AF710518

Chain Rule

$$\frac{df}{dx} = \frac{df}{du}\frac{du}{dx}$$

Derivatives of Exponential Functions

$$\frac{d}{dx}(e^u) = e^u \frac{du}{dx}$$

$$\frac{d}{dx}(b^u) = (\ln b)b^u \frac{du}{dx}$$

Derivatives of Logarithmic Functions

$$\frac{d}{dx}(\ln u) = \frac{1}{u}\frac{du}{dx}$$

$$\frac{d}{dx}(\log_b u) = \frac{1}{\ln b}\frac{1}{u}\frac{du}{dx}$$

Derivatives of Trigonometric Functions

$$\frac{d}{dx}(\sin u) = \cos u \frac{du}{dx}$$

$$\frac{d}{dx}(\cos u) = -\sin u \frac{du}{dx}$$

$$\frac{d}{dx}(\tan u) = \sec^2 u \frac{du}{dx}$$

$$\frac{d}{dx}(\csc u) = -\csc u \cot u \frac{du}{dx}$$

$$\frac{d}{dx}(\sec u) = \sec u \tan u \frac{du}{dx}$$

$$\frac{d}{dx}(\cot u) = -\csc^2 u \frac{du}{dx}$$

Brief Calculus
for
Management and the Life and Social Sciences

The Irwin Series in Quantitative Analysis for Business

Consulting Editor Robert B. Fetter Yale University

Brief Calculus for Management and the Life and Social Sciences

Donald L. Stancl   Mildred L. Stancl
both of *St. Anselm College*

1988

IRWIN

Homewood, Illinois 60430

Chapter opening photos: Marilyn Gartman Agency, Inc.
Chapter One *(New York),* © Everett C. Johnson, 2
Chapter Two *(San Francisco),* © Spenser Grant, 68
Chapter Three *(Chicago),* © Eric Futran, 132
Chapter Four *(St. Louis),* © Everett C. Johnson, 208
Chapter Five *(Seattle),* © Milt and Joan Mann, 264
Chapter Six *(Boston),* © Michael Phillip Manheim, 304
Chapter Seven *(Miami Beach),* © Michael Phillip Manheim, 350
Appendix *(Houston),* © Milt and Joan Mann, 412

Acquisitions editor: Richard T. Hercher, Jr.
Project editor: Waivah Clement
Production manager: Bette Ittersagen
Designer: Lucy Lesiak
Artist: Tom Mallon
Compositor: Arcata Graphics/Kingsport
Typeface: 10/12 Times Roman
Printer: Arcata Graphics/Hawkins County

ISBN 0-256-06303-6

Library of Congress Catalog Card No. 87–82174

Printed in the United States of America
1 2 3 4 5 6 7 8 9 0 K 5 4 3 2 1 0 9 8

The more comprehensive text *Calculus for Management and the Life and Social Sciences* by Donald and Mildred Stancl is also available from Irwin. All of these chapters are included along with chapters on Differential Equations, Probability, and Sequences and Series.

To each other

Preface

Brief Calculus for Management and the Life and Social Sciences is a textbook for one-semester calculus courses designed for students of management, economics, and the life and social sciences. It is a shortened version of our text *Calculus for Management and the Life and Social Sciences,* consisting of Chapters 1 through 7 and most of Chapter 9 of the longer book. The only prerequisite for this book is intermediate algebra.

Goals: Our goals in this text are:

- To present the basic concepts and techniques of calculus at a level of sophistication appropriate to students who are for the most part mathematically inexperienced.
- To demonstrate how calculus can be used to build models and solve problems in the disciplines of management, economics, the life sciences, and the social sciences.
- To help students develop a mastery of the concepts and techniques of calculus and an ability to apply these to model-building and problem-solving in their chosen disciplines.

General Approach: The general approach of our book emphasizes the use of functions to construct models and the use of calculus to analyze models in the subject areas of management, economics, and the sciences. Whenever possible we motivate our development of the concepts of calculus with a prior discussion of specific situations in these subject areas. We have avoided the formal theorem-proof method of presentation, but we do furnish proofs of our results when we feel that they will be both understood by and instructive to the student. When proofs are not appropriate, we attempt to provide convincing justifications for our results through less formal means: for instance, by appealing to suggestive special cases or geometric figures. Finally, in both the text and the exercises we place heavy emphasis on drawing and interpreting the graphs of functions.

Features: Some features of the book are:

Review: In Chapter 1 we offer a comprehensive review of the material necessary for an understanding of the remainder of the text. Topics covered include exponents and roots, functions and graphing, polynomial and rational expressions, linear functions, and quadratic functions. Students who have a solid mathematical background may be able to skip some or all of Chapter 1.

Supplements: Each chapter is followed by a supplement containing additional material not usually covered in a textbook of this type. The supplements to Chapters 2 and 5 are historical in nature, being brief descriptions of the development of the derivative and the integral, respectively. Each of the other supplements consists of an application of a topic covered in the preceding chapter to a specific situation in management, economics, or the sciences. The supplements can be covered in class, but they are also appropriate for assignment as outside reading or (an approach we have had success with) as a basis for student presentations to the class. There is an exercise set at the end of each supplement; answers to the odd-numbered supplementary exercises can be found at the back of the book.
Exercises: The book contains over 1600 exercises, many of which have several parts. Each numbered section is followed by an exercise set. The exercises are graded by difficulty with the easier ones at the beginning of the set. Nearly every set contains applied exercises and, for the convenience of the instructor and the students, these are clearly marked as to the particular subject area they address (management, life science, or social science). In addition to the section exercises, there is at the end of each chapter a set of review exercises designed to provide a thorough test of the student's mastery of the concepts and techniques of the chapter. Answers to odd-numbered section exercises and to all review exercises can be found at the back of the book.
Examples: Every newly introduced concept or technique is immediately followed by one or more illustrative examples. There are nearly 300 such examples in the text, many of them pertaining to business, economics, the life sciences, or the social sciences.
Figures: Whenever possible we have utilized figures to illustrate mathematical concepts and to illuminate examples and problems. The book contains more than 350 figures.
Computer Programs: Many of the topics covered in the text are illustrated by computer programs that are described in a separate Computer Manual and are available on disk (see below).

Organization: The organization of the book is straightforward. Chapter 1 is an introductory chapter that contains a review of algebra and prepares students for the calculus to come. Chapters 2 through 6 form the heart of the text: Here we present the derivative and its applications (Chapters 2 and 3), exponential and logarithmic functions and their applications (Chapter 4), and the indefinite and definite integrals with their applications (Chapters 5 and 6). Chapter 7 is devoted to functions of several variables. For those who wish to cover the trigonometric functions, they are presented in the Appendix. A dependency diagram follows:

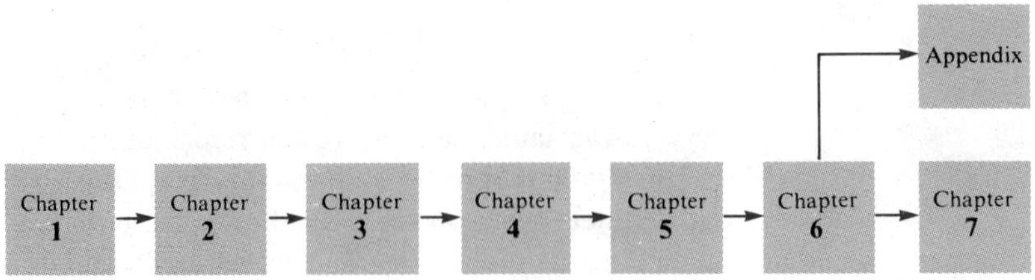

Computer Manual: We have also written a Computer Manual to accompany this text. The Manual includes an IBM PC formatted disk containing BASIC programs that illustrate the material in the text, as well as descriptions of these programs and directions and suggestions for their use. The Computer Manual is correlated with the text: At the end of each section of the text, reference is made to the programs that are relevant to that section.

Additional Supplements: An Instructor's Manual and a Student Solutions Manual and Study Guide for students are available from the publisher. The Instructor's Manual contains answers to all exercises in the text and a test bank; the Student Solutions Manual and Study Guide contains answers to all exercises, detailed solutions of many exercises, study tips for students, and self-tests.

Acknowledgments: We would like to thank the following reviewers, whose valuable suggestions and criticisms have contributed greatly to this book: Robert Fetter, Yale University, Consulting Editor; Ray Boersema, Front Range Community College; Genelle Beck, University of Texas, Austin; Larry Bowen, University of Alabama; Joseph Sukta, Moraine Valley Community College; Howard Frisinger, Colorado State University; John Hodges, University of Arkansas, Little Rock; John Hornsby, University of New Orleans; Gail Earles, St. Cloud State University; Donald Trahan, Naval Postgraduate School; Thalia Papageorgiou, Bentley College; Thomas Caplinger, Memphis State University; David Pentico, Duquesne University; Enrique Venta, Loyola University of Chicago; Robert L. Moore, University of Alabama; Jean Clarke, Virginia Commonwealth University.

We also gratefully acknowledge the suggestions, assistance, and patience of our editors and the rest of the staff at Richard D. Irwin, Inc. Finally, our thanks to our colleagues at St. Anselm College and all others who have supported us in the writing of this book.

Donald L. Stancl
Mildred L. Stancl

Contents

Brief Calculus for Management and the Life and Social Sciences

1

Numbers and Functions

A function is a rule that describes a relationship. For example, a function might describe the relationship between

- The number of units of product a firm sells and its profits.
- The size of a city's population and the amount of pollution in its air.
- A society's intake of cholesterol and its incidence of heart disease.

As these examples suggest, functions are often used to model practical situations, so we must be able to analyze them. Calculus is a powerful tool for analyzing functions. As we proceed through this book, we will show how calculus can be used to study functions and to determine the rate of change of one quantity with respect to another, the maximum or minimum value that a quantity can attain, and many other interesting and important things. However, before we begin, we must develop some familiarity with functions and their properties. This is our goal in Chapter 1.

Since our functions will be described in terms of real numbers, we begin this chapter with an introduction to the system of real numbers and some of its properties. We then provide a general discussion of functions and their graphs, consider the algebra of functions, and conclude with brief studies of linear, quadratic, polynomial, and rational functions.

1.1 REAL NUMBERS

We will be working with real numbers throughout this book, so we must know what they are and have an understanding of some of their properties. In this section we will discuss the set of real numbers, the real line, inequalities among real numbers, intervals, the concept of the absolute value of a real number, and exponents and roots.

The Set of Real Numbers

A **set** is a collection of objects; the objects are called **elements** of the set. If object x is an element of set A, we write $x \in A$ and say that "x belongs to A." If object x is not an element of set A, we write $x \notin A$ and say that "x does not belong to A."

Often it is possible to specify a set by listing all its elements between braces. For example, if A is the set whose elements are the numbers -1, 0, and 1, then we may write $A = \{-1, 0, 1\}$. Note that $1 \in A$, but $2 \notin A$. If it is not convenient or possible to specify a set by listing its elements, then we may use the **set-builder notation** $\{x \mid x \text{ has property P}\}$. This is read as "the set of all x such that x has property P." For instance, the set $B = \{5\}$ may be written as $B = \{x \mid 2x - 10 = 0\}$, since the only object x that satisfies the equation $2x - 10 = 0$ is the number 5.

Two important sets of numbers are

the **positive integers** $\mathbf{N} = \{1, 2, 3, \ldots\}$

and

the **integers** $\mathbf{Z} = \{0, 1, -1, 2, -2, 3, -3, \ldots\}$.

(The three dots . . . mean "and so on.") Another important set of numbers is the set **Q,** which consists of all quotients of integers. Some elements of **Q** are

$$\frac{0}{1} = 0, \quad \frac{1}{2}, \quad -\frac{3}{5}, \quad -\frac{22}{33}, \quad \frac{319}{100}, \quad \frac{466}{13}, \quad \text{and} \quad \frac{14}{7} = 2.$$

Numbers that can be written as quotients of integers are called **rational numbers.** Numbers that cannot be written as quotients of integers, such as $\sqrt{2}$ and π, for example, are called **irrational numbers.** The set **Q** is thus referred to as the set of rational numbers, and we may write **Q** in set-builder notation as

$$\mathbf{Q} = \left\{\frac{m}{n} \,\middle|\, m \in Z, n \in Z, n \neq 0\right\};$$

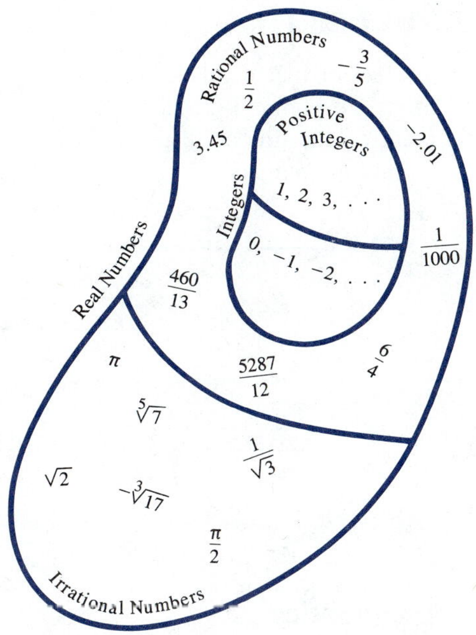

FIGURE 1–1 The Real Number System

note that $n \neq 0$, because division by 0 is not defined. Note also that every integer is an element of **Q,** since every integer m may be written as $m/1$.

The set **R** of **real numbers** consists of all rational numbers together with all irrational numbers. Thus **R** contains all integers, all fractions of integers, and all irrational numbers such as $\sqrt{2}$, π, $-\sqrt[3]{5}$, and so on. Both rational and irrational numbers have decimal representations, such as

$$\frac{1}{2} = 0.5, \quad \frac{2}{3} = 0.6666\ .\ .\ .\ , \quad \text{and} \quad \sqrt{2} = 1.414213\ .\ .\ .\ ,$$

so every real number has a decimal representation. On the other hand, it can be shown that every decimal number represents either a rational number or an irrational number; hence we may think of the set **R** as consisting of all decimal numbers. Figure 1–1 shows the relationships among the number sets we have discussed here. From now on when we write "number," we shall mean "real number," and a set written in the form

$$\{x \mid x \text{ has property P}\}$$

will be understood to mean the set of all real numbers x such that x has property P.

The Real Line

The set of real numbers may be represented by a horizontal line called the **real line.** See Figure 1–2. Each number corresponds to a point on the line, and each point on the line corresponds to a number. For example, in Figure 1–3 we have plotted the numbers $-7/4$, -1, $-1/2$, 0, $2/3$, 1.75, 2, and 2.8 on the real line.

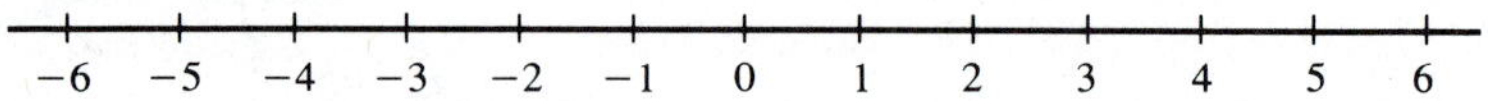

FIGURE 1–2 The Real Line

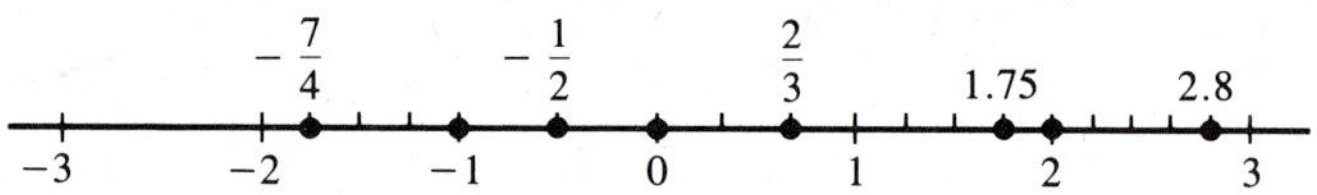

FIGURE 1–3

Inequalities

The real line makes it possible for us to compare numbers using inequalities. If a and b are numbers and a is to the left of b on the real line, then we say that ***a* is less than *b*** (written $a < b$) or that ***b* is greater than *a*** (written $b > a$). See Figure 1–4. If $a < 0$, then a is a **negative** number; if $a > 0$, then a is a **positive** number. An expression of the form $a \leq b$ means that a is less than b or equal to b. Similarly, $b \geq a$ means that b is greater than a or equal to a.

FIGURE 1–4 $a < b$ **or** $b > a$

EXAMPLE 1 The following are true:

$$-3 < -2, \qquad -1 < -\frac{2}{3}, \qquad 1 \leq 1, \qquad 1 < 6.5, \qquad 2 \geq 0, \qquad \frac{5}{4} > \frac{9}{8}.$$

The following are false:

$$-3 < -4, \qquad 1 < \frac{2}{3}, \qquad 2 \leq -3.25, \qquad 2 < 2, \qquad 1.01 > 1.1, \qquad \frac{5}{4} > \frac{11}{8}.$$

An inequality of the form $a < b < c$ is a **double inequality.** It is true if both $a < b$ and $b < c$ are true, that is, if b is between a and c on the real line.

EXAMPLE 2 The following are true:

$$-3 < -\frac{3}{2} < -1, \qquad -\frac{2}{3} < 0.02 \leq 0.021, \qquad \frac{1}{2} < \frac{3}{4} < \frac{7}{8}.$$

The following are false:

$$-2 < -3 < -4, \qquad 2 < \frac{5}{2} < 2.25, \qquad 1 \leq 2 < 2.$$

Do not write statements such as $2 < 4 > 0$. All inequality symbols in a double inequality should point in the same direction, so the correct way to write the inequality relation among 0, 2, and 4 is either $0 < 2 < 4$ or $4 > 2 > 0$.

Intervals

It is often convenient to consider the set of all numbers that lie between two given numbers, or the set of all numbers that are greater than or less than a given number. Such sets of numbers are called **intervals.** For example,

$$\{x \mid 0 \leq x < 1\}$$

is an interval because it consists of all the numbers between 0 and 1, including 0 but not including 1. Figure 1–5 shows a **line diagram** for this interval; the closed circle at 0 indicates that 0 belongs to the interval; the open circle at 1 indicates that 1 does not belong to the interval. We write

$$[0, 1) = \{x \mid 0 \leq x < 1\};$$

the expression $[0, 1)$ is called **interval notation.** Figure 1–6 depicts all possible forms of intervals and gives their interval notation. The symbols $+\infty$ and $-\infty$ are read as "plus infinity" and "minus infinity," respectively.

FIGURE 1–5 The Interval $\{x \mid 0 \leq x < 1\} = [0, 1)$

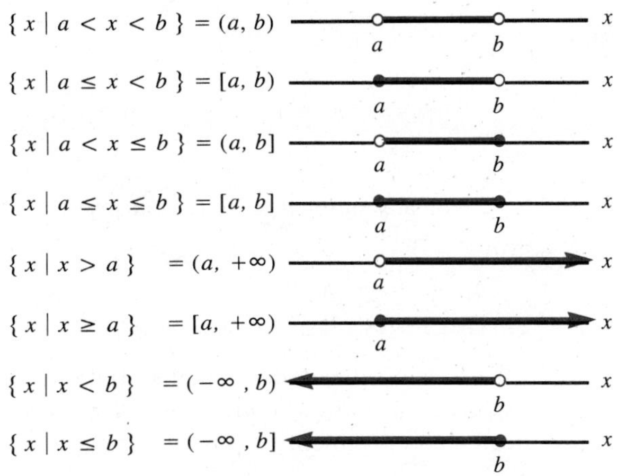

FIGURE 1–6 Intervals

EXAMPLE 3 The line diagrams corresponding to the intervals $[1, 3]$, $(-1, 2)$, $(-\infty, 1)$, and $(3, +\infty)$ are shown in Figure 1–7.

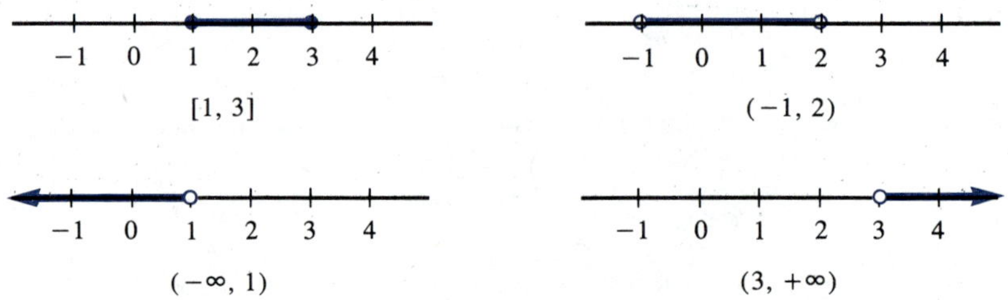

FIGURE 1–7

Absolute Value

The **absolute value** of a number is its distance from 0 on the real line, regardless of whether the number is greater than or less than 0. For example, the numbers 3 and -3 are both 3 units from 0, so both have an absolute value of 3. A more formal definition of the absolute value of a number is as follows:

Absolute Value

The **absolute value** of a number x, denoted $|x|$, is defined by

$$|x| = \begin{cases} x & \text{if } x \geq 0, \\ -x & \text{if } x < 0. \end{cases}$$

EXAMPLE 4

$$|2| = 2, \qquad |-2| = 2, \qquad |0| = 0, \qquad |\sqrt{2}| = \sqrt{2}, \qquad \left|\frac{2}{3} - \frac{5}{3}\right| = 1.$$

For any number x, note that $|x| = 0$ if $x = 0$, $|x|$ is positive if $x \neq 0$, and $|x| = |-x|$.

Exponents and Roots

We conclude this section with a brief review of exponents and roots. In an expression of the form x^r, where x and r are numbers, x is called the **base** and r is called the **exponent** or **power.**

Nonnegative Integral Exponents

Let x be a number and n a positive integer. We define

$$x^n = \underbrace{x \cdot x \cdots x}_{n \text{ factors}}$$

Also, if $x \neq 0$, we define $x^0 = 1$.

EXAMPLE 5 $2^1 = 2 \qquad 2^3 = 2 \cdot 2 \cdot 2 = 8 \qquad 3^0 = 1 \qquad (-5)^3 = (-5)(-5)(-5) = -125.$

If x is a number and n a positive integer, any number r such that $r^n = x$ is called an ***n*th root** of x. If x has just one nth root r, then r is the **principal *n*th root** of x; if x has two nth roots, its principal nth root is the positive one.

EXAMPLE 6 Since $(-3)^3 = -27$, -3 is a third root (or cube root) of -27; since $2^4 = 16$ and also $(-2)^4 = 16$, 2 and -2 are both fourth roots of 16. Since -3 is the only cube root of -27, it is the principal cube root of -27; since 2 is the positive fourth root of 16, it is the principal fourth root of 16.

We denote the principal nth root of x by $\sqrt[n]{x}$. If $n = 2$, we write $\sqrt{x}$ instead of $\sqrt[2]{x}$ and call $\sqrt{x}$ the **square root** of x. Note that if n is even, then $\sqrt[n]{x}$ does not exist when x is negative and is positive when x is positive.

EXAMPLE 7 From Example 6,

$$\sqrt[3]{-27} = -3 \qquad \text{and} \qquad \sqrt[4]{16} = 2.$$

Note also that

$$\sqrt[2]{25} = \sqrt{25} = 5$$

is positive, but $\sqrt{-25}$ does not exist.

Now we can define rational exponents.

Rational Exponents

If x is a number, m and n are positive integers, and $\sqrt[n]{x}$ exists, then

$$x^{m/n} = (\sqrt[n]{x})^m.$$

EXAMPLE 8

$$2^{1/2} = (\sqrt{2})^1 = \sqrt{2}, \qquad 25^{1/2} = (\sqrt{25})^1 = 5,$$
$$(-27)^{4/3} = (\sqrt[3]{-27})^4 = (-3)^4 = 81.$$

Lastly, we define negative exponents.

Negative Exponents

If x is a nonzero number and r is a positive number, then

$$x^{-r} = \frac{1}{x^r}.$$

EXAMPLE 9

$$2^{-1} = \frac{1}{2}, \qquad (-3)^{-4} = \frac{1}{(-3)^4} = \frac{1}{81}, \qquad 4^{-3/2} = \frac{1}{4^{3/2}} = \frac{1}{8}.$$

Now that we have defined x^r for rational numbers r, we state the rules of exponents.

Rules of Exponents

Let x, y, r, and s be numbers and assume that the following expressions are all defined. Then

a. $x^r x^s = x^{r+s}$

b. if $x \neq 0$, $\dfrac{x^r}{x^s} = x^{r-s}$

c. $(x^r)^s = x^{rs}$

d. $(xy)^r = x^r y^r$

e. if $y \neq 0$, $\left(\dfrac{x}{y}\right)^r = \dfrac{x^r}{y^r}$

EXAMPLE 10

a. $3^2 3^{-4} = 3^{2+(-4)} = 3^{-2} = \dfrac{1}{9}$

b. $\dfrac{3^2}{3^{-4}} = 3^{2-(-4)} = 3^6 = 729$

c. $(3^2)^3 = 3^6 = 729$

d. $(3 \cdot 5)^2 = 3^2 \cdot 5^2 = 9 \cdot 25 = 225$

e. $\left(\dfrac{3}{5}\right)^2 = \dfrac{3^2}{5^2} = \dfrac{9}{25}$

EXAMPLE 11 Simplify $\sqrt[3]{\dfrac{x^3y^{-2}z^5x^{-6}}{y^4(z^2)^{-2}}}$.

We have

$$\begin{aligned}\sqrt[3]{\frac{x^3y^{-2}z^5x^{-6}}{y^4(z^2)^{-2}}} &= (x^{3+(-6)}y^{-2-4}z^{5-(-4)})^{1/3}\\ &= (x^{-3}y^{-6}z^9)^{1/3}\\ &= x^{-1}y^{-2}z^3\\ &= \frac{z^3}{xy^2}.\end{aligned}$$

Computer Manual: Program INTERVAL.

1.1 EXERCISES

The Real Line

1. Plot the following numbers on the real line:

-5 $\quad -4.5 \quad -\frac{7}{4} \quad -1 \quad -\frac{2}{5} \quad 0 \quad \frac{1}{3}$

$\frac{3}{4} \quad 1 \quad \frac{3}{2} \quad \frac{17}{8} \quad \frac{19}{6} \quad 4.2 \quad 5$

2. What numbers are represented by the plotted points?

(a)

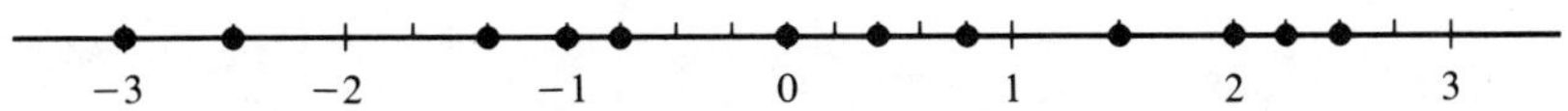

(b)

Inequalities

In Exercises 3 through 16 state whether the inequality is true or false.

3. $-3 < 2$

4. $-6 < -8$

5. $0 \le 2$

6. $4 \ge 2$

7. $-3 > -4$

8. $2 < 2$

9. $\frac{2}{3} < \frac{3}{5}$

10. $-\frac{2}{3} < -\frac{3}{5}$

11. $-1 < 0 < 1$

12. $-3 < -4 < -5$

13. $-4 \le 4 \le 10$

14. $\frac{2}{3} < \frac{4}{7} \le 1$

15. $\frac{4}{7} < \frac{2}{3} \le 1$

16. $5 < 5.1 < 5.01$

Intervals

In Exercises 17 through 24 draw the line diagram corresponding to the given interval.

17. $(-2, 7]$ **18.** $(4, 8)$ **19.** $[-1, 4]$

20. $\left[\frac{1}{2}, \frac{7}{4}\right)$ **21.** $(-\infty, -2]$ **22.** $(2.5, +\infty)$

23. $[1, +\infty)$ **24.** $(-\infty, 6)$

Absolute Value

In Exercises 25 through 33 evaluate the expression.

25. $|4|$ **26.** $|-4|$ **27.** $-|-8|$

28. $\left|\frac{2}{3}\right|$ **29.** $\left|-\frac{4}{5}\right|$ **30.** $-|\sqrt{2}|$

31. $|7 - 12|$ **32.** $2|8 - 11|$ **33.** $2-|-\sqrt{2}|$

Exponents and Roots

In Exercises 34 through 45 evaluate the expression.

34. 2^6 **35.** 3^{-4} **36.** π^0

37. $\left(\frac{2}{5}\right)^3$ **38.** $\left(-\frac{3}{4}\right)^2$ **39.** $36^{1/2}$

40. $\sqrt[3]{64}$ **41.** $81^{3/4}$ **42.** $\sqrt{-16}$

43. $100^{-1/2}$ **44.** $32^{-3/5}$ **45.** $(-8)^{1/3}$

In Exercises 46 through 54 simplify the expression. Use only positive exponents in your answer.

46. $\dfrac{5^2y^3z^8}{x^2y^3z^{-2}}$ **47.** $(3x^2y^3z^{-1})^3$ **48.** $(4x^2y^{-2}z^5)^2(2x^{-2}yz^4)^{-1}$

49. $\dfrac{x^2y^{2/3}z^{-5/4}}{x^{1/2}y^{-1/3}z}$ **50.** $\dfrac{(x^4y^2z^6)^{-3/2}}{xyz}$ **51.** $\dfrac{(27x^3y^{-3})^{1/3}}{(3x^{2/3}y^{4/3})^3}$

52. $\sqrt{x^2y^3}$ **53.** $\sqrt[4]{32x^4y^9}$ **54.** $\sqrt[3]{\dfrac{27x^5y^3}{x^3z^2}}$

1.2 FUNCTIONS AND GRAPHING

As we noted at the beginning of this chapter, functions describe relationships. For instance, a function might describe the relationship between a company's sales and its profits, or the relationship between the average daily temperature and the amount of heating oil used to heat a building. Throughout this book we will be creating functions, describing them, graphing them, and examining their behavior. In this section we begin our study of functions.

Functions

A **function** is a rule that associates with each element of one set of numbers a single element of another set of numbers. Figure 1–8 depicts this situation symbolically: the rule shown there assigns to each number in the set X a single number in the set Y, and thus it is a function from X to Y. On the other hand, Figure 1–9 depicts a rule that assigns more than one number in the set Y to some of the numbers in the set X; hence it is not a function from X to Y.

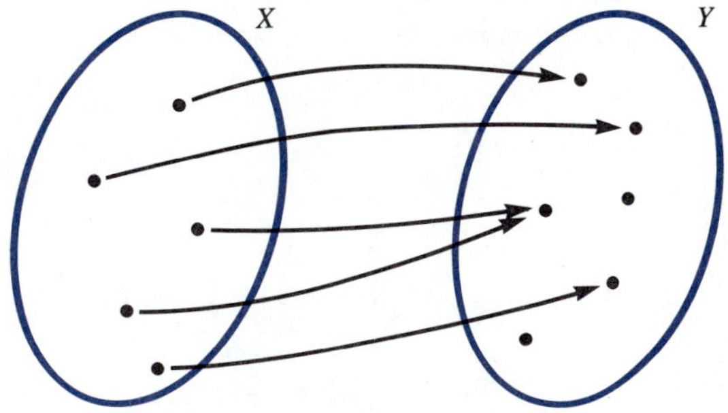

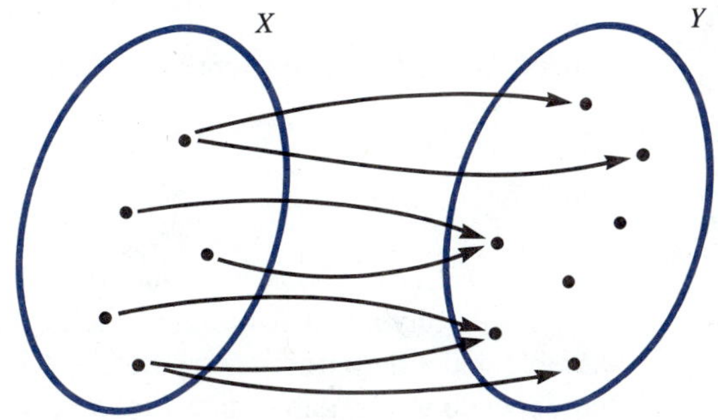

FIGURE 1–9 This Is Not a Function from X to Y

Functions are often denoted by the letters f, g, or h and specified by a **defining equation** in **functional notation.** For example, the equation

$$f(x) = 3x - 2$$

defines a function f using functional notation. The letter x is the **independent variable** of f, and the notation $f(x)$ is read "f of x." If we substitute a particular real number for x in the defining equation, we can see what value the function assigns to that number. For instance, since

$$f(0) = 3(0) - 2 = -2,$$

$$f(1) = 3(1) - 2 = 1,$$

and

$$f(-2) = 3(-2) - 2 = -8,$$

we see that f assigns -2 to 0, 1 to 1, and -8 to -2. The numbers $f(0)$, $f(1)$, and $f(-2)$ are called the **functional values** of f at 0, 1, and -2, respectively. Thus $f(x)$ denotes the functional value of f at x. Notice that a function f and a functional value $f(x)$ are not the same thing, since f is a *rule* but $f(x)$ is a *number*. Throughout this book we will specify functions by their defining equations, using phrases such as "let f be the function defined by the equation $f(x) = 3x - 2$," or "let f be the function given by $f(x) = 3x - 2$."

It is possible, and often useful, to substitute algebraic expressions for the independent variable of a function. For instance, consider the function g defined by the equation $g(x) = 5x + 2$: we have

$$g(t) = 5t + 2,$$

$$g(a^2) = 5a^2 + 2,$$

and

$$g(b - 1) = 5(b - 1) + 2 = 5b - 5 + 2 = 5b - 3.$$

EXAMPLE 1 Suppose it costs a firm $C(x) = x^2 + 5x + 100$ dollars to produce x units of its product. How much will it cost the firm to produce 20 units?

It will cost

$$C(20) = 20^2 + 5(20) + 100 = 600$$

dollars. If the firm is currently producing b units, its cost is

$$C(b) = b^2 + 5b + 100$$

dollars.

We will not always use functional notation to describe functions. Sometimes it is more convenient to define functions by equations involving the independent variable and a **dependent variable.** For example, the equation $y = 3x - 2$ specifies the dependent variable y as a function of the independent variable x. Substituting numerical values for x will yield numerical values for y. Note that the function defined by $y = 3x - 2$ is the same as the function defined by $f(x) = 3x - 2$. In fact, $s = 3t - 2$ and $g(u) = 3u - 2$ also define this same function. The letters used are not important; what is important is the relationship between the variables. It is common practice in the applications of mathematics to use letters that are suggestive of the things they represent. Thus, if a firm sells quantity q of its product at a price of \$20 per unit, its revenue will be given by the equation $R = 20q$. Here q is the independent variable, representing quantity sold; R is the dependent variable, representing revenue in dollars. If the firm sells 5000 units, its revenue is

$$R = 20(5000) = 100{,}000$$

dollars.

The set of values which the independent variable of a function is allowed to assume is called the **domain** of the function. Sometimes the domain is stated explicitly along with the defining equation of the function. Thus, if a function f is defined by the statement

$$f(x) = 5x^2 + 2, \qquad x > 1,$$

this indicates that the independent variable x is only allowed to take on values greater than 1; hence the domain of this function is $\{x|x > 1\} = (1, +\infty)$. Often the domain of the function is not stated explicitly. It is then understood that the domain consists of all numbers that may legitimately be substituted for the independent variable in the defining equation.

EXAMPLE 2 Consider the square-root function g defined by $g(x) = \sqrt{x}$. We cannot substitute negative numbers for x in this equation because the square root of a negative number is not defined as a real number. We can substitute any nonnegative number for x, however, and thus the domain of g is the interval $[0, +\infty)$.

Similarly, suppose the function h is defined by

$$h(x) = \frac{2}{x - 1}.$$

Since division by 0 is not defined, we cannot substitute 1 for x in this equation, but we can substitute any other number for x. Hence the domain of h is $\{x|x \neq 1\}$.

Not every defining equation yields a function. For example, the equation $x^2 + y^2 = 1$ does not define y as a function of x, because it sometimes assigns more than one value of y to a single value of x. (For instance, if $x = 0$, then $y = 1$ and also $y = -1$.) A function must assign a *single* value of the dependent variable to each value of the independent variable.

Graphing

It is usually possible to depict a function geometrically by **graphing** it. A graph gives us information about the function in a visual way that is easy to understand. For instance, the graph in Figure 1–10 shows the number of bacteria present in a

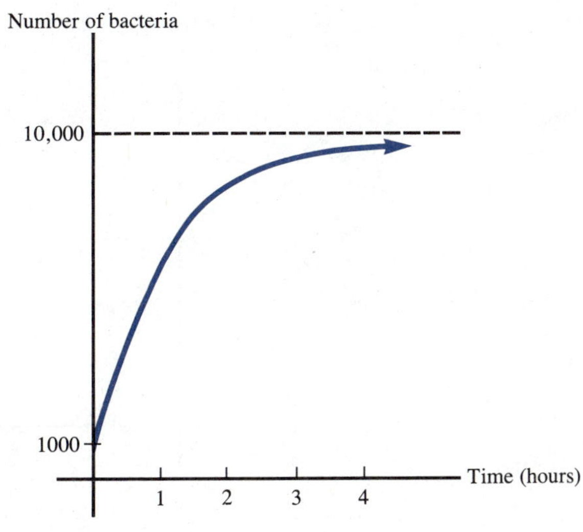

FIGURE 1–10

culture as a function of time. The horizontal scale represents time in hours, the vertical scale the number of bacteria present. We immediately see from the graph that the initial number of bacteria present was 1000, that as time passed the bacteria population increased, but at a slower and slower rate, and that the number of bacteria was limited by some factor (the supply of nutrients, perhaps) to fewer than 10,000.

The process of graphing a function begins with two copies of the real line, one horizontal and one vertical, that intersect at their zero points, as in Figure 1–11. The horizontal real line and the vertical real line are called **axes:** the horizontal axis is often called the ***x*-axis,** the vertical one the ***y*-axis.** The two together form the **cartesian coordinate system.** If a and b are any real numbers, the point (a, b) is located on the cartesian coordinate system opposite a on the x-axis and opposite b on the y-axis, as shown in Figure 1–12. The number a is called the ***x*-coordinate** of the point (a, b), and the number b is its ***y*-coordinate.** The point (0, 0) where the axes intersect is called the **origin.**

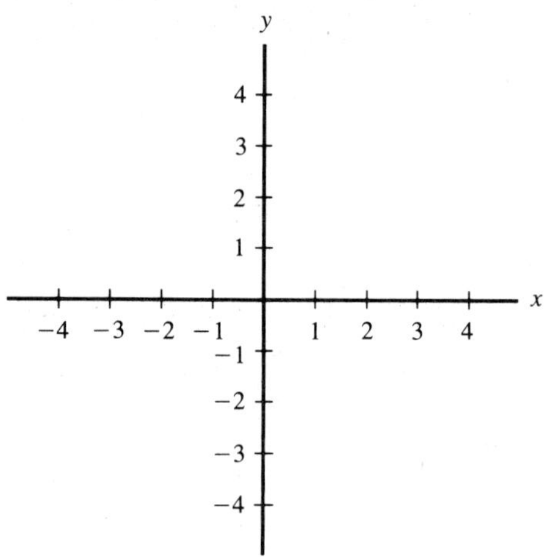

FIGURE 1–11 The Cartesian Coordinate System

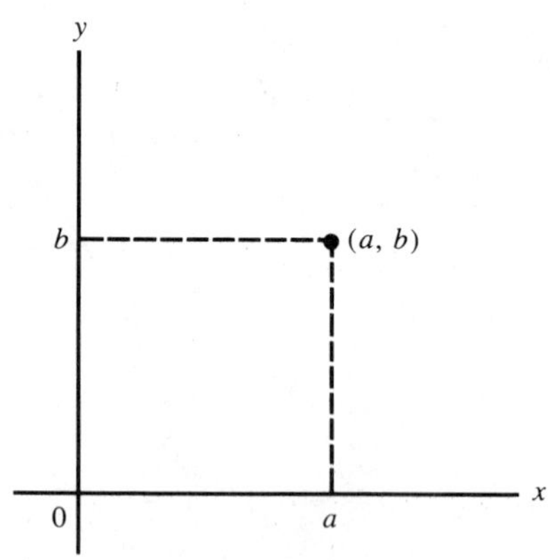

FIGURE 1–12

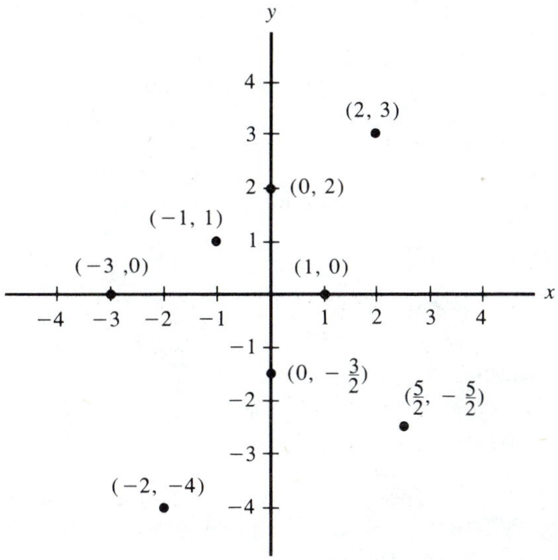

FIGURE 1–13

EXAMPLE 3 In Figure 1–13 we have plotted the points (1, 0), (2, 3), (0, 2), (−1, 1), (−3, 0), (−2, −4), $\left(0, -\frac{3}{2}\right)$), and $\left(\frac{5}{2}, -\frac{5}{2}\right)$.

The **graph** of a function f is the set of all points $(x, f(x))$ for which x is in the domain of f. To draw the graph of a function, we proceed as follows:

Graphing Procedure

To graph a function f:

1. Determine its domain.
2. Substitute representative numbers from the domain into the defining equation to obtain points $(x, f(x))$ that lie on the graph.
3. Plot the points obtained in step 2 on the cartesian coordinate system. When enough points have been plotted to make the shape of the graph clear, connect them with a curve.

It is also frequently useful to find the **intercepts** of a function. The ***y*-intercept** of a function f is the point where its graph crosses the y-axis; this is the point $(0, f(0))$. The ***x*-intercepts** of the function are the points where the graph crosses or touches the x-axis; we find them by solving the equation $f(x) = 0$ for x, if possible.

EXAMPLE 4 Let us graph the function f given by $f(x) = 2x - 1$.

1. Since any number can be substituted for x in the defining equation for f, the domain of f is the set **R** of real numbers.

2. We select some representative numbers from the domain of f and find the corresponding points on the graph:

x	$f(x) = 2x - 1$	$(x, f(x))$
-2	$f(-2) = -5$	$(-2, -5)$
0	$f(0) = -1$	$(0, -1)$
1	$f(1) = 1$	$(1, 1)$
3	$f(3) = 5$	$(3, 5)$

The y-intercept of f is the point $(0, f(0)) = (0, -1)$. Since the equation $2x - 1 = 0$ has the solution $x = \frac{1}{2}$, there is one x-intercept, namely the point $(\frac{1}{2}, 0)$. The graph of f is shown in Figure 1–14.

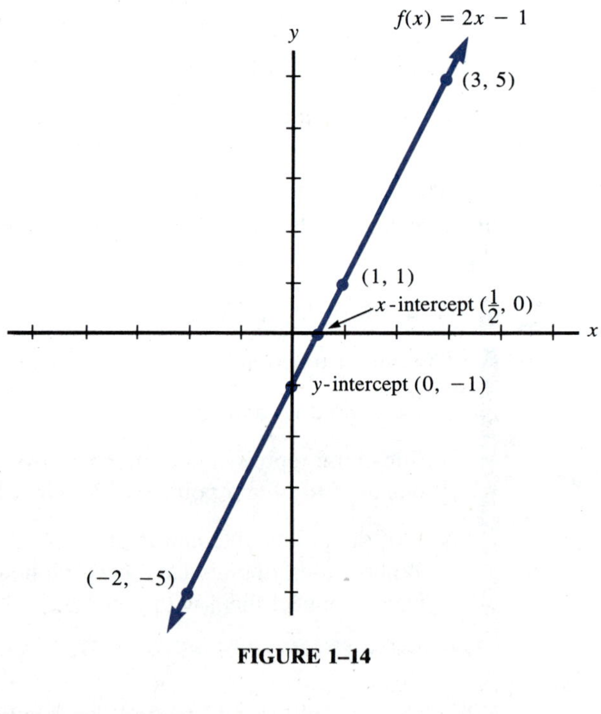

FIGURE 1–14

EXAMPLE 5 Let us graph the **absolute value function** $y = |x|$. The domain of this function is the set of real numbers (why?). We have

x	$y = \lvert x \rvert$	(x, y)
-3	$y = 3$	$(-3, 3)$
-2	$y = 2$	$(-2, 2)$
-1	$y = 1$	$(-1, 1)$
0	$y = 0$	$(0, 0)$
1	$y = 1$	$(1, 1)$
2	$y = 2$	$(2, 2)$
3	$y = 3$	$(3, 3)$

If $x = 0$, then $y = 0$, so the y-intercept is the point $(0, 0)$. If $y = 0$, then the equation $0 = |x|$ has the single solution $x = 0$, so $(0, 0)$ is also the only x-intercept. The graph of the absolute value function is shown in Figure 1–15.

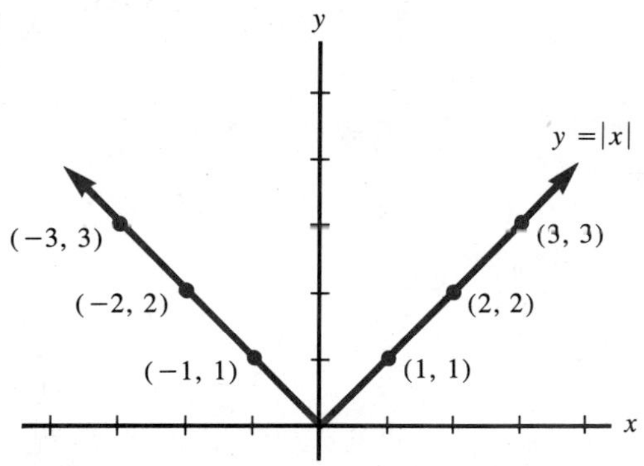

FIGURE 1–15

It is important to realize that a point (a, b) lies on the graph of a function f if and only if $f(a) = b$. Therefore if (a, b) is on the graph of f, then the number a must be in the domain of f; on the other hand, if the number a is *not* in the domain of f, then (a, b) cannot be on the graph of f for any number b, and therefore a vertical line through a on the x-axis will not intersect the graph of f.

EXAMPLE 6 Let us graph the function h given by

$$h(x) = \frac{2}{x - 1}.$$

The domain of h is $\{x \mid x \neq 1\}$, so there is no point on the graph of the form $(1, b)$. We indicate this by drawing a dotted vertical line through 1 on the x-axis. (This line is called a **vertical asymptote** for the function.) The graph cannot touch or cross the dotted line. See Figure 1–16, where the graph of h has been drawn. (Check that the labeled points in Figure 1–16 are indeed on the graph of h.)

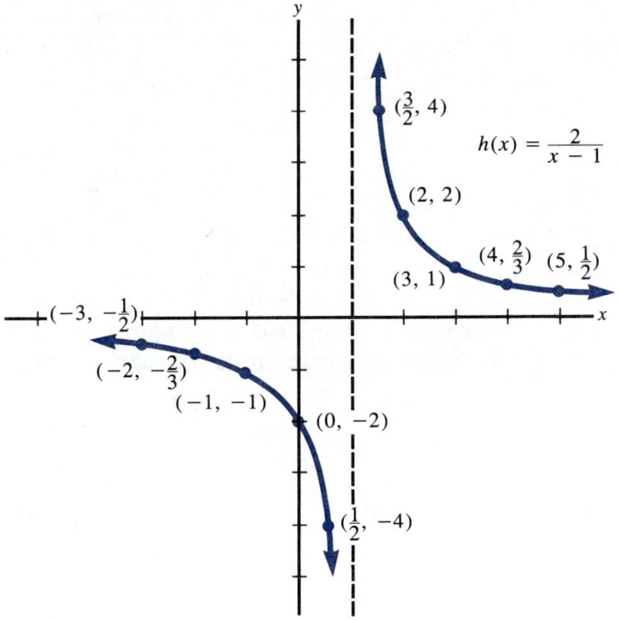

FIGURE 1–16

EXAMPLE 7 Suppose a firm's profits in year t, where $t = 0$ represents 1970, are given by the function $P = t^2 - 16t + 48$, where P is in millions of dollars. The graph of this function is shown in Figure 1–17. From the graph we see that in 1970 the firm had a profit of \$48 million. Profits declined after 1970 until in 1974 ($t = 4$) they were zero. Between 1974 and 1982 the firm had a negative profit (that is, a loss) each year; the largest loss was \$16 million in 1978 ($t = 8$). The firm broke even again in 1982 ($t = 12$), and since then profits have been increasing.

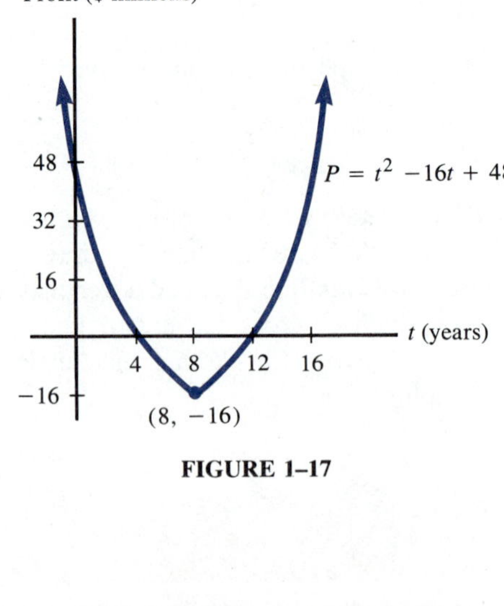

FIGURE 1–17

Our graphing procedure relies on finding points on the graph and then plotting them. Unless the function under consideration is a simple one, this is usually quite tedious. It is also potentially unreliable, for we cannot always be certain that we have plotted enough points to give us a true picture of the graph. Later we will learn how to use calculus to sketch the graphs of functions quickly and reliably.

It is possible to graph equations containing two variables even when they do not represent functions. For example, the graph of $x^2 + y^2 = 1$ is a circle of radius 1 (Figure 1–18). Note that it is possible for a vertical line to intersect the graph in more than one point: the y-axis, for instance, intersects the circle at the points (0, 1) and (0, −1). But this means that the single value $x = 0$ is assigned the two numbers 1 and −1 by the equation, and hence the equation does not define y as a function of x. Thus an examination of the graph of an equation can tell us whether or not the equation defines a function.

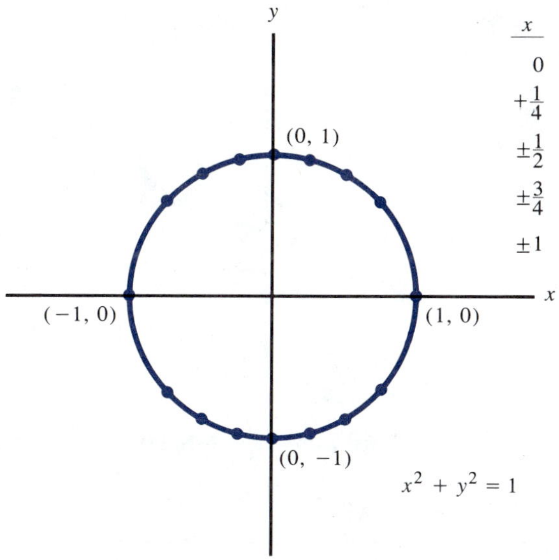

x	$y = \pm\sqrt{1-x^2}$	(x, y)
0	± 1	$(0, \pm 1)$
$\pm\frac{1}{4}$	$\pm\frac{\sqrt{15}}{4}$	$(\pm\frac{1}{4}, \pm\frac{\sqrt{15}}{4})$
$\pm\frac{1}{2}$	$\pm\frac{\sqrt{3}}{2}$	$(\pm\frac{1}{2}, \pm\frac{\sqrt{3}}{2})$
$\pm\frac{3}{4}$	$\pm\frac{\sqrt{7}}{4}$	$(\pm\frac{3}{4}, \pm\frac{\sqrt{7}}{4})$
± 1	0	$(\pm 1, 0)$

FIGURE 1–18

The Vertical Line Test

A graph is the graph of a function if and only if no vertical line intersects it in more than one point.

EXAMPLE 8 The graph of Figure 1–19 is the graph of a function, but that of Figure 1–20 is not.

Computer Manual: Program FNVALUE.

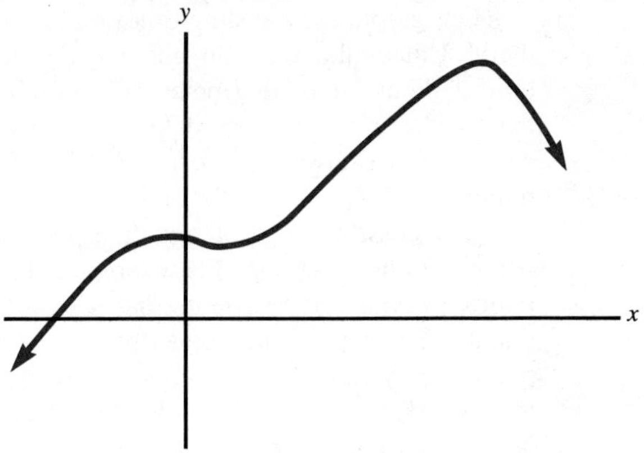

FIGURE 1–19 This Is the Graph of a Function

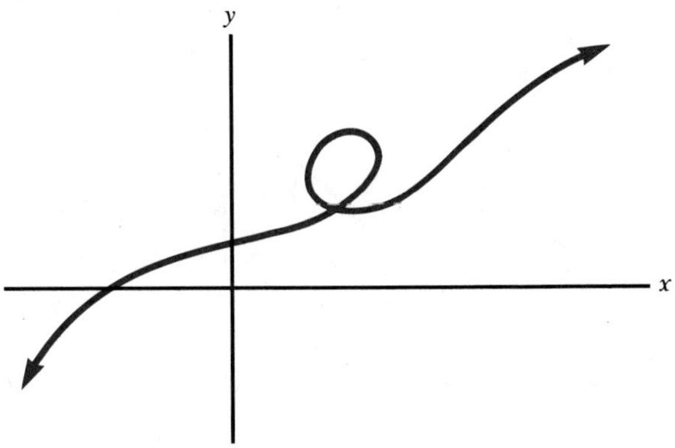

FIGURE 1–20 This Is Not the Graph of a Function

1.2 EXERCISES

Functions

1. Let f be the function whose defining equation is $f(x) = 4x - 2$. Find
   (a) $f(0)$ **(b)** $f(1)$ **(c)** $f(-3)$ **(d)** $f(0.25)$
   (e) $f(1.5)$ **(f)** $f(b)$ **(g)** $f(b + 2)$ **(h)** the domain of f
2. Let $y = x^2 - 4x + 3$. Find y when
   (a) $x = 0$ **(b)** $x = 1$ **(c)** $x = -3$ **(d)** $x = t^2$
   (e) $x = \frac{1}{2}$ **(f)** $x = -\frac{3}{4}$

   What is the domain of this function?

3. Let k be the function defined by $k(x) = 2\sqrt{x+1}$. Find

(a) $k(-1)$ (b) $k(0)$ (c) $k(3)$ (d) $k(2)$
(e) $k(b-1)$ if $b > 0$ (f) the domain of k

4. Let $y = \dfrac{2}{\sqrt{x}}$. Find y when

(a) $x = 1$ (b) $x = 4$ (c) $x = \dfrac{1}{4}$ (d) $x = c, \quad c > 0$

(e) $x = c^2$ if $c > 0$

What is the domain of this function?

Management

5. If a firm produces and sells x units of its product, its profit is

$$P(x) = -x^2 + 30x - 200$$

thousand dollars. Find the firm's profit if it makes and sells

(a) 0 units (b) 5 units (c) 10 units
(d) 15 units (e) 20 units (f) 25 units

6. If the price for a product is p dollars per unit, the demand for the product, that is, the number of units consumers will be willing to purchase, is given by the equation $d = 200{,}000 - 8000p$. Find the demand d for the product if its price per unit is

(a) \$1 (b) \$5 (c) \$15 (d) \$25

Life Science

Temperature may be measured in degrees Fahrenheit (°F) or degrees Celsius (°C). The equations

$$F = \frac{9}{5}C + 32 \qquad \text{and} \qquad C = \frac{5(F-32)}{9}$$

are used to convert from one temperature scale to the other. Use these equations in Exercises 7 and 8.

7. Convert to °C:

(a) 5°F (b) 32°F (c) 77°F (d) 212°F

8. Convert to °F:

(a) −10°C (b) 0°C (c) 20°C (d) 100°C

9. The size of a city's population in year t is given by

$$p(t) = \frac{210t + 400}{t + 2};$$

here $t = 0$ represents 1980 and $p(t)$ is in thousands of individuals. Find the size of the city's population in 1980 and 1985 and forecast the size of its population in 1992.

Social Science

10. The average score of high school seniors on a standardized mathematics test over a 10-year period is given by

$$A(t) = 480 - 12t + t^2, \qquad 0 \le t \le 10.$$

Find the average score on the test at the beginning, the middle ($t = 5$), and the end of the 10-year period.

Consumers are frequently asked to rank products or services on a numerical scale such as

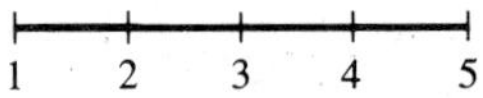

where 1 is the lowest ranking and 5 is the highest ranking. If students are asked to evaluate a professor's course using such a scale, the evaluation can be thought of as a function f from the set of all possible judgments about the course to the interval $[1, 5]$. Thus, f will assign to each judgment about the course a single numerical ranking between 1 and 5, inclusive. Exercises 11 and 12 refer to this evaluation function f.

11. Evaluate

(a) f(very poor) **(b)** f(poor) **(c)** f(average)
(d) f(above average) **(e)** f(excellent)

12. How would you evaluate

(a) f(halfway between average and above average)?
(b) f(two thirds of the way from poor toward average)?
(c) f(three fourths of the way from very poor toward poor)?

Graphing

In Exercises 13 through 18 graph the function defined by the equation.

13. $f(x) = 4 - 3x$ **14.** $g(x) = x^2$ **15.** $y = x^3$

16. $y = \sqrt{x - 1}$ **17.** $y = \dfrac{1}{x + 1}$ **18.** $h(x) = \begin{cases} x & \text{if } 0 < x < 5, \\ 2 & \text{if } x \ge 5. \end{cases}$

In Exercises 19 through 24 state whether or not the graph shown is the graph of a function; if it is, find the domain of the function.

19.

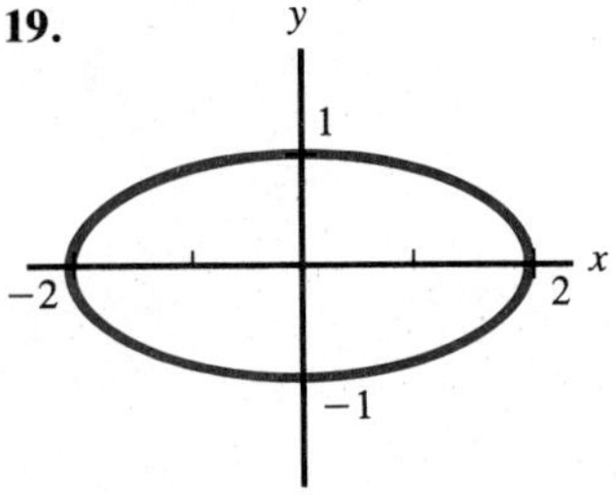

20.

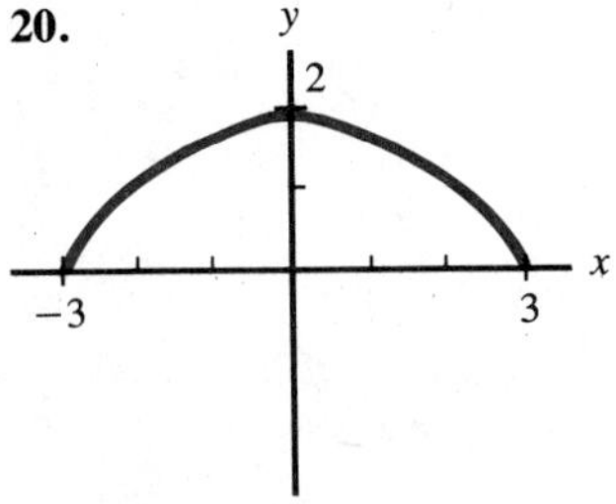

21.

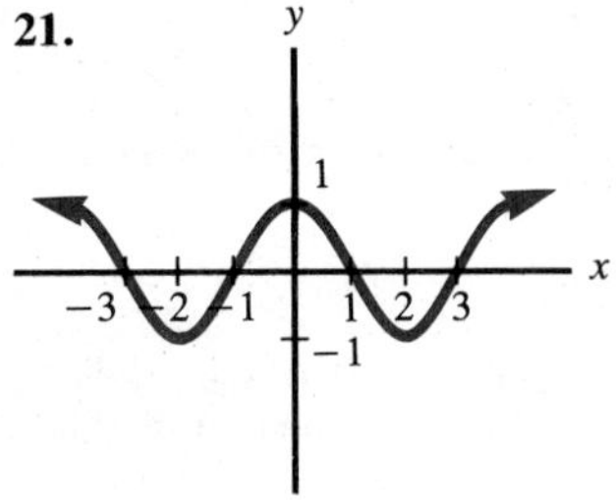

22.

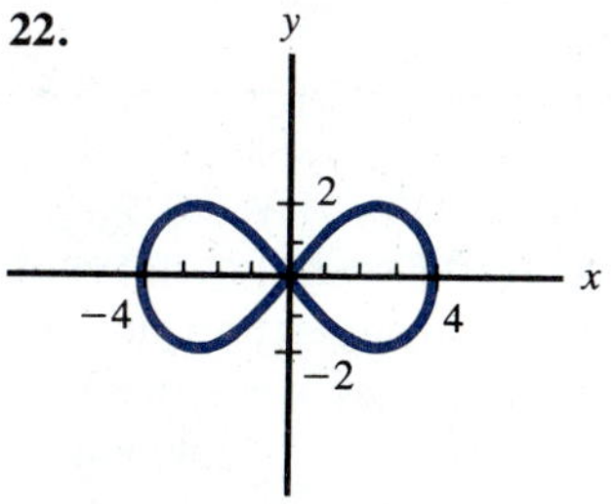

23.

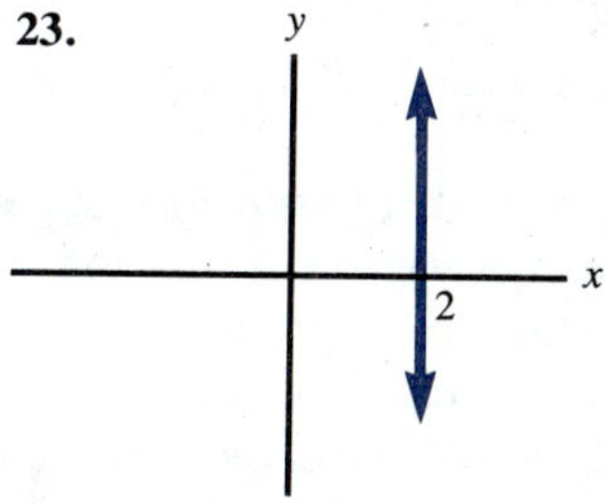

24.

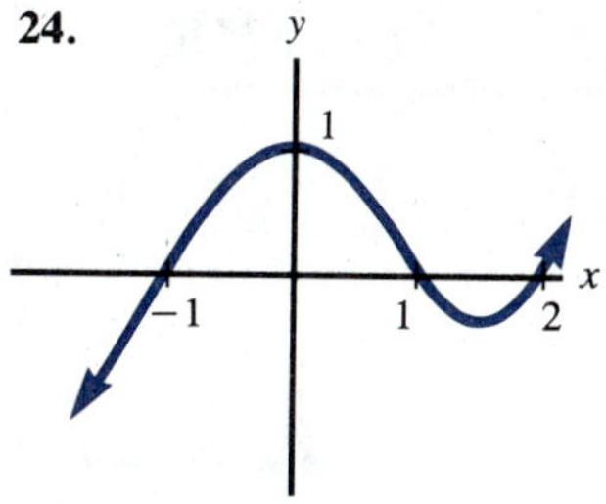

Management

25. Alpha Company's annual profits are described by the equation

$$P = -t^2 + 8t - 12,$$

where $t = 0$ represents 1980 and P is profit in year t in tens of thousands of dollars.

(a) Graph P as a function of t.

(b) When is the company making money? When is it losing money? What is its maximum profit and when does it occur?

26. The demand d for a product is given by the equation $d = -10p^2 + 100{,}000$, where p is the unit price of the product in dollars.

(a) Graph the function defined by this equation.

(b) Does demand increase or decrease as the unit price increases? How many units of the product could be given away free? At what price would demand for the product be zero?

Life Science

27. The time required for a patient to become unconscious after being injected with x units of an anesthetic, $x \geq 1$, is y minutes, where

$$y = \frac{3x}{x - 0.5}.$$

(a) Graph the function defined by this equation.

(b) What is the maximum time required for a patient injected with the anesthetic to become unconscious? Is there a minimum time?

28. Graph the function of Exercise 9. Is the size of the city's population limited?

Social Science

29. A psychologist performing an experiment with rats finds that they forget a task quickly unless they are given constant reinforcement. It appears that if no reinforcement is given, then t days after a rat learns a task it will remember only

$$y = \frac{100}{t + 1}$$

percent of the task. Graph this function. In theory, will a rat ever entirely forget a task it has once learned?

30. Graph the function of Exercise 10. When did the average score reach its lowest level?

1.3 ALGEBRA

We will frequently find it useful to combine functions to create new ones. For instance, we will soon see that if functions that describe a firm's cost and revenue are known, then the firm's profit function can be found by subtracting its cost function from its revenue function. In this section we show how new functions may be created from old ones, and we briefly review factorization of polynomial functions.

The Arithmetic of Functions

Functions can be added, subtracted, multiplied, and divided to form new functions. We illustrate with examples.

EXAMPLE 1 Let

$$f(x) = 2x^3 + 3x^2 + 5x - 3$$

and

$$g(x) = 4x^3 - 3x - 6.$$

We find the defining equation of the function $h = f + g$ by adding the corresponding terms of $f(x)$ and $g(x)$:

$$\begin{aligned} h(x) = f(x) + g(x) &= (2x^3 + 3x^2 + 5x - 3) + (4x^3 - 3x - 6) \\ &= (2x^3 + 4x^3) + 3x^2 + (5x - 3x) + (-3 - 6) \\ &= 6x^3 + 3x^2 + 2x - 9. \end{aligned}$$

Thus the defining equation for $h = f + g$ is

$$h(x) = 6x^3 + 3x^2 + 2x - 9.$$

Notice that both f and g have the set of real numbers as their domain and that $h = f + g$ does also.

If we can add functions, then we can also subtract one function from another. This is often useful when we are dealing with cost, revenue, and profit, since profit equals revenue minus cost.

Cost, Revenue, and Profit Functions

A **cost function** is a function that gives the cost of making x units of a product. A **revenue function** is a function that gives the revenue gained from selling x units of a product. A **profit function** is a function that gives the profit obtained from making and selling x units of a product. If C is the cost function, R the revenue function, and P the profit function for a product, then

$$P = R - C.$$

EXAMPLE 2 Suppose that if a firm makes and sells x units of its product, its cost function C is defined by the equation

$$C(x) = 20x + 1000$$

and its revenue function R by the equation

$$R(x) = 35x.$$

Here $x \geq 0$, and cost and revenue are in dollars. We find the firm's profit function P by subtracting C from R:

$$P(x) = R(x) - C(x) = 35x - (20x + 1000) = (35x - 20x) - 1000 = 15x - 1000.$$

Thus P is given by

$$P(x) = 15x - 1000.$$

The profit earned by making and selling 100 units is therefore

$$P(100) = 15(100) - 1000 = 500$$

dollars. Note that the domain of the function P is the interval $[0, +\infty)$, which is also the domain of R and C. Figure 1–21 shows the graphs of the cost function C, the revenue function R, and the profit function P on the same set of axes.

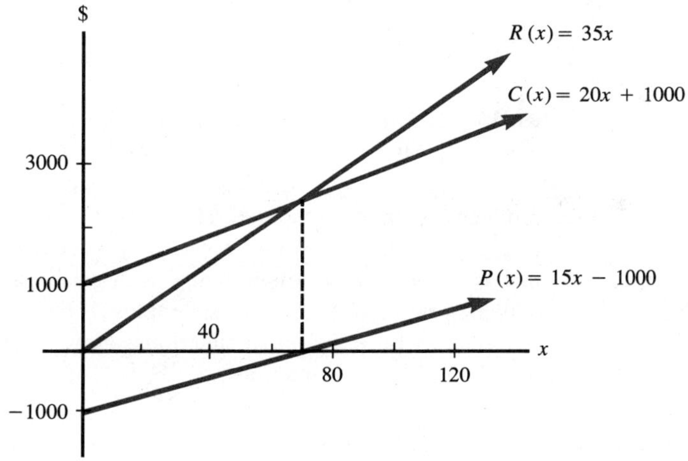

FIGURE 1–21

EXAMPLE 3 Suppose that

$$f(x) = 4x - 7 \qquad \text{and} \qquad g(x) = 2x + 6.$$

We find the defining equation of the function $h = fg$ by multiplying each term of $f(x)$ by each term of $g(x)$, using the rules of exponents, and then adding like terms:

$$\begin{aligned} h(x) = f(x)g(x) &= (4x - 7)(2x + 6) \\ &= (4x)(2x) + (4x)(6) + (-7)(2x) + (-7)(6) \\ &= 8x^2 + 24x - 14x - 42 \\ &= 8x^2 + 10x - 42. \end{aligned}$$

Therefore $h = fg$ has the defining equation

$$h(x) = 8x^2 + 10x - 42.$$

Notice that f, g, and $h = fg$ all have the set of real numbers as their domain.

EXAMPLE 4 If $p(x) = 2x + 1$ and $q(x) = x - 3$, we may form the function $h = p/q$ by dividing $p(x)$ by $q(x)$:

$$h(x) = \frac{p(x)}{q(x)} = \frac{2x + 1}{x - 3}.$$

Notice that the domain of p is the set of real numbers, as is that of q, but that the domain of $h = p/q$ is $\{x \mid x \neq 3\}$.

A **polynomial function** is a function whose defining equation is of the form

$$p(x) = a_n x^n + a_{n-1}x^{n-1} + \cdots + a_1 x + a_0,$$

where n is a nonnegative integer and the **coefficients** $a_n, a_{n-1}, \ldots, a_1, a_0$ are numbers. All the functions in Examples 1, 2, and 3 are polynomials, and, as these examples suggest, adding, subtracting, and multiplying polynomials always produces another polynomial. However, as Example 4 shows, dividing one polynomial by another does not necessarily produce another polynomial. A function that is a quotient of polynomials, such as the function $h = p/q$ in Example 4, is called a **rational function.** Examples 1 through 4 showed us how to perform arithmetic with polynomial functions. Now let us consider how to perform arithmetic with rational functions.

EXAMPLE 5 Suppose that

$$f(x) = \frac{x + 4}{x - 1} \qquad \text{and} \qquad g(x) = \frac{2x - 3}{x + 2}.$$

Let us find $f + g$, $f - g$, fg, and f/g.

To add f and g, we must find a common denominator for them: the product $(x - 1)(x + 2)$ of their denominators will serve as a common denominator. To write f and g as rational functions with the common denominator $(x - 1)(x + 2)$, we must multiply the numerator and denominator of $f(x)$ by $x + 2$ and the numerator and denominator of $g(x)$ by $x - 1$:

$$f(x) = \frac{x+4}{x-1}\frac{x+2}{x+2}, \qquad g(x) = \frac{2x-3}{x+2}\frac{x-1}{x-1}.$$

Therefore,

$$\begin{aligned} f(x) + g(x) &= \frac{x+4}{x-1}\frac{x+2}{x+2} + \frac{2x-3}{x+2}\frac{x-1}{x-1} \\ &= \frac{(x+4)(x+2) + (2x-3)(x-1)}{(x-1)(x+2)} \\ &= \frac{x^2+6x+8+2x^2-5x+3}{x^2+x-2} \\ &= \frac{3x^2+x+11}{x^2+x-2}. \end{aligned}$$

Similarly,

$$\begin{aligned} f(x) - g(x) &= \frac{x+4}{x-1}\frac{x+2}{x+2} - \frac{2x-3}{x+2}\frac{x-1}{x-1} \\ &= \frac{(x+4)(x+2) - (2x-3)(x-1)}{(x-1)(x+2)} \\ &= \frac{(x^2+6x+8) - (2x^2-5x+3)}{x^2+x-2} \\ &= \frac{-x^2+11x+5}{x^2+x-2}. \end{aligned}$$

To multiply f by g, we multiply their numerators and their denominators:

$$f(x)g(x) = \frac{x+4}{x-1}\frac{2x-3}{x+2} = \frac{(x+4)(2x-3)}{(x-1)(x+2)} = \frac{2x^2+5x-12}{x^2+x-2}.$$

To divide f by g, we invert the expression for $g(x)$ and multiply $f(x)$ by the result:

$$\frac{f(x)}{g(x)} = \frac{x+4}{x-1}\frac{x+2}{2x-3} = \frac{(x+4)(x+2)}{(x-1)(2x-3)} = \frac{x^2+6x+8}{2x^2-5x+3}.$$

Let us summarize the rules of arithmetic for rational functions:

The Arithmetic of Rational Functions

Let f and g be rational functions, with

$$f(x) = \frac{p(x)}{q(x)} \qquad \text{and} \qquad g(x) = \frac{r(x)}{s(x)}.$$

Then

$$f(x) + g(x) = \frac{p(x)s(x) + r(x)q(x)}{q(x)s(x)},$$

$$f(x) - g(x) = \frac{p(x)s(x) - r(x)q(x)}{q(x)s(x)},$$

$$f(x)g(x) = \frac{p(x)r(x)}{q(x)s(x)},$$

$$\frac{f(x)}{g(x)} = \frac{p(x)s(x)}{q(x)r(x)}.$$

Composition of Functions

There is another method for creating new functions from old ones, called **composition of functions.** To illustrate composition of functions, let f and g be defined by

$$f(x) = x^2 \qquad \text{and} \qquad g(x) = x + 1.$$

We can substitute the expression $g(x)$ for x in the equation for $f(x)$ to obtain

$$f(g(x)) = g(x)^2 = (x + 1)^2.$$

The equation

$$f(g(x)) = (x + 1)^2$$

then defines a new function called the **composite of g with f.** This procedure can be carried out using any functions f and g as long as the substitution of $g(x)$ for x in $f(x)$ makes sense.

Composite Functions

If f and g are functions, then the function whose defining equation is given by $f(g(x))$ is called the **composite of g with f.**

EXAMPLE 6 Let

$$f(x) = 3x + 7 \qquad \text{and} \qquad g(x) = -2x.$$

The composite of g with f is defined by $f(g(x))$, where

$$f(g(x)) = 3(g(x)) + 7 = 3(-2x) + 7 = -6x + 7.$$

Thus,

$$f(g(0)) = -6(0) + 7 = 7,$$
$$f(g(1)) = -6(1) + 7 = 1,$$
$$f(g(2)) = -6(2) + 7 = -5,$$

and so on.

Given two functions f and g, the composite of g with f is the function whose defining equation is given by $f(g(x))$. It is also possible to consider the composite of f with g: this is the function whose defining equation is given by $g(f(x))$.

EXAMPLE 7 Let

$$f(x) = 3x + 7 \qquad \text{and} \qquad g(x) = -2x.$$

The composite of f with g is defined by $g(f(x))$, where

$$g(f(x)) = -2(f(x)) = -2(3x + 7) = -6x - 14.$$

Note that $g(f(x)) \neq f(g(x))$.

EXAMPLE 8 Let

$$f(x) = x^2 - 2x + 2 \qquad \text{and} \qquad g(x) = 2x - 1.$$

Then

$$\begin{aligned} f(g(x)) &= (g(x))^2 - 2g(x) + 2 \\ &= (2x - 1)^2 - 2(2x - 1) + 2 \\ &= 4x^2 - 4x + 1 - 4x + 2 + 2 \\ &= 4x^2 - 8x + 5 \end{aligned}$$

and

$$\begin{aligned} g(f(x)) &= 2f(x) - 1 \\ &= 2(x^2 - 2x + 2) - 1 \\ &= 2x^2 - 4x + 4 - 1 \\ &= 2x^2 - 4x + 3. \end{aligned}$$

Note that again $f(g(x)) \neq g(f(x))$.

Composite functions are often useful when one variable is a function of a second and the second is a function of a third. We can then use composition of functions to write the first variable as a function of the third. Thus, if

$$y = x^2 + 1 \qquad \text{and} \qquad x = t + 5,$$

we may write y as a function of t by forming the composite function:

$$y = (t + 5)^2 + 1 = t^2 + 10t + 26.$$

EXAMPLE 9 Suppose the cats eat the rats and the rats eat the bats. Suppose also that if c is the number of cats, r the number of rats, and b the number of bats, then

$$c = 0.1r^{2/3} \qquad \text{and} \qquad r = \sqrt{b}.$$

How many cats can b bats support? The answer is

$$c = 0.1(\sqrt{b})^{2/3} = 0.1\sqrt[3]{b}.$$

For instance, 1000 bats can support 1 cat.

Factoring

Polynomial functions can sometimes be **factored** into a product of simpler polynomials. Factoring can be important if we need to solve an equation of the form $p(x) = 0$, where p is a polynomial. We illustrate with examples.

EXAMPLE 10 Let us factor the polynomial expression

$$p(x) = 3x^3 + 9x^2 - 30x.$$

Factorization begins with a search for the **common factors** of the terms of the polynomial. In this example each term has the common factor x, and the coefficients 3, 9, and -30 have the common factor 3. Such common factors can be **factored out** of the polynomial. Thus,

$$p(x) = 3x(x^2 + 3x - 10).$$

Now we seek to factor the polynomial expression in parentheses. If this expression does factor, it will do so as

$$x^2 + 3x - 10 = (x + a)(x + b),$$

where a and b are numbers. But since

$$(x + a)(x + b) = x^2 + (a + b)x + ab,$$

we see by comparing

$$x^2 + 3x - 10$$

with

$$x^2 + (a + b)x + ab$$

that we must have

$$ab = -10 \qquad \text{and} \qquad a + b = 3;$$

that is, we need to find a pair of numbers whose product is -10 and whose sum is 3. Such a pair is 5 and -2. Therefore

$$x^2 + 3x - 10 = (x + 5)(x - 2),$$

and

$$p(x) = 3x(x + 5)(x - 2).$$

We can check the factorization by multiplying out the right side to see that it is equivalent to the original expression for $p(x)$.

Now suppose we wish to find the x-intercepts of the polynomial function p whose defining equation is $p(x) = 3x^3 + 9x^2 - 30x$. Then we must solve the equation $p(x) = 0$ for x. If we use the factored form of $p(x)$, we must solve

$$3x(x + 5)(x - 2) = 0$$

for x. But

a product is zero if and only if at least one of its factors is zero.

Therefore the equation

$$3x(x + 5)(x - 2) = 0$$

tells us that either

$$x = 0, \quad x + 5 = 0, \quad \text{or} \quad x - 2 = 0;$$

that is, that either

$$x = 0, \quad x = -5, \quad \text{or} \quad x = 2.$$

Hence the x-intercepts of p are $(0, 0)$, $(-5, 0)$, and $(2, 0)$.

EXAMPLE 11 An ecologist studying a herd of minke whales has estimated that the population P of the herd in year t, where $t = 0$ represents 1984, is given by

$$P = -t^2 - 14t + 220.$$

Suppose the ecologist wants to know when the population will be 100 individuals. Then she must set $P = 100$ and solve the polynomial equation

$$100 = -t^2 - 14t + 220,$$

or

$$-t^2 - 14t + 120 = 0,$$

or

$$t^2 + 14t - 120 = 0.$$

To factor this polynomial, she must find two numbers whose product is -120 and whose sum is 14; 20 and -6 are such numbers. Therefore

$$t^2 + 14t - 120 = (t + 20)(t - 6),$$

so the equation in factored form is

$$(t + 20)(t - 6) = 0.$$

Setting each factor equal to zero, she finds that $t = -20$ and $t = 6$. Thus the population was 100 in 1964 ($t = -20$) and will be 100 again in 1990 ($t = 6$).

Computer Manual: Programs POLYRITH, RATRITH, COMPOSIT.

1.3 EXERCISES

The Arithmetic of Functions

1. If $f(x) = 2x - 5$ and $g(x) = 3x + 9$, find the defining equations for $f + g$, $f - g$, fg, and f/g. Also find $f(2) + g(2)$, $f(0) - g(0)$, $f(3)g(3)$, and $f(-1)/g(-1)$. What is the domain of $f + g$? of f/g?

2. If $f(x) = 7x + 2$ and $g(x) = x^2 + x - 1$, find the defining equations for $f + g$, $f - g$, fg, and f/g. Also find $f(3) - g(3)$ and $f(2)/g(2)$.

3. If $f(x) = x^2 + 6x - 10$ and $g(x) = -3x^2 + 2x - 13$, find the defining equations for fg and f/g. Also find $f(1)g(1)$ and $f(0)/g(0)$.

4. If $f(x) = 5x^4 - 6x^2 + 2x$ and $g(x) = -3x^3 + 7x^2 - 8x + 11$, find the defining equations for $f + g$ and $f - g$. Also find $f(2) + g(2)$ and $f(-1) - g(-1)$.

5. If $f(x) = \dfrac{x - 2}{x + 3}$ and $g(x) = \dfrac{2x + 4}{x + 1}$, find the defining equations for $f + g$, $f - g$, fg, and f/g. Also find $f(1) + g(1)$ and $f(0)/g(0)$. What is the domain of $f + g$? of fg?

6. If $f(x) = \dfrac{x^2 + 9}{x + 4}$ and $g(x) = \dfrac{x^2 + x}{x^2 - 3}$, find the defining equations for $f + g$, $f - g$, fg, and f/g. Also find $f(2) - g(2)$ and $f(3)g(3)$. What is the domain of fg?

7. If $f(x) = \dfrac{3x - 5}{x + 1}$ and $g(x) = \dfrac{(x - 3)(x + 2)}{4 - 2x}$, find the defining equations for $f + g$, fg, and f/g. What is the domain of $f + g$? of fg? of f/g?

8. If $f(x) = \dfrac{(x + 6)(2x - 8)}{(x - 1)(2x + 5)}$ and $g(x) = \dfrac{x(x + 6)}{x - 1}$, find the defining equations for $f - g$, fg, and f/g. What is the domain of $f - g$? of fg? of f/g?

Management

9. If a firm produces and sells x units of its product, its cost is $C(x)$ dollars and its revenue $R(x)$ dollars, where

$$C(x) = 200x + 10{,}000 \qquad \text{and} \qquad R(x) = 230x.$$

(a) Find the firm's profit function P.

(b) Find the firm's profit if it makes and sells 100 units. If the firm is currently producing and selling q units, what will be the additional profit if it makes and sells one more unit?

10. Each freezer unit is made by installing one cooling assembly in one insulated metal shell. If the cost of the raw materials used to make x cooling assemblies is given by

$$f(x) = 200x + 5000$$

and the cost of raw materials used to make x insulated metal shells is given by

$$g(x) = 400x + 2000,$$

find the function h that describes the total cost of raw materials used in making x freezer units. Graph h, f, and g on the same set of axes. Find the cost of raw materials if 20 freezer units are made.

Life Science

11. An ecologist studying wolves and their prey estimates that over an 8-year period the sizes of the wolf population and the rabbit population are given by

$$W(t) = 2000 + 400t - 50t^2$$

and

$$R(t) = 100{,}000 + 600t - 150t^2$$

respectively. Here $0 \le t \le 8$. Graph the functions W, R, and $W + R$ on the same set of axes and estimate when the peak population of wolves, the peak population of rabbits, and the peak population of wolves and rabbits combined occur.

Composition of Functions

In Exercises 12 through 19 find $f(g(x))$ and $g(f(x))$.

12. $f(x) = 4x$, $g(x) = 5 - 2x$.

13. $f(x) = 3x + 6$, $g(x) = x^2 + 4$.

14. $f(x) = x^2 - 3x + 12$, $g(x) = \sqrt{x + 1}$.

15. $f(x) = x^4 + 5x^2 + 1$, $g(x) = x + h$.

16. $f(x) = x^3 + x$, $g(x) = x^{2/3}$.

17. $f(x) = \dfrac{1}{x}$, $g(x) = x + h$.

18. $f(x) = \dfrac{x + 3}{x - 4}$, $g(x) = \dfrac{2}{x}$.

19. $f(x) = \dfrac{2x - 3}{7 + 2x}$, $g(x) = \dfrac{5x + 1}{x^2 - 1}$.

20. If $y = 5x + 3$ and $x = t^2 - 4$, express y as a function of t.

21. If $y = 4x - 16$ and $x = t + 4$, express y as a function of t.

22. If $u = 17v^2 + 2\sqrt{v} + 3$ and $v = w^{3/4}$, express u as a function of w.

23. If y and x are as in Exercise 21, express t as a function of y.

Management

24. The demand for a product depends on its price: if the product sells for p dollars per unit, demand for it is

$$d = 500p^2 + 3000p + 10{,}000$$

units. It is estimated that the price of the product will fall during the near future according to the equation

$$p = \frac{60}{t + 10},$$

where t is the time in days from the present. Find a function that expresses the demand for the product as a function of time. What is the demand for the product now? What will it be 5 days from now?

Life Science

25. The carbon monoxide concentration in the air of a large city depends on the number of cars driven in the city each day. The number of cars driven in the city is expected to increase with time. Suppose that if x cars are driven in the city each day, then the concentration of carbon monoxide in the atmosphere is
$$c(x) = 10^{-9}x^2 + 10^{-5}x + 1$$
parts per million (ppm); suppose also that the number of cars driven in the city per day is forecast to be
$$x = 10{,}000t + 100{,}000$$
in year t, where $t = 0$ represents 1987. Find a function that describes the carbon monoxide concentration as a function of time, and forecast this concentration in the year 1995.

Factoring

In Exercises 26 through 37 solve the equation $f(x) = 0$ for x.

26. $f(x) = x^2 - x - 12$
27. $f(x) = x^2 + 8x + 15$
28. $f(x) = x^2 - 16$
29. $f(x) = 18x - 2x^3$
30. $f(x) = x^2 + 4x + 4$
31. $f(x) = 4x^4 - 24x^3 + 36x^2$
32. $f(x) = 3x^2 - 9x - 12$
33. $f(x) = -x^3 + 9x^2 - 14x$
34. $f(x) = x^3 + x^2 - 2x$
35. $f(x) = x^6 - 2x^5 + x^4$
36. $f(x) = 2x^2 + x - 3$
37. $f(x) = \frac{1}{2} + \frac{1}{2}x - x^2$

Management

38. A firm is said to break even if its profit from making and selling its product is zero, since if this happens the firm is neither making nor losing money. Suppose Acme Company's profit is
$$P(x) = x^2 + 20x - 1500$$
dollars if it makes and sells x units. How many units must Acme make and sell in order to break even?

Life Science

39. It is estimated that the concentration of sulfur dioxide in the air over a large city will average
$$y = 100 + 6t - t^2$$
parts per million in year t, where $t = 0$ represents 1990. In what year will the concentration average 28 ppm?

Social Science

40. The number of riders on a city's transit system is given by
$$y = t^2 - 40t + 800,$$
where t is in years, with $t = 0$ representing 1970, and y is thousands of riders. When will ridership reach one half of the level of 1970?

1.4 LINEAR FUNCTIONS

A **linear function** is a polynomial function that can be written in the form

$$y = mx + b,$$

where m and b are numbers. A knowledge of linear functions will be required when we begin our study of calculus in the next chapter. In addition, linear functions are important because they can be used to model many practical situations. In this section we discuss graphing linear functions, finding the equation of a line, and finding the point where two lines intersect. We also illustrate linear models and their uses with descriptions of linear break-even analysis and supply–demand analysis.

Graphing Linear Functions

The graph of a linear function $y = mx + b$ is a line. (Throughout this book "line" means "straight line.") Since two points determine a line, when graphing a linear function we need only find two points on its graph.

EXAMPLE 1 The graph of the linear function

$$y = 2x + 4$$

is shown in Figure 1–22. Check that the points $(-1, 2)$ and $(1, 6)$ are on the line.

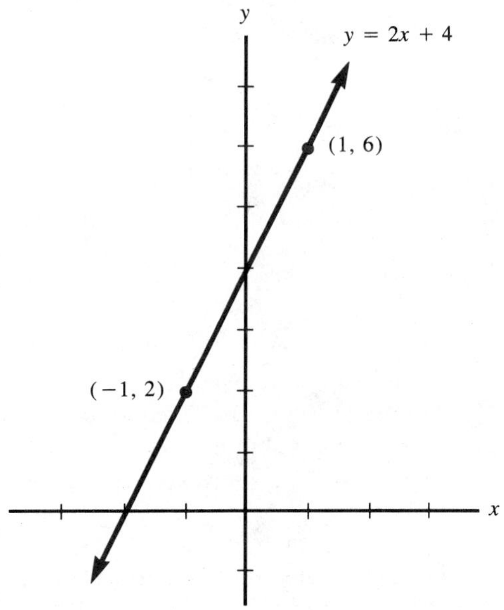

FIGURE 1–22

EXAMPLE 2 The equation

$$6x + 2y = 4$$

may be solved for y to obtain the linear function

$$y = -3x + 2;$$

thus the graph of the equation is a line. The line is depicted in Figure 1–23: here we used the x-intercept $(\frac{2}{3}, 0)$ and the y-intercept $(0, 2)$ as the two points we needed to draw the graph.

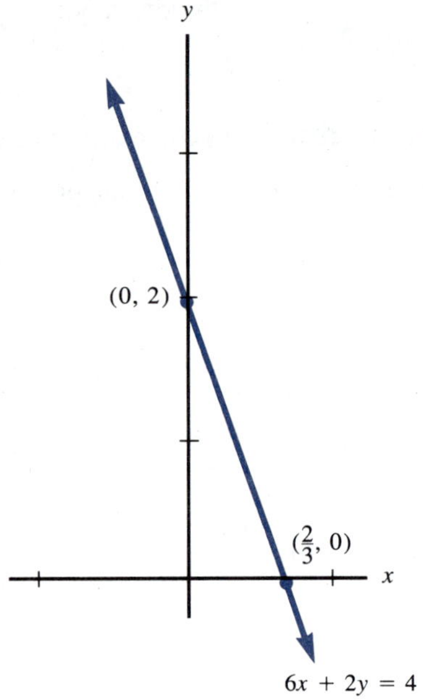

FIGURE 1–23

For a linear function $y = mx + b$, every increase of 1 in the value of x causes a change of m in the value of y. See Figure 1–24. Therefore a linear function $y = mx + b$ increases (if m is positive) or decreases (if m is negative) at the constant rate m. The number m is called the **slope** of the line. For example, the line of Figure 1–22 has slope $m = 2$: thus every increase of 1 in the value of x causes the value of y to increase by 2. Similarly, the line of Figure 1–23 has slope $m = -3$, so every increase of 1 in the value of x causes the value of y to decrease by 3. Note also that $y = mx + b$ has y-intercept $(0, b)$.

A vertical line such as that pictured in Figure 1–25 is *not* the graph of a function. (Why not?) A vertical line has no slope.

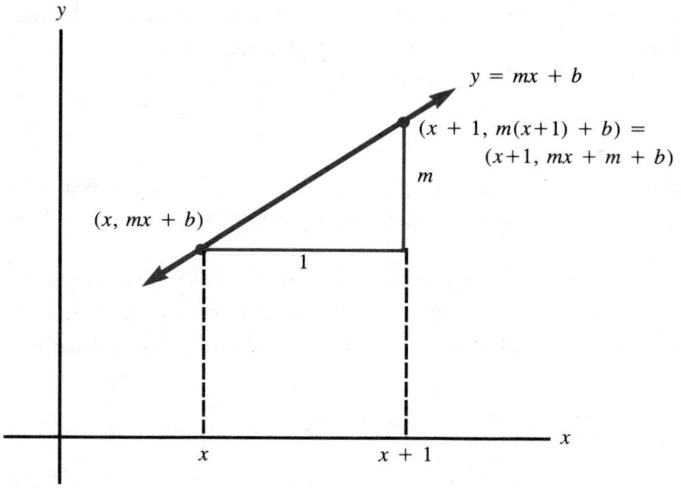

FIGURE 1–24

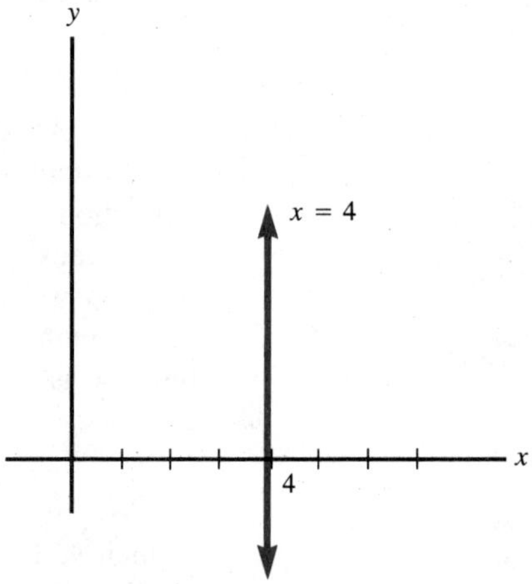

FIGURE 1–25

Finding the Equation of a Line

Every line is the graph of some equation, called the **equation of the line.** If a line is vertical, its equation is of the form $x = c$, where $(c, 0)$, is the x-intercept of the line. Thus the line of Figure 1–25 is the graph of the equation $x = 4$. If a nonvertical line has slope m and y-intercept $(0, b)$, then its equation is $y = mx + b$.

EXAMPLE 3 Let us find the equation of the line that has slope -4 and passes through the point $(0, 3)$. Since we are told the slope and y-intercept of the line, we can immediately write its equation $y = -4x + 3$.

EXAMPLE 4 Let us find the equation of the line that has slope 2 and passes through the point $(3, 5)$. Here we know the slope of the line but not its y-intercept. However, since $m = 2$, the equation must be of the form

$$y = 2x + b.$$

Since the point $(3, 5)$ is on the line, it must satisfy this equation. Hence

$$5 = 2(3) + b = 6 + b,$$

so $b = -1$. Therefore the equation is

$$y = 2x - 1.$$

We have noted that if a function is linear, then it changes at a constant rate equal to the slope of the line. The converse of this statement is also true: if a function changes at a constant rate, it is linear and the rate is the slope of the line.

EXAMPLE 5 Suppose Alpha Company's sales were \$100,000 in 1980 and have increased by the constant amount of \$20,000 each year. Let us find Alpha's annual sales S as a function of time t.

Because the change in Alpha's sales is constant at \$20,000 each year, the function we seek will be linear with slope $m = 20{,}000$. If we measure time in years, with $t = 0$ representing 1980, then we have $b = 100{,}000$. Hence the function is

$$S = 20{,}000t + 100{,}000.$$

We often want to find the equation of a line when its slope is not given explicitly. However, we can find the slope of a line if we know two points on the line. Thus, if the equation of the line is $y = mx + b$ and (x_1, y_1), (x_2, y_2) are points on the line, then we must have

$$y_1 = mx_1 + b \qquad \text{and} \qquad y_2 = mx_2 + b.$$

Subtracting the first equation from the second, we obtain

$$y_2 - y_1 = mx_2 - mx_1 = m(x_2 - x_1).$$

Therefore

$$m = \frac{y_2 - y_1}{x_2 - x_1}.$$

The Slope of a Line through Two Points

A nonvertical line passing through the points (x_1, y_1) and (x_2, y_2) has slope

$$m = \frac{y_2 - y_1}{x_2 - x_1}.$$

Geometrically, this result says that we can find the slope of a nonvertical line by choosing any two points (x_1, y_1), (x_2, y_2) on the line, using them to draw a right triangle as in Figure 1–26, and setting

$$m = \frac{\text{vertical ``rise''}}{\text{horizontal ``run''}} = \frac{y_2 - y_1}{x_2 - x_1}.$$

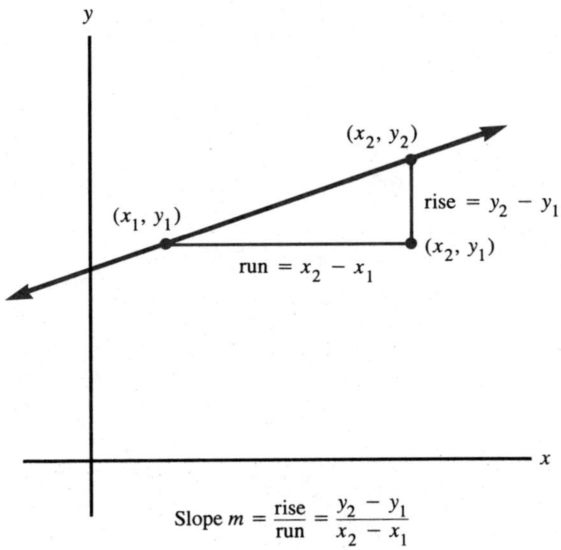

FIGURE 1–26

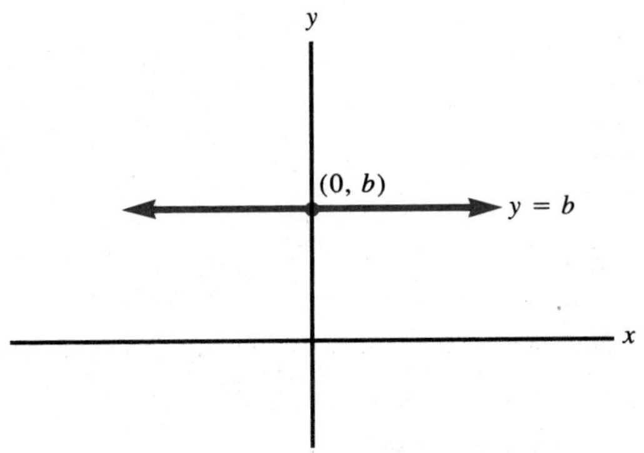

FIGURE 1–27 Graph of the Constant Function $y = b$

EXAMPLE 6 Let us find the equation of the line passing through the points (2, 5) and (−2, 13). Our strategy will be to find the slope m first, then substitute one of the points into the resulting equation to find b. Setting

$$(x_1, y_1) = (2, 5) \qquad \text{and} \qquad (x_2, y_2) = (-2, 13),$$

we have

$$m = \frac{y_2 - y_1}{x_2 - x_1} = \frac{13 - 5}{-2 - 2} = -2.$$

Therefore $y = -2x + b$. But since the point (2, 5) is on the line, we must have $5 = -2(2) + b$, from which we find that $b = 9$. Hence the equation of the line is

$$y = -2x + 9.$$

If the slope of a line is zero, the line is horizontal (why?) and its equation reduces to $y = b$. See Figure 1–27. A function of the form $y = b$ is called a **constant function.** Also, if a line has no slope then it is vertical and will have an equation of the form $x = c$.

Sometimes it is convenient to write the equation of a line in the form

$$Ax + By = C.$$

This is called the **general form** of the equation of a line.

EXAMPLE 7 Beta Company produces large and small boilers. Each large boiler uses 20 kilograms (kg) of steel, and each small one uses 10 kg of steel. If Beta has 1000 kg of steel on hand and wants to use it all up, it can make any combination of x large boilers and y small boilers as long as

$$20x + 10y = 1000, \qquad x \geq 0, \qquad y \geq 0.$$

The graph of this equation is the line segment shown in Figure 1–28.

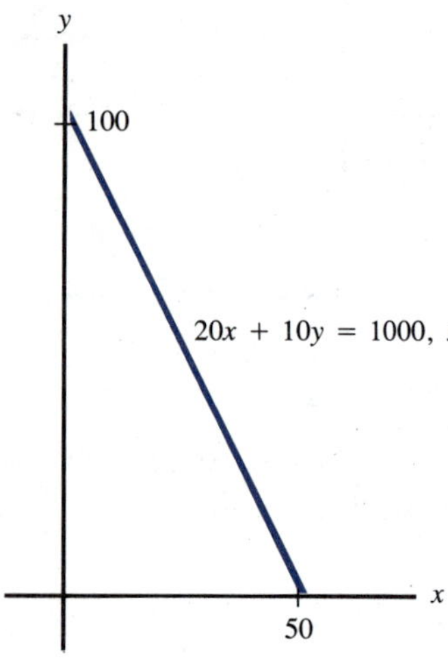

FIGURE 1–28

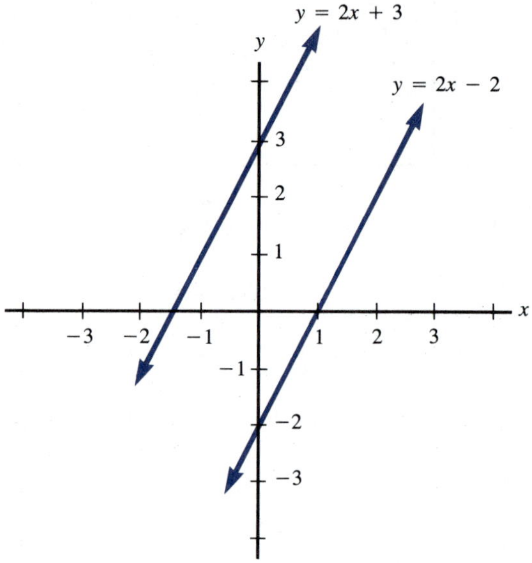

FIGURE 1–29 Parallel Lines

Intersection of Lines

Lines with the same slope are **parallel;** parallel lines do not intersect. The lines whose equations are

$$y = 2x - 2 \qquad \text{and} \qquad y = 2x + 3$$

are parallel and do not intersect. See Figure 1–29.

Lines whose slopes are not equal are not parallel, and they intersect in a point. Since the y-coordinate of the point of intersection must be the same for both lines, we can find the point of intersection by setting the right sides of their equations equal, solving for x, and then back-substituting to find y. We illustrate with an example.

EXAMPLE 8 Find the point of intersection of the lines whose equations are

$$y = 2x - 9 \qquad \text{and} \qquad y = x + 3.$$

We set

$$2x - 9 = x + 3$$

and solve for x to obtain $x = 12$. Substituting 12 for x in either $y = 2x - 9$ or $y = x + 3$ yields $y = 15$. Therefore the point of intersection of the lines is (12, 15). See Figure 1–30.

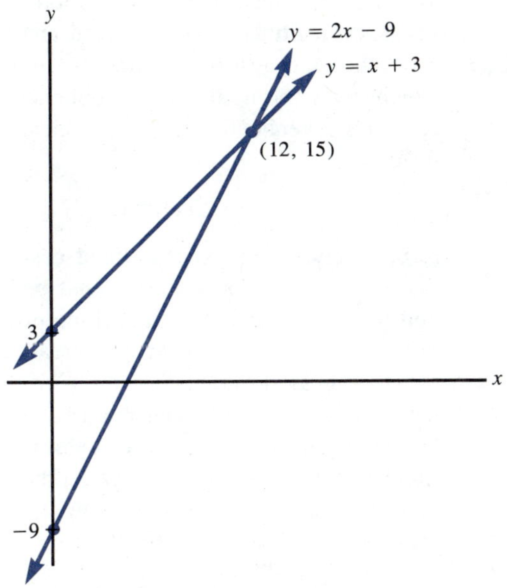

FIGURE 1–30

Finding the point of intersection of two lines can often be of practical importance.

EXAMPLE 9 Suppose each Brand X vitamin pill contains 10 milligrams (mg) of vitamin A and 20 mg of vitamin C, and each Brand Y pill contains 10 mg of A and 30 mg of C. If you need exactly 50 mg of A and 120 mg of C each day, how many pills of each brand should you take?

Let x denote the number of pills of Brand X and y the number of pills of Brand Y that you should take. Then your vitamin requirements imply that

$$\text{(vitamin A)} \qquad 10x + 10y = 50,$$

$$\text{(vitamin C)} \qquad 20x + 30y = 120.$$

Solving these equations for y, we have

$$y = -x + 5 \qquad \text{and} \qquad y = -\frac{2}{3}x + 4.$$

Therefore

$$-x + 5 = -\frac{2}{3}x + 4,$$

or

$$-\frac{1}{3}x = -1,$$

so $x = 3$. Hence $y = -3 + 5 = 2$.

Break-Even Analysis

We often use linear functions to model a firm's cost and revenue when performing **break-even analysis.** Break-even analysis begins with a **cost function** C, which gives the total cost of producing x units of a product, and a **revenue function** R, which gives the total revenue obtained from selling x units of the product. If the cost and revenue functions are linear, they will have defining equations of the form

$$C(x) = vx + F \qquad \text{and} \qquad R(x) = px,$$

respectively. Here F is the **fixed cost,** which is the cost of producing 0 units of the product, v is the **variable cost per unit,** which is the cost of producing each unit of product, and p is the selling price per unit. Note that since x is the number of units produced and sold, $x \geq 0$.

Figure 1–31 shows the graphs of typical linear cost and revenue functions on the same set of axes. The point of intersection of the cost line and the revenue line is called the firm's **break-even point,** and its x-coordinate is the **break-even quantity.** As Figure 1–31 shows, if the firm makes and sells its break-even quantity, then its cost will equal its revenue and it will break even. If the firm makes and sells fewer units than its break-even quantity, it will suffer a loss; if it makes and sells more, it will enjoy a gain.

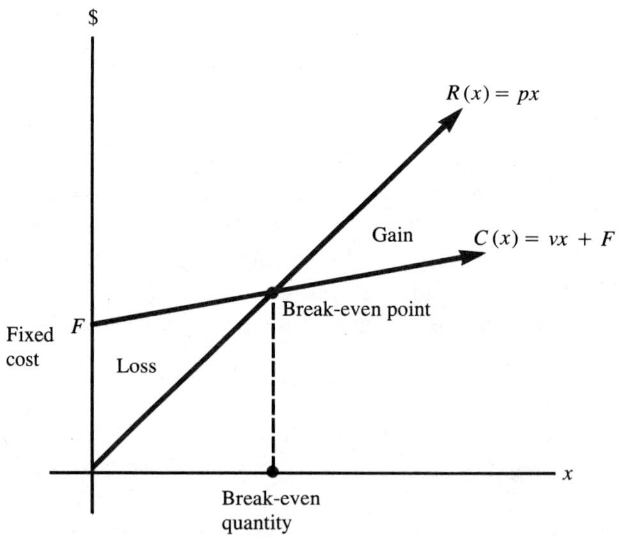

FIGURE 1–31

EXAMPLE 10 Suppose that Gamma Company's cost and revenue functions are given by

$$C(x) = 5x + 100{,}000 \qquad \text{and} \qquad R(x) = 7x,$$

respectively. Then Gamma's fixed cost is \$100,000, its variable cost per unit is \$5, and it sells each unit for \$7. We find Gamma's break-even quantity by setting cost equal to revenue and solving for x:

$$5x + 100{,}000 = 7x$$

yields $x = 50{,}000$. Therefore the break-even point is (50,000, 350,000). Thus if Gamma makes and sells 50,000 units, its cost and its revenue will both equal \$350,000 and the company will break even.

Note also that Gamma's profit function P is given by

$$P(x) = R(x) - C(x) = 2x - 100{,}000.$$

From this linear function we see that each unit produced and sold contributes \$2 toward profit. Note that we can also obtain the break-even quantity of 50,000 units by solving the equation $P(x) = 0$. Gamma Company's cost, revenue, and profit functions are graphed on the same set of axes in Figure 1–32.

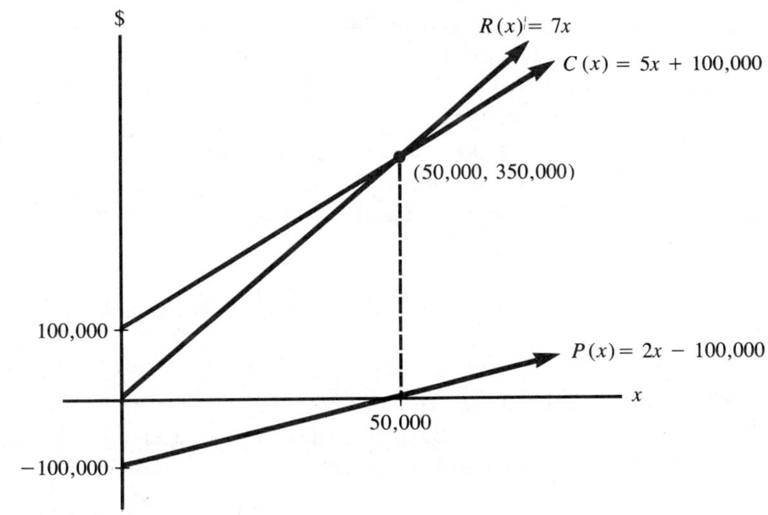

FIGURE 1–32

Supply and Demand

Linear functions are also often used to model situations of supply and demand. A **supply function** is a function that gives the number of units of a commodity that producers will supply to the market at any given price per unit. A **demand function** gives the number of units of the commodity that consumers will purchase at any given price per unit. Figure 1–33 shows typical linear supply and demand

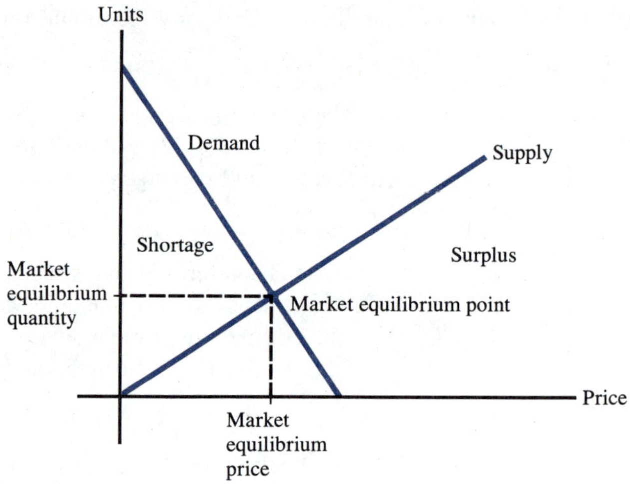

FIGURE 1–33

functions graphed on the same set of axes. (The graph in Figure 1–33 indicates that some units are supplied at every price greater than zero; this need not always be the case.) Note that as the unit price of the commodity increases, the quantity supplied to the market by producers increases while the quantity demanded by consumers decreases. The point of intersection of the supply and demand functions is called the **market equilibrium point,** and its price coordinate is the **market equilibrium price.** At the market equilibrium price, quantity supplied and quantity demanded are equal, so there will be neither a shortage nor a surplus of the product at this price.

EXAMPLE 11 Suppose the supply and demand functions for microcomputer stands are given by

$$s = 500p \qquad \text{and} \qquad d = -1000p + 120{,}000,$$

respectively. Here p is the unit price of the stands, in dollars, s is the quantity of stands supplied to the market by producers, and d is the quantity of stands demanded by consumers. If we set $s = d$, we have

$$500p = -1000p + 120{,}000,$$

or

$$1500p = 120{,}000,$$

which yields $p = 80$. It then follows that the market equilibrium point is (80, 40,000). See Figure 1–34. If the stands sell for \$80 each, both quantity supplied and quantity demanded will be 40,000 units. If they sell for less than \$80, quantity demanded will exceed quantity supplied, and there will be a shortage of the stands; if they sell for more than \$80 each, quantity supplied will exceed quantity demanded and there will be a surplus.

Computer Manual: Programs TWOPOINT and INTRSECT.

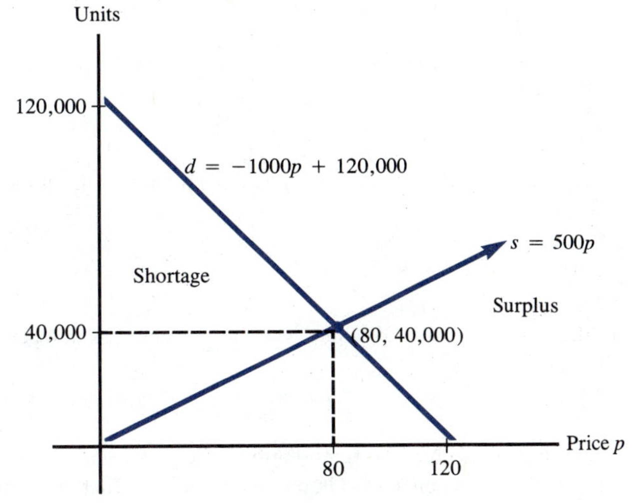

FIGURE 1–34

1.4 EXERCISES

Graphing Linear Functions

In Exercises 1 through 8 graph the equation.

1. $y = 4x - 8$ **2.** $y = -x + 12$ **3.** $y = 2x$

4. $2x + 3y = 12$ **5.** $2x - 5y = 10$ **6.** $2x + 4y = 0$

7. $y = 4$ **8.** $x = -2$

Management

9. Acme Company's sales are described by the function

$$S = 200t + 500,$$

where t is time in years with $t = 0$ representing 1985 and S is annual sales in thousands of dollars. Graph this function. What were Acme's sales in 1985? Are sales increasing or decreasing each year? by how much?

10. Baker Corporation makes regular and deluxe model radios. The numbers of regular and deluxe radios the firm can produce are related by the equation

$$20x + 30y = 60{,}000, \qquad x \geq 0, \quad y \geq 0,$$

where x is the number of regular radios and y is the number deluxe radios. Graph this equation. Can Baker produce 1200 regular and 1200 deluxe radios? 1800 regular and 800 deluxe? 1600 regular and 1000 deluxe? What is the maximum number of each kind of radio that the firm can produce?

Life Science

11. If the concentration of pollutants in a lake is x mg/L, then the proportion of fish eggs that will hatch is

$$y = 0.25 - 0.005x.$$

Graph this equation. For every increase of 1 mg/L in pollution, what additional proportion of fish eggs fail to hatch? At what level of pollution will all the eggs fail to hatch?

Social Science

12. In a certain state the proportion of voters who are registered as independents is given by the equation

$$-\frac{1}{5}t + 20y = 9,$$

where t is time in years with $t = 0$ representing 1980 and y is the proportion of registered voters who are independents. Graph this equation. What proportion of registered voters were independents in 1970? in 1980? in 1984? What is the change in the proportion of independents each year?

Finding the Equation of a Line

In Exercises 13 through 20 find the equation of the line.

13. The line with slope -3 and y-intercept $(0, 2)$.
14. The line with slope 5 and y-intercept $(0, -3)$.
15. The line passing through the point $(4, 8)$ with slope 6.
16. The line passing through the point $(-5, -9)$ with slope -2.
17. The line passing through the points $(2, 6)$ and $(3, 8)$.
18. The line passing through the points $(5, -2)$ and $(0, -7)$.
19. The line passing through the points $(6, 2)$ and $(4, 2)$.
20. The line passing through the points $(5, -2)$ and $(5, 8)$.
21. Let F be temperature in degrees Fahrenheit, and let C be temperature in degrees Celsius. When $F = 32$, $C = 0$; when $F = 212$, $C = 100$. The relation between the two temperature scales is linear. Find an equation that relates F and C.

Management

22. Cramer Company's profits have been declining at the constant rate of \$7000 per year and were \$58,000 in 1982. Find a linear equation describing Cramer's profits as a function of time, and then find

    (a) the firm's profit in 1986; **(b)** the year in which the firm will lose \$5,000.

23. Simple interest is interest earned on the amount originally invested. Suppose you invest \$10,000 at a simple interest rate of 10% per year. Find a function that describes the value of your investment t years after you make it. Then find

    (a) the value of the investment 6 years after you make it;
    (b) the time required for the value of the investment to grow to \$17,000.

24. Farson Corporation produces small and large clocks. Their production cost is \$25 for each small clock and \$35 for each large one. Farson's production budget for the clocks is \$105,000, and management intends to use the entire amount. Find and graph an equation describing the number of each size clock the company can produce. How many small clocks can the firm produce if it wants to make 1250 large clocks? What is the maximum number of each size clock the firm can produce?

25. The advertising director of Theta Corporation knows that when the firm spent \$100,000 on broadcasting TV ads the ads were seen by 2,500,000 people and when the firm spent \$300,000 broadcasting the ads, they were seen by 3,450,000 people. Find a function that expresses audience size as a function of amount spent, assuming that these quantities are linearly related. What is the significance of the slope of this linear function? If the ads must be seen by at least 5,000,000 people, how much must be budgeted for them? What size audience can be reached if the budget for the ads is \$1,000,000?

Life Science

26. A pediatrician studying babies finds that, on average, newborns are 53 centimeters (cm) tall and grow at the rate of 2.1 cm per month until the age of 9 months. Find

(a) a function that describes the relationship between average height and age in months;
(b) the average height of babies at age 4.5 months;
(c) the number of months required for the average height to reach 64 cm.

27. The concentration of a chemical accidentally spilled into a river is decreasing at a constant rate and was 600 ppm 1 month after the spill and 510 ppm 3 months after the spill.

(a) Find a function that describes the concentration of the chemical t months after the spill.
(b) What was the initial concentration of the chemical?
(c) How long will it take until all traces of the chemical are gone?

Social Science

28. Newly hired salespersons were given an aptitude test, and one year later their test scores were correlated with the total dollar value of the sales they made during their first year of employment. It was found that those who scored 52 on the test averaged \$119,000 in sales their first year, and those who scored 74 averaged \$163,000 in first-year sales. Assuming a linear relationship between score on the test and average first-year sales, find a function describing this relationship. What is the significance of the slope of this linear function? What is the significance of its y-intercept?

Intersection of Lines

In Exercises 29 through 34 find the point of intersection of the given lines.

29. $y = 2x - 5,\ y = 5x + 7$

30. $y = -3x + 1,\ y = -2x + 2$

31. $-2x + 4y = 8,\ 2x - 3y = 12$

32. $3x + y = 4,\ -5x + 4y = 16$

33. $-3x + 6y = 2,\ y = 0.5x - 5$

34. $5x + 4y = 10,\ 15x + 12y = 21$

Management

35. Triangle Company manufactures small and large wheelbarrows. Two worker-hours are needed to produce each small wheelbarrow, and three worker-hours are required to make each large one. Each wheelbarrow has one rubber wheel. There are 126 worker-hours and 52 rubber wheels available today. How many of each size wheelbarrow will Triangle produce today?

36. An investor has $10,000 and wants to invest part of the money in stock A and part in stock B. Stock A pays a dividend of 6 percent per year, stock B a dividend of 8 percent per year. How much should be invested in each stock if the annual dividends must amount to $750?

Life Science

37. A certain habitat currently supports 1618 individuals of animal species A and none of species B. Biologists plan to introduce 2 individuals of species B into the area. They estimate that if this is done the population of species A will thereafter decrease by 50 individuals per year while that of species B will increase by 30 individuals per year. How long will it take after the introduction of species B until the populations of the two species in the habitat are the same?

Social Science

38. National mathematics test scores for first-year college students have been decreasing linearly at a rate of 4.5 points per year. The average score in 1985 was 478. A certain college knows that its students' mathematics test scores have been increasing linearly, with the average score being 385 in 1983 and 397 in 1986. When will the average test score for this college reach the national average?

39. In 1980, 48.6 percent of registered voters in a certain state were Democrats, 36.4 percent were Republicans, and the rest were independents. The percentage of registered Democrats has been declining at a constant rate of 2 percent per year, the percentage of registered Republicans has been declining at a constant rate of 0.8 percent per year, and the percentage of registered independents has been increasing at a constant rate of 1.9 percent per year. When will the number of independents equal the number of

(a) Republicans?
(b) Democrats?
(c) Republicans and Democrats combined?

Break-Even Analysis

40. Ajax Company's fixed cost is $100,000, its variable cost per unit is $6, and it sells each unit it makes for $10. Find Ajax's cost and revenue functions and its break-even point.

41. Boodle, Inc. has cost and revenue functions given by

$$C(x) = 15x + 35{,}000 \qquad \text{and} \qquad R(x) = 20x,$$

respectively. Graph these functions on the same set of axes. What are the significances of the slopes and intercepts of these functions? Find Boodle's break-even point. Find Boodle's profit function. What are the significances of its slope and intercepts?

42. Chaco Corporation's profit function is given by $P(x) = 7x - 280{,}000$. Graph this function and find Chaco's break-even quantity.

Supply and Demand

43. The supply and demand functions for a certain product are

$$s = 8000p \qquad \text{and} \qquad d = -12{,}000p + 200{,}000,$$

respectively. Graph these functions on the same set of axes. Find the market equilibrium point for the product. What are the significances of the slopes and intercepts of these functions?

44. Repeat Exercise 43 if $s = 8000p - 6000$.

45. When book bags sold for \$10 each, the quantity supplied to the market was 15,000 units and the quantity demanded by consumers was 90,000 units. When they sold for \$20 each, the quantity supplied was 40,000 units and the quantity demanded was 60,000 units. Assuming linearity of supply and demand, find the supply and demand functions for book bags and their market equilibrium point.

1.5 QUADRATIC AND RATIONAL FUNCTIONS

It will be helpful for us if we have some knowledge of the characteristics of quadratic and rational functions. In this section we discuss these functions and their graphs.

Quadratic Functions

A polynomial function whose defining equation is of the form

$$y = ax^2 + bx + c,$$

where a, b, c are numbers and $a \neq 0$, is called a **quadratic function.** The graph of a quadratic function is called a **parabola:** if $a > 0$, the parabola opens upward, as in Figure 1–35; if $a < 0$, it opens downward, as in Figure 1–36. The point labeled V in each figure is the **vertex** of the parabola. Every parabola which is the graph of a function is symmetric with respect to a vertical line through its vertex; in other words, the two halves of the graph on either side of such a vertical line are mirror images of one another.

When drawing the graph of a quadratic function we should find its intercepts.

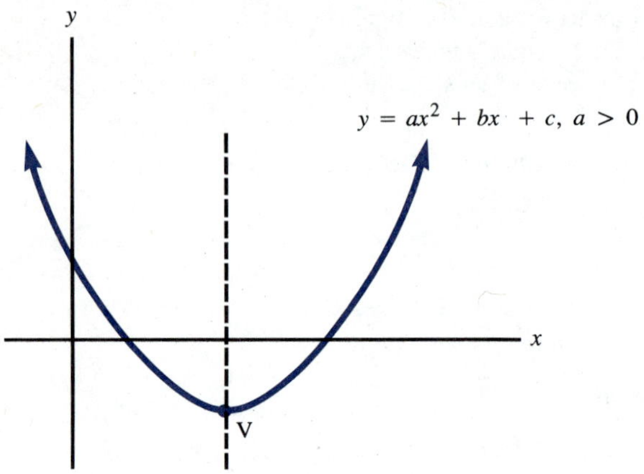

FIGURE 1–35

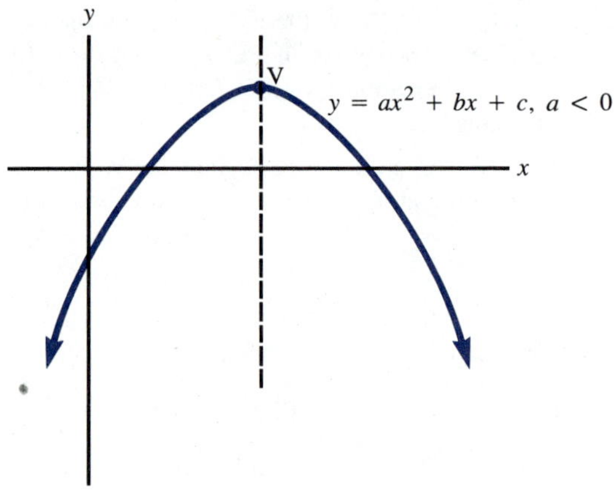

FIGURE 1–36

The y-intercept of the quadratic function $y = ax^2 + bx + c$ is the point $(0, c)$ (why?), and its x-intercepts may be found by solving the equation

$$ax^2 + bx + c = 0$$

for x. We can solve this equation by factoring, as discussed in Section 1.3, or by means of the **quadratic formula.**

The Quadratic Formula

Given a quadratic equation $ax^2 + bx + c = 0$:

1. If $b^2 - 4ac \geq 0$, the solutions of the equation are

$$x = \frac{-b \pm \sqrt{b^2 - 4ac}}{2a}.$$

2. If $b^2 - 4ac < 0$, the equation has no real solutions.

The quantity $b^2 - 4ac$ is called the **discriminant** of the quadratic.

(A derivation of the quadratic formula is outlined in Exercise 15 at the end of this section.)

We illustrate the process of graphing quadratic functions with examples.

EXAMPLE 1 Let us graph the quadratic function whose defining equation is

$$y = x^2 - 2x - 3.$$

Here $a = 1 > 0$, so the graph is a parabola that opens upward. The y-intercept is the point $(0, -3)$. To find the x-intercepts (if any), we must solve the equation

$$x^2 - 2x - 3 = 0$$

for x. Since this can be factored as

$$(x - 3)(x + 1) = 0,$$

its solutions are $x = 3$ and $x = -1$. Hence the x-intercepts are $(3, 0)$ and $(-1, 0)$. Because of the symmetry of the parabola, the x-coordinate of the vertex lies halfway between the x-intercepts and hence in this case is $x = 1$. But if $x = 1$,

$$y = 1^2 - 2(1) - 3 = -4.$$

Therefore the vertex is $(1, -4)$. The graph is shown in Figure 1–37.

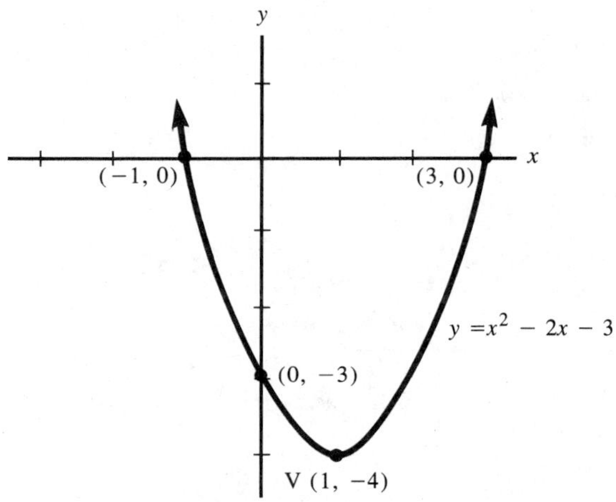

FIGURE 1–37

EXAMPLE 2 Let us graph the quadratic function defined by

$$y = -x^2 + 2x + 4.$$

Here $a = -1$, $b = 2$, and $c = 4$. Since $a < 0$, the parabola opens downward. The y-intercept is $(0, 4)$. We find the x-intercepts by using the quadratic formula to solve

$$-x^2 + 2x + 4 = 0.$$

$$x = \frac{-2 \pm \sqrt{2^2 - 4(-1)(4)}}{2(-1)} = \frac{-2 \pm \sqrt{20}}{-2}$$

$$= \frac{-2 \pm 2\sqrt{5}}{-2} = 1 \pm \sqrt{5}.$$

Thus the x-intercepts are $(1 + \sqrt{5}, 0)$ and $(1 - \sqrt{5}, 0)$. [Since $\sqrt{5} \approx 2.24$, these points are approximately $(3.24, 0)$ and $(-1.24, 0)$.] Again, the x-coordinate of the vertex will be halfway between the x-intercepts; hence it is $x = 1$, which implies that

$$y = -(1)^2 + 2(1) + 4 = 5.$$

Thus the vertex is $(1, 5)$. The graph is pictured in Figure 1–38.

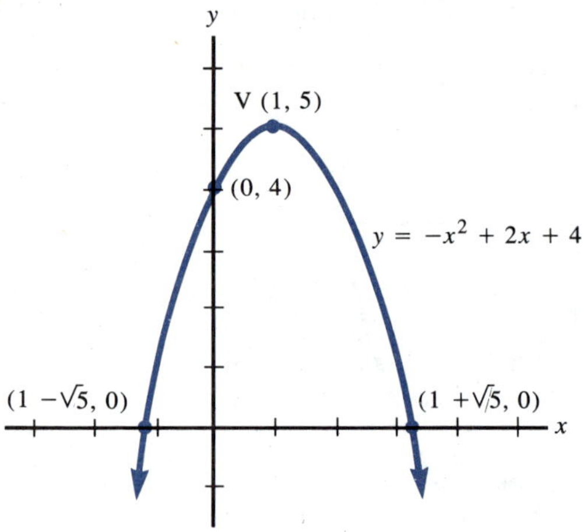

FIGURE 1–38

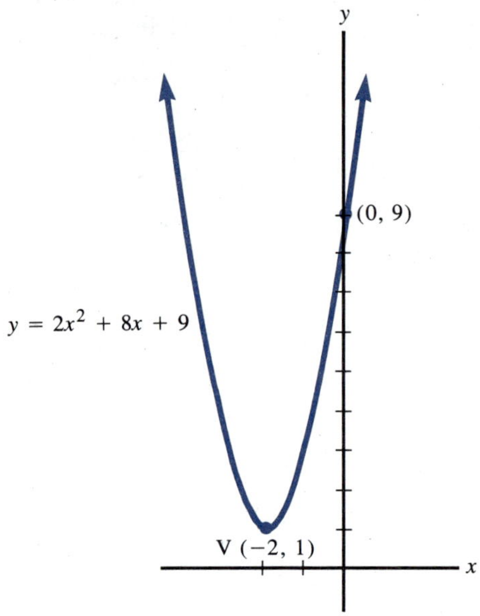

FIGURE 1–39

EXAMPLE 3 The graph of the quadratic function given by

$$y = 2x^2 + 8x + 9$$

is depicted in Figure 1–39. The parabola opens upward because $a = 2 > 0$, the y-intercept is $(0, 9)$, and there are no x-intercepts because

$$b^2 - 4ac = 8^2 - 4(2)(9) = -8 < 0.$$

In this case the vertex $(-2, 1)$ was found by trial and error. In Chapter 3 we will show how to use calculus to sketch the graph of a function; this will allow us to find the vertex of any quadratic function quite easily.

Rational Functions

In Section 1.3 we defined a **rational function** as one that is a quotient of polynomials. An important characteristic of many rational functions is their possession of **asymptotes:** an asymptote is a line that the graph approaches arbitrarily closely. In Figure 1–40 the line $x = 1$ is a **vertical asymptote,** and the line $y = 2$ is a **horizontal asymptote.**

The graph of a rational function $f = p/q$ has a vertical asymptote at $x = c$ if $q(c) = 0$ while $p(c) \neq 0$. Thus, to find the vertical asymptotes of f, we reduce the fraction $p(x)/q(x)$ to its lowest terms, set the denominator equal to zero, and solve for x. For the time being we will find horizontal asymptotes by trial and error. In the next chapter we will show how to find horizontal asymptotes using calculus.

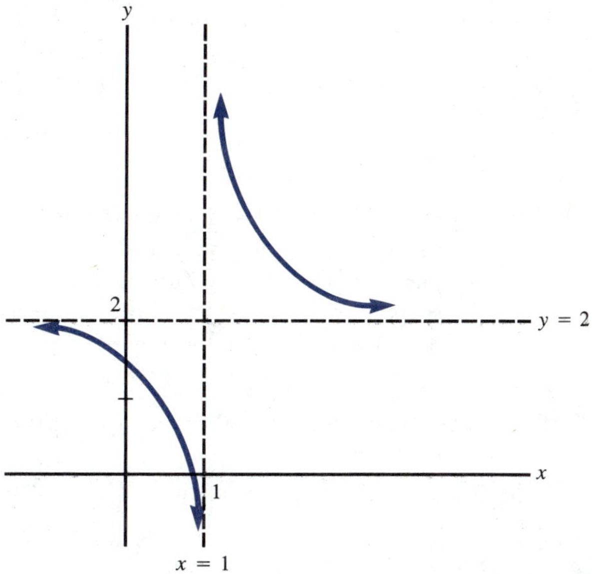

FIGURE 1–40

EXAMPLE 4 Let us graph the rational function f given by

$$f(x) = \frac{x + 2}{x - 2}.$$

The fraction $(x + 2)/(x - 2)$ is already in lowest terms, so we set $x - 2 = 0$ and solve to obtain $x = 2$. Therefore there is a vertical asymptote at $x = 2$, which we indicate on the cartesian coordinate system by a dotted vertical line. See Figure 1–41. The y-intercept of the function is $(0, f(0)) = (0, -1)$. Since the value of a fraction is zero if and only if the value of its numerator is zero

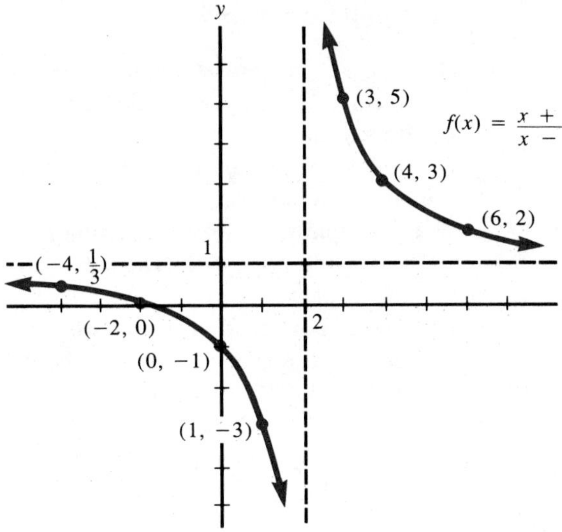

FIGURE 1–41

while the value of its denominator is not, we find the x-intercepts of $f(x)$ by solving $x + 2 = 0$ to obtain $x = -2$. Now that we have the vertical asymptotes and the intercepts for the function, we plot some additional points to determine the shape of the graph. These points are indicated in Figure 1–41. Note that the line $y = 1$ is a horizontal asymptote for the function.

EXAMPLE 5 Let us graph the rational function g defined by

$$g(x) = \frac{2x^2 - 8}{x^2 - 1}.$$

Factoring the numerator and denominator of the expression for $g(x)$, we obtain

$$g(x) = \frac{2(x - 2)(x + 2)}{(x - 1)(x + 1)}.$$

Since this fraction is in lowest terms, we can find the vertical asymptotes of g by setting its denominator equal to zero and solving for x. Thus the vertical asymptotes are the lines $x = 1$ and $x = -1$. The y-intercept is the point $(0, g(0)) = (0, 8)$. We find the x-intercepts by setting the numerator of $g(x)$ equal to zero and solving for x; they are $(2, 0)$ and $(-2, 0)$. The graph of g is shown in Figure 1–42. Note that the line $y = 2$ is a horizontal asymptote.

Computer Manual: Programs QUADFORM and PARABOLA.

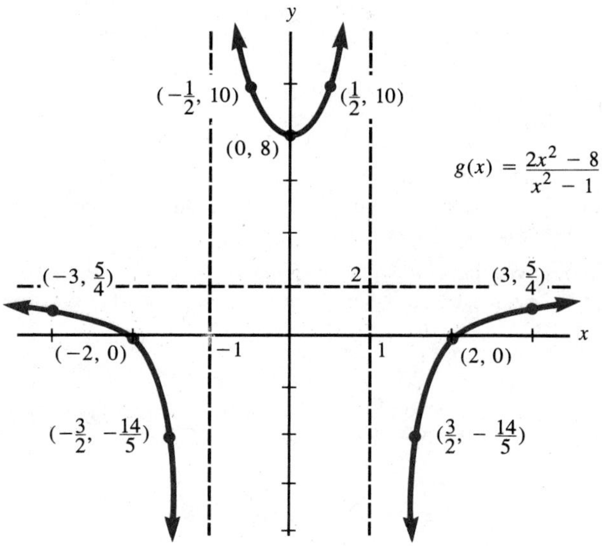

FIGURE 1–42

1.5 EXERCISES

Quadratic Functions

In Exercises 1 through 8 graph the quadratic function whose defining equation is given.

1. $y = x^2 + x - 30$ **2.** $y = -x^2 + 6x - 8$ **3.** $y = -x^2 - 10x - 25$

4. $y = x^2 - 6x + 9$ **5.** $y = x^2 - 5x + 3$ **6.** $y = -2x^2 + 6x - 1$

7. $y = -2x^2 + 7x + 8$ **8.** $y = 0.5x^2 - 3x + 5$

Management

9. Abelard's cost function is given by

$$C(x) = 2x^2 - 20x + 3200,$$

and its revenue function by

$$R(x) = x^2 + 100x.$$

Draw the graphs of C and R on the same set of axes. Find Abelard's break-even quantities by setting cost equal to revenue and solving for x.

10. Referring to Exercise 9, find Abelard's profit function and draw its graph. What is Abelard's maximum profit? How many units must the firm make and sell to attain its maximum profit?

11. It costs Brandon Camera Company \$100 to make each camera, with a fixed cost of \$0. If the firm charges $200 - x$ dollars for each camera, it will sell x of them.

(a) Find Brandon's profit function.
(b) Graph Brandon's profit function.
(c) What is Brandon's maximum profit?

Life Science

12. The concentration of sulfur dioxide in the air of a city from 7 A.M. to 7 P.M. is given by

$$y = -0.06t^2 + 0.9t + 4.2,$$

where y is the concentration in ppm and t is time in hours, with $t = 0$ representing 7 A.M. Graph this function. At what time of day is the concentration of sulfur dioxide greatest? What is the maximum concentration?

13. The furbish lousewort plant was placed on the endangered species list in 1985, and as a result it has been estimated that the number of plants that will germinate in year t will be

$$y = 0.02t^2 - 0.08t + 0.16,$$

where $t = 0$ represents 1985 and y is in thousands. Graph this function. When will annual germinations of the furbish lousewort be at a minimum?

Social Science

14. Voter interest, as measured by the percentage of eligible voters who state that they intend to vote, usually peaks before election day. A political scientist finds that t days before an election, the percentage y of eligible voters who intend to vote is given by

$$y = -0.07t^2 + 1.26t + 60.$$

Here $0 \le t \le 30$. Graph this function. When is voter interest at its peak?

15. Derive the quadratic formula by showing that

$$ax^2 + bx + c = a\left(x + \frac{b}{2a}\right)^2 + c - \frac{b^2}{4a}$$

and then solving

$$a\left(x + \frac{b}{2a}\right)^2 + c - \frac{b^2}{4a} = 0$$

for x.

Rational Functions

In Exercises 16 through 23 graph the rational function whose defining equation is given.

16. $y = \dfrac{x}{x - 1}$

17. $y = \dfrac{x + 4}{x + 1}$

18. $y = \dfrac{(x + 4)(x - 6)}{(x - 2)(x + 3)}$

19. $y = \dfrac{x}{x^2 - 1}$

20. $y = \dfrac{x^2 - 9}{x^2 - 4}$

21. $f(x) = \dfrac{x^2}{x^2 - 1}$

22. $f(x) = \dfrac{x^2 + x - 6}{x^2 - 6x + 8}$

23. $f(x) = \dfrac{x^2 + 1}{x^2 + 2}$

Management

24. A firm's cost function is given by

$$C(x) = 5x + 50.$$

Find and graph a function that describes the firm's average cost per unit produced. (Average cost per unit produced equals total cost divided by the number of units produced.) What happens to the average cost per unit as more and more units are produced?

25. Repeat Exercise 24 if

$$C(x) = 2x^2 + 8x + 20.$$

Life Science

26. During an experiment there are y million bacteria present at time t, where

$$y = \frac{t - 40}{t - 10} + 10.$$

Here t is in days, $0 \le t \le 9$. Graph this function and find the number of bacteria present at the start of the experiment and at the end of the experiment.

Social Science

27. The cost of providing x million elementary school children with a free glass of milk at lunch is estimated to be

$$y = \frac{2x}{10 + x}$$

million dollars. Graph this function. What happens to the cost as more and more children are served?

1 REVIEW EXERCISES

Answers to all review exercises are at the back of the book.

In Exercises 1 through 12 indicate whether the given statement is true or false.

1. $2 < 3$ **2.** $|-5| = -5$ **3.** $-3 > -2$

4. $4 < 6 \le 5$ **5.** $|-2| = 2$ **6.** $-|-3| = 3$

7. $-3 < -3$ **8.** $-4 < -2 < 1$ **9.** $|8 - 2| = 6$

10. $-4 \ge -5$ **11.** $3 \le 6 < 8$ **12.** $-2|5 - 9| = 8$

In Exercises 13 through 16 draw the line diagram corresponding to the given interval.

13. $(0, 1)$ **14.** $[0, 1]$ **15.** $(-\infty, 1]$ **16.** $(0, +\infty)$

17. Simplify the following expressions. Write your answers using only positive exponents.

(a) $(x^3y^{-3}z^{3/2})^{2/3}$ **(b)** $\sqrt[5]{\dfrac{x^4y^{3/4}z^{-2}y^{1/4}z^{-3}}{x^{-1}y^7z^5}}$

In Exercises 18 through 29 let

$$f(x) = 5 - 3x, \quad g(x) = 2 - x^2, \quad \text{and} \quad h(x) = \frac{x+1}{x-2}.$$

Find the indicated functional value if it exists.

18. $f(0)$ **19.** $f(2)$ **20.** $f(-1)$ **21.** $f(-a)$ **22.** $g(4)$
23. $g(\frac{1}{2})$ **24.** $g(\sqrt{2})$ **25.** $g(\sqrt{b}),\ b \geq 0$ **26.** $h(3)$ **27.** $h(0)$
28. $h(-1)$ **29.** $h(2)$

30. Graph the function k whose defining equation is $k(x) = \dfrac{x}{x+1}$.

31. Let $f(x) = -2x^3 + 5x^2 + 3x - 6$ and $g(x) = 4x^3 - 6x + 4$. Find defining equations for $f + g$, $f - g$, fg, and f/g.

32. Let $f(x) = \dfrac{x^2 - 2}{2x - 6}$ and $g(x) = \dfrac{x+5}{3-x}$. Find defining equations for $f + g$, $f - g$, fg, and f/g.

33. Let $f(x) = 2x + 3$, $g(x) = 5x + 2$, and $h(x) = \dfrac{2}{x-1}$. Find $f(g(x))$, $g(f(x))$, $f(h(x))$, and $h(g(x))$.

34. It costs a firm $C(x) = 0.5(x + 2)^2 + 100$ dollars to produce x units of one of its products, and the number of units produced depends on the number of worker-hours w devoted to production according to the equation $x = 3\sqrt{w + 1}$. Express the cost of making the product in terms of the worker-hours devoted to it.

In Exercises 35 through 40 solve the equation $f(x) = 0$ for x.

35. $f(x) = 64 - x^2$ **36.** $f(x) = x^2 + 4x - 32$ **37.** $f(x) = 3x^2 - 24x + 48$
38. $f(x) = x^3 - x$ **39.** $f(x) = 2x^4 - 2x^3 - 24x^2$ **40.** $f(x) = 4x^2 + 4x - 3$

In Exercises 41 through 44 graph the equation and find the slope of the line if possible.

41. $y = 4x - 3$ **42.** $2x + 5y = 20$ **43.** $y = 6$
44. $x = 6$

In Exercises 45 through 48 find the equation of the given line.

45. The line with slope $\frac{1}{3}$ passing through the point $(2, 6)$.

46. The line passing through the points $(2, -7)$ and $(-9, 26)$.

47. The horizontal line passing through the point $(5, 12)$.

48. The vertical line passing through the point $(5, 12)$.

49. The number of bacteria alive in a culture that has been treated with a bactericide declines at a constant rate of 6000 bacteria per hour. If 120,000 bacteria were alive in the culture at the time the bactericide was applied, find an equation describing the number of living bacteria present as a function of time. How long will it take until all the bacteria are dead?

50. A firm's sales were \$368,000 in 1983 and \$740,000 in 1986. Assuming linearity, find an equation describing the firm's sales as a function of time, predict the firm's sales in 1991, and predict the year in which sales will reach \$1,484,000.

51. Find the point of intersection of the lines

(a) $y = 3x + 6$ and $y = 5x - 12$;

(b) $4x - 3y = 9$ and $2x + 5y = 6$.

52. A farmer wants to feed his cattle a diet supplement containing both calcium and an antibiotic. One brand of diet supplement contains 100 units of calcium and 50 units of antibiotic per pound, and another contains 140 units of calcium and 25 units of antibiotic per pound. If the cattle require 446 units of calcium and 160 units of antibiotic per week, how much of each brand should be fed to them?

53. A firm's fixed cost is \$92,000, its variable cost per unit is \$46, and it sells each unit it makes for \$62. Find its cost, revenue, and profit functions and its break-even point.

54. The supply and demand functions for a commodity are given by

$$s = 6250p - 15{,}625 \qquad \text{and} \qquad d = -8500p + 183{,}500,$$

respectively. What are the significances of the slopes and intercepts of these functions? Find the market equilibrium price for the commodity.

In Exercises 55 through 60 graph the function whose defining equation is given.

55. $y = x^2 - 4x - 12$ **56.** $y = -x^2 + 2x + 26$ **57.** $y = x^2 + 2x + 4$

58. $y = \dfrac{x}{x + 3}$ **59.** $y = \dfrac{2x - 1}{x - 1}$ **60.** $y = \dfrac{x^2 - 1}{x + 2}$

61. A manufacturer who intends to put out a line of electronic desk calendars estimates that the calendars can be produced for \$10 each, that a selling price of \$20 per calendar would result in annual sales of 30,000 calendars, and that each rise of \$5 in the selling price would result in the sale of 5000 fewer calendars.

(a) Find the demand function for the calendars.

(b) Find the profit function for the calendars and sketch its graph. (*Hint:* Total profit equals profit per unit times demand.)

(c) What is the maximum profit the manufacturer could make on the calendars? At what selling price would the maximum profit occur?

62. It is estimated that if x million dollars is spent on research concerning automobile emissions, then y percent of the pollutants in such emissions could be eliminated, where

$$y = \frac{100x}{x + 10}.$$

Sketch the graph of this function. How much would the research cost if it were desired to eliminate 80 percent of the pollutants? 90 percent? 99.9 percent?

1 SUPPLEMENT: A LINEAR PROGRAMMING MODEL

In this supplement we demonstrate how to solve a problem by constructing a mathematical model for it and then analyzing the model. The problem we will use as our illustrative example is a **linear programming** problem.

The Problem: Zeta Company makes watercoolers in two sizes, small and large. It makes a profit of \$6 on each small one and \$10 on each large one. Each small watercooler requires 2 kg of steel and 2 kg of plastic; each large one requires 3 kg of steel and 9 kg of plastic. Zeta Company has 1200 kg of steel and 2700 kg of plastic available. How many of each size of watercooler should Zeta produce in order to maximize its profit?

How does one go about constructing a model for such a problem? The first step is to decide what it is we must find. The problem asks, "How many of each size of watercooler should Zeta produce . . . ," so we must find the number of each size that the firm should make. Therefore we need to name these quantities. Let us set

$$x = \text{number of small watercoolers Zeta should produce}$$

and

$$y = \text{number of large watercoolers Zeta should produce.}$$

Note that by the nature of the problem neither x nor y can be negative, so we have $x \geq 0$ and $y \geq 0$.

Are we interested in any particular relationship between the quantities we must find? Yes: we want to find the combination of x small and y large watercoolers that will maximize Zeta's profit. Therefore we must find an expression for Zeta's profit that involves x and y, so that we can then try to find the values of x and y that will make this expression as large as possible. Since Zeta will produce x small and y large watercoolers, and since each small one will contribute \$6 and each large one \$10 to profit, Zeta's total profit P will be given by

$$P = 6x + 10y.$$

Now we must ask if there are any limitations on x and y other than the ones we have noted so far. Again the answer is yes: x and y cannot take on just any values because they are constrained by the amount of raw materials available. For instance, the total steel used in making x small and y large watercoolers will be $2x + 3y$ kg, and this must be no more than 1200 kg, since that is all the steel Zeta has available. Thus we have a

$$\text{steel constraint:} \qquad 2x + 3y \leq 1200.$$

Similarly, since there are only 2700 kg of plastic available, we have a

$$\text{plastic constraint:} \qquad 2x + 9y \leq 2700.$$

Are there any other relationships between x and y that might be of interest to us? Are there any other limitations on x and y? Is there any information in the problem that we have not used? The answer to all these questions is no, so our model is complete. Let us collect the pieces of the model in one place.

The Model: Let

$$x = \text{number of small watercoolers Zeta should produce}$$

and

$$y = \text{number of large watercoolers Zeta should produce.}$$

Maximize

$$P = 6x + 10y$$

subject to the constraints

$$2x + 3y \leq 1200,$$

$$2x + 9y \leq 2700,$$

$$x \geq 0,$$

$$y \geq 0.$$

Notice how we constructed our model. We began by deciding which quantities we had to find, and we named them. We next sought a relationship between these quantities that was of interest to us and we created a mathematical expression for this relationship. Then we looked for limitations on the quantities and expressed them mathematically. Finally, we checked to see if we had missed any relationships between the quantities or any of the limitations on them. These four steps form a general approach to mathematical model building, and they can be used to create models for many types of problems other than the kind we have applied them to here.

Now that we have created our model, let us use it to solve the problem. The constraints tell us that only certain combinations of values for x and y are possible: we will call these **feasible combinations.** If we could somehow identify the feasible combinations, then perhaps we could test them in the profit equation to see which one yields the largest profit. Therefore we will try to identify the feasible combinations of x and y.

Consider one of the constraints, say $2x + 3y \leq 1200$. Does any part of this look familiar? Yes: the "equal to" part, that is, $2x + 3y = 1200$, is the equation of a line. If we graph this line, as in Figure 1–43, then the line is the set of all points (x, y) such that $2x + 3y = 1200$; in other words, the points on the line represent feasible combinations of x and y for this constraint.

What about the $2x + 3y < 1200$ part of the constraint? Substituting points that lie below and above the line $2x + 3y = 1200$ into the inequality, we see that points lying below the line satisfy the inequality, but points lying above the line do not. For instance, (0, 0) and (100, 100) satisfy the inequality, since

$$2(0) + 3(0) = 0 < 1200 \qquad \text{and} \qquad 2(100) + 3(100) = 500 < 1200;$$

but (400, 400) does not, because

$$2(400) + 3(400) = 2000 \not< 1200.$$

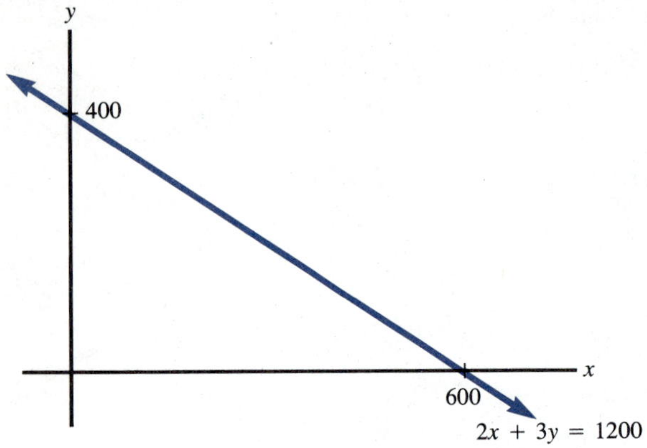

FIGURE 1–43

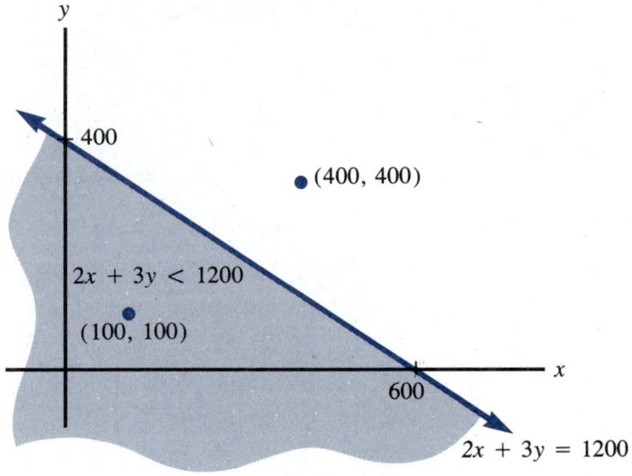

FIGURE 1–44

Thus the points below the line represent feasible combinations for this constraint, but those above it do not. Hence the set of all feasible combinations for the constraint $2x + 3y \leq 1200$ is represented by the line and the shaded region below it in Figure 1–44. (Here we tested the inequality with the three points (0, 0), (100, 100), and (400, 400), but in practice one test point is sufficient: if a test point satisfies the inequality, so will all points on the same side of the line as the test point; if a test point does not satisfy the inequality, then all points on the other side of the line will satisfy the inequality.)

A similar analysis for the constraint $2x + 9y \leq 2700$ shows that the feasible combinations for this constraint are represented by the points on and below the line $2x + 9y = 2700$. See Figure 1–45. The combinations that are feasible for the constraints $x \geq 0$ and $y \geq 0$ are shown in Figures 1–46 and 1–47, respectively. Thus the feasible combinations for all four constraints are represented by the points that are in and on the boundary of the shaded region in Figure 1–48. These points make up the **feasibility region** for the problem: if a point (x, y) lies in this feasibility region, then Zeta Company can produce the combination of x small and y large watercoolers; if the point does not lie within the feasibility region, the combination cannot be produced.

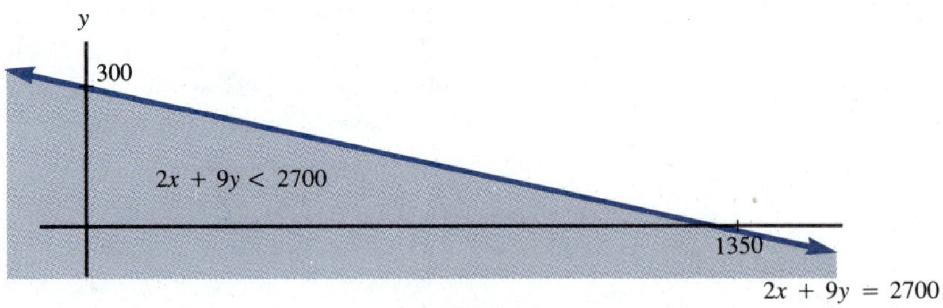

FIGURE 1–45

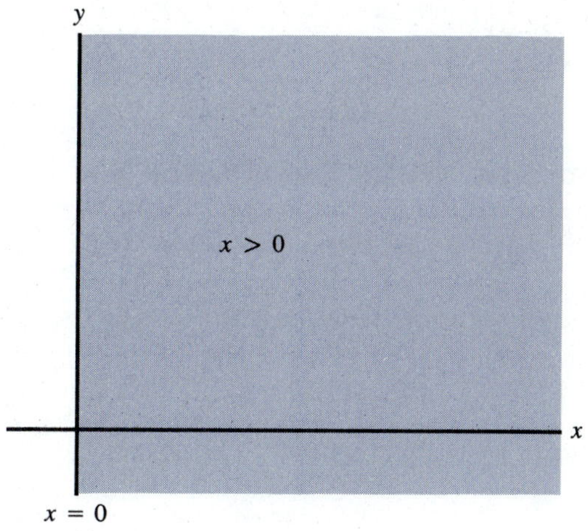

FIGURE 1–46

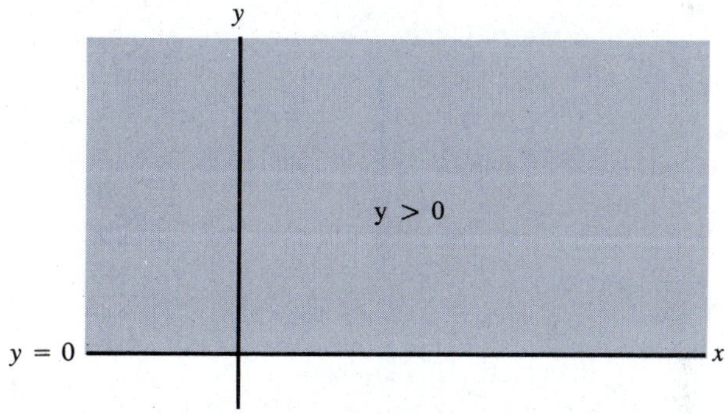

FIGURE 1–47

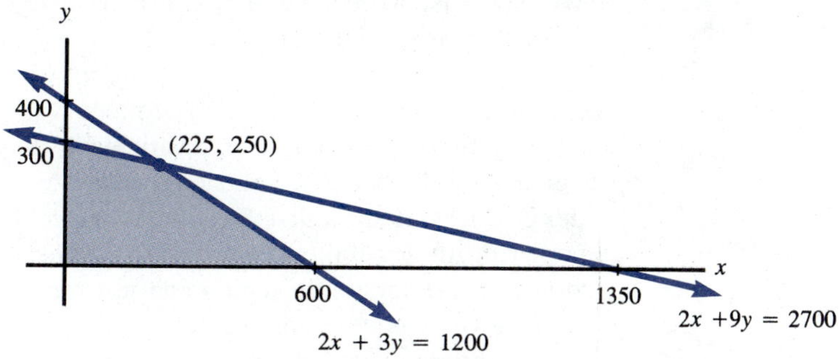

FIGURE 1–48

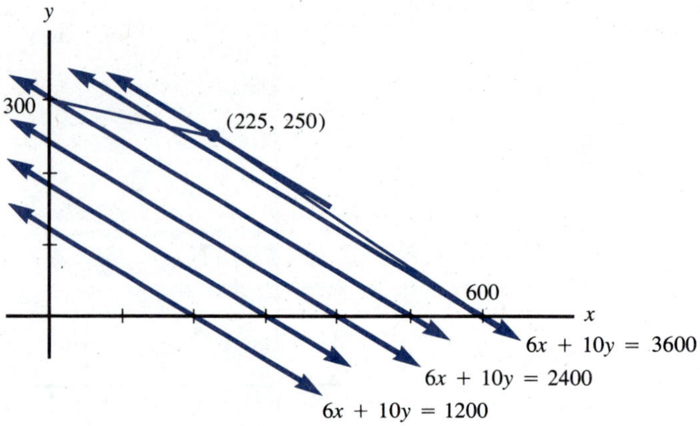

FIGURE 1–49

Now that we have identified the feasible combinations, how can we test them in the profit equation to see which one gives the maximum profit? There are certainly too many possible combinations to substitute them all into the equation, so we will have to figure out another method. Is there anything familiar about the profit equation? Yes: it is linear. Each different value of P gives rise to a different line, but since the lines all have the same slope (namely, $-6/10$) they are all parallel. Figure 1–49 shows some of these lines for various value of P. We want to maximize P, and the figure shows that the largest value of P will occur at the corner of the feasibility region where the lines

$$2x + 3y = 1200 \qquad \text{and} \qquad 2x + 9y = 2700$$

intersect. This point is (225, 250). Therefore Zeta should make $x = 225$ small watercoolers and $y = 250$ large watercoolers for a maximum profit of

$$P = 6(225) + 10(250) = 3850$$

dollars.

We have just solved a typical linear programming problem in two variables x and y. Every linear programming problem resembles ours, in that it consists of a linear **objective function** which must be maximized or minimized and a set of linear constraints on the variables. The method illustrated here will work for any linear programming problem in two variables. Thus, to solve such a problem, set up a model for it and graph the constraints to find the feasibility region. The reasoning we used to show that the optimum value of the objective function occurred at a corner of the feasibility region always holds; thus it is only necessary to find the corners of the feasibility region and test them in the objective function to see which one gives the optimum value.

Linear programming problems are very important in business, engineering, and the sciences. Many of them have more than two variables; indeed, hundreds of variables are not uncommon. Linear programming problems with more than two variables cannot be solved by graphing; other methods are available, the most popular being the **simplex method** invented by George Dantzig in 1947. Recently two other methods for solving linear programming problems have been developed. They are being tested to see if they are more efficient than the simplex method.

Computer Manual: Program INTRSECT.

SUPPLEMENTARY EXERCISES

Management

1. A manufacturer produces two kinds of electronic calculators: a regular model and a deluxe model. Each regular model has 1 plastic case and 10 microcircuits and contributes \$3 to profit. Each deluxe model has 1 plastic case and 20 microcircuits and contributes \$4 to profit. If the manufacturer has 24 plastic cases and 320 microcircuits available, find the maximum profit and the quantity of each model that must be produced in order to attain it.

2. Do Exercise 1 if each regular model contributes \$3 to profit and each deluxe model contributes \$10 to profit.

3. Knifty-Knit Sweaters is about to schedule a week's production run of ski sweaters. The sweaters come in two sizes, medium and large, and Knifty-Knit has orders for 2000 medium and 1000 large sweaters. These orders must be filled from the week's production run. Each medium sweater requires 8 oz of wool, and each large one requires 10 oz of wool. It takes 10 minutes to knit a sweater, regardless of size, and each sweater is packed in a plastic envelope for shipping. Knifty-Knit has purchased 72,000 oz of wool and 8000 plastic envelopes. Also, the firm's contract with the Knitter's Union guarantees each of its 15 knitters at least 40 hours of work per week. If Knifty-Knit makes a profit of \$4 on each medium sweater and \$6 on each large sweater, how many of each size must it make in order to maximize its profit?

Life Science

4. Each Brand X vitamin pill contains 50 mg of vitamin A, 10 mg of vitamin B, and 20 mg of vitamin C. Each Brand Y pill contains 40 mg of A, 50 mg of B, and 60 mg of C. Each Brand X pill costs 3.5 cents, and each Brand Y pill costs 4.2 cents. If you need at least 440 mg of A, 320 mg of B, and 440 mg of C per day, how many pills of each brand should you take in order to minimize your cost?

Social Science

5. A judge has ordered a town to desegregrate its high schools. Currently, East School has 1800 minority students, 1000 nonminority students, and a capacity of 3000 students. West School has 1200 minority students, 2200 nonminority students, and a capacity of 3500 students. The judge has ruled that East School can have at most 60% minority enrollment and West school can have at most 60% nonminority enrollment. Only minority students will be transferred from East School to West School, and only nonminority students will be transferred from West School to East School. Minimize the total number of students who must be transferred.

2

The Derivative

Calculus is a method for studying functions. For instance, suppose a demographer concludes that a certain function will describe a nation's population size as a function of time over the next 100 years. The techniques of calculus can be applied to this function to answer questions such as: At what rate will the population be changing at any given time? When will the population be increasing and when will it be decreasing? What will be the maximum and minimum sizes of the population over the next century? The particular concept of calculus needed to answer questions like these is called the **derivative.**

In this chapter we study the derivative. The derivative is defined in terms of another concept of calculus called the **limit,** so we will begin with a brief intuitive introduction to limits. We will then define the derivative, develop some rules for finding derivatives of functions, show how to use the derivative to find rates of change, exhibit some other useful applications of the derivative, and conclude with a discussion of the various notations for the derivative and for derivatives of the derivative.

2.1 LIMITS

This section introduces the concept of the limit of a function f as its independent variable x approaches the number a. The notion of a limit will be needed for the definition of the derivative in the next section and is important in its own right. We will illustrate the idea of the limit of a function geometrically, state the properties of limits, show how to use these properties to find limits of polynomial and rational functions, and consider limits at infinity and asymptotes of functions.

The Limit of a Function

Consider the linear function f whose defining equation is $f(x) = 2x + 1$. We ask the following question: if we let x approach the number 1 on the x-axis, from either side of 1 and without ever letting $x = 1$, what can we say about the behavior of the corresponding functional values $f(x)$? For instance, Table 2–1 shows some functional values $f(x) = 2x + 1$ as x approaches 1 from the left and from the right:

TABLE 2–1

	x approaches 1 from the left				
x	0.8	0.9	0.99	0.999	0.9999
$f(x) = 2x + 1$	2.6	2.8	2.98	2.998	2.9998
	x approaches 1 from the right				
x	1.2	1.1	1.01	1.001	1.0001
$f(x) = 2x + 1$	3.4	3.2	3.02	3.002	3.0002

Notice that as x gets closer and closer to 1, from either side of 1, the values $f(x)$ get closer and closer to the single number 3. Hence we say that the **limit of $f(x)$ as x approaches 1 is 3,** written

$$\lim_{x \to 1} f(x) = 3,$$

or

$$\lim_{x \to 1} (2x + 1) = 3.$$

We can also tell that $\lim_{x\to 1} (2x + 1) = 3$ from the graph of the function. Figure 2–1 shows the graph of f. As the figure indicates, as x approaches 1 from either side on the x-axis the functional values $f(x)$ approach 3 on the y-axis.

Having illustrated the idea of the limit of a function, we now present an informal definition of the concept.

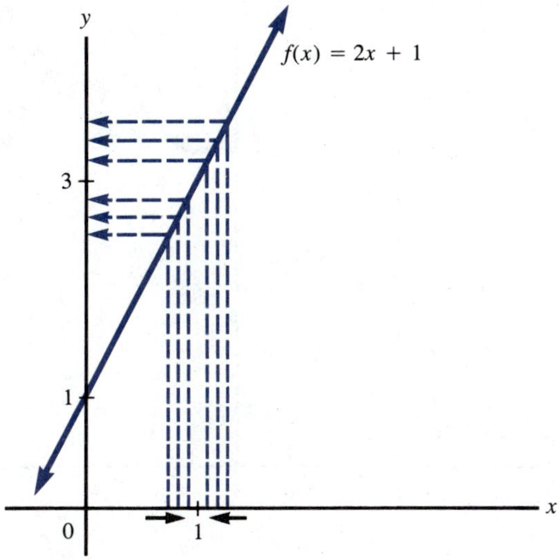

FIGURE 2–1 $\lim_{x\to 1}(2x+1) = 3$

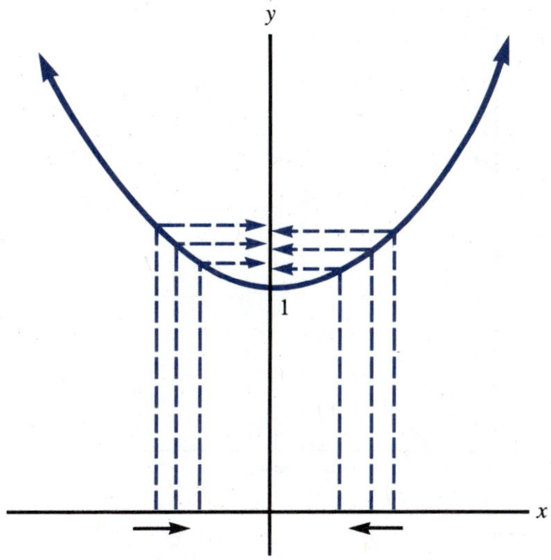

FIGURE 2–2 $\lim_{x\to 0}(x^2+1) = 1$

The Limit of a Function

If f is a function and a is a number, the **limit of f as x approaches a is L,** written

$$\lim_{x\to a} f(x) = L,$$

provided that as x comes arbitrarily close to a from either side, without ever being equal to a, the corresponding functional values $f(x)$ come arbitrarily close to the single number L. If there is no such single number L, the limit does not exist.

EXAMPLE 1 If $f(x) = x^2 + 1$ and $a = 0$, we claim that $\lim_{x\to 0} f(x) = 1$. Figure 2–2 verifies this: it shows that as x approaches 0 on the x-axis, from either side and without ever being equal to 0, the corresponding functional values approach the single number 1 on the y-axis.

EXAMPLE 2 If f is defined by

$$f(x) = \begin{cases} x - 1 & \text{if} \quad 0 \le x \le 3, \\ x + 1 & \text{if} \quad x > 3, \end{cases}$$

then $\lim_{x\to 3} f(x)$ does not exist. To see this, look at Figure 2–3, which depicts the graph of this function. Notice that as x approaches 3 on the x-axis from the

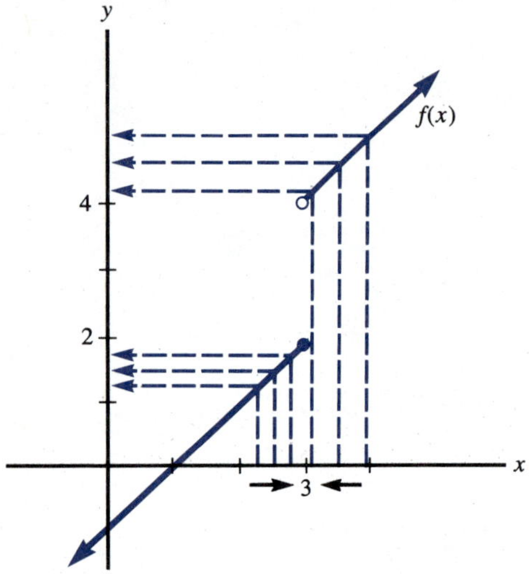

FIGURE 2–3 $\lim_{x \to 3} f(x)$ Does Not Exist

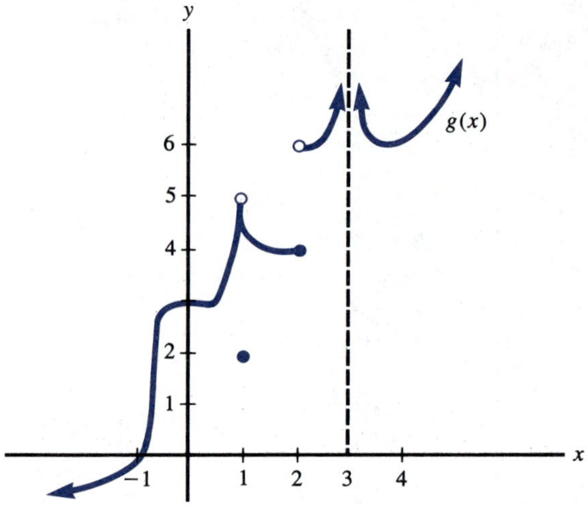

FIGURE 2–4

left, the corresponding functional values approach 2 on the y-axis, but as x approaches 3 from the right, the corresponding functional values approach 4 on the y-axis. Therefore, there is no single number L that the functional values approach as x approaches 3, so the limit does not exist.

EXAMPLE 3 Figure 2–4 shows the graph of a function g. From the graph, we see that

a. $\text{Lim}_{x \to -1}\, g(x) = 0$, because as x approaches -1 from either side, without ever being equal to -1, the functional values $g(x)$ approach 0;

b. $\text{Lim}_{x \to 0}\, g(x) = 3$, because as x approaches 0 from either side, without ever being equal to 0, the functional values actually become equal to 3;

c. $\text{Lim}_{x \to 1}\, g(x) = 5$, because as x approaches 1 from either side, without ever being equal to 1, the functional values approach 5; note that this is so even though the functional value $g(1) \neq 5$;

d. $\text{Lim}_{x \to 2}\, g(x)$ does not exist, because as x approaches 2 from the left the functional values approach 4, but as x approaches 2 from the right they approach 6;

e. $\text{Lim}_{x \to 3}\, g(x)$ does not exist, because as x approaches 3 from the left (and also from the right), the functional values grow larger and larger without bound and hence do not approach a single number L.

We should remark that our definition of limit is highly intuitive and not mathematically precise. A more rigorous definition of the limit of a function is beyond the scope of this book.

Properties of Limits

The following properties are useful when it is necessary to evaluate limits. They may be proved using the rigorous definition of limit referred to in the preceding paragraph.

Properties of Limits

Let a be any number and assume that both $\text{Lim}_{x\to a} f(x)$ and $\text{Lim}_{x\to a} g(x)$ exist.

1. If $f(x) = g(x)$ whenever $x \neq a$, then $\text{Lim}_{x\to a} f(x) = \text{Lim}_{x\to a} g(x)$.
2. For any number c, $\text{Lim}_{x\to a} c = c$.
3. For any number r, $\text{Lim}_{x\to a} x^r = a^r$, provided that a^r is defined; if a^r is not defined as a real number, the limit does not exist.
4. For any real number c, $\text{Lim}_{x\to a} cf(x) = c \cdot \text{Lim}_{x\to a} f(x)$.
5. $\text{Lim}_{x\to a} [f(x) \pm g(x)] = \text{Lim}_{x\to a} f(x) \pm \text{Lim}_{x\to a} g(x)$.
6. $\text{Lim}_{x\to a} [f(x)g(x)] = [\text{Lim}_{x\to a} f(x)][\text{Lim}_{x\to a} g(x)]$.
7. If $\text{Lim}_{x\to a} g(x) \neq 0$, then

$$\text{Lim}_{x\to a} \frac{f(x)}{g(x)} = \frac{\text{Lim}_{x\to a} f(x)}{\text{Lim}_{x\to a} g(x)}.$$

8. If $\text{Lim}_{x\to a} g(x) = 0$ and $\text{Lim}_{x\to a} f(x) \neq 0$, then $\text{Lim}_{x\to a} \dfrac{f(x)}{g(x)}$ does not exist.

EXAMPLE 4

a. The functions f and g defined by

$$f(x) = x^2 + 1 \qquad \text{and} \qquad g(x) = \frac{x^3 + x}{x}$$

are equal except when $x = 0$, and we know from Example 1 that $\text{Lim}_{x\to 0} f(x) = 1$. Therefore by Property 1 we have

$$\text{Lim}_{x\to 0} g(x) = \text{Lim}_{x\to 0} f(x) = 1.$$

b. By Property 2, we have $\text{Lim}_{x\to a} 4 = 4$ for any number a.

c. Property 3 shows that $\text{Lim}_{x\to -2} x = -2$, $\text{Lim}_{x\to 3} x^2 = 3^2 = 9$, and

$$\text{Lim}_{x\to 4} \sqrt{x} = \text{Lim}_{x\to 4} x^{1/2} = 4^{1/2} = 2.$$

But also by Property 3,

$$\lim_{x\to 0} x^{-1} \text{ does not exist, since } 0^{-1} = 1/0 \text{ is not defined,}$$

and

$$\lim_{x\to -1} \sqrt[4]{x} \text{ does not exist, since } \sqrt[4]{-1} \text{ is not defined as a real number.}$$

d. By Properties 4 and 3,

$$\lim_{x\to 2} 3x^2 = 3 \cdot \lim_{x\to 2} x^2 = 3(2^2) = 12.$$

e. By Properties 2 through 5, we have

$$\begin{aligned}\lim_{x\to 3} (2x^3 + 4x^2 - x + 5) &= \lim_{x\to 3} (2x^3) + \lim_{x\to 3} (4x^2) - \lim_{x\to 3} x + \lim_{x\to 3} 5 \\ &= 2 \cdot \lim_{x\to 3} x^3 + 4 \cdot \lim_{x\to 3} x^2 - \lim_{x\to 3} x + 5 \\ &= 2(3^3) + 4(3^2) - 3 + 5 \\ &= 92.\end{aligned}$$

f. Properties 2 through 6 imply that

$$\begin{aligned}\lim_{x\to 3} [(2x^{1/2})(3x - 1)] &= \left[\lim_{x\to 3} 2\sqrt{x}\right]\left[\lim_{x\to 3} (3x - 1)\right] \\ &= [2\sqrt{3}][3(3) - 1] \\ &= 16\sqrt{3}.\end{aligned}$$

g. Properties 2 through 7 show that

$$\begin{aligned}\lim_{x\to 0} \frac{x^2 + 2}{2x + 3} &= \frac{\lim_{x\to 0} (x^2 + 2)}{\lim_{x\to 0} (2x + 3)} \\ &= \frac{0^2 + 2}{2(0) + 3} \\ &= \frac{2}{3}.\end{aligned}$$

h. By Property 8,

$$\lim_{x\to 1} \frac{x^2 + 1}{x - 1}$$

does not exist because

$$\lim_{x\to 1} (x^2 + 1) = 2$$

while

$$\lim_{x\to 1} (x - 1) = 0.$$

As Example 4(e) suggests, Properties 2 through 5 imply that we can find the limit of any polynomial function as x approaches a simply by substituting a for x in the defining equation of the function.

> **The Limit of a Polynomial Function**
> If p is a polynomial function, then $\text{Lim}_{x \to a}\, p(x) = p(a)$.

EXAMPLE 5 We have

$$\lim_{x \to 2} (x^3 - 5x^2 + 3x - 12) = 2^3 - 5(2^2) + 3(2) - 12 = -18.$$

If $f = p/q$ is a rational function, where the polynomials p and q have a common factor, Property 1 implies that we can cancel this common factor from p/q without affecting the limit of f. Therefore to find the limit of a rational function, we first reduce it to its lowest terms by canceling common factors from its numerator and denominator. We then use our result for the limit of polynomials and apply either Property 7 or Property 8. We illustrate with an example.

EXAMPLE 6 **a.** Consider

$$\lim_{x \to 3} \frac{x^2 + x - 12}{x^2 - x - 6} = \lim_{x \to 3} \frac{(x + 4)(x - 3)}{(x + 2)(x - 3)}.$$

We cancel the common factor $x - 3$ from the rational expression. Then by Property 7 we have

$$\lim_{x \to 3} \frac{x^2 + x - 12}{x^2 - x - 6} = \frac{\lim_{x \to 3} (x + 4)}{\lim_{x \to 3} (x + 2)} = \frac{3 + 4}{3 + 2} = \frac{7}{5}.$$

b. We use the same rational function as in part (a), but this time take its limit as x approaches -2. Now

$$\lim_{x \to -2} \frac{x^2 + x - 12}{x^2 - x - 6} = \frac{\lim_{x \to -2} (x + 4)}{\lim_{x \to -2} (x + 2)} = \frac{-2 + 4}{-2 + 2} = \frac{2}{0}.$$

By Property 8, the limit does not exist.

Limits at Infinity

Consider the function f whose graph is shown in Figure 2–5. This function has the lines $y = L$ and $y = K$ as horizontal asymptotes. Note that as x increases without bound on the x-axis, the corresponding functional values approach the number L on the y-axis. Similarly, as x decreases without bound on the x-axis, the corresponding functional values approach K on the y-axis. We express these facts by writing the **limits at infinity**

$$\lim_{x\to+\infty} f(x) = L \qquad \text{and} \qquad \lim_{x\to-\infty} f(x) = K,$$

respectively. The notation $x \to +\infty$ means that x is allowed to increase without bound, that is, move to the right along the x-axis without bound; the notation $x \to -\infty$ means that x is allowed to decrease without bound, that is, move to the left along the x-axis without bound. We can use limits at infinity to find the horizontal asymptotes of a function.

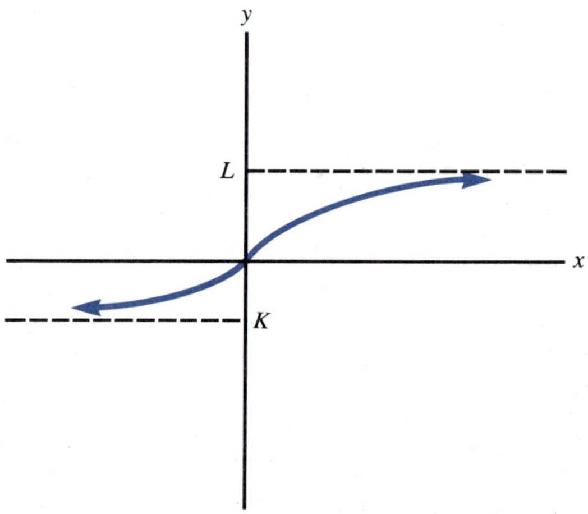

FIGURE 2–5

EXAMPLE 7 Let us find the horizontal asymptotes of the function f defined by

$$f(x) = \frac{1}{x}.$$

When x is a large positive number, $1/x$ is a small positive fraction; we can make $1/x$ as close to 0 as we want by taking x sufficiently large. Therefore as $x \to +\infty$, $1/x$ approaches 0:

$$\lim_{x\to+\infty} \frac{1}{x} = 0.$$

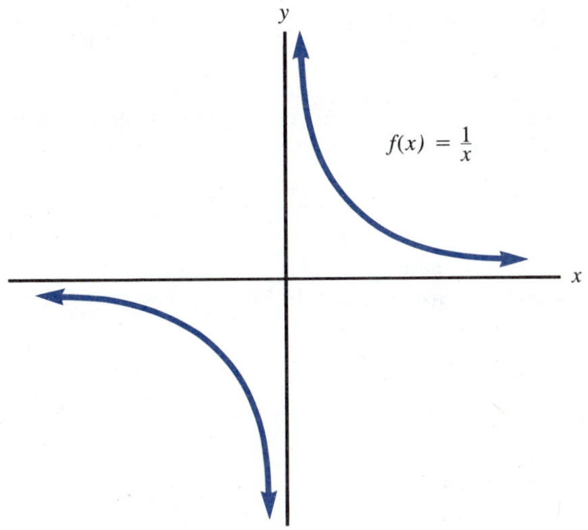

FIGURE 2–6

This shows that the x-axis is a horizontal asymptote for the function f as $x \to +\infty$. Similar reasoning as $x \to -\infty$ shows that the x-axis is also a horizontal asymptote for f as $x \to -\infty$. The graph of f is shown in Figure 2–6.

In Example 7 we showed that as $x \to +\infty$, $1/x$ approaches 0. The reasoning used there holds also for $1/x^r$, where r is any positive number. On the other hand, as x becomes large without bound, so does x^r, for any positive number r. Therefore we have the following facts about limits at infinity for powers of x.

Limits at Infinity for Powers of x

For any positive number r:

1. $\text{Lim}_{x\to+\infty} \dfrac{1}{x^r} = 0$ and $\text{Lim}_{x\to-\infty} \dfrac{1}{x^r} = 0$.

2. $\text{Lim}_{x\to+\infty} x^r$ and $\text{Lim}_{x\to-\infty} x^r$ do not exist.

It follows from fact 2 that the limits at infinity for any polynomial function do not exist; therefore polynomial functions do not have horizontal asymptotes.

When evaluating limits at infinity for rational functions, we divide each term of the numerator and denominator by the largest power of x that appears in the function and then use the above results.

EXAMPLE 8 Let us evaluate

$$\lim_{x\to+\infty} \frac{2x^2 + 1}{x^2 - 5x + 4}.$$

The largest power of x present is x^2, so we rewrite the limit as

$$\lim_{x\to+\infty} \frac{2x^2/x^2 + 1/x^2}{x^2/x^2 - 5x/x^2 + 4/x^2} = \lim_{x\to+\infty} \frac{2 + 1/x^2}{1 - 5(1/x) + 4(1/x^2)}.$$

But $\lim_{x\to+\infty} (1/x) = 0$ and $\lim_{x\to+\infty} (1/x^2) = 0$ by our results for limits at infinity for powers of x. Hence

$$\lim_{x\to+\infty} \frac{2x^2 + 1}{x^2 - 5x + 4} = \frac{2 + 0}{1 - 5(0) + 4(0)} = \frac{2}{1} = 2.$$

EXAMPLE 9 Since

$$\lim_{x\to-\infty} \frac{x^3 + 8}{x^2 - 5} = \lim_{x\to-\infty} \frac{x^3/x^3 + 8/x^3}{x^2/x^3 - 5/x^3}$$

$$= \lim_{x\to-\infty} \frac{1 + 8(1/x^3)}{1/x - 5(1/x^3)}$$

$$= \frac{1 + 8(0)}{0 - 5(0)}$$

$$= \frac{1}{0},$$

the limit does not exist.

Finding the limits at infinity of a function can help identify its asymptotes and thus be useful in drawing its graph.

EXAMPLE 10 Let us graph the function f defined by

$$f(x) = \frac{2x^2}{x^2 - 4}.$$

Since

$$f(x) = \frac{2x^2}{(x - 2)(x + 2)},$$

the lines $x = 2$ and $x = -2$ are vertical asymptotes for f. Since

$$\lim_{x\to\pm\infty} f(x) = \lim_{x\to\pm\infty} \frac{2x^2/x^2}{x^2/x^2 - 4/x^2}$$

$$= \lim_{x\to\pm\infty} \frac{2}{1 - 4(1/x^2)}$$

$$= \frac{2}{1 - 4(0)}$$

$$= 2,$$

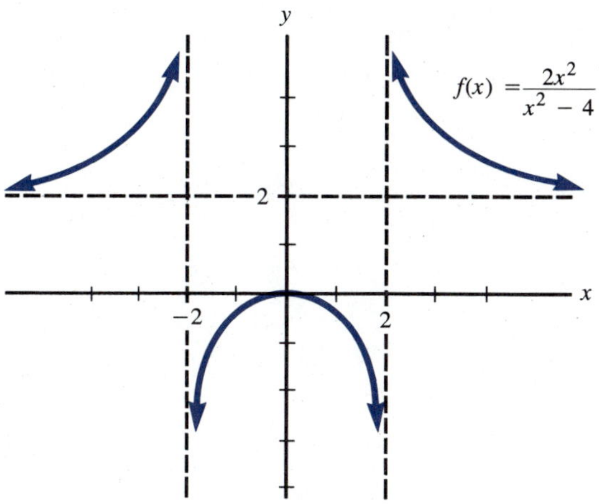

FIGURE 2–7

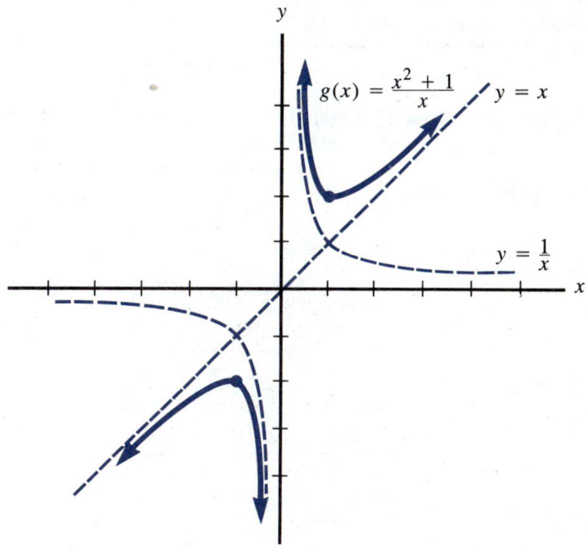

FIGURE 2–8

the line $y = 2$ is a horizontal asymptote for f as x approaches plus infinity and a horizontal asymptote for f as x approaches minus infinity. Plotting points now yields the graph of f shown in Figure 2–7.

EXAMPLE 11 Graph the function defined by

$$g(x) = \frac{x^2 + 1}{x}.$$

Note that $x = 0$ (the y-axis) is a vertical asymptote for g. Since

$$\frac{x^2 + 1}{x} = \frac{x^2}{x} + \frac{1}{x} = x + \frac{1}{x},$$

we may rewrite g in the form

$$g(x) = x + \frac{1}{x}$$

But $\text{Lim}_{x \to +\infty} (1/x) = 0$ implies that when x is very large

$$g(x) = x + \frac{1}{x} \approx x + 0 = x;$$

in other words, as x approaches plus infinity, $g(x)$ approaches x, so the graph of g will approach the line $y = x$. Therefore we say that the line $y = x$ is an **oblique asymptote** for g as x approaches plus infinity. Similarly, the line $y = x$ is an oblique asymptote for g as x approaches minus infinity. Since $g(x) = x + 1/x$, it follows that we can obtain the graph of g by visually adding the graphs of $y = x$ and $y = 1/x$, as in Figure 2–8.

Computer Manual: Programs LIMIT and INFLIMIT.

2.1 EXERCISES

The Limit of a Function

In Exercises 1 through 4 complete the table and find the limit if it exists; if the limit does not exist, so state.

1. $\lim_{x \to 3} f(x)$, where $f(x) = x^2 - 9$.

x approaches 3 from the left

x	2.8	2.9	2.99	2.999	2.9999
$f(x)$					

x approaches 3 from the right

x	3.2	3.1	3.01	3.001	3.0001
$f(x)$					

2. $\lim_{x \to 3} f(x)$, where $f(x) = \dfrac{x^2 - 9}{x - 3}$.

x approaches 3 from the left

x	2.8	2.9	2.99	2.999	2.9999
$f(x)$					

x approaches 3 from the right

x	3.2	3.1	3.01	3.001	3.0001
$f(x)$					

3. $\lim_{x \to 0} f(x)$, where $f(x) = \dfrac{1}{x}$.

x approaches 0 from the left

x	−0.1	−0.01	−0.001	−0.0001	−0.00001
$f(x)$					

x approaches 0 from the right

x	0.1	0.01	0.001	0.0001	0.00001
$f(x)$					

4. $\lim_{x \to 1} f(x)$, where $f(x) = \begin{cases} x & \text{if} \quad 0 \le x < 1, \\ 2x & \text{if} \quad 1 \le x < 2. \end{cases}$

x approaches 1 from the left

x	1/2	3/4	7/8	15/16	31/32	63/64
$f(x)$						

x approaches 1 from the right

x	3/2	5/4	9/8	17/16	33/32	65/64
$f(x)$						

The following is the graph of a function f:

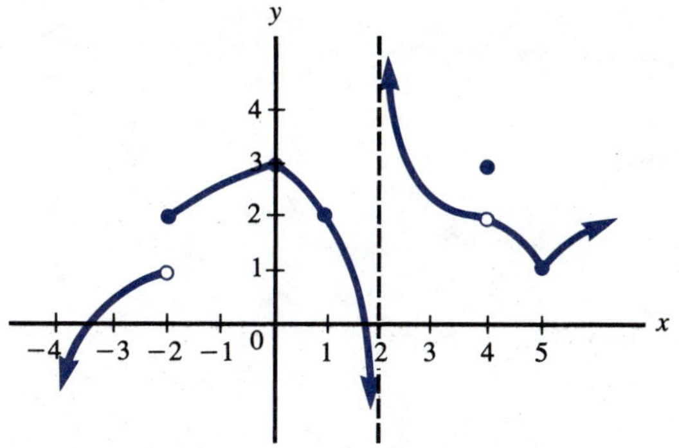

In Exercises 5 through 10 find the indicated limit if it exists; if it does not exist, so state.

5. $\lim_{x\to -2} f(x)$ **6.** $\lim_{x\to 0} f(x)$ **7.** $\lim_{x\to 1} f(x)$

8. $\lim_{x\to 2} f(x)$ **9.** $\lim_{x\to 4} f(x)$ **10.** $\lim_{x\to 5} f(x)$

The following is the graph of a function g:

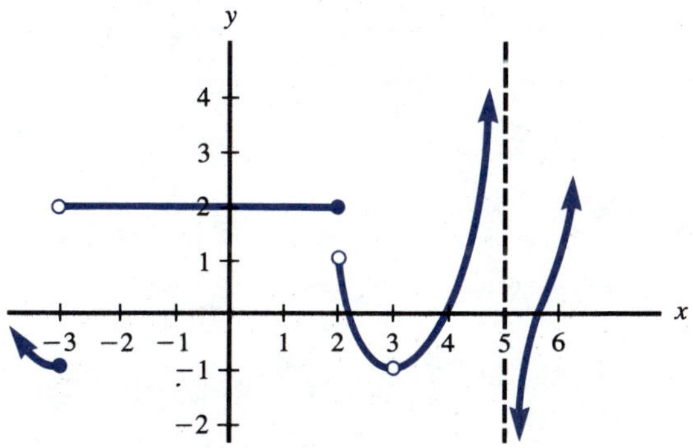

In Exercises 11 through 16 find the indicated limit if it exists; if it does not exist, so state.

11. $\lim_{x\to -3} g(x)$ **12.** $\lim_{x\to -1} g(x)$ **13.** $\lim_{x\to 0} g(x)$

14. $\lim_{x\to 2} g(x)$ **15.** $\lim_{x\to 3} g(x)$ **16.** $\lim_{x\to 5} g(x)$

Properties of Limits

In Exercises 17 through 36 find the indicated limit if it exists; if it does not exist, so state.

17. $\lim_{x\to 2} 5$ **18.** $\lim_{x\to -4} \sqrt{3}$ **19.** $\lim_{x\to 0} x$

20. $\lim_{x\to 2} x^3$ **21.** $\lim_{x\to -3} 3x^4$ **22.** $\lim_{x\to 4} -5x^{3/2}$

23. $\lim_{x\to 3} (x^2 + 3x)$ **24.** $\lim_{x\to 0} (2x^3 - 5x^2 + 3x - 17)$

25. $\lim_{x\to -2} (-3x^3 + 2x^2 + 5x + 3)$ **26.** $\lim_{x\to 8} (x^{2/3} + 4x^{-1/3} - 5)$

27. $\lim_{x\to 2} \left(\frac{2}{x} - \frac{3}{x^2} + 1\right)$ **28.** $\lim_{x\to 0} (4x^{-1} + 3x^2 + 8)$

29. $\lim_{x\to -1} (2x + 4)(3x - 2)$ **30.** $\lim_{x\to 2} (x^2 - 4x - 6)(x^3 + x + 2)$

31. $\lim_{x\to 1} \frac{4x + 7}{5x - 2}$ **32.** $\lim_{x\to 3} \frac{x^2 - 9}{x + 2}$

33. $\lim\limits_{x\to 3} \dfrac{x^2 - 5x + 6}{x^2 + x - 12}$

34. $\lim\limits_{x\to 3} \dfrac{x^2 - 5x + 6}{x^2 + 10x + 24}$

35. $\lim\limits_{x\to 2} \dfrac{x^2 - 5x - 14}{x^2 - 4}$

36. $\lim\limits_{x\to 7} \dfrac{x^2 - 5x - 14}{x^2 - 4}$

Limits at Infinity

In Exercises 37 through 40 find the indicated limits at infinity for the function f whose graph is shown. If a limit does not exist, so state.

37. $\lim\limits_{x\to +\infty} f(x)$

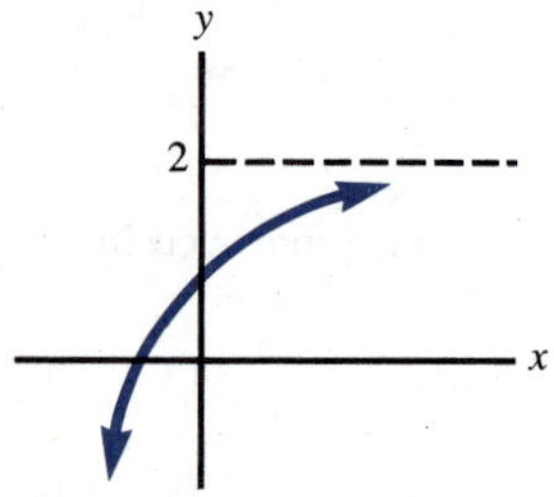

38. $\lim\limits_{x\to -\infty} f(x)$

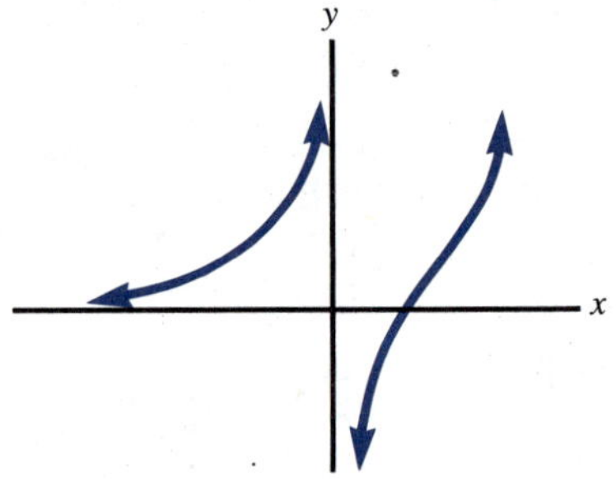

39. $\lim\limits_{x\to +\infty} f(x)$, $\lim\limits_{x\to -\infty} f(x)$

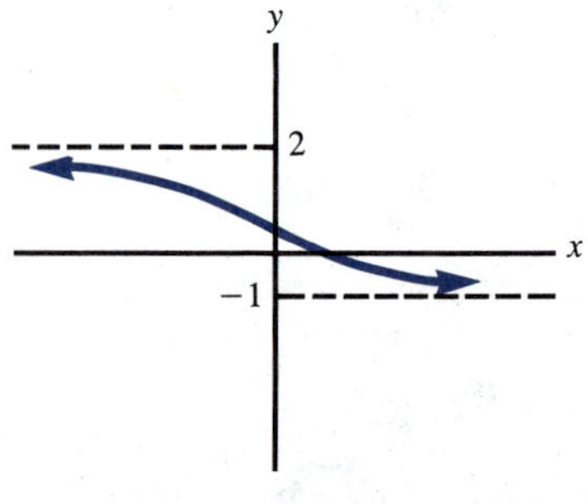

40. $\lim\limits_{x\to +\infty} f(x)$, $\lim\limits_{x\to -\infty} f(x)$

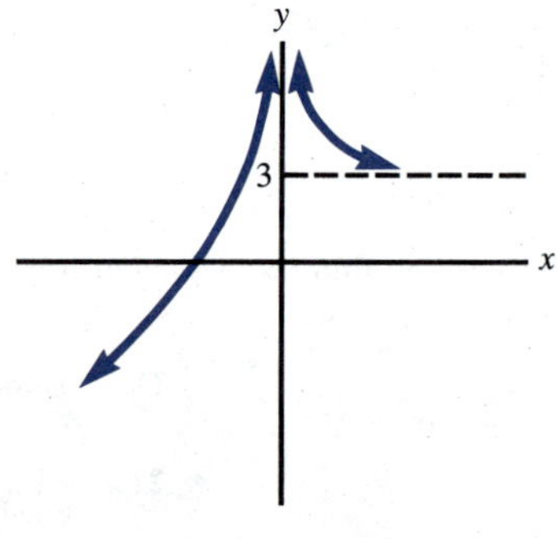

In Exercises 41 through 48 find the indicated limits at infinity. If a limit does not exist, so state.

41. $\lim\limits_{x\to +\infty} \dfrac{4}{x}$

42. $\lim\limits_{x\to -\infty} \dfrac{-3}{x^2}$

43. $\lim\limits_{x\to -\infty} \dfrac{3x^2}{x}$

44. $\lim\limits_{x\to +\infty} \dfrac{2x^2}{x^2}$

45. $\lim\limits_{x\to +\infty} \dfrac{2x^2 + 5x - 7}{3x^2 + 8x + 2}$

46. $\lim\limits_{x\to +\infty} \dfrac{-2x^3 + x^2 + 17x - 1}{4x^2 + 3x + 2}$

47. $\lim\limits_{x\to -\infty} \dfrac{-8x^3 + 15x + 2}{9x^4 - 16}$

48. $\lim\limits_{x\to -\infty} \dfrac{x^2 - 5x + 9}{3 - 2x + x^4}$

In Exercises 49 through 54 find all asymptotes of the given function and sketch its graph.

49. $f(x) = \dfrac{2x + 1}{x - 3}$ **50.** $f(x) = \dfrac{5x + 2}{2x - 6}$ **51.** $f(x) = \dfrac{2x - 2}{x^2}$

52. $f(x) = \dfrac{3x}{x^2 - 1}$ **53.** $f(x) = \dfrac{2x^2 - 2}{x}$ **54.** $f(x) = \dfrac{x^3 + 1}{2x}$

Management

55. Acme Company's average cost per unit produced is given by

$$A(x) = \frac{5x + 80}{x + 1},$$

where x is the number of units produced and $A(x)$ is the average cost per unit in dollars. Find and interpret $\lim_{x \to +\infty} A(x)$.

56. A computer company is introducing a new model of computer. The firm estimates that it will take $T(x)$ worker-hours to build the xth unit of the new computer, where

$$T(x) = \frac{300x^3 + 50x + 100}{x^3}.$$

Find and interpret $\lim_{x \to +\infty} T(x)$.

Life Science

57. If a person who suffers a bite by a poisonous snake receives an immediate shot of antivenin, then t seconds after the shot is given there will be

$$y = \frac{0.5t + 2000}{2t + 3}$$

ppm of poison in the victim's blood. Find and interpret $\lim_{t \to +\infty} y$.

2.2 THE DERIVATIVE

The concept of the derivative is the basis of calculus. The derivative of a function tells us many things about the function. For instance, as we shall see later in this chapter, we can use the derivative to examine the rate of change of a function. Thus, suppose a certain function describes the concentration of an antibiotic in a patient's blood as a function of time; then the derivative of the function will tell us the rate at which the concentration of the antibiotic is changing with respect to time. Derivatives can also help us graph functions, find their maximum and minimum values, and in general analyze them in many useful ways. Before we can do any of these things, however, we must first develop a thorough understanding of what the derivative is. In this section we introduce the derivative, discuss its geometric significance, and briefly consider the concepts of differentiability and continuity of a function.

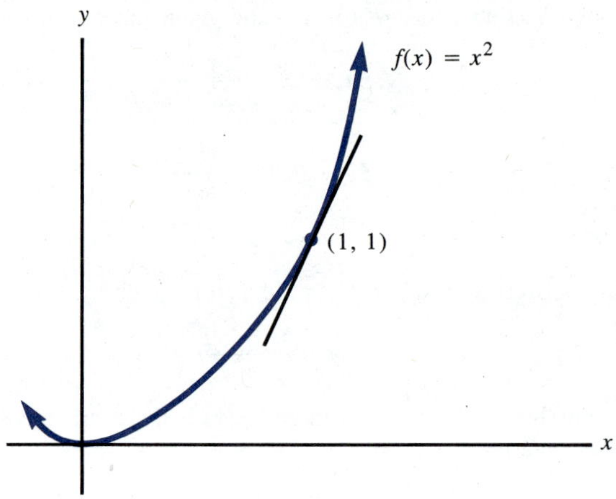

FIGURE 2–9

Definition of the Derivative

In order to make our initial discussion of the derivative specific, we choose to consider the function f defined by $f(x) = x^2$ and the point (1, 1) on its graph. We may construct a unique line that just touches the graph at the point (1, 1). See Figure 2–9. This line is called the **line tangent to f at (1, 1).** If we knew the slope of this tangent line, we could write its equation; we will now try to find its slope.

Suppose we add a number h to 1. The point

$$(1 + h, f(1 + h)) = (1 + h, (1 + h)^2)$$

will lie on the graph of f, and we can draw a line through the points (1, 1) and $(1 + h, (1 + h)^2)$. This line is called a **secant line.** Figure 2–10 depicts the situation if h is positive. The secant line has

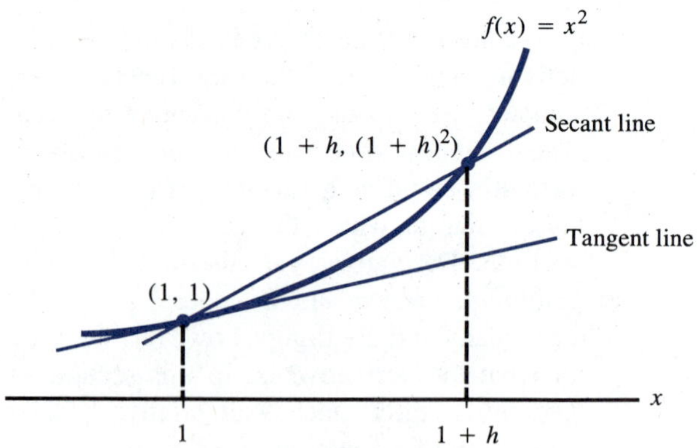

FIGURE 2–10

$$\text{slope} = \frac{(1+h)^2 - 1}{1+h-1} = \frac{1 + 2h + h^2 - 1}{h} = \frac{h(2+h)}{h} = 2 + h.$$

Now let h approach 0. As h gets smaller and smaller, $1 + h$ approaches 1 on the x-axis, the point $(1 + h, (1 + h)^2)$ "slides" along the graph toward the point (1, 1), and the secant line approaches the tangent line at (1, 1). Figure 2–11 shows how as h approaches 0 the secant line "pivots" at (1, 1) and approaches the tangent line. Since the secant line approaches the tangent line, it follows that the slope of the secant line must approach the slope of the tangent line. In other words,

$$\lim_{h \to 0} (\text{slope of secant line}) = \text{slope of tangent line},$$

or

$$\lim_{h \to 0} (2 + h) = 2 + 0 = 2.$$

Thus the slope of the line tangent to f at (1, 1) is equal to 2. This slope is called the **derivative of f at $x = 1$.**

The preceding analysis can be applied to any function f at any number in its domain as long as the function has a unique nonvertical tangent line at the point $(x, f(x))$ on its graph. Figure 2–12 depicts the process. We begin by adding h to x to obtain the point $(x + h, f(x + h))$ on the graph. We then form the secant line through $(x, f(x))$ and $(x + h, f(x + h))$. The secant line will have

$$\text{slope} = \frac{f(x+h) - f(x)}{x + h - x} = \frac{f(x+h) - f(x)}{h},$$

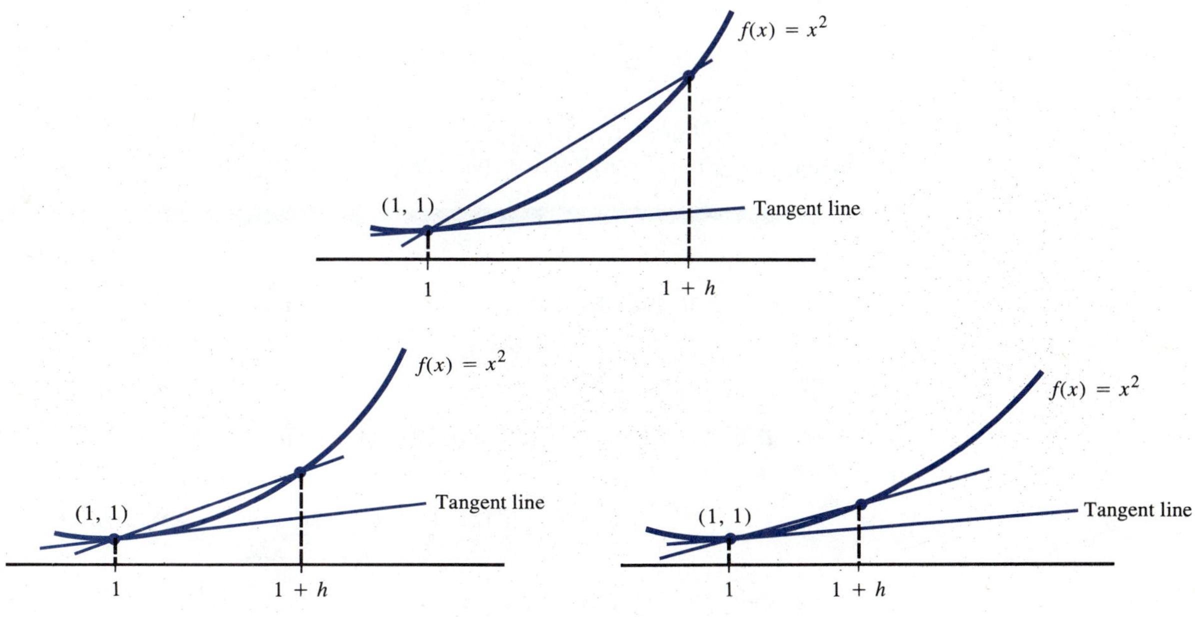

FIGURE 2–11 $h \to 0$

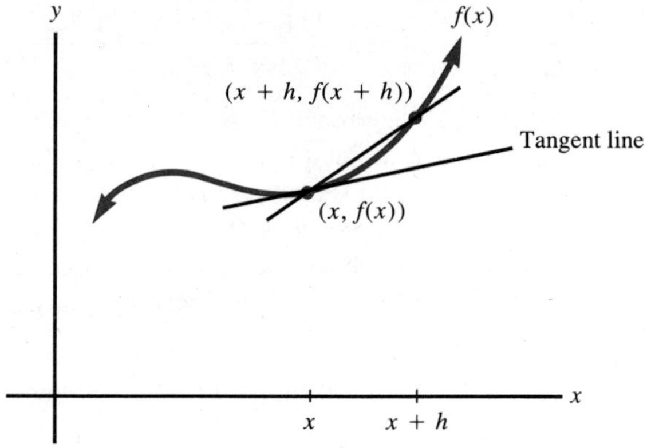

FIGURE 2–12

and thus
$$\text{slope of tangent line} = \lim_{h \to 0} \frac{f(x+h) - f(x)}{h}.$$

The limit on the right side of this equation is called the **derivative of f at x.**

> **The Derivative of a Function**
>
> Let f be a function. The **derivative of f at x,** denoted by $f'(x)$, is defined to be
>
> $$f'(x) = \lim_{h \to 0} \frac{f(x+h) - f(x)}{h},$$
>
> provided this limit exists. The derivative $f'(x)$ is the slope of the line tangent to f at the point $(x, f(x))$. The notation $f'(x)$ is read "f prime of x."

EXAMPLE 1 Let $f(x) = x^2$. Let us find the derivative $f'(x)$. We begin by forming the quotient

$$\frac{f(x+h) - f(x)}{h},$$

which is called the **difference quotient of f at x.** Since $f(x) = x^2$, we have $f(x+h) = (x+h)^2$, so

$$\frac{f(x+h) - f(x)}{h} = \frac{(x+h)^2 - x^2}{h}.$$

Next we simplify the difference quotient, our goal being the removal of the factor h from the denominator:

$$\frac{f(x+h)-f(x)}{h} = \frac{(x+h)^2 - x^2}{h}$$
$$= \frac{x^2 + 2xh + h^2 - x^2}{h}$$
$$= \frac{h(2x+h)}{h}$$
$$= 2x + h.$$

Therefore the derivative of $f(x) = x^2$ at x is

$$f'(x) = \lim_{h\to 0}\frac{f(x+h)-f(x)}{h} = \lim_{h\to 0}(2x+h) = 2x + 0 = 2x.$$

Notice that the derivative of the function f defined by $f(x) = x^2$ is another function f' defined by $f'(x) = 2x$. If we let $x = a$ in the equation $f'(x) = 2x$, we obtain the slope of the line tangent to f at the point $(a, f(a))$. For instance, if we take $x = 1$, then $f'(1) = 2(1) = 2$, so the slope of the line tangent to f at $(1, f(1)) = (1, 1)$ is equal to 2. This is the result we obtained at the beginning of this section.

The procedure we used in Example 1 to find the derivative is known as the **Three-Step Rule.**

The Three-Step Rule

To find the derivative $f'(x)$ of $f(x)$:

1. Form the difference quotient $\dfrac{f(x+h)-f(x)}{h}$.
2. Simplify the difference quotient to eliminate the factor h from the denominator.
3. Then $$f'(x) = \lim_{h\to 0}\frac{f(x+h)-f(x)}{h}$$ if this limit exists.

EXAMPLE 2 Let f be given by $f(x) = x^2 + 2x + 4$. We will use the Three-Step Rule to find the defining equation for the derivative f'. We have

$$\frac{f(x+h)-f(x)}{h} = \frac{[(x+h)^2 + 2(x+h) + 4] - [x^2 + 2x + 4]}{h}$$
$$= \frac{x^2 + 2xh + h^2 + 2x + 2h + 4 - x^2 - 2x - 4}{h}$$

$$= \frac{h(2x + h + 2)}{h}$$

$$= 2x + h + 2.$$

Therefore

$$f'(x) = \lim_{h \to 0} \frac{f(x + h) - f(x)}{h} = \lim_{h \to 0} (2x + h + 2) = 2x + 2.$$

EXAMPLE 3 If

$$g(x) = \frac{1}{x},$$

find $g'(x)$ and then find the equation of the line tangent to g at $(2, g(2))$.

We have

$$\frac{g(x + h) - g(x)}{h} = \frac{1/(x + h) - 1/x}{h}$$

$$= \frac{\dfrac{x - (x + h)}{x(x + h)}}{h}$$

$$= \frac{-h}{hx(x + h)}$$

$$= \frac{-1}{x(x + h)}.$$

Thus

$$g'(x) = \lim_{h \to 0} \frac{g(x + h) - g(x)}{h} = \lim_{h \to 0} \frac{-1}{x(x + h)} = \frac{-1}{x(x + 0)} = \frac{-1}{x^2}.$$

Hence if $g(x) = 1/x$, then $g'(x) = -1/x^2$. Thus the line tangent to g at $(2, g(2)) = (2, ½)$ has slope equal to

$$g'(2) = -\frac{1}{2^2} = -\frac{1}{4},$$

and the equation of the tangent line is

$$y = -\frac{1}{4}x + b.$$

But the point $(2, ½)$ is on the tangent line, so

$$\frac{1}{2} = -\frac{1}{4}(2) + b,$$

which implies that $b = 1$. Hence the equation of the tangent line is

$$y = -\frac{1}{4}x + 1.$$

See Figure 2–13.

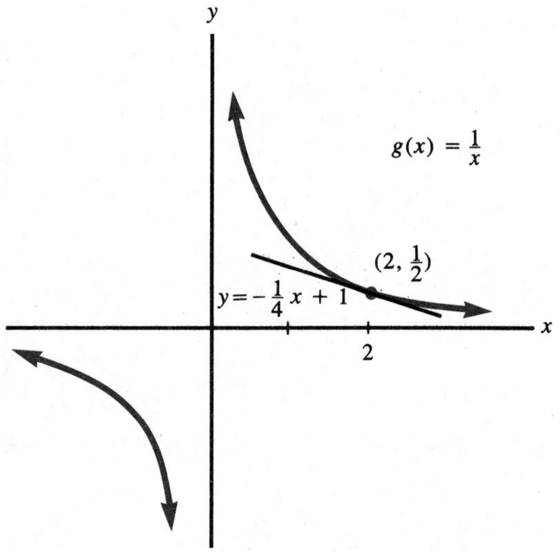

FIGURE 2–13

Differentiability and Continuity

If a function f has a derivative $f'(a)$ at $x = a$, we say that the function is **differentiable at *a*;** if it does not have a derivative at a, we say it is **not differentiable at *a*.**

EXAMPLE 4 The function f defined by $f(x) = \sqrt{x}$ is differentiable at a if $a > 0$, but it is not differentiable at 0. To see this, suppose $a \geq 0$ and note that

$$\lim_{h \to 0} \frac{\sqrt{a+h} - \sqrt{a}}{h} = \lim_{h \to 0} \frac{\sqrt{a+h} - \sqrt{a}}{h} \frac{\sqrt{a+h} + \sqrt{a}}{\sqrt{a+h} + \sqrt{a}}$$

$$= \lim_{h \to 0} \frac{a + h - a}{h(\sqrt{a+h} + \sqrt{a})}$$

$$= \lim_{h \to 0} \frac{1}{\sqrt{a+h} + \sqrt{a}}.$$

If $a > 0$, this limit exists and is equal to $\frac{1}{2\sqrt{a}}$. Thus $f'(a) = \frac{1}{2\sqrt{a}}$ exists if $a > 0$. However, if $a = 0$, the limit does not exist, and therefore f is not differentiable at 0.

If the derivative $f'(a)$ exists, it is the slope of the unique nonvertical line tangent to f at $(a, f(a))$. Therefore if f does not have a unique tangent line at $(a, f(a))$, as in Figures 2–14 and 2–15, or if the tangent line is vertical, as in

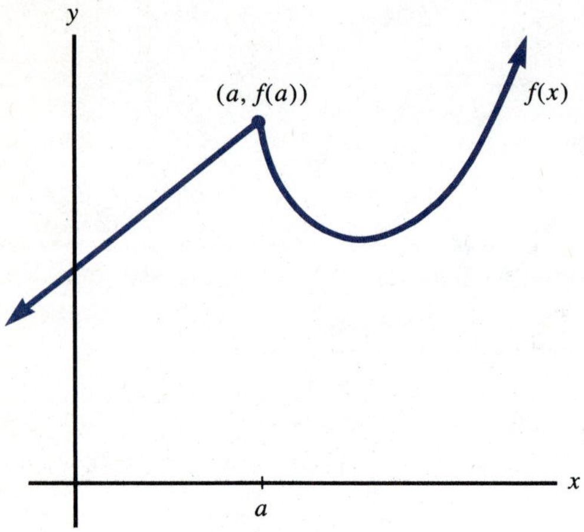

FIGURE 2–14

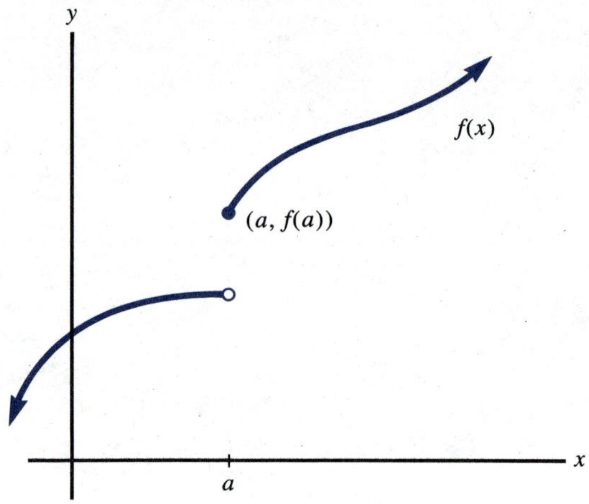

FIGURE 2–15

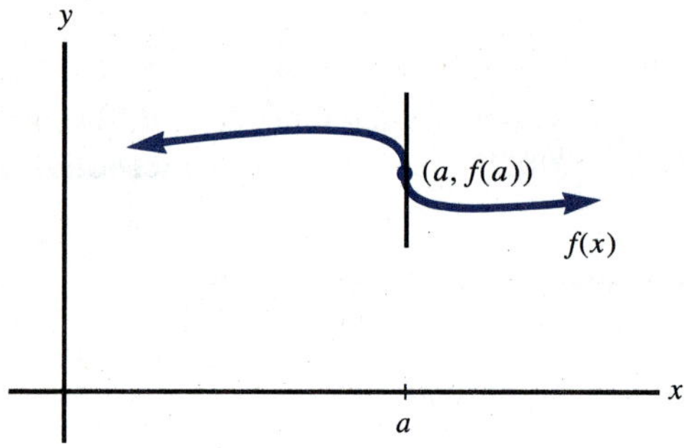

FIGURE 2–16

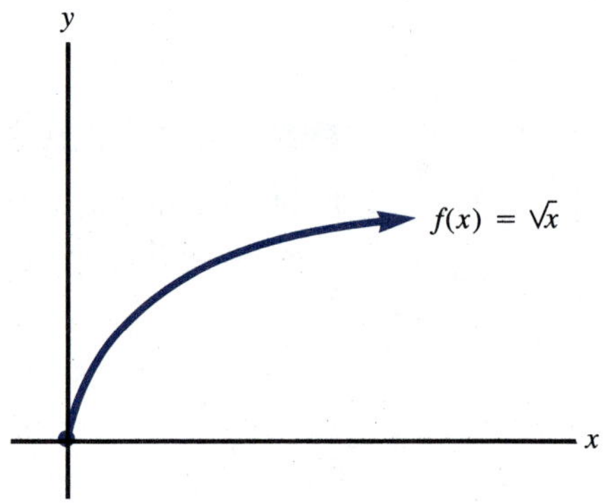

FIGURE 2–17

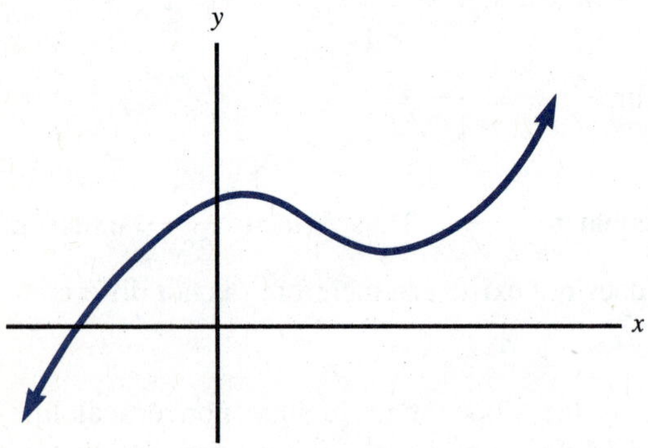

FIGURE 2–18

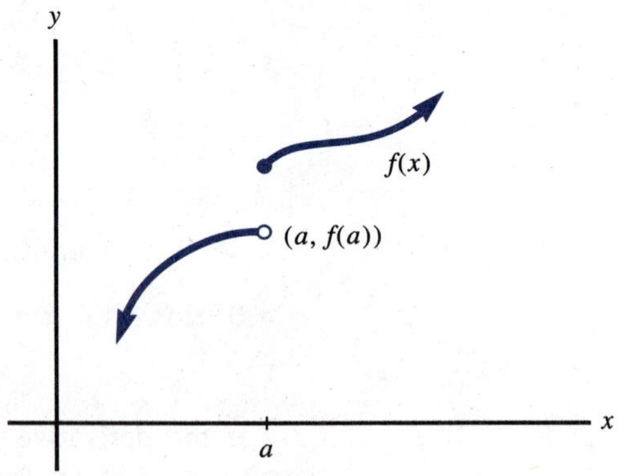

FIGURE 2–19

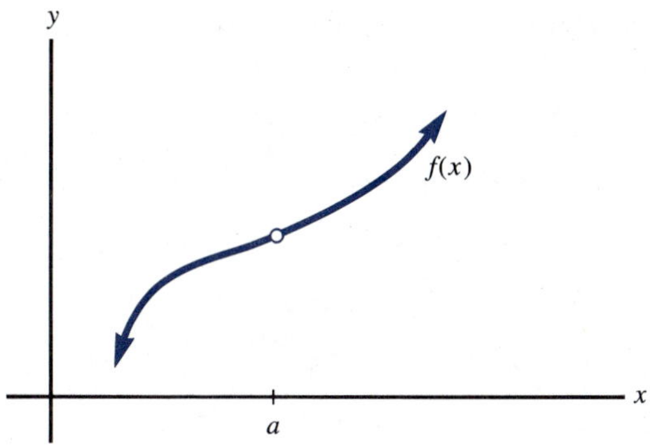

FIGURE 2–20

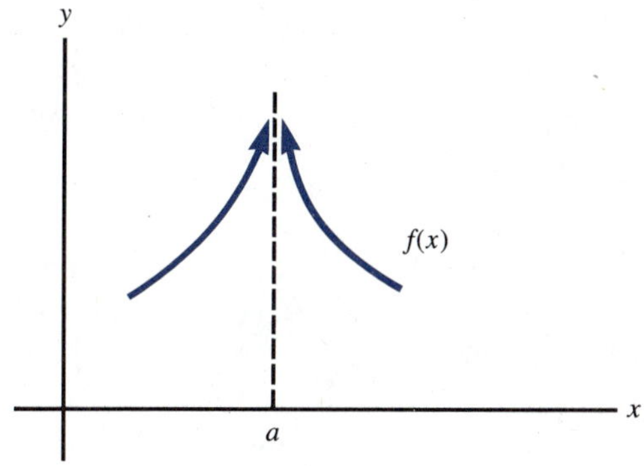

FIGURE 2–21

Figure 2–16, then the derivative $f'(a)$ cannot exist. For instance, the square-root function of Example 4 has a vertical tangent line at (0, 0) (see Figure 2–17), so it is not differentiable at $x = 0$.

Closely allied to the notion of differentiability is the notion of continuity. A function f is **continuous at $x = a$** if we can trace its graph through the point $(a, f(a))$ without lifting our pencil from the paper. This means that f is continuous at $x = a$ if there is neither a jump nor a gap in its graph at $x = a$. Figure 2–18 shows the graph of a function that is continuous at every real number; such a function is said to be **continuous everywhere.** If f is not continuous at $x = a$, we say it is **discontinuous at $x = a$.** Figures 2–19, 2–20, and 2–21 show different ways in which a function can be discontinuous at $x = a$: in Figure 2–19 the function has a jump in its graph at $x = a$; in Figures 2–20 and 2–21 it has a gap in its graph at $x = a$.

We can state the definition of continuity of a function at $x = a$ in terms of limits.

The First Limit Definition of Continuity

A function f is continuous at $x = a$ if $f(a)$ exists and

$$\lim_{x \to a} f(x) = f(a).$$

This definition of continuity agrees with our previous geometric one: if $f(a)$ exists, then the point $(a, f(a))$ is on the graph, so there can be no gap in the graph of f at $x = a$; if $\lim_{x \to a} f(x) = f(a)$, then there can be no jump in the graph at $x = a$. (Why?)

We have already remarked in the previous section that if p is a polynomial function, then $\lim_{x \to a} p(x) = p(a)$. Thus polynomial functions are continuous at every real number. Rational functions that have been reduced to lowest terms are continuous except at numbers where they have vertical asymptotes.

Note that if we set $x = a + h$, then we may write

$$\lim_{x \to a} f(x) \qquad \text{as} \qquad \lim_{h \to 0} f(a + h).$$

Hence we may rewrite the first limit definition of continuity as

The Second Limit Definition of Continuity

A function f is continuous at $x = a$ if $f(a)$ exists and

$$\lim_{h \to 0} f(a + h) = f(a).$$

This second version of the limit definition of continuity will be useful to us later.

The relationship between continuity and differentiability is this: a function differentiable at $x = a$ is continuous at $x = a$. To see this, suppose the function f has a derivative at $x = a$. If f were not continuous at $x = a$, it would have either a jump or a gap in its graph at $x = a$. But if this were the case, it could not have a unique tangent line at $(a, f(a))$, and hence it could not have a derivative at $x = a$. Thus f must be continuous at $x = a$.

Differentiability Implies Continuity

If the function f is differentiable at $x = a$, then it is continuous at $x = a$.

The converse of the preceding statement is *not* true: continuity does *not* imply differentiability. In other words, it is possible for a function to be continuous at $x = a$ but not differentiable at $x = a$. For example, the absolute value function defined by $f(x) = |x|$ has the graph shown in Figure 2–22; from the graph we see that this function is continuous at $x = 0$, but it is not differentiable at $x = 0$ because it does not have a unique tangent line there.

Computer Manual: Program DFQUOTNT.

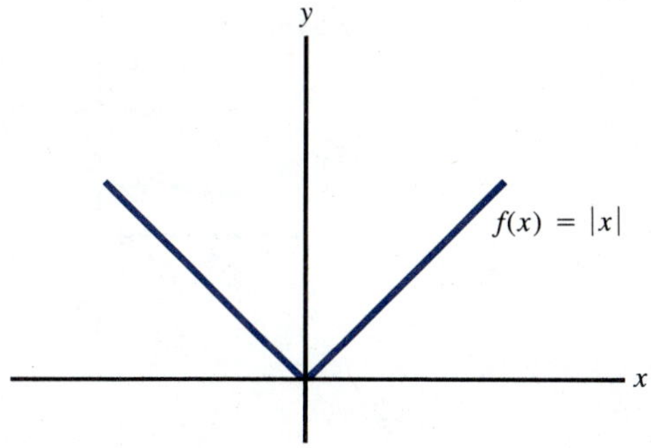

FIGURE 2–22

2.2 EXERCISES

Definition of the Derivative

In Exercises 1 through 6 use the Three-Step Rule to find $f'(x)$.

1. $f(x) = 3x - 2$

2. $f(x) = 2x^2 + x - 1$

3. $f(x) = x^3$

4. $f(x) = \dfrac{x}{x - 1}$

5. $f(x) = \dfrac{1}{x^2}$

6. $f(x) = \sqrt{x + 1}$

In Exercises 7 through 12 find the equation of the line tangent to f at $(a, f(a))$. You may use the results of Exercises 1 through 6.

7. $f(x) = 3x - 2, \quad a = 2$

8. $f(x) = 2x^2 + x - 1, \quad a = 1$

9. $f(x) = x^3, \quad a = -1$

10. $f(x) = \dfrac{x}{x - 1}, \quad a = -2$

11. $f(x) = \dfrac{1}{x^2}, \quad a = 3$

12. $f(x) = \sqrt{x + 1}, \quad a = 3$

Differentiability and Continuity

Find all values of x at which the functions of Exercises 13 through 16 are not differentiable.

13.

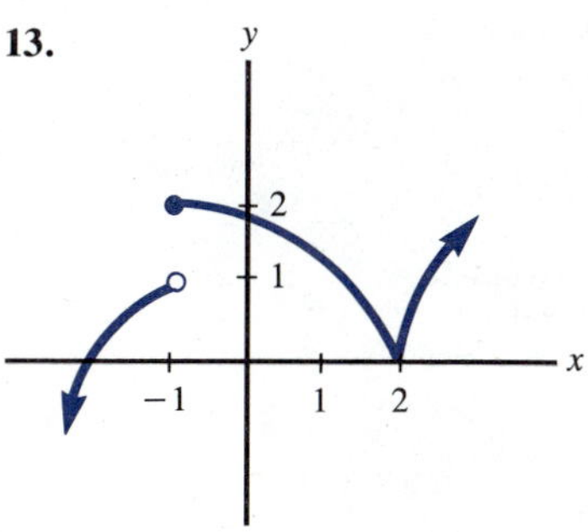

14.

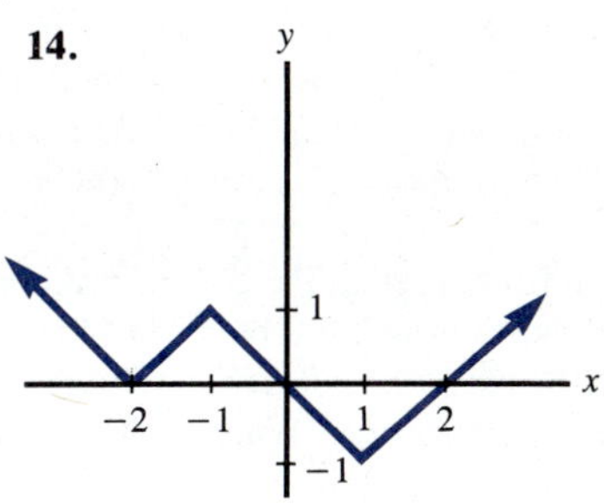

15.

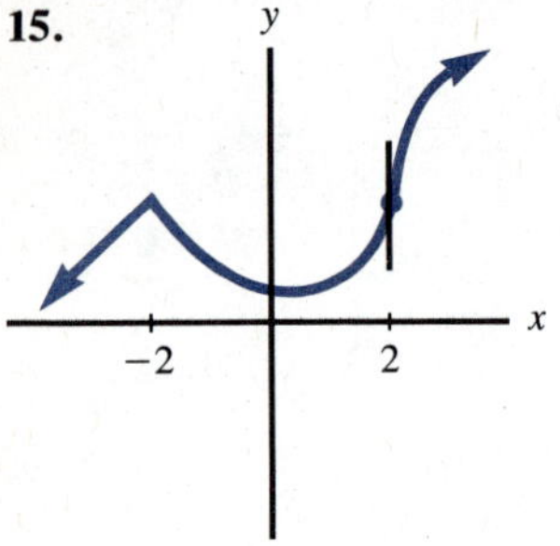

16.

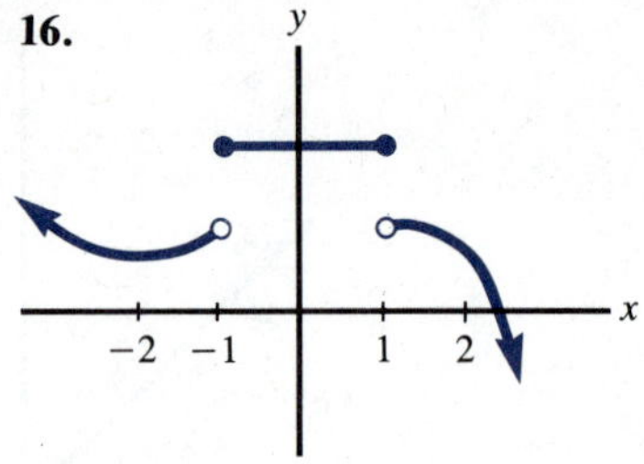

Find all values of x at which the functions of Exercises 17 through 20 are discontinuous.

17.

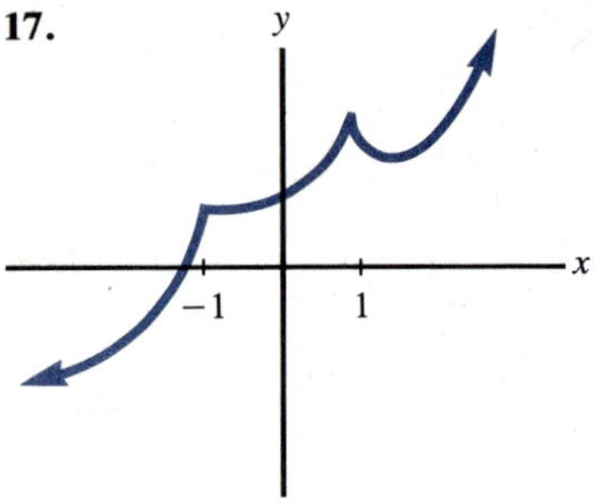

18.

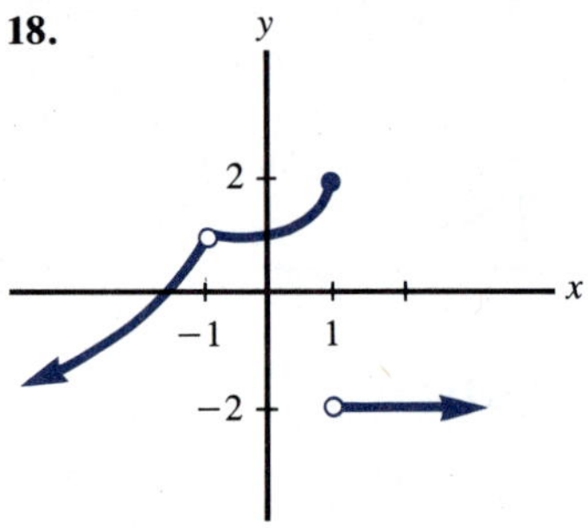

19.

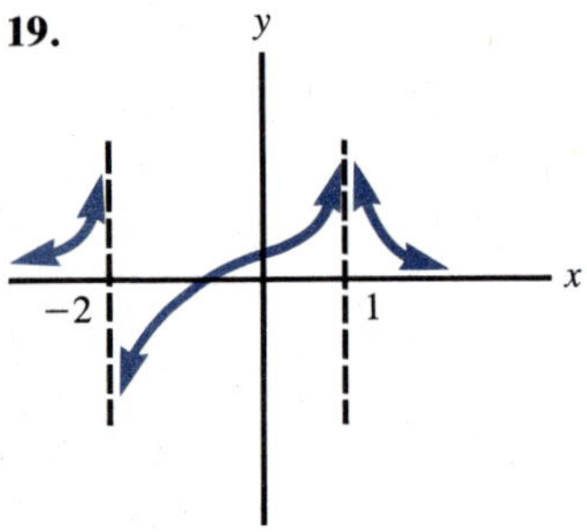

20.

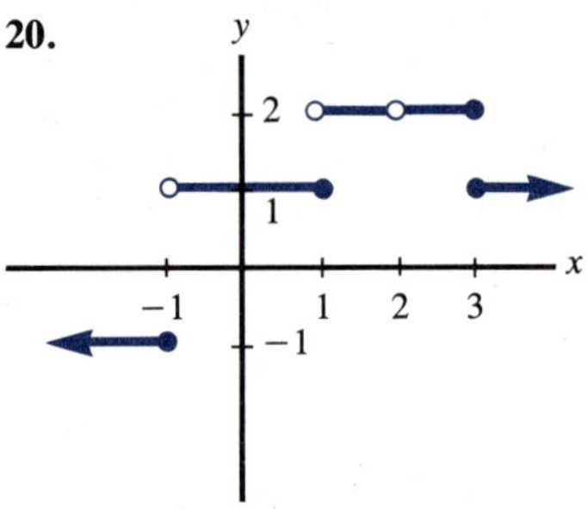

Management

21. A candy company has three machines that can make chocolate bonbons. Machine A, which is new, can make 2000 bonbons per hour at a cost of 1.8 cents each. Machine B, which is older, can make 1500 bonbons per hour at a cost of 2.2 cents each. Machine C, which is ancient, can make 800 bonbons per hour at a cost of 3.5 cents each.

(a) Find a function that expresses the cost of making x bonbons per hour. (*Hint:* If $x \leq 2000$, the company will use only Machine A; if $2000 < x \leq 3500$, they will use A and B, etc.)

(b) Graph the function of part (a). Where is it not differentiable?

22. A firm that sells computer supplies charges \$9.00 per ribbon for printer ribbons on orders of fewer than 10 ribbons, \$8.60 per ribbon on orders of 10 through 24 ribbons, and \$8.20 per ribbon on orders of 25 or more ribbons. Find and graph a function that expresses the cost of an order of x ribbons. Where is the function not differentiable?

2.3 RULES OF DIFFERENTIATION

The process of finding the derivative of a function is called **differentiation.** As of now, we can only differentiate functions by means of the Three-Step Rule, which is often both time-consuming and tedious to apply. Fortunately, it is possible to use the Three-Step Rule to prove some very general rules of differentiation. Once we have these rules of differentiation available and understand how to use them, we will be able to differentiate entire classes of functions (such as the polynomial functions) almost automatically. In this section we prove some of the most important rules of differentiation and illustrate their usage.

Let c be any number and consider the **constant function** f defined by

$$f(x) = c \qquad \text{for all numbers } x.$$

By the Three-Step Rule,

$$f'(x) = \lim_{h\to 0} \frac{f(x+h) - f(x)}{h} = \lim_{h\to 0} \frac{c - c}{h} = \lim_{h\to 0} 0 = 0.$$

Thus we have proved the **constant rule.**

The Constant Rule

If c is a number and $f(x) = c$ for all x, then

$$f'(x) = 0.$$

That is, the derivative of a constant function is zero.

EXAMPLE 1 Let $f(x) = 2$ and $g(x) = \sqrt{2}$ for all x. Then $f'(x) = 0$ and $g'(x) = 0$ for all x; thus $f'(5) = 0$, $f'(\pi) = 0$, $g'(0) = 0$, and so on.

Now let n be a positive integer and consider the **power function** defined by

$$f(x) = x^n \qquad \text{for all numbers } x.$$

We have

$$f'(x) = \lim_{h\to 0} \frac{f(x+h) - f(x)}{h} = \lim_{h\to 0} \frac{(x+h)^n - x^n}{h}.$$

But
$$(x+h)^n = x^n + nx^{n-1}h + \frac{n(n-1)}{2}x^{n-2}h^2 + \cdots + h^n,$$

where all the terms after the first two on the right side of the equality have h^2 as a factor. Therefore if we factor h out of all terms after the first one, we will obtain

$$(x+h)^n = x^n + h\left[nx^{n-1} + \frac{n(n-1)}{2}x^{n-2}h + \cdots + h^{n-1}\right],$$

where now all the terms inside the brackets after the first one still have h as a factor, and hence will approach 0 as $h \to 0$. Thus

$$\begin{aligned} f'(x) &= \lim_{h\to 0} \frac{x^n + h[nx^{n-1} + (n(n-1)/2)x^{n-2}h + \cdots + h^{n-1}] - x^n}{h} \\ &= \lim_{h\to 0}\left[nx^{n-1} + \frac{n(n-1)}{2}x^{n-2}h + \cdots + h^{n-1}\right] \\ &= nx^{n-1} + 0 + \cdots + 0 \\ &= nx^{n-1}. \end{aligned}$$

We have shown that if $f(x) = x^n$, then $f'(x) = nx^{n-1}$. This fact, which is known as the **power rule,** remains true if we replace n by any nonzero number (though we shall not prove this).

The Power Rule

If r is any number except 0 and $f(x) = x^r$, then

$$f'(x) = rx^{r-1}.$$

That is, the derivative of x raised to a nonzero power is the power times x raised to the power minus 1.

EXAMPLE 2

a. If $f(x) = x$, then $f(x) = x^1$, and $f'(x) = 1x^{1-1} = 1x^0 = 1 \cdot 1 = 1$. Thus if $f(x) = x$, then $f'(x) = 1$.
b. If $g(x) = x^2$, then $g'(x) = 2x^{2-1} = 2x^1 = 2x$.
c. If $h(x) = x^5$, then $h'(x) = 5x^{5-1} = 5x^4$.
d. If $k(x) = \sqrt{x}$, then $k(x) = x^{1/2}$, so

$$k'(x) = \frac{1}{2}x^{(1/2)-1} = \frac{1}{2x^{1/2}} = \frac{1}{2\sqrt{x}}.$$

e. If $s(x) = 1/x = x^{-1}$, then $s'(x) = (-1)x^{-1-1} = -x^{-2} = -1/x^2$.

The power rule shows how to differentiate $f(x) = x^r$ when r is any nonzero number. We would like to be able to differentiate functions whose defining equations are of the form $f(x) = cx^r$, where c is a number. But if g is any function for which $g'(x)$ exists, and $f(x) = c \cdot g(x)$, then

$$f'(x) = \lim_{h\to 0} \frac{f(x+h) - f(x)}{h} = \lim_{h\to 0} \frac{c \cdot g(x+h) - c \cdot g(x)}{h}$$

$$= c \cdot \lim_{h\to 0} \frac{g(x+h) - g(x)}{h} = c \cdot g'(x),$$

and we have proved the **constant-multiple rule.**

The Constant-Multiple Rule

If c is a number and g is a function such that $f(x) = c \cdot g(x)$ and $g'(x)$ exists, then

$$f'(x) = c \cdot g'(x).$$

That is, the derivative of a constant times a function is the constant times the derivative of the function.

EXAMPLE 3 **a.** If $f(x) = 2x$, then $f'(x) = (2x)' = 2(x)' = 2 \cdot 1 = 2$.
b. If $h(x) = 3x^2$, then $h'(x) = (3x^2)' = 3(x^2)' = 3(2x) = 6x$.
c. If $k(x) = \frac{1}{3}x^{-3}$, then $k'(x) = -x^{-4}$.

Now we can differentiate functions whose defining equations are of the form $f(x) = cx^r$. Polynomials are formed by taking sums and differences of such functions, and we would like to be able to differentiate polynomials. But if $k(x) = f(x) + g(x)$, where $f'(x)$ and $g'(x)$ exist, then

$$k'(x) = \lim_{h\to 0} \frac{k(x+h) - k(x)}{h}$$

$$= \lim_{h\to 0} \frac{[f(x+h) + g(x+h)] - [f(x) + g(x)]}{h}$$

$$= \lim_{h\to 0} \frac{f(x+h) - f(x)}{h} + \lim_{h\to 0} \frac{g(x+h) - g(x)}{h}$$

$$= f'(x) + g'(x).$$

Similarly, if $k(x) = f(x) - g(x)$, then $k'(x) = f'(x) - g'(x)$. Thus we have the **sum rule.**

The Sum Rule

If $f'(x)$ and $g'(x)$ both exist and $k(x) = f(x) \pm g(x)$, then

$$k'(x) = f'(x) \pm g'(x).$$

That is, the derivative of a sum (difference) is the sum (difference) of the derivatives.

EXAMPLE 4 If $k(x) = -2x^2 + 3$, then

$$k'(x) = (-2x^2 + 3)' = -2(x^2)' + (3)' = -2(2x) + 0 = -4x.$$

If $s(x) = x^4 - 5x^3$, then

$$s'(x) = 4x^3 - 5(3x^2) = 4x^3 - 15x^2.$$

We note that the sum rule also holds if there are more than two summands. Thus, if

$$k(x) = f_1(x) \pm f_2(x) \pm \cdots \pm f_n(x),$$

then

$$k'(x) = f_1'(x) \pm f_2'(x) \pm \cdots \pm f_n'(x).$$

The four rules of differentiation we have proved in this section allow us to differentiate polynomials and also functions that are formed by taking sums and differences of nonintegral powers of the variable.

EXAMPLE 5 **a.** If $f(x) = 2x^3 + 5x^2 - 12x + 6$, then

$$f'(x) = 2(3x^2) + 5(2x) - 12(1) + 0 = 6x^2 + 10x - 12.$$

b. If $g(x) = 2\sqrt{x} - 4x^{6/5} + 6x - 10/x^2 = 2x^{1/2} - 4x^{6/5} + 6x - 10x^{-2}$, then

$$\begin{aligned} g'(x) &= 2\left(\frac{1}{2}x^{-1/2}\right) - 4\left(\frac{6}{5}x^{1/5}\right) + 6(1) - 10(-2x^{-3}) \\ &= x^{-1/2} - \frac{24}{5}x^{1/5} + 6 + 20x^{-3} \\ &= \frac{1}{\sqrt{x}} - \frac{24}{5}\sqrt[5]{x} + \frac{20}{x^3} + 6. \end{aligned}$$

EXAMPLE 6 Let $f(x) = x^5 - 2x^3 + 7x + 2$. Find the equation of the line tangent to f at the point $(1, f(1)) = (1, 8)$.

We find the derivative

$$f'(x) = 5x^4 - 6x^2 + 7$$

and evaluate it at $x = 1$ to obtain

$$f'(1) = 5(1^4) - 6(1^2) + 7 = 6.$$

Since $f'(1) = 6$, the tangent line has slope equal to 6. Hence its equation is

$$y = 6x + b.$$

But since the point (1, 8) is on the line, we have

$$8 = 6(1) + b = 6 + b.$$

Thus $b = 2$. Therefore the equation of the tangent line is

$$y = 6x + 2.$$

EXAMPLE 7 A patient's white blood cell count t hours after being admitted to a hospital is

$$f(t) = 5t^2 - 80t + 500$$

cells per cubic millimeter. The graph of this quadratic function is the parabola of Figure 2–23. Note that the line tangent to the graph at the vertex of the parabola is horizontal and therefore has slope zero. But the slope of the line tangent to f at $(t, f(t))$ is $f'(t)$. Thus if

$$f'(t) = 10t - 80 = 0,$$

then $t = 8$. Since

$$f(8) = 5(8)^2 - 80(8) + 500 = 180,$$

the vertex of the parabola is $(8, f(8)) = (8, 180)$. Therefore the patient's minimum white blood cell count occurs 8 hours after admission and is equal to 180 cells per cubic millimeter.

Computer Manual: Program DIFFPOLY.

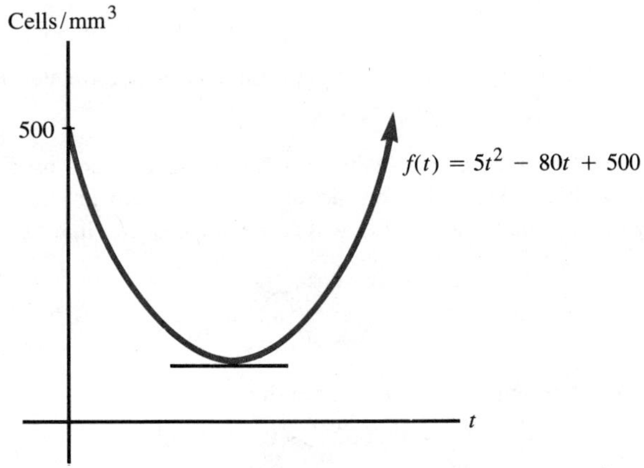

FIGURE 2–23

2.3 EXERCISES

In Exercises 1 through 16 use the rules of differentiation to find $f'(x)$.

1. $f(x) = 2$ **2.** $f(x) = -4$ **3.** $f(x) = 3x^2$ **4.** $f(x) = 6x + 2$

5. $f(x) = 8x^2 - 12x + 3$ **6.** $f(x) = -3x^4 + 7x^3 - 5x^2$

7. $f(x) = x^{100} - x^{99} + 1$ **8.** $f(x) = \frac{1}{5}x^5 + \frac{1}{3}x^3$

9. $f(x) = -5x^{12} + \frac{3}{21}x^7 - 6.5x^4 + 2200x^3 + 0.02x^2 + 200{,}000x + 3$

10. $f(x) = 2\sqrt{x}$ **11.** $f(x) = 4\sqrt[3]{x}$ **12.** $f(x) = \frac{2}{x}$ **13.** $f(x) = -\frac{32}{x^2}$

14. $f(x) = 3\sqrt{x} + 4x^{2/3} - 5x^{-4} + 12$ **15.** $f(x) = \frac{1}{x} - \frac{1}{x^2} + \frac{1}{x^3} - \frac{1}{x^4}$

16. $f(x) = x^{-1/2} + \frac{4}{x^{3/2}} - 5x^{-5} + 12x^{-4/3}$

In Exercises 17 through 24 find the equation of the line tangent to f at $(a, f(a))$.

17. $f(x) = 5x - 3, \quad a = 2$ **18.** $f(x) = x^2 - 3x + 1, \quad a = 0$ **19.** $f(x) = x^3 - 1, \quad a = -2$

20. $f(x) = x^4 - 3x^3 + 2x, \quad a = 1$ **21.** $f(x) = 4\sqrt{x} - 5x, \quad a = 9$ **22.** $f(x) = 3x^{4/3} - 2x^{2/3}, \quad a = 27$

23. $f(x) = \frac{2}{x} - \frac{4}{x^2}, \quad a = 2$ **24.** $f(x) = 6x^{-2} + x^{-3/4}, \quad a = 16$

Management

25. A firm's profit function is given by $P(x) = -x^2 + 40x - 350$.
(a) Graph this function.
(b) Find the firm's maximum profit by finding the value of x for which the slope of the tangent line is 0.

Life Science

26. The number of black bear cubs born in New England was estimated to be

$$f(t) = -0.5t^2 + 5t + 100$$

in year t, where $t = 0$ represents 1980. Graph this function and then find the maximum number of black bear cubs born annually and the year in which this maximum occurred by finding the value of t for which the slope of the line tangent to f is zero.

Social Science

27. A politician's voter approval rating is given by

$$y = 0.0002t^2 - 0.01t + 0.6,$$

where t represents the number of months since the last election and y is the proportion of voters who approve of the politician's performance. Graph this function and find the politician's minimum approval rating and the time at which it occurred.

2.4 APPLICATION: RATES OF CHANGE

We have several times referred to the fact that the derivative of a function describes the rate of change of the function with respect to its independent variable. In this section we see why this is true, and we demonstrate how to use this property of the derivative in some practical applications.

Suppose we make a journey in a small airplane. We fly in a straight line, and after t hours in the air we have traveled a distance of

$$d(t) = 4t^2 + 150t$$

miles from our starting point, where $0 \leq t \leq 4$. What is the average velocity of our plane during the second hour of the trip?

Average velocity is distance traveled divided by elapsed time. During the second hour of the trip the time changes from $t = 1$ to $t = 2$, so the distance traveled during the second hour is $d(2) - d(1)$ miles, and therefore the average velocity during the second hour is

$$\frac{d(2) - d(1)}{2 - 1} = \frac{316 - 154}{2 - 1} = 162$$

miles per hour. The same reasoning holds for any time interval: the average velocity of the plane from time t to time $t + h$ is

$$\frac{d(t + h) - d(t)}{t + h - t} = \frac{d(t + h) - d(t)}{h}$$

miles per hour. Notice that the right side of the equation is the difference quotient of the **distance function** $d = d(t)$. If we take the limit of this difference quotient as h approaches 0, we obtain the derivative $d'(t)$. On the other hand, as h approaches 0 the difference quotient gives the average velocity over shorter and shorter intervals of time. Hence, if we take the limit as h approaches 0, we obtain the **instantaneous velocity** at time t. Thus the derivative

$$d'(t) = 8t + 150$$

of $d(t)$ gives the instantaneous velocity of the airplane at time t. For example, when $t = 2$ we have

$$d'(2) = 8(2) + 150 = 16 + 150 = 166.$$

Therefore at the instant when $t = 2$ the velocity of the airplane is 166 miles per hour.

A distance function such as the one we have just considered is simply a means of specifying the position of an object at time t relative to its starting point. The analysis we performed for $d(t) = 4t^2 + 150t$ can be repeated for any function $s = s(t)$ that gives the position of an object relative to some reference point, as long as the motion is in a straight line.

Velocity

If an object moves in a straight line so that its position at time t relative to some reference point is given by the function $s = s(t)$, then its instantaneous velocity at time t is

$$v(t) = s'(t).$$

If we take the reference point to be the zero point on the line of motion, then one direction along the line is positive and the other is negative. Thus if the velocity is positive at time $t = a$, the object is moving along the line in the positive direction; if the velocity is negative, it is moving in the negative direction.

EXAMPLE 1 A rocket is launched straight up into the atmosphere. The altitude of the rocket t minutes after it is launched is

$$s(t) = -2t^2 + 100t$$

kilometers. We have chosen the rocket's zero point to be its launch pad and its upward motion to be positive motion. Figure 2–24 shows the graph of the position function $s = s(t)$. Note that this is the graph of the rocket's distance from its launch pad (that is, its altitude) at time t, and not a picture of its motion; the rocket moves in a straight line.

The instantaneous velocity of the rocket at time t is

$$v(t) = s'(t) = -4t + 100$$

kilometers per minute. For instance, exactly 10 minutes after launching, the rocket is at an altitude of

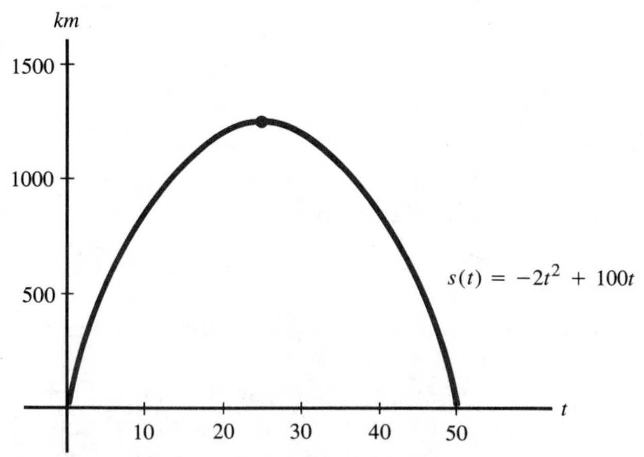

FIGURE 2–24

$$s(10) = 800$$

kilometers and its velocity at the instant when $t = 10$ is

$$v(10) = 60$$

kilometers per minute. Therefore at the instant when $t = 10$, the rocket's altitude is increasing at a rate of 60 kilometers per minute; we can tell that the altitude is increasing because $v(10)$ is positive. See Figure 2–25.

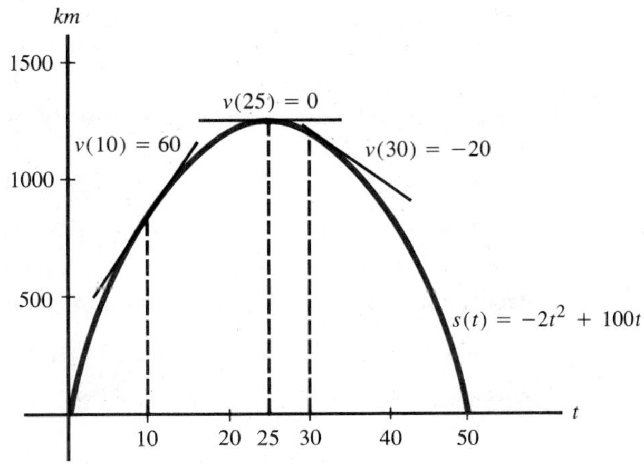

FIGURE 2–25

Similarly, 30 minutes after launching, the rocket is at an altitude of

$$s(30) = 1200$$

kilometers and has a velocity of

$$v(30) = -20$$

kilometers per minute. Hence when $t = 30$ the rocket's altitude is decreasing at the rate of 20 kilometers per minute; the fact that the velocity is negative tells us that the altitude is decreasing. See Figure 2–25.

Notice that $v(25) = 0$. Thus at the instant when $t = 25$, the rocket's altitude is neither increasing nor decreasing. Therefore the rocket reaches its peak altitude 25 hours after launching. Again, see Figure 2–25.

We have seen that the derivative of a position function is a velocity function. But velocity is just the name we give to the rate of change of position with respect to time. Thus the derivative of a position function is the rate of change of the position function with respect to time. We can apply the same reasoning to any function to conclude that its derivative represents the rate of change of the function with respect to the independent variable.

The Derivative as Rate of Change

For any function f, the derivative f' gives the instantaneous rate of change of f with respect to its independent variable.

EXAMPLE 2 A stone dropped from a height falls

$$s(t) = 16t^2$$

feet in t seconds. The velocity of the stone at time t is thus

$$v(t) = s'(t) = 32t$$

feet per second. For instance, the velocity of the stone exactly 2 seconds after it is dropped is

$$v(2) = 64$$

feet per second. The derivative of $v(t)$ will give the stone's rate of change of velocity with respect to time; this is known as its **acceleration.** If we let $a(t)$ denote the stone's acceleration at time t, then

$$a(t) = v'(t) = 32$$

feet per second per second. Note that the acceleration of the stone is constant.

In the preceding example we said that the acceleration of the stone was the derivative of its velocity. This is true in general.

Acceleration

If $v(t)$ is the velocity of an object at time t and $a(t)$ its acceleration at time t, then

$$a(t) = v'(t).$$

The fact that the derivative of a function can be viewed as its rate of change with respect to its independent variable has many applications in addition to finding velocity and acceleration.

EXAMPLE 3 Alpha Company has studied the productivity of its employees and has been found that after t years of experience an employee's monthly productivity can be expected to be

$$f(t) = -2t^2 + 120t + 100$$

units. The derivative

$$f'(t) = -4t + 120$$

gives the rate of change of monthly productivity with respect to years of experience. For instance, when $t = 10$ we have $f'(10) = 80$. Thus employees with 10 years

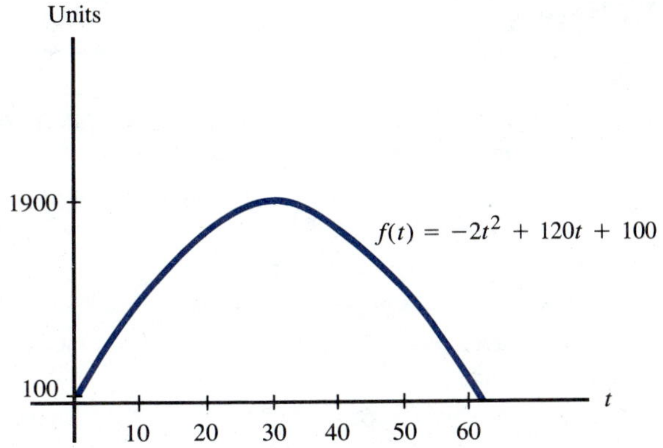

FIGURE 2–26

of experience are increasing their monthly productivity at a rate of 80 units per year. Similarly, since $f'(20) = 40$, employees with exactly 20 years of experience are also increasing their monthly productivity, but only at the rate of 40 units per year.

Note that $f'(30) = 0$, and hence employees with exactly 30 years of experience will be neither increasing nor decreasing their productivity. Since $f'(t) = -4t + 120$ is negative for $t > 30$, employee productivity decreases when $t > 30$. It follows that greatest productivity is reached when $t = 30$, and hence that maximum monthly productivity is $f(30) = 1900$ units. See Figure 2–26 for a graph of the productivity function.

EXAMPLE 4 If Beta Corporation makes x units of product, its average cost per unit is

$$A(x) = 0.10x + \frac{40}{x}$$

dollars. Thus, if Beta produces 10 units, its average cost per unit is

$$A(10) = 0.10(10) + \frac{40}{10} = 5$$

dollars. The derivative

$$A'(x) = 0.10 - \frac{40}{x^2}$$

gives the rate of change of Beta's average cost per unit. For instance, since

$$A'(10) = 0.10 - \frac{40}{10^2} = -0.30,$$

when Beta produces 10 units its average cost per unit is decreasing by 30 cents per unit.

Computer Manual: Program DIFFPOLY.

2.4 EXERCISES

1. In t hours an automobile travels $d(t) = 40t$ miles in a straight line from its starting point. Find its velocity and acceleration t hours after it started.

2. A vehicle travels $d(t) = 40t^2 - 3t^3$ kilometers in a straight line in t hours, where $0 \leq t \leq 4$. Find the vehicle's

(a) velocity 1, 2, and 3.5 hours after it started;
(b) acceleration 1, 2, and 3.5 hours after it started.

3. A drag racer's car has traveled $d(t) = 12t^2 + 110t$ feet from the starting line t seconds after the start of a race.

(a) What is the car's velocity 1 second after the start of the race? 4 seconds after the start?
(b) What is the car's acceleration 1 second after the start of the race? 4 seconds after the start?

4. You throw a ball straight up. The altitude of the ball t seconds after you have thrown it is $s(t) = -2t^2 + 20t$ feet.

(a) What is the altitude of the ball 2 seconds after you have thrown it? What is its velocity at this time? Is the ball going up or coming down? What is its acceleration?
(b) Answer the questions of part (a) for $t = 5$, $t = 7$, and $t = 10$ seconds.
(c) What is the maximum altitude the ball will reach? When will the ball attain its maximum altitude?

5. On the moon the velocity of a stone dropped from a height is $v(t) = 1.6t$ meters per second. Find the acceleration of the stone at time t. If the stone was dropped from rest, find the distance the stone will fall in t seconds. (Hint: Find a function whose derivative is $v(t)$.)

Management

6. Total monthly sales of Gamma Corporation t months after the founding of the company are

$$S(t) = t^3 - 7t^2 + 10t + 100$$

thousand dollars.

(a) What were total monthly sales 1 month after the founding of the corporation? At what rate were sales changing then? Were sales increasing or decreasing?
(b) Answer the questions of part (a) for $t = 3$ months, $t = 6$ months, and $t = 1$ year.

7. Delta Company's cost function is given by $C(x) = 5x^2 + 2000$, where x is in thousands of units produced. Delta's average cost per thousand units is therefore given by $A(x) = C(x)/x$. At what rate is Delta's average cost per thousand units changing when production is 5000 units? 10,000 units? 20,000 units? 30,000 units? At each of these production levels, is average cost per thousand units increasing or decreasing?

8. A firm's profit function is given by

$$P(x) = -0.02x^3 + 4.5x^2 - 300x - 5000.$$

At what rate is profit changing when 40 units are being produced? when 80 units are being produced? when 150 units are being produced? At each of these production levels, is profit per unit increasing or decreasing?

9. If the demand function for a product is given by $d = -p^2 - 30p + 4000$, where d is in thousands of units, find the rate at which demand for the product is changing when the price is \$10 per unit; when the price is \$50 per unit.

Life Science

10. A reagent is added to a chemical solution containing an enzyme, and the enzyme precipitates out of solution. The amount of enzyme precipitated out t seconds after the reagent has been added is $y = -0.04t^2 + 0.20t$ grams, where $0 \leq t \leq 2.5$. Find the rate at which the enzyme is precipitating out of solution 0.5 second, 1 second, 1.5 seconds, and 2 seconds after the reagent is added.

11. The number of bacteria present in a culture t hours after the start of an experiment is given by $f(t) = t^3 + 6t^2 + 100$. Find the rate at which the size of the bacteria population in the culture is changing when $t = 2$, $t = 5$, and $t = 10$.

12. A field is sprayed with weed killer, and t days later it contains an average of

$$W = t^3 - 3t^2 - 144t + 900$$

weeds per acre. Find the rate at which the average number of weeds per acre is changing 1 day after spraying; 5 days after spraying; 10 days after spraying; 12 days after spraying. Is the average number of weeds per acre increasing or decreasing at these times?

Social Science

13. Students who take an intensive foreign language course in which only the language being taught is spoken to them typically understand

$$P = 0.005t^3 - 0.05t^2 + 2t$$

percent of what is being said to them by the end of day t of the course, $0 \leq t \leq 25$. Find the rate at which the comprehension of a typical student is changing 5 days, 10 days, and 15 days into the course.

14. In a memory experiment, subjects memorize a list of 20 numbers and then watch rock videos for various lengths of time. After t minutes of watching the videos, the typical subject can remember approximately

$$y = -0.02t^2 + 20$$

of the numbers. Find the rate at which the typical subject forgets the numbers when $t = 5$, $t = 10$, $t = 15$, and $t = 25$ minutes.

15. A politician's campaign manager estimates that the proportion of voters who support the politician t days before an election is y, where

$$y = 0.45 + \frac{0.10}{t}$$

and $1 \leq t \leq 10$. Find the rate at which the politician's support is changing 10 days, 5 days, and 1 day before the election.

2.5 APPLICATION: THE THEORY OF THE FIRM

The **theory of the firm** is concerned with questions such as

- How many units of its product must a firm make and sell in order to maximize its profit?
- What price should the firm charge for its product in order to attain its maximum profit?
- How will changing the product's price affect the firm's revenue?

In this section we show how to use calculus to answer these questions. We do this by exploiting both viewpoints of the derivative: that it is the slope of the tangent line, and that it gives the rate of change with respect to the independent variable.

Marginality

If a firm produces and sells x units of its product, then, as we have seen, it has a **cost function** C, a **revenue function** R, and a **profit function**

$$P = R - C.$$

These functions need not be linear, but as in the linear case we can find the firm's **break-even quantities** by setting $R(x) = C(x)$ and solving for x, or, alternatively, by setting $P(x) = 0$ and solving for x.

EXAMPLE 1 Suppose Acme Company's cost and revenue functions are given by

$$C(x) = x^2 + 500{,}000 \qquad \text{and} \qquad R(x) = 1500x,$$

respectively. Then Acme's profit function is given by

$$P(x) = R(x) - C(x) = -x^2 + 1500x - 500{,}000.$$

Since

$$P(x) = -(x - 500)(x - 1000),$$

setting $P(x) = 0$ and solving for x yields $x = 500$ and $x = 1000$. Therefore the firm's break-even quantities are 500 units and 1000 units. Figures 2–27 and 2–28 show the graphs of C, R, and P. Note that Acme will make a profit if it makes and sells between 500 and 1000 units, but will suffer a loss if it makes and sells fewer than 500 or more than 1000 units.

The derivatives of the cost, revenue, and profit functions are called **marginal** functions. Thus C' is the **marginal cost function,** R' the **marginal revenue function,** and P' the **marginal profit function.** The economic interpretation of the marginal cost function is as follows: since the derivative $C'(a)$ is the instantaneous rate of change of the cost when $x = a$, $C'(a)$ represents the approximate additional cost of producing one more unit. In other words, $C'(a)$ is the approximate additional

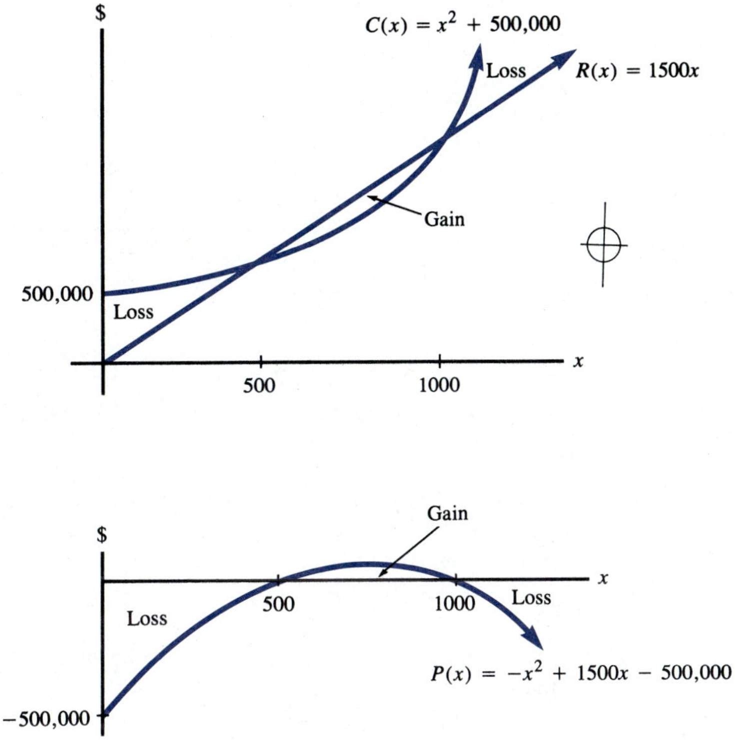

FIGURE 2–28

cost of producing the $(a + 1)$st unit. Similarly, $R'(a)$ is the approximate additional revenue obtained from selling the $(a + 1)$st unit, and $P'(a)$ is the approximate additional contribution to profit made by the $(a + 1)$st unit.

EXAMPLE 2 Let Acme Company's cost, revenue, and profit functions be as in Example 1. Then Acme's marginal cost, revenue, and profit functions are given by

$$C'(x) = 2x, \quad R'(x) = 1500, \quad \text{and} \quad P'(x) = -2x + 1500,$$

respectively. Since $C'(600) = 2(600) = 1200$, it will cost Acme approximately \$1200 to make the 601st unit. Similarly, since $R'(600) = 1500$, Acme's revenue from selling the 601st unit will be approximately \$1500. (Actually, in this example Acme's revenue from selling the 601st unit will be *exactly* \$1500, since every unit sells for \$1500.) Also, the approximate additional contribution to profit made by the 601st unit will be $P'(600) = \$300$.

Figure 2–29 shows the graph of a typical profit function P. Notice that the maximum profit occurs when $x = a$ units are made and sold. Since the tangent line to P at $(a, P(a))$ is horizontal, its slope will be zero; hence we must have $P'(a) = 0$. (Why?) But since $P(x) = R(x) - C(x)$,

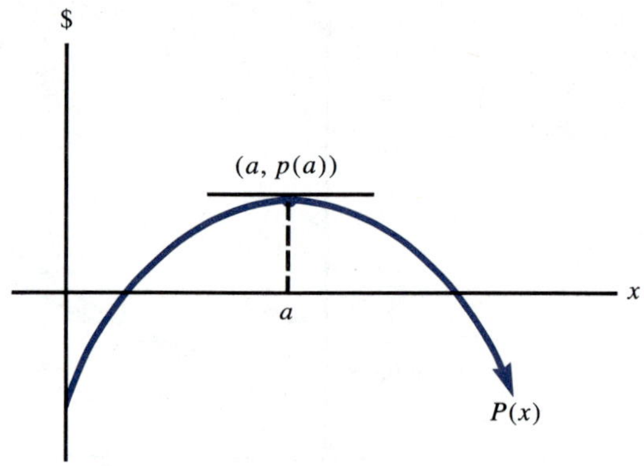

FIGURE 2–29

$$0 = P'(a) = R'(a) - C'(a),$$

and thus $R'(a) = C'(a)$. Therefore profit is maximized at the value of x for which

$$R'(x) = C'(x),$$

or, in the language of economics, at the quantity where marginal revenue equals marginal cost.

Profit Maximization

Profit is maximized at the quantity for which marginal revenue equals marginal cost.

EXAMPLE 3 Consider Acme Company of Examples 1 and 2 again. If we set marginal revenue equal to marginal cost and solve for x, we have

$$R'(x) = C'(x)$$

$$1500x = 2x$$

$$750 = x.$$

Therefore, as Figure 2–30 indicates, Acme's profit is maximized when it makes and sells 750 units, and its maximum profit is $P(750) = \$62{,}500$.

What if a firm has not yet determined the price it should charge for its product? For a single firm, the price that must be charged in order to obtain a given revenue depends on the demand for the product: if the demand is large, the firm can sell many units at a low price; if demand is small, it cannot sell very many units, and therefore must charge a higher price per unit in order to obtain the same revenue. If we let p denote the unit price of the product and x the quantity demanded,

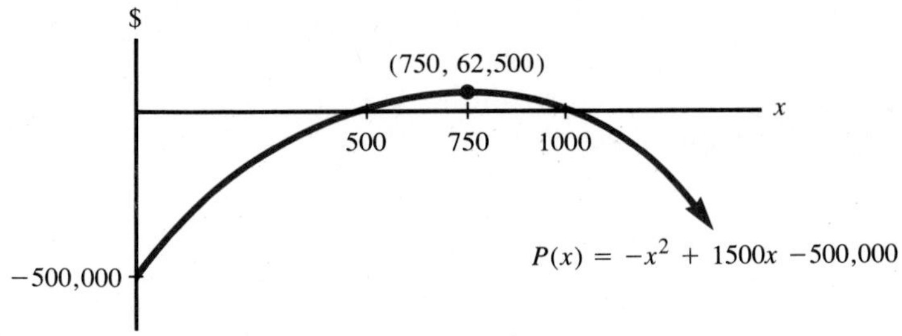

FIGURE 2–30

then p depends on x by means of a demand function $p = d(x)$. Since revenue is quantity sold times unit price, it follows that the firm's revenue is given by

$$R(x) = xp = xd(x).$$

EXAMPLE 4 As in the previous examples, suppose Acme Company's cost function is given by

$$C(x) = x^2 + 500{,}000,$$

but now suppose that the firm has not yet established a price for its product. Let the demand function for the product be defined by

$$p = 2100 - 0.5x.$$

Then Acme's revenue function is given by

$$R(x) = xp = x(2100 - 0.5x) = 2100x - 0.5x^2.$$

As before, we maximize profit by setting $R'(x) = C'(x)$ and solving for x: we have

$$R'(x) = C'(x)$$
$$2100 - x = 2x$$
$$700 = x.$$

Therefore Acme will maximize its profit by making and selling 700 units. In order to sell 700 units, the firm must charge

$$p = 2100 - 0.5(700) = \$1750$$

per unit. We have thus determined that if Acme prices its product at \$1750 per unit, it will sell 700 units for a maximum profit of $P(700) = \$235{,}000$.

Compare Example 4 with Example 3. In Example 3 we assumed that Acme had set a price of \$1500 per unit and found that under this condition the firm should make and sell 750 units for a maximum profit of \$62,500. In Example 4 it turned out that if Acme's demand function is defined by $p = 2100 - 0.5x$, then the firm should make fewer units, sell them at a higher price, and make a larger profit. A different choice of demand function could have led to the opposite conclusion: that Acme would be better off selling more units at a lower price. The property of the demand function that leads to such results is called **elasticity of demand.**

Elasticity of Demand

A demand function for a product relates quantity demanded q to unit price p. Here we consider the price p to be the independent variable so that the demand function is defined by an equation of the form $q = d(p)$. Figure 2–31 shows the graph of a typical demand function. Note that as price p increases, quantity demanded q decreases.

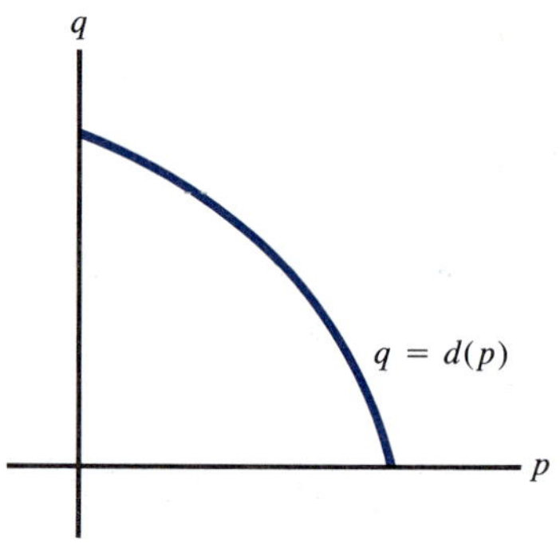

FIGURE 2–31

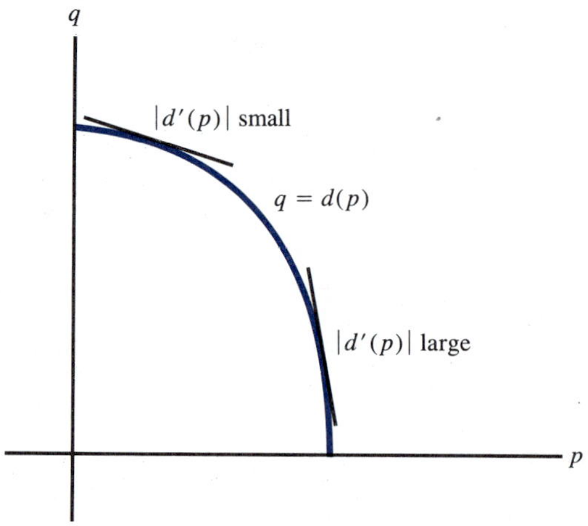

FIGURE 2–32

The derivative $q' = d'(p)$ gives the rate of change of quantity demanded with respect to price. Since the line tangent to the demand curve at any point will have negative slope, as indicated in Figure 2–32, $d'(p)$ will always be negative. If $|d'(p)|$ is large, then quantity demanded is decreasing quite rapidly as price increases; if $|d'(p)|$ is small, then quantity demanded is decreasing quite slowly as price increases. In other words, if $|d'(p)|$ is large, a small change in price results in a large change in quantity demanded; if $|d'(p)|$ is small, a small change in price results in a small change in quantity demanded.

EXAMPLE 5 Suppose

$$d(p) = \frac{10{,}000}{p}.$$

Then

$$d'(p) = -\frac{10{,}000}{p^2},$$

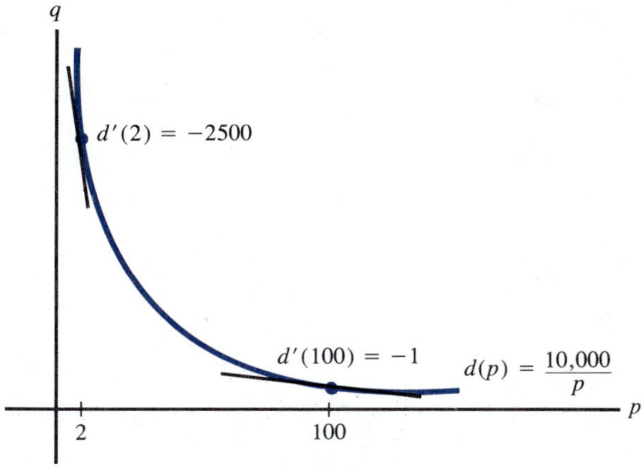

FIGURE 2–33

so $d'(2) = -2500$ and $d'(100) = -1$. Thus when the price is \$2, a small increase in price will cause a large decrease in quantity demanded (in fact, we can say that when $p = \$2$, an increase of \$1 in the price will cause quantity demanded to decrease by approximately 2500 units). On the other hand, when the price is \$100, a small increase in price will cause only a small decrease in quantity demanded. See Figure 2–33.

When a small change in price results in a large change in quantity demanded, we say that demand is **elastic.** When a small change in price results in only a small change in quantity demanded, we say that demand is **inelastic.** In order to make these notions precise, we define the elasticity of demand for a commodity.

Elasticity of Demand

If the demand function for a commodity is defined by an equation of the form $q = d(p)$, the commodity's **elasticity of demand at price p** is $E(p)$, where

$$E(p) = \frac{p[d'(p)]}{d(p)}.$$

Exercise 15 at the end of this section outlines the derivation of the elasticity formula.

The elasticity of demand $E(p)$ measures the proportional change in quantity demanded relative to price. If $|E(p)| > 1$, demand is elastic; if $|E(p)| < 1$, demand is inelastic. A product can have elastic demand at one price and inelastic demand at another.

EXAMPLE 6 If the demand function for a product is defined by

$$q = -1000p + 10{,}000,$$

then the elasticity of demand for the product is

$$E(p) = \frac{p[d'(p)]}{d(p)} = \frac{-1000p}{-1000p + 10{,}000}.$$

Suppose that $p = \$2$. Then

$$|E(p)| = |E(2)| = |-0.25| = 0.25 < 1,$$

so demand for the product is inelastic at a price of \$2. Note that when $p = \$2$, quantity demanded $q = 8000$ units and revenue is thus \$2(8000) = \$16,000. If the price is increased by 10 percent to \$2.20, the quantity demanded decreases to 7800 and the revenue increases to \$2.20(7800) = \$17,160. Hence, when $p = \$2$, a 10 percent increase in price results in a small decrease in quantity demanded and an increase in revenue.

Now suppose that $p = \$8$. Then

$$|E(p)| = |E(8)| = |-4| = 4 > 1,$$

so demand for the product is elastic at a price of \$8. When $p = \$8$, quantity demanded $q = 2000$ units and revenue is \$8(2000) = \$16,000 again. But now a 10 percent increase in the price, from \$8 to \$8.80, results in a decrease in quantity demanded to 1200 units, which in turn yields revenue of \$10,560. Hence when $p = \$8$, a 10 percent increase in price leads to a large decrease in quantity demanded and a large decrease in revenue.

When $|E(p)| = 1$, demand is said to be of **unit elasticity.** For instance, in this example demand for the product is of unit elasticity when $p = \$5$. This means that small changes in price around \$5 will produce moderate changes in demand that will leave revenue approximately constant.

We conclude this section by remarking that instead of considering quantity demanded to be a function of price, as we have done here, economists usually consider price to be a function of quantity demanded. Therefore they define demand functions by equations of the form $p = d(q)$. When price is thus considered to be a function of quantity demanded, the formula for elasticity of demand has the alternative form

$$E(q) = \frac{d(q)}{q[d'(q)]}.$$

Computer Manual: Program DIFFPOLY.

2.5 EXERCISES

Marginality

1. Zeta Company's cost function is given by $C(x) = 0.5x^2 + 100{,}000$, and the company sells each unit it makes for \$600.
   - **(a)** Find Zeta's total cost, revenue, and profit from making and selling 100 units.
   - **(b)** Find Zeta's approximate additional cost, revenue, and profit from making and selling the 101st unit.
   - **(c)** Find Zeta's maximum profit and the number of units the firm must make and sell in order to attain its maximum profit.
2. Let Zeta Company's cost function be as in Exercise 1, and suppose that the demand function for Zeta's product is given by $p = 1200 - 2x$.
   - **(a)** Find Zeta's approximate additional cost, revenue, and profit from making and selling the 101st unit; the 201st unit.
   - **(b)** Find Zeta's maximum profit and the number of units the company must make and sell in order to attain it.
   - **(c)** Find the price per unit the firm must charge in order to attain the maximum profit.
3. Repeat Exercise 2 if the demand function is given by $p = 700 - 0.5x$.
4. Eta Corporation's cost function is given by $C(x) = 20x^2 + 40{,}000$, and the demand function for its product by $p = 6000 - 100x$. Find Eta's maximum profit, and the number of units the company must make and sell and the price per unit it must charge in order to attain its maximum profit.
5. Repeat Exercise 4 if $C(x) = 2.3x^2 + 4000x$, $p = 10{,}000 - 0.10x$, and x is in thousands of units.
6. Theta, Inc. knows the following:
   —its fixed cost is \$20,000;
   —its marginal cost function is given by $C'(x) = 10x$;
   —its demand function is linear;
   —at a price of \$2650 per unit, it sold 90 units;
   —at a price of \$2800 per unit, it sold 80 units.
   - **(a)** Find Theta's cost, revenue, and profit functions.
   - **(b)** Find Theta's maximum profit and the number of units the company must make and sell and the price per unit it must charge in order to attain its maximum profit.

Elasticity of Demand

7. The demand function for a product is given by $q = -3000p + 120{,}000$.
   - **(a)** Find the elasticity of demand for the product when $p = \$15$. Find quantity demanded and revenue when $p = \$15$, $p = \$13.50$, and $p = \$16.50$.
   - **(b)** Find the elasticity of demand for the product when $p = \$20$. Find quantity demanded and revenue when $p = \$20$, $p = \$19$, and $p = \$21$.
   - **(c)** Find the elasticity of demand for the product when $p = \$30$. Find quantity demanded and revenue when $p = \$30$, $p = \$27$, and $p = \$33$.

8. The demand function for a product is defined by $q = -5p + 3000$.
   (a) Find the elasticity of demand for the product when $p = \$100$ and when $p = \$500$.
   (b) Find quantity demanded and revenue when $p = \$100$ and when $p = \$500$, and also when the price is increased by 10 percent in each case.
   (c) Find the price at which the product has unit elasticity of demand. Find quantity demanded and revenue when the price is increased or decreased by 5 percent from this value.

9. Cutter Cotter Pin Company needs to increase its revenue. The marketing department has recommended a 5 percent decrease in the price the company charges for its cotter pins, arguing that this would increase the quantity demanded so much that revenue would actually increase. The accounting department has recommended a 5 percent increase in the price, arguing that this would increase revenue because the resulting decrease in quantity demanded would be more than offset by the higher price. Suppose the demand function for cotter pins is given by $q = -1250p + 25{,}000$, where p is price per dozen cotter pins and q is in dozens of cotter pins. Whose recommendation should be taken if the current price per dozen of cotter pins is **(a)** \$6? **(b)** \$10? **(c)** \$12?

10. The demand for gasoline in a certain country is given by

$$q = \begin{cases} -10p + 50, & 0 \le p < 1.20, \\ -35p + 80, & 1.20 \le p \le 2.00, \end{cases}$$

where p is the price in dollars per gallon and q is in millions of gallons per day.
   (a) Find the elasticity of demand for gasoline when $p = 0.80$, $p = 1.15$, $p = 1.25$, and $p = 1.50$.
   (b) Graph the demand function. Can you explain why the price per gallon must rise to at least \$1.20 before significant conservation of gasoline occurs?

11. Show that a commodity whose demand function is given by an equation of the form $q = a/p$, where a is a positive number, will have unit elasticity of demand at every price and that revenue from sales of the commodity will therefore remain constant no matter what the price.

12. The demand function for a product is given by $p = 80 - 0.005q^2$, where q is in thousands of units. Use the alternative form of the elasticity of demand formula to find the elasticity of demand for the product when $q = 100$ and when $q = 50$.

13. If a commodity has a demand function whose equation is of the form $q = c$, where c is a constant, show that its elasticity of demand is 0. This is called **perfect inelasticity;** when it occurs, revenue depends only on price. (Why?)

14. If the demand function for a commodity has an equation of the form $p = c$, where c is a constant, show that its elasticity of demand does not exist. This is known as **perfect elasticity:** when it occurs, revenue depends only on quantity demanded. (Why?)

15. If the price of a commodity changes by some percentage, then in response the quantity of the commodity demanded by consumers will change by some percentage. Elasticity of demand is defined to be the ratio of the percentage change in quantity demanded to the percentage change in price. If $d(p)$ is a demand function and the price changes from p to $p + h$, then the quantity demanded changes from $d(p)$ to $d(p + h)$, and hence

$$\text{percentage change in price} = \frac{(p + h) - p}{p} 100 = \frac{h}{p} 100$$

and

$$\text{percentage change in quantity demanded} = \frac{d(p + h) - d(p)}{d(p)} 100.$$

Take the limit as h approaches 0 of the ratio

$$\frac{\text{percentage change in quantity demanded}}{\text{percentage change in price}}$$

and interpret the result.

2.6 APPLICATION: THE NEWTON–RAPHSON METHOD

We have encountered equations of the form $f(x) = 0$ when attempting to find the x-intercepts of functions in order to graph them, when trying to find break-even quantities by setting the profit function equal to zero, and in several other contexts. The solutions of the functional equation $f(x) = 0$ are called the **roots** or **zeros** of the function f. Until now we have been careful to choose our functions so that it was a relatively simple procedure to find their roots. However, in many practical situations the roots of the functions involved are not easily found; often the best we can do is approximate them to some desired degree of accuracy. In this section we discuss the Newton–Raphson method for approximating the roots of a function. This method uses the function and its derivative to produce successive approximations to a root. Each approximation depends on the preceding one, and in most circumstances the approximations get closer and closer to the true value of the root.

Suppose the continuous function f has a root r. By sketching the graph of f, or by some other method, we make a reasonable initial guess at the value of r: we call this guess r_0. See Figure 2–34. The line tangent to f at $(r_0, f(r_0))$ has slope $f'(r_0)$, so its equation is

$$y = f'(r_0)x + b.$$

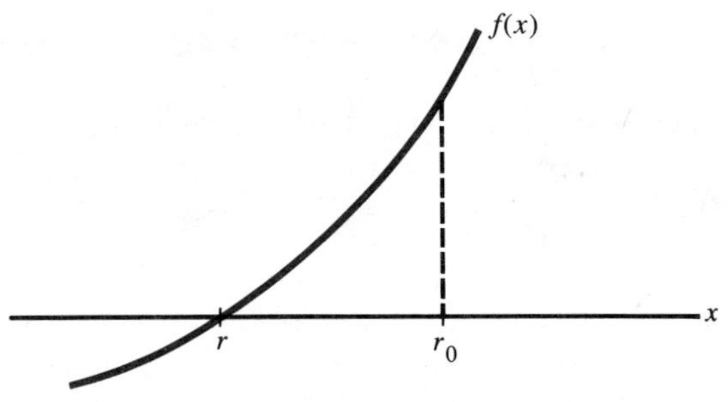

FIGURE 2–34

But since the point $(r_0, f(r_0))$ is on the line,

$$f(r_0) = f'(r_0)r_0 + b,$$

so

$$b = f(r_0) - f'(r_0)r_0.$$

Thus the equation of the tangent line at $(r_0, f(r_0))$ is

$$y = f'(r_0)x + f(r_0) - f'(r_0)r_0.$$

The x-intercept of this line is obtained by setting $y = 0$ and solving for x. If we do this, we find that the x-intercept is $(r_1, 0)$, where

$$r_1 = r_0 - \frac{f(r_0)}{f'(r_0)}.$$

(Check this.) As Figure 2–35 indicates, r_1 is usually closer to the true root r than the initial guess r_0 is.

Next we repeat the procedure with r_1 as our initial guess, to obtain another tangent line to the curve whose x-intercept is the number

$$r_2 = r_1 - \frac{f(r_1)}{f'(r_1)}.$$

Again, r_2 will usually be closer to r than r_1 is.

Now we iterate the process again with r_2 as our initial guess. In this way we construct a sequence of numbers $r_0, r_1, r_2, \ldots$ that we hope will be successively better approximations to the true root r.

The Newton–Raphson Method

Let f be a function with root r. Let r_0 be an initial guess at the value of r, and for $n = 1, 2, 3, \ldots$ set

$$r_n = r_{n-1} - \frac{f(r_{n-1})}{f'(r_{n-1})}.$$

Under most circumstances the numbers $r_0, r_1, r_2, \ldots$ will be successively better approximations to r.

At the end of this section we discuss the circumstances under which the Newton–Raphson method fails.

EXAMPLE 1 Let us use the Newton–Raphson method to approximate the positive root of the quadratic equation $x^2 - 3 = 0$. The positive root of this equation is $\sqrt{3}$, which has approximate value 1.7320508, accurate to seven decimal places. As we shall

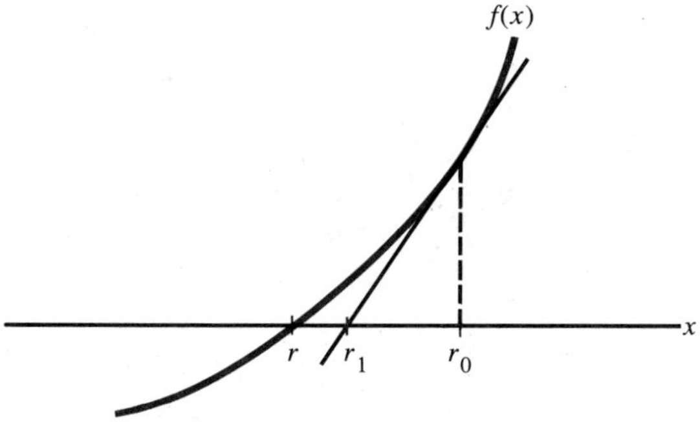

FIGURE 2–35

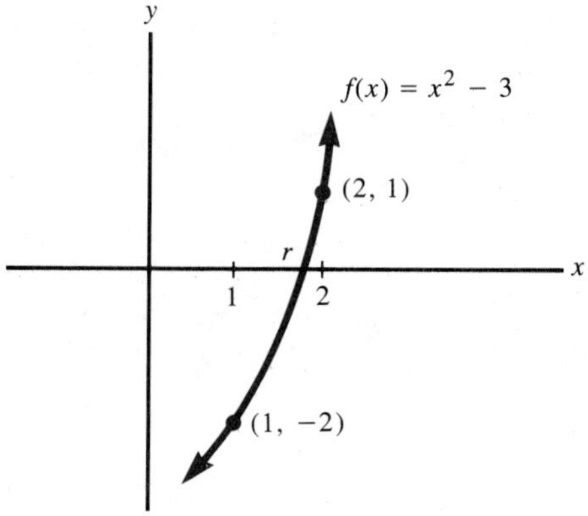

FIGURE 2–36

see, with a reasonable initial guess the Newton–Raphson method will obtain this approximate value with only four iterations.

Let $f(x) = x^2 - 3$; we seek the positive root of f. Note that $f(1) = -2$ and $f(2) = 1$, so that as x goes from 1 to 2, $f(x)$ goes from a negative value to a positive one. But since f is a polynomial function it is continuous, and it follows that the graph of f must cross the x-axis at some number r between 1 and 2. See Figure 2–36. Thus we have concluded that the positive root of f lies between 1 and 2. Therefore let us use $r_0 = 1$ as our initial guess for the root. Since

$$f(x) = x^2 - 3 \qquad \text{and} \qquad f'(x) = 2x,$$

we have

a. First iteration: $$r_1 = r_0 - \frac{f(r_0)}{f'(r_0)} = 1 - \frac{f(1)}{f'(1)} = 1 - \frac{-2}{2} = 2.$$

b. Second iteration: $$r_2 = r_1 - \frac{f(r_1)}{f'(r_1)} = 2 - \frac{f(2)}{f'(2)} = 2 - \frac{1}{4} = 1.75.$$

c. Third iteration: $$r_3 = r_2 - \frac{f(r_2)}{f'(r_2)} = 1.75 - \frac{f(1.75)}{f'(1.75)}$$

$$= 1.75 - \frac{0.0625}{3.5} = 1.7321429.$$

d. Fourth iteration: $$r_4 = r_3 - \frac{f(r_3)}{f'(r_3)} = 1.7321429 - \frac{f(1.7321429)}{f'(1.7321429)}$$

$$= 1.7321429 - \frac{0.000319}{3.4642858} = 1.7320508.$$

EXAMPLE 2 A firm's profit function is given by $P(x) = x^3 + x - 1$, where profit is in millions of dollars and x is in thousands of units. Estimate the firm's break-even quantity.

Since $P(0) = -1$ and $P(1) = 1$, and since the polynomial P is continuous, its graph must cross the x-axis between 0 and 1. Let us take $r_0 = 1$ as our initial guess for the root. Since x is in thousands, we need four-decimal place accuracy in order to estimate x to the nearest unit. Successive iterations yield the following approximations to the root:

$$r_0 = 1,$$
$$r_1 = 0.75,$$
$$r_2 = 0.6860465,$$
$$r_3 = 0.6831228,$$
$$r_4 = 0.6823283,$$
$$r_5 = 0.6823254.$$

Thus, to four decimal places, the root is 0.6823, so the firm's break-even quantity is 682 units.

The Newton–Raphson method sometimes fails because one of the derivatives $f'(r_{n-1})$ does not exist or is 0; when this occurs, it is not possible to continue the method to get the next approximation r_n. Figure 2–37 shows how it is possible for $f'(r_1)$ to be 0. This type of failure can usually be circumvented by choosing a different value for r_0. Sometimes, however, the Newton–Raphson method will fail no matter how we choose r_0. Figure 2–38 shows how this can happen: no matter how we choose r_0 in Figure 2–38, the numbers r_0, r_1, r_2, . . . will actually get further and further from the true value of the root.

Computer Manual: Program NEWTON.

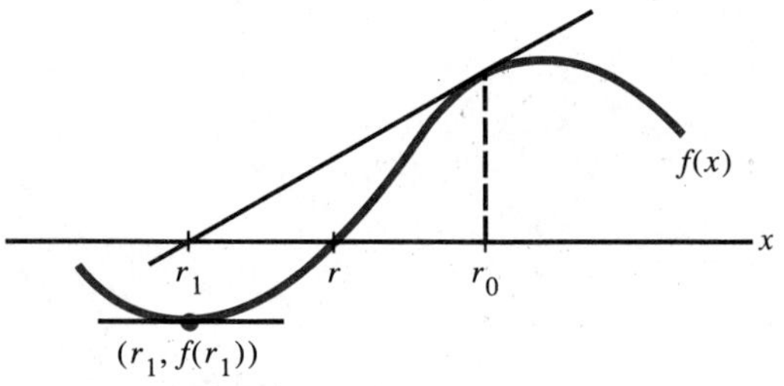

FIGURE 2–37

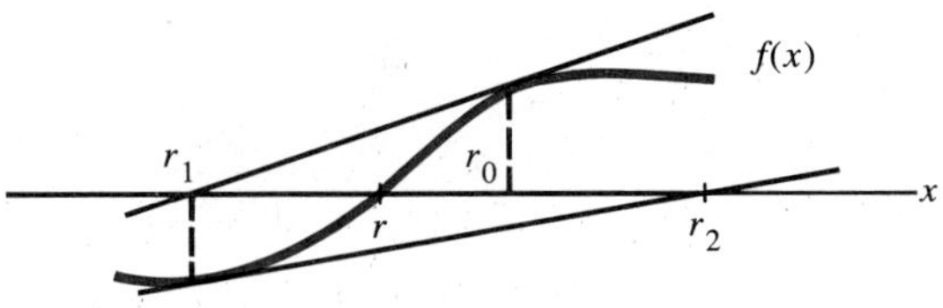

FIGURE 2–38

2.6 EXERCISES

In Exercises 1 through 6 use the Newton–Raphson method to approximate the positive roots of f.

1. $f(x) = x^2 - 2$
2. $f(x) = x^2 + 2x - 6$
3. $f(x) = 2x^2 - 4x + 1$
4. $f(x) = x^3 + 3x^2 - 24x + 1$
5. $f(x) = x^4 - 2$
6. $f(x) = x^5 - 100$
7. The function f defined by $f(x) = -x^3 + 3x^2 - 3x + \frac{1}{2}$ has a root between 0 and 1. What happens if you attempt to use the Newton–Raphson method with initial guess $r_0 = 1$ to approximate this root? What happens if you use $r_0 = 0$?
8. The function defined by $f(x) = x^{1/3}$ has 0 as its unique root. Show that applying the Newton–Raphson method with any initial guess r_0 will not work.

Management

9. A firm's profit function is given by
$$P(x) = -4x^3 + 150x^2 - 1200x - 400,$$
where x is in thousands of units and $P(x)$ is in thousands of dollars. Use the Newton–Raphson method to find the firm's break-even quantities. (*Hint:* Use $r_0 = 12$ and $r_0 = 25$.)
10. If the supply and demand functions for a commodity are given by
$$s(p) = 20\sqrt{p} \qquad \text{and} \qquad d(p) = 200 - 5.2p,$$
respectively, find the equilibrium price for the product. (*Hint:* $d(20) - s(20) > 0$ and $d(25) - s(25) < 0$.)

Life Science

11. In a commercial biochemical process a reaction that liberates heat takes place in a vat, the outside of which is simultaneously being cooled. Equilibrium occurs when the outside temperature of the vat equals the temperature of the solution inside the vat. The temperature of the solution t minutes after the start of the reaction is T degrees Celsius, where $T = 2t^4$; the outside temperature of the vat is 100° C initially and decreases linearly at a rate of 20° C per minute. How long will it take until equilibrium is reached?

Social Science

12. The proportion of eligible voters who are registered with a political party in year t is given by
$$y = 0.30 + \frac{0.35}{t}, \qquad t \geq 1.$$
The proportion who are registered as independents is
$$y = 0.32 + 0.04\sqrt[3]{t}, \qquad t \geq 1.$$
How long will it be until the proportion of those registered with a party and the proportion registered as independents are equal? Compute your answer accurate to two decimal places.

2.7 DIFFERENTIAL NOTATION AND HIGHER DERIVATIVES

Thus far we have written the derivative of a function f as f'. This is known as the **prime notation** for derivatives. Another common way of writing derivatives, known as **differential notation,** is sometimes more convenient and useful than prime notation. In this section we discuss differential notation for derivatives. We also consider the higher derivatives of a function: higher derivatives are derivatives of derivatives. For instance, as we saw in Section 2.4, velocity is the derivative of position, and acceleration is the derivative of velocity; thus acceleration is a higher derivative of position.

Differential Notation

Consider the function f defined by the equation

$$f(x) = x^3 + 4x + 1.$$

Using prime notation, we write the derivative of this function as

$$f'(x) = 3x^2 + 4.$$

If the function is given in the form $y = x^3 + 4x + 1$, we can still use prime notation and write the derivative as $y' = 3x^2 + 4$.

Another notation often used for the derivative of a function is **differential notation:** if f is a function of x, its derivative is written in differential notation as

$$\frac{df}{dx}$$

and referred to as the **derivative of f with respect to x.** The notation

$$\left.\frac{df}{dx}\right|_{x=a}$$

indicates that the derivative is to be evaluated at $x = a$.

EXAMPLE 1 Let $f(x) = x^3 + 2x^2 + 3x$ and $y = \dfrac{2}{x} + 3x$.

Then the derivative of f with respect to x is

$$\frac{df}{dx} = 3x^2 + 4x + 3$$

and

$$\left.\frac{df}{dx}\right|_{x=1} = 3(1)^2 + 4(1) + 3 = 10.$$

Similarly, the derivative of y with respect to x is

$$\frac{dy}{dx} = -\frac{2}{x^2} + 3$$

and
$$\left.\frac{dy}{dx}\right|_{x=2} = \frac{5}{2}.$$

At this time we will not consider df/dx and dy/dx to be fractions, but merely symbols for the derivative. Later we will show how these symbols can indeed be interpreted as fractions.

Differential notation may be used with independent variables other than x. It is especially convenient when we must make explicit which variable is to be regarded as the independent one.

EXAMPLE 2 Let $f(t) = 3\sqrt{t}$. Then the derivative of f with respect to t is

$$\frac{df}{dt} = \frac{3}{2\sqrt{t}}.$$

Now let $y = x^3$. Then dy/dx is the derivative of y with respect to x. As usual, x is the independent variable and y is the dependent one. Thus

$$\frac{dy}{dx} = \frac{d}{dx}(x^3) = 3x^2.$$

On the other hand, dx/dy is the derivative of x with respect to y, so here y is to be regarded as the independent variable and x as the dependent one. If we solve the equation $y = x^3$ for x in terms of y to obtain $x = y^{1/3}$, we find that

$$\frac{dx}{dy} = \frac{d}{dy}(y^{1/3}) = \frac{1}{3}y^{-2/3}.$$

Higher Derivatives

If f' is the derivative of f, then f' may itself have a derivative $(f')'$, which we write as f''. Similarly, f'' may have a derivative $(f'')' = f'''$, and so on. The derivative f' is the **first derivative of f**, f'' is the **second derivative of f**, and f''' is the **third derivative of f**. If $n > 3$, we denote the nth derivative of f by $f^{(n)}$.

EXAMPLE 3 Let $f(x) = x^4 + 5x^3 + 3x^2 + 4x - 1$. Then we have

first derivative: $f'(x) = 4x^3 + 15x^2 + 6x + 4$

second derivative: $f''(x) = 12x^2 + 30x + 6$

third derivative: $f'''(x) = 24x + 30$

fourth derivative: $f^{(4)}(x) = 24$

fifth derivative: $f^{(5)}(x) = 0$

Note that $f^{(n)}(x) = 0$ for $n \geq 5$.

When using differential notation, we write the second derivative as

$$\frac{d^2f}{dx^2},$$

with the obvious extension to higher derivatives.

EXAMPLE 4 Let $y = 1/x$. Then

$$\frac{dy}{dx} = -\frac{1}{x^2}, \quad \frac{d^2y}{dx^2} = \frac{2}{x^3}, \quad \frac{d^3y}{dx^3} = -\frac{6}{x^4}, \quad \frac{d^4y}{dx^4} = \frac{24}{x^5}.$$

Computer Manual: Program HIGHPOLY.

2.7 EXERCISES

Differential Notation

In Exercises 1 through 12 find the indicated derivatives.

1. $\frac{dy}{dx}$, $y = 2x^3 - 3x + 12$

2. $\frac{dy}{dx}$, $y = 5\sqrt{x} - 2x^{2/3}$

3. $\frac{df}{dt}$, $f(t) = 4t^{-2} + 3t^2 + 2t$

4. $\frac{dW}{dq}$, $W = 14q^2 + 10q - q^{-1}$

5. $\frac{dg}{dx}$, $g(x) = \frac{2}{x} - \frac{4}{x^3}$

6. $\left.\frac{df}{dx}\right|_{x=2}$, $f(x) = x^2 + 5x - 2$

7. $\left.\frac{dy}{dx}\right|_{x=4}$, $y = x^5 - 5x^{3/2} + 2$

8. $\left.\frac{dh}{dv}\right|_{v=8}$, $h(v) = 2v^{-3} + 2v^{-1/3}$

9. $\left.\frac{dq}{dp}\right|_{p=1}$, $q = 400p^2 - 1200p + 100$

10. $\frac{dy}{dx}, \frac{dx}{dy}$, $3x + 2y = 5$

11. $\frac{dy}{dt}, \frac{dt}{dy}$, $4ty = 1$

12. $\frac{du}{dv}, \frac{dv}{du}$, $u^3 = 27v$

Higher Derivatives

In Exercises 13 through 18 find the indicated higher derivatives.

13. $f''(x), f'''(x)$, $f(x) = x^3 + x^2 + x + 1$

14. $f''(x), f^{(4)}(x)$, $f(x) = x^5 - 2x^3 + 4x$

15. $\frac{d^2f}{dx^2}, \frac{d^3f}{dx^3}$, $f(x) = x^4 - 4x^3 + 3x + 12$

16. $\frac{d^2y}{dx^2}, \frac{d^4y}{dx^4}$, $y = \frac{2}{x^2} - \frac{1}{x}$

17. $g''(x), g'''(4)$, $g(x) = x^4 + 2\sqrt{x}$

18. $\left.\frac{d^2p}{dq^2}\right|_{q=1}, \left.\frac{d^3p}{dq^3}\right|_{q=1}$, $p = 3q^{-1} + q^2$

2 REVIEW EXERCISES

Answers to all review exercises are at the back of the book.

1. Let the function f have the following graph:

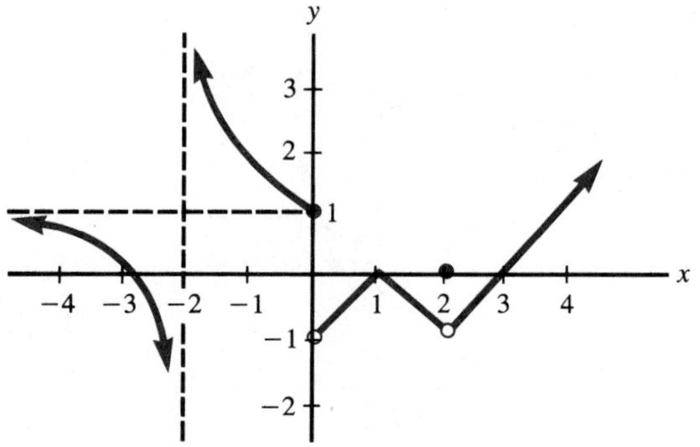

Find the indicated limits:

(a) $\lim_{x\to-2} f(x)$ **(b)** $\lim_{x\to-1} f(x)$ **(c)** $\lim_{x\to0} f(x)$ **(d)** $\lim_{x\to1} f(x)$

(e) $\lim_{x\to2} f(x)$ **(f)** $\lim_{x\to3} f(x)$ **(g)** $\lim_{x\to+\infty} f(x)$ **(h)** $\lim_{x\to-\infty} f(x)$

In Exercises 2 through 17 evaluate the given limit.

2. $\lim_{x\to2} (3x^2 - 5x - 2)$

3. $\lim_{x\to1} (x^4 + x^3 - 2x^2 + 5x - 1)$

4. $\lim_{x\to0} (x^{1000} - 157x^{501} + 2)$

5. $\lim_{x\to9} \left(4\sqrt{x} - \frac{2}{x}\right)$

6. $\lim_{x\to1} \left(\frac{1}{x} - \frac{2}{x^2}\right)$

7. $\lim_{x\to-1} (x^4 + x + 1)(x^2 - x - 1)$

8. $\lim_{x\to1} \frac{2}{x-1}$

9. $\lim_{x\to2} \frac{x+3}{x+2}$

10. $\lim_{x\to2} \frac{x+3}{x-2}$

11. $\lim_{x\to0} \frac{3x+1}{2x^2+5}$

12. $\lim_{x\to5} \frac{x^2 - 7x + 10}{x^2 + 2x - 15}$

13. $\lim_{x\to5} \frac{x^2 + 7x + 10}{x^2 + 2x - 15}$

14. $\lim_{x\to+\infty} \frac{4x^2 - 3x}{x+2}$

15. $\lim_{x\to-\infty} \frac{x^2 + 2x}{4x^2 - 3x}$

16. $\lim_{x\to+\infty} \frac{x+2}{4x^2 - 3x}$

17. $\lim_{x\to+\infty} \frac{\sqrt{x} + \sqrt[3]{x} - 1}{2 + 4\sqrt{x}}$

18. Find all asymptotes and graph

(a) $f(x) = \dfrac{3x - 7}{4 - 2x}$ **(b)** $f(x) = \dfrac{x^2 - 3x + 2}{x}$

19. Use the Three-Step Rule to find $f'(x)$ if $f(x) = x^3 + x^2$. Find the equation of the line tangent to f at $(2, f(2))$.

20. Use the Three-Step Rule to find $g'(x)$ if $g(x) = \sqrt[3]{x}$. For what values of x does $g'(x)$ not exist?

The following is the graph of a function h:

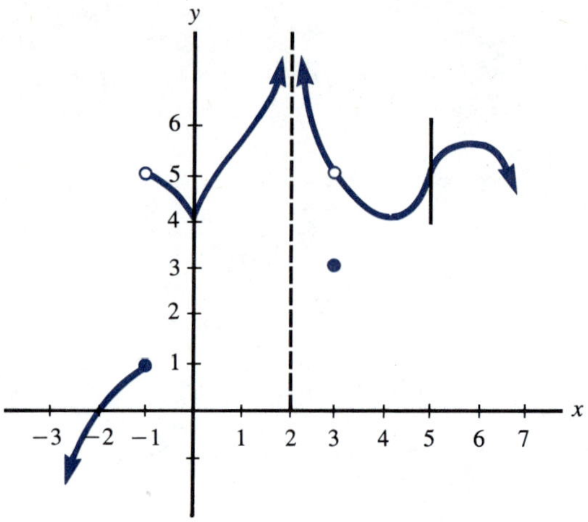

21. At what values of x is h not differentiable?

22. At what values of x is h discontinuous?

In Exercises 23 through 31 find $f'(x)$.

23. $f(x) = \sqrt{3}$

24. $f(x) = \sqrt{3}\,x$

25. $f(x) = 3x^2 + 6x + 12$

26. $f(x) = -x^6 + 4x^5 - 12x^3 + 16x^2 - 8x - 3$

27. $f(x) = \dfrac{2}{x^3} - \dfrac{1}{x} + 3$

28. $f(x) = 4x^{-2} + 2x^{-3}$

29. $f(x) = 4\sqrt{x}$

30. $f(x) = 4x^{3/2} - 3\sqrt{x}$

31. $f(x) = 5x^{4/5} - 2x^{-1/2} + \dfrac{3}{x^{2/3}}$

32. Find the equation of the line tangent to f at $(3, f(3))$ if

$$f(x) = 4x^2 - 6x + 2$$

33. Find the equation of the line tangent to g at $(1, g(1))$ if

$$g(x) = 3\sqrt{x} + \frac{2}{x} - 2.$$

34. A mortar shall is fired straight up into the air. The altitude of the shell t seconds after firing is

$$s(t) = -5t^2 + 150t$$

meters. Find the shell's velocity and acceleration 3 seconds and 6 seconds after firing. Find the shell's maximum altitude. How long is the shell in the air?

35. A firm estimates that the value of a computer t years after it is purchased new will be

$$V(t) = 576 - 192\sqrt{t}$$

thousand dollars. Here $0 \le t \le 9$. Find the rate at which the value of the computer is changing at the end of 1 year; at the end of 4 years; at the end of 8 years.

36. A firm's cost function is given by $C(x) = x^2 + 320$, and the demand function for its product by $p = 56 - x$. Find the firm's maximum profit, the number of units that must be made and sold in order to attain the maximum profit, and the price per unit the firm must charge in order to attain the maximum profit.

37. The demand function for a commodity is given by $q = -6000p + 240{,}000$. Find

(a) the elasticity of demand for the commodity when $p = \$10$, $p = \$20$, and $p = \$30$;
(b) the revenue when $p = \$10$, $p = \$20$, and $p = \$30$.
(c) the revenue when $p = \$11$, $p = \$22$, and $p = \$33$.

38. Use the Newton–Raphson method to approximate the cube root of 10.

39. The demand function for a product is given by

$$d = -2p^3 - 90p^2 - 1200p + 25{,}000.$$

Here price p is in hundreds of dollars. Find the price, to the nearest cent, at which demand for the product is 0.

In Exercises 40 through 44 find the indicated derivatives.

40. $f(x) = 2x^4 - 5x^2 + 12,\quad \left.\dfrac{df}{dx}, \dfrac{d^2f}{dx^2}, \dfrac{d^3f}{dx^3}\right|_{x=2}$

41. $y = 2\sqrt{x} + \dfrac{2}{\sqrt{x}},\quad \dfrac{dy}{dx}, \dfrac{d^2y}{dx^2}$

42. $g(x) = 3x^3 - 4x^{4/5} + 2,\quad g'(x), g''(x), g'''(1)$

43. $w = 2u^2 + 3bu + 2\sqrt{u} + 1,\quad b$ a constant, $\dfrac{dw}{du}, \dfrac{d^2w}{du^2}$

44. $\dfrac{8x}{y} = 11,\quad \dfrac{dy}{dx}, \dfrac{dx}{dy}$

2 SUPPLEMENT: NEWTON AND LEIBNIZ

No one person "invented" calculus. The development of calculus was a lengthy process contributed to by many mathematicians over many centuries. Nevertheless, there was a short period of time during which the partial results and incomplete methods previously discovered were greatly extended and then organized into a coherent mathematical theory that was thereafter known as "the calculus." This rapid development of the subject occurred in the latter part of the 17th century and was the independent work of Isaac Newton (1643–1727) of England and Gottfried Wilhelm Leibniz (1646–1716) of Germany.

Isaac Newton was one of the greatest of human intellects. He made monumental contributions to physics by enunciating the law of universal gravitation and the three laws of motion that bear his name, working out many of the consequences of these laws as they apply to astronomy and mechanics, experimenting in optics, and proposing a particle theory of light. His development of calculus went hand-in-hand with his work in physics as he created new mathematical tools to solve problems in celestial and earthly mechanics. In addition to these activities, Newton also experimented in chemistry and alchemy, studied and wrote on theology, and served as Master of the Royal Mint.

Newton apparently developed what he called his "fluxional calculus" during the period 1665–1666. Quantities that changed with time he referred to as **fluents;** he called their rates of change with respect to time **fluxions.** For instance, if a particle moves along a curve its x- and y-coordinates change with time and hence are fluents, and therefore their rates of change with respect to time, which Newton denoted by $\dot{x}$ and $\dot{y}$, respectively, are fluxions. Thus the fluxions $\dot{x}$ and $\dot{y}$ are what we would now write as dx/dt and dy/dt, the derivatives of x and y with respect to time. The notation $\dot{x}$ for the derivative with respect to time survives today in physics and engineering.

The method by which Newton found derivatives was rather cumbersome, proceeding much in the manner of our Three-Step Rule but using the idea of infinitesimal quantities rather than limits. His reasoning was as follows: if h is the duration of an infinitely short time period, then $\dot{x}h$ and $\dot{y}h$ are infinitesimals, that is, infinitely small nonzero amounts by which x and y change during the time period. If we substitute $x + \dot{x}h$ for x and $y + \dot{y}h$ for y in the relationship between x and y, subtract the original equation from the result, and then drop all terms containing powers of h that are 2 or greater because they are infinitesimally small compared with the remaining terms, we can obtain an expression for $\dot{y}$ in terms of $\dot{x}$. To illustrate, suppose

$$y = x^2.$$

Then

$$y + \dot{y}h = (x + \dot{x}h)^2,$$

or

$$y + \dot{y}h = x^2 + 2x\dot{x}h + (\dot{x})^2h^2.$$

Subtracting $y = x^2$ from this last equation, and dropping the term containing h^2 because it is infinitesimal compared with the rest of the terms, we have

$$\dot{y}h = 2x\dot{x}h,$$

or

$$\dot{y} = 2x\dot{x}.$$

This last equation can be rewritten in the form

$$\frac{\dot{y}}{\dot{x}} = 2x,$$

which expresses what we today call the derivative of y with respect to x as a ratio of the time derivatives $\dot{y}$ and $\dot{x}$.

Newton seems to have been a secretive man who was not anxious to share his discoveries with others. He did not communicate his discoveries in calculus and physics to the world at large for many years, but instead revealed them in piecemeal fashion to a few selected colleagues. Finally, at the instigation of his friend Edmund Halley, he published in 1687 a book entitled *Philosophiae Naturalis Principia Mathematica,* known today as the *Principia.* In the *Principia* Newton presented for the first time his work on gravitation and celestial mechanics, as well as the fluxional calculus that made it possible. Unfortunately, Newton recast all the calculus into geometrical form, in approved classical style, and this served to obscure the power and utility of his method of fluxions.

During the years 1673–1676, after Newton had developed his version of the calculus but before he had published it, Leibniz independently created a slightly different version. He, too, utilized infinitesimals, but he called his infinitesimals **differentials:** the differentials for x and y he denoted by dx and dy, respectively. These were defined geometrically as shown in Figure 2–39: dx is an arbitrary infinitesimal increment to x, and dy is determined by dx and the tangent line at (x, y). Leibniz defined the derivative of y with respect to x to be the ratio dy/dx of the differentials dy and dx.

Leibniz's terminology and notation have survived to the present day. This is why we use the differential notation dy/dx for the derivative. We will discuss differentials in more detail in Section 3.7.

The method by which Leibniz found derivatives was much like Newton's. In the relationship between x and y, x was replaced by $x + dx$, y was replaced by $y + dy$, and the original equation was subtracted from the result. Then all terms

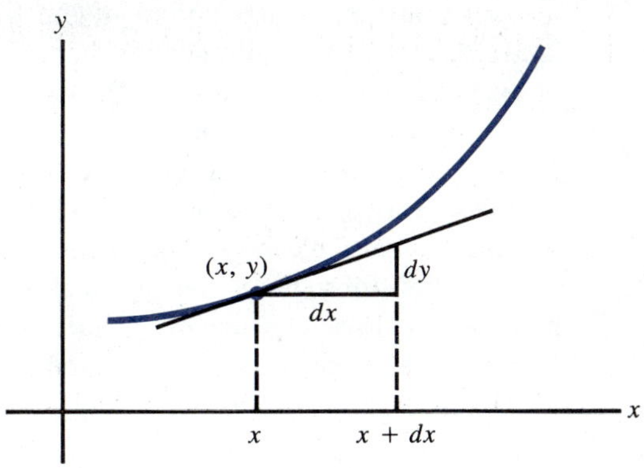

FIGURE 2–39

containing powers of dx or dy that were 2 or greater, or all terms with a factor of the form $dx\,dy$, were eliminated on the grounds that they were infinitesimally small compared with the remaining terms. The end result was an expression relating dy and dx. However, Leibniz also worked out and used many of the rules of differentiation, such as (in his notation)

$$d(cx) = c \cdot dx$$

and

$$d(x^n) = nx^{n-1}\,dx.$$

Unlike Newton, Leibniz was eager to let others know of his discoveries. Even before the formal publication of his results in 1684, knowledge of his methods was quickly spread through the European mathematical community by personal correspondence and word of mouth.

In 1676 Newton learned that Leibniz was working on a version of calculus; in an attempt to assert his priority, he sent Leibniz a letter in which he mentioned his earlier work in the field. Leibniz replied with a letter outlining his own methods. When Leibniz published his results in 1684, he did not mention Newton's claim of priority; in response Newton, in the *Principia,* told of this exchange of letters. This started a quarrel, not so much between Newton and Leibniz but between their followers, over who had priority of discovery. Accusations of plagiarism were made by both sides, and the situation became, particularly in England, one of national pride. As a result, English mathematicians for many years insisted on Newton as the ultimate mathematical authority and adopted his terminology and notation exclusively. Since the terminology and notation used by Leibniz were generally superior to those of Newton, this had an unfortunate stultifying effect on English mathematics.

Newton and Leibniz both used infinitesimals in their development of calculus. Such "infinitely small" but nonzero quantities seemed illogical to mathematicians and made them uneasy. Newton himself was not satisfied with his reliance on infinitesimals, and in his later work he tried to get by without them; in so doing, he nearly arrived at the notion of the limit of a function. However, it was not until 1821, when the French mathematician Augustin Louis Cauchy introduced the idea of limit in its modern form, that it was possible for calculus to be put on a firm logical foundation.

Calculus is one of the greatest tools ever discovered for the investigation of nature. It was developed as a method for solving problems about the movement of objects, particularly astronomical bodies, but its applications soon began to expand, first throughout the physical sciences and engineering, and then to the biological and social sciences, including management and economics. The usefulness of calculus continues to grow today as it is used to investigate new problems as they arise in these and other fields.

Suggestion

Many books describe the development of calculus in much greater detail than we have been able to do here. If you are interested in this topic and would like to read more about it, here are some places to start:

Andrade, E. N. Da C. "Isaac Newton." In *The World of Mathematics,* vol. 1, ed. James R. Newman. New York: Simon and Schuster, 1956.

Bell, E. T. *Men of Mathematics.* New York: Simon and Schuster, 1937.

Boyer, Carl B. *History of the Calculus and Its Conceptual Development.* New York: Dover, 1959.

Boyer, Carl B. *A History of Mathematics.* New York: John Wiley & Sons, 1968.

Eves, Howard. *An Introduction to the History of Mathematics.* New York: Holt, Rinehart, and Winston, 1961.

Eves, Howard. *Great Moments in Mathematics after 1650.* Dolciani Mathematical Expositions No. 7, The Mathematical Association of America, 1983.

3

More about Differentiation

The graph of a function is usually of considerable help in analyzing the behavior of the function. For instance, Figure 3–1 shows the graph of a function that predicts high-school enrollment in a large city for the years 1990 through 2020. From the graph we see the following:

- From 1990 to 1996 enrollment is predicted to decline, with the rate of decline decreasing as 1996 approaches.
- The minimum enrollment will be 200,000 students in 1996.
- From 1996 to 2004 enrollment will grow at an increasing rate;
- From 2004 to 2010 enrollment will continue to grow, but at a decreasing rate;
- The maximum enrollment will be 250,000 students in 2010;
- From 2010 to 2020 enrollment will decline at an increasing rate.

To analyze the behavior of a function as we have just done, we must be able to draw the graph accurately enough to show

- Where the function is increasing and where it is decreasing.
- What its maximum and minimum values are and where they occur.
- The general shape of the curve.

Our major task in this chapter will be to show how to use the first and second derivatives of a function to determine these things. Before we embark upon this program, however, we develop three more rules of differentiation: the product rule, the quotient rule, and the chain rule. These rules will greatly enlarge the class of functions we can differentiate. At the end of the chapter we briefly discuss the differentials dy and dx, the technique of implicit differentiation, and related rates problems.

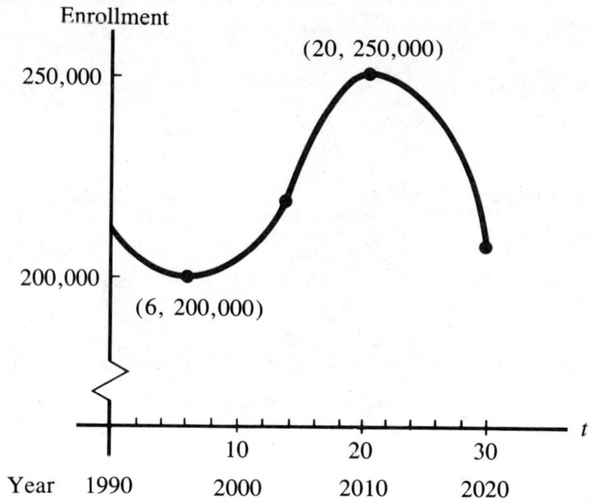

3.1 THE PRODUCT RULE AND THE QUOTIENT RULE

In Section 2.3 we presented some of the basic rules of differentiation: the constant rule, the power rule, the constant-multiple rule, and the sum rule. These rules are not adequate to handle all functions of interest, however. In particular, they do not tell us how to differentiate functions that are products or quotients of other functions, such as

$$f(x) = (x^2 + 2x + 3)(x^5 + 4x^3 - 12)$$

and

$$g(x) = \frac{3x^2 + 2x - 5}{5x + 3}.$$

In this section we develop two new rules of differentiation, the product rule and the quotient rule, which can be used to find the derivatives of such functions, and we illustrate their usage.

The Product Rule

Let f and g be functions and suppose that $f'(x)$ and $g'(x)$ both exist. We would like to find the derivative of the product function $k = fg$ at x. Using the Three-Step Rule, we form the difference quotient of k at x, then rewrite it in terms of the difference quotients of f and g by adding and subtracting $f(x)g(x + h)$ in its numerator:

$$\frac{k(x+h)-k(x)}{h}=\frac{f(x+h)g(x+h)-f(x)g(x)}{h}$$

$$=\frac{f(x+h)g(x+h)-f(x)g(x+h)+f(x)g(x+h)-f(x)g(x)}{h}$$

$$=\frac{[f(x+h)-f(x)]g(x+h)+f(x)[g(x+h)-g(x)]}{h}$$

$$=\frac{f(x+h)-f(x)}{h}g(x+h)+f(x)\frac{g(x+h)-g(x)}{h}.$$

Because $f'(x)$ and $g'(x)$ exist, we have

$$\operatorname*{Lim}_{h\to 0}\frac{f(x+h)-f(x)}{h}=f'(x) \qquad \text{and} \qquad \operatorname*{Lim}_{h\to 0}\frac{g(x+h)-g(x)}{h}=g'(x).$$

Also, since $g'(x)$ exists, $g(x)$ is differentiable and hence continuous at x; therefore by the second limit definition of continuity (Section 2.2),

$$\operatorname*{Lim}_{h\to 0} g(x+h)=g(x).$$

Thus

$$k'(x)=\operatorname*{Lim}_{h\to 0}\frac{f(x+h)-f(x)}{h}\operatorname*{Lim}_{h\to 0} g(x+h)$$

$$+\operatorname*{Lim}_{h\to 0} f(x)\operatorname*{Lim}_{h\to 0}\frac{g(x+h)-g(x)}{h}$$

$$=f'(x)g(x)+f(x)g'(x).$$

This is the **product rule** for derivatives:

The Product Rule

If $k=fg$ and $f'(x)$ and $g'(x)$ both exist, then

$$k'(x)=f(x)g'(x)+f'(x)g(x).$$

EXAMPLE 1 If $k(x)=(x^2-3x+4)(x^3+x)$, then

$$k'(x)=(x^2-3x+4)(x^3+x)'+(x^2-3x+4)'(x^3+x)$$

$$=(x^2-3x+4)(3x^2+1)+(2x-3)(x^3+x)$$

$$=5x^4-12x^3+15x^2-6x+4.$$

[We could have found $k'(x)$ without using the product rule by multiplying out the right side of $k(x)$ and then taking the derivative. Check that this approach yields the same expression for $k'(x)$.]

EXAMPLE 2 Let

$$f(x) = \frac{x^3 + 1}{x^4}.$$

If we rewrite $f(x)$ as

$$f(x) = (x^3 + 1)x^{-4},$$

we may use the product rule to obtain

$$\begin{aligned} f'(x) &= (x^3 + 1)(x^{-4})' + (x^3 + 1)'x^{-4} \\ &= (x^3 + 1)(-4x^{-5}) + (3x^2)x^{-4}, \end{aligned}$$

which simplifies to

$$f'(x) = -\frac{1}{x^2} - \frac{4}{x^5}.$$

(Check this.)

The Quotient Rule

For functions f and g, suppose that $f'(x)$ and $g'(x)$ both exist and that $g(x) \neq 0$. We would like to find the derivative of the quotient function $k = f/g$ at x. Using a trick similar to that employed in deriving the product rule, we write

$$\begin{aligned} \frac{k(x + h) - k(x)}{h} &= \frac{f(x + h)/g(x + h) - f(x)/g(x)}{h} \\ &= \frac{f(x + h)g(x) - f(x)g(x + h)}{hg(x)g(x + h)} \\ &= \frac{f(x + h)g(x) - f(x)g(x) + f(x)g(x) - f(x)g(x + h)}{hg(x)g(x + h)} \\ &= \frac{f(x + h) - f(x)}{h} \, \frac{g(x)}{g(x)g(x + h)} \\ &\quad - \frac{f(x)}{g(x)g(x + h)} \, \frac{g(x) - g(x + h)}{h}. \end{aligned}$$

As was the case for the product rule, the existence of $f'(x)$ and $g'(x)$ implies that

$$\lim_{h \to 0} \frac{f(x + h) - f(x)}{h} = f'(x) \qquad \text{and} \qquad \lim_{h \to 0} \frac{g(x + h) - g(x)}{h} = g'(x).$$

Also, the existence of $g'(x)$ implies that g is differentiable and hence continuous at x; therefore by the second limit definition of continuity (Section 2.2),

$$\lim_{h\to 0} g(x+h) = g(x).$$

Hence

$$k'(x) = \lim_{h\to 0}\frac{f(x+h)-f(x)}{h}\lim_{h\to 0}\frac{g(x)}{g(x)g(x+h)}$$

$$-\lim_{h\to 0}\frac{f(x)}{g(x)g(x+h)}\lim_{h\to 0}\frac{g(x)-g(x+h)}{h}$$

$$= f'(x)\frac{g(x)}{g(x)g(x)} - \frac{f(x)}{g(x)g(x)}g'(x)$$

$$= \frac{g(x)f'(x)-f(x)g'(x)}{[g(x)]^2}.$$

This is the **quotient rule** for derivatives.

The Quotient Rule

If $k = f/g$, $f'(x)$ and $g'(x)$ both exist, and $g(x) \neq 0$, then

$$k'(x) = \frac{g(x)f'(x)-f(x)g'(x)}{[g(x)]^2}.$$

EXAMPLE 3 If $k(x) = 1/x^2$, then by the power rule, $k'(x) = -2/x^3$. We can obtain the same result using the quotient rule:

$$k'(x) = \frac{x^2(1)' - 1(x^2)'}{[x^2]^2} = \frac{x^2(0)-1(2x)}{x^4} = -\frac{2x}{x^4} = -\frac{2}{x^3}.$$

EXAMPLE 4 If $f(x) = (x^2+x+1)/(2x^3+1)$, then by the quotient rule,

$$f'(x) = \frac{(2x^3+1)(x^2+x+1)' - (x^2+x+1)(2x^3+1)'}{[2x^3+1]^2}$$

$$= \frac{(2x^3+1)(2x+1) - (x^2+x+1)(6x^2)}{[2x^3+1]^2}$$

$$= \frac{4x^4+2x^3+2x+1-6x^4-6x^3-6x^2}{[2x^3+1]^2}$$

$$= \frac{-2x^4-4x^3-6x^2+2x+1}{4x^6+4x^3+1}.$$

Computer Manual: Programs PRODPOLY and QUOPOLY.

3.1 EXERCISES

In Exercises 1 through 24 use the product rule or the quotient rule to find $k'(x)$.

1. $k(x) = (3x + 5)(2x - 1)$

2. $k(x) = (5x^2 + 1)(2x^2)$

3. $k(x) = (4x^3 + x^2)(2x^4)$

4. $k(x) = (3x^5 + 2x)(x^2 + 4x - 1)$

5. $k(x) = \dfrac{1}{x}$

6. $k(x) = \dfrac{3}{x^3}$

7. $k(x) = \dfrac{2x + 4}{5x - 1}$

8. $k(x) = \dfrac{4 - 3x}{7 - x}$

9. $k(x) = \dfrac{x^2}{2x + 1}$

10. $k(x) = \dfrac{x^2 + 3x - 2}{9 - 3x}$

11. $k(x) = (5x + 3)^2$

12. $k(x) = (x^2 + 4x)^2$

13. $k(x) = (x^4 + 2)x^{-3}$

14. $k(x) = 2x^{-3}(x^3 + 4)$

15. $k(x) = \dfrac{3x^3 - 2}{x^4}$

16. $k(x) = \dfrac{x^2 + 3x + 7}{x^3}$

17. $k(x) = (3x^{1/3} - 2\sqrt{x})(x^2 - 3x + 1)$

18. $k(x) = \dfrac{2x^{-1} + 3x^{-2}}{x^{-2} - 1}$

19. $k(x) = \dfrac{x^{3/2} - x^{-1}}{x^{2/3}}$

20. $k(x) = x(2x + 1)(3x + 2)$

21. $k(x) = x(x^2 + 1)(x^3 + x^2 + 1)$

22. $k(x) = (1 - 2x)(x^2 - 1)(3 + 4x)(x^3 + 2x)$

23. $k(x) = \dfrac{(x^2 + 1)(x^3 + 2)}{x^2 - 2x}$

24. $k(x) = \dfrac{x^2 - 2x}{(x^2 + 1)(x^3 + 2)}$

Management

25. A firm's cost function is given by

$$C(x) = \frac{x^2}{x + 1} - 1, \qquad x \geq 1.$$

Find the firm's marginal cost when $x = 1$, $x = 10$, and $x = 20$, and interpret your results.

26. If Acme Company makes $x \geq 1$ units of its product, its average cost per unit is given by

$$A(x) = \frac{20x - 4}{5x - 0.5}.$$

Find the rate at which Acme's average cost per unit is changing when $x = 1$, $x = 10$, and $x = 100$. Also find $\text{Lim}_{x \to +\infty} A(x)$ and interpret your result.

Life Science

27. Suppose that t hours after the introduction of a single bacterium into a culture there are

$$y = \frac{200t}{t + 10}$$

bacteria in the culture. Find the rate at which the number of bacteria is growing when $t = 1$, $t = 5$, and $t = 10$. Also find $\lim_{t \to +\infty} y$ and interpret your result.

28. A drug is given to reduce a benign tumor. A dosage of x units will shrink the tumor by y cm, where

$$y = \frac{x^2 + 1}{2 + 3x}.$$

Find the rate at which the tumor will shrink if the dosage is 5 units; if the dosage is 10 units.

Social Science

29. A psychologist asks the subjects of an experiment to perform a series of complex tasks, and finds that after x repetitions of the series they can, on average, complete

$$y = \frac{100x + 30}{x + 1}$$

percent of the tasks without mistakes. Find the rate at which learning takes place as a function of the number of repetitions if the number of repetitions is 0, 2, and 5.

3.2 THE CHAIN RULE

As we noted in Section 1.3, we can create new functions from old ones by addition, subtraction, multiplication, and division. Now the sum rule, product rule, and quotient rule for derivatives show us how to differentiate such new functions in terms of the old ones. However, in Section 1.3 we also learned another method of creating new functions from old ones: the method of composition of functions. Recall that if f and g are functions, then the composite of g with f is the function defined by $f(g(x))$. To complete our rules of differentiation, we need a rule that will tell us how to differentiate a composite function in terms of its component functions; that is, we need a rule that will tell us how to find the derivative of $f(g(x))$ in terms of the derivatives $f'(x)$ and $g'(x)$. In this section we introduce such a rule and show how to use it. This rule of differentiation for composite functions is extremely useful because many functions can be written as composites of simpler ones. We give an example to show how this is done.

EXAMPLE 1 Let $k(x) = (2x + 3)^{20}$. We seek to write the function k as a composite function by finding f and g such that $k(x) = f(g(x))$. The way to do this is to choose $g(x)$ to be the "inside" expression in $k(x)$ and $f(x)$ to be the "outside" expression in $k(x)$. Thus, if we take $g(x) = 2x + 3$ and $f(x) = x^{20}$, then

$$f(g(x)) = f(2x + 3) = (2x + 3)^{20} = k(x).$$

Similarly if

$$h(x) = \sqrt{2x^3 + 4x},$$

and we take

$$g(x) = 2x^3 + 4x \qquad \text{and} \qquad f(x) = \sqrt{x},$$

then

$$f(g(x)) = f(2x^3 + 4x) = \sqrt{2x^3 + 4x} = h(x).$$

Now we develop the rule for differentiating composite functions. Suppose $k(x) = f(g(x))$, where $f'(g(x))$ and $g'(x)$ both exist. We form the difference quotient

$$\frac{k(x + h) - k(x)}{h} = \frac{f(g(x + h)) - f(g(x))}{h}.$$

If $g(x + h) - g(x) \neq 0$, we may rewrite this difference quotient as

$$\frac{k(x + h) - k(x)}{h} = \frac{f(g(x + h)) - f(g(x))}{g(x + h) - g(x)} \frac{g(x + h) - g(x)}{h}.$$

Since $g'(x)$ exists, g is continuous at x. Therefore

$$\lim_{h \to 0} g(x + h) = g(x),$$

or

$$\lim_{h \to 0} [g(x + h) - g(x)] = 0.$$

Let us set $r = g(x + h) - g(x)$. As h approaches 0, r approaches 0, and $g(x + h) = g(x) + r$. Thus

$$\begin{aligned} k'(x) &= \lim_{h \to 0} \frac{f(g(x + h)) - f(g(x))}{g(x + h) - g(x)} \lim_{h \to 0} \frac{g(x + h) - g(x)}{h} \\ &= \lim_{r \to 0} \frac{f(g(x) + r) - f(g(x))}{r} g'(x) \\ &= f'(g(x))g'(x). \end{aligned}$$

We have thus shown that if $k(x) = f(g(x))$, then $k'(x) = f'(g(x))g'(x)$, a result known as the **chain rule.** We have proved the chain rule under the assumption that $g(x + h) - g(x) \neq 0$, but it is true in general.

The Chain Rule: First Formulation

If $k(x) = f(g(x))$ and $f'(g(x))$ and $g'(x)$ both exist, then

$$k'(x) = f'(g(x))g'(x).$$

EXAMPLE 2 Suppose $k(x) = (2x + 3)^{20}$. We will use the chain rule to find $k'(x)$. From Example 1,

$$k(x) = f(g(x)),$$

where

$$f(x) = x^{20} \qquad \text{and} \qquad g(x) = 2x + 3.$$

Since

$$f'(x) = 20x^{19} \qquad \text{and} \qquad g'(x) = 2,$$

the chain rule tells us that

$$\begin{aligned} k'(x) &= f'(g(x))g'(x) \\ &= f'(2x + 3)g'(x) \\ &= 20(2x + 3)^{19}(2) \\ &= 40(2x + 3)^{19}. \end{aligned}$$

EXAMPLE 3 Let $h(x) = \sqrt{2x^3 + 4x}$. In Example 1 we showed that

$$h(x) = f(g(x)),$$

where

$$f(x) = \sqrt{x} \qquad \text{and} \qquad g(x) = 2x^3 + 4x.$$

Since

$$f'(x) = \frac{1}{2\sqrt{x}} \qquad \text{and} \qquad g'(x) = 6x^2 + 4,$$

we have

$$\begin{aligned} h'(x) &= f'(g(x))g'(x) \\ &= f'(2x^3 + 4x)g'(x) \\ &= \frac{1}{2\sqrt{2x^3 + 4x}}(6x^2 + 4) \\ &= \frac{3x^2 + 2}{\sqrt{2x^3 + 4x}}. \end{aligned}$$

We do not really need to write out $f(x)$, $g(x)$, $f'(x)$, and $g'(x)$ every time we differentiate a composite function. Roughly speaking, what the chain rule says is that to differentiate a composite expression $f(g(x))$ we must differentiate from the "outside in." In other words, we first differentiate the "outside" expression $f(x)$, then multiply the result by the derivative of the "inside" expression $g(x)$. With a little practice this becomes easy to do.

EXAMPLE 4 Let us differentiate $k(x) = (2x + 3)^{20}$ again. This time we will not write out $f(x)$, $g(x)$, $f'(x)$, and $g'(x)$, but instead will differentiate from the outside in as follows:

$$\begin{aligned} k'(x) &= [(2x + 3)^{20}]' = 20(2x + 3)^{19}(2x + 3)' \\ &= 20(2x + 3)^{19}(2) \\ &= 40(2x + 3)^{19}. \end{aligned}$$

EXAMPLE 5 Let us differentiate $h(x) = \sqrt{2x^3 + 4x}$ again. This time we will differentiate from the outside in. We have

$$\begin{aligned} h'(x) &= (\sqrt{2x^3 + 4x})' \\ &= [(2x^3 + 4x)^{1/2}]' \\ &= \frac{1}{2}(2x^3 + 4x)^{-1/2}(2x^3 + 4x)' \\ &= \frac{1}{2\sqrt{2x^3 + 4x}}(6x^2 + 4) \\ &= \frac{3x^2 + 2}{\sqrt{2x^3 + 4x}}. \end{aligned}$$

The same differentiation using differential notation would be written as follows:

$$\begin{aligned} \frac{dh}{dx} = \frac{d}{dx}(\sqrt{2x^3 + 4x}) &= \frac{1}{2\sqrt{2x^3 + 4x}}\frac{d}{dx}(2x^3 + 4x) \\ &= \frac{1}{2\sqrt{2x^3 + 4x}}(6x^2 + 4) \\ &= \frac{3x^2 + 2}{\sqrt{2x^3 + 4x}}. \end{aligned}$$

The chain rule may be used with the other rules of differentiation.

EXAMPLE 6 Let

$$p(x) = (x^2 + 1)^3(x^4 - 1)^5.$$

Then by the product rule we have

$$p'(x) = (x^2 + 1)^3[(x^4 - 1)^5]' + [(x^2 + 1)^3]'(x^4 - 1)^5.$$

Now we apply the chain rule to find

$$\begin{aligned} [(x^4 - 1)^5]' &= 5(x^4 - 1)^4(x^4 - 1)' = 5(x^4 - 1)^4(4x^3) \\ &= 20x^3(x^4 - 1)^4 \end{aligned}$$

and

$$[(x^2+1)^3]' = 3(x^2+1)^2(x^2+1)' = 3(x^2+1)^2(2x)$$
$$= 6x(x^2+1)^2.$$

Therefore

$$\begin{aligned} p'(x) &= (x^2+1)^3[20x^3(x^4-1)^4] \\ &\quad + [6x(x^2+1)^2](x^4-1)^5 \\ &= 2x(x^2+1)^2(x^4-1)^4[10x^2(x^2+1)+3(x^4-1)] \\ &= 2x(x^2+1)^2(x^4-1)^4[13x^4+10x^2-3]. \end{aligned}$$

EXAMPLE 7 Let $q(x) = [(x+1)/(x-1)]^{2/3}$. To differentiate $q(x)$, we use the chain rule followed by the quotient rule:

$$\begin{aligned} q'(x) &= \left[\left(\frac{x+1}{x-1}\right)^{2/3}\right]' = \frac{2}{3}\left(\frac{x+1}{x-1}\right)^{-1/3}\left(\frac{x+1}{x-1}\right)' \\ &= \frac{2}{3}\left(\frac{x+1}{x-1}\right)^{-1/3}\frac{(x-1)(x+1)' - (x+1)(x-1)'}{(x-1)^2} \\ &= \frac{2}{3}\left(\frac{x+1}{x-1}\right)^{-1/3}\frac{(x-1)(1) - (x+1)(1)}{(x-1)^2} \\ &= -\frac{4}{3(x-1)^2}\left(\frac{x+1}{x-1}\right)^{-1/3}. \end{aligned}$$

The functions to which we have thus far applied the chain rule have all been defined by equations of the form

$$k(x) = [g(x)]^r.$$

But the chain rule tells us that the derivative of such a function is

$$k'(x) = r[g(x)]^{r-1}g'(x).$$

(Why?) This special case of the chain rule is known as the **general power rule.**

The General Power Rule

If

$$k(x) = [g(x)]^r, \qquad r \neq 0,$$

then

$$k'(x) = r[g(x)]^{r-1}g'(x).$$

EXAMPLE 8 If

$$k(x) = (x^3 - 2x^2 + 17)^{12},$$

then

$$\begin{aligned} k'(x) &= 12(x^3 - 2x^2 + 17)^{11}(x^3 - 2x^2 + 17)' \\ &= 12(3x^2 - 4x)(x^3 - 2x^2 + 17)^{11}. \end{aligned}$$

We conclude this section with another formulation of the chain rule that is often useful. If $y = f(x)$ defines y as a function of x and $x = g(t)$ defines x as a function of t, then

$$y = f(g(t))$$

defines y as a function of t. Differentiating with respect to t by means of the chain rule, we have

$$\frac{dy}{dt} = \frac{df}{dx}\frac{dg}{dt},$$

which we may write as

$$\frac{dy}{dt} = \frac{dy}{dx}\frac{dx}{dt}.$$

This is the second formulation of the chain rule.

The Chain Rule: Second Formulation

If $y = f(x)$ and $x = g(t)$ and dy/dx and dx/dt both exist, then

$$\frac{dy}{dt} = \frac{dy}{dx}\frac{dx}{dt}.$$

EXAMPLE 9 Suppose $y = x^2 + 1$ and $x = 3t - 2$. Then by the second formulation of the chain rule,

$$\frac{dy}{dt} = \frac{dy}{dx}\frac{dx}{dt} = \frac{d}{dx}(x^2 + 1)\frac{d}{dt}(3t - 2) = 2x \cdot 3 = 6x.$$

If we wish, we may write dy/dt as a function of t by substituting $3t - 2$ for x:

$$\frac{dy}{dt} = 6x = 6(3t - 2) = 18t - 12.$$

In the previous example it would have been easy to substitute $3t - 2$ for x in the equation that defines y and then find dy/dt directly:

$$y = x^2 + 1 = (3t - 2)^2 + 1 = 9t^2 - 12t + 5,$$

so

$$\frac{dy}{dt} = \frac{d}{dt}(9t^2 - 12t + 5) = 18t - 12.$$

Nevertheless, as we shall see later in this chapter, we often need to find dy/dt without writing y explicitly as a function of t. We then must use the second formulation of the chain rule.

The second formulation of the chain rule may be used with variables other than x, y, and t.

EXAMPLE 10 Let $u = 3v^2 - 2v + 2$ and $v = 2\sqrt{w}$. Then

$$\frac{du}{dw} = \frac{du}{dv}\frac{dv}{dw}$$

$$= \frac{d}{dv}(3v^2 - 2v + 2)\frac{d}{dw}(2\sqrt{w})$$

$$= \frac{6v - 2}{\sqrt{w}}.$$

Computer Manual: Program POWRPOLY.

3.2 EXERCISES

In Exercises 1 through 6 write k as a composite function.

1. $k(x) = (3x + 5)^{10}$ **2.** $k(x) = (2 - 3x)^{12}$ **3.** $k(x) = (x^2 - 2x)^5$

4. $k(x) = (2x^{-2} + 5x - 1)^{-3}$ **5.** $k(x) = \sqrt{4x - 1}$ **6.** $k(x) = \sqrt[3]{x^2 + 1}$

In Exercises 7 through 28 use the chain rule to find $k'(x)$.

7. $k(x) = (3x + 5)^{10}$ **8.** $k(x) = (2 - 3x)^{12}$ **9.** $k(x) = (x^2 - 2x)^5$

10. $k(x) = (2x^{-2} + 5x - 1)^{-3}$ **11.** $k(x) = \sqrt{4x - 1}$ **12.** $k(x) = \sqrt[3]{x^2 + 1}$

13. $k(x) = (5x^2 - 3x + 1)^{100}$ **14.** $k(x) = (2x^{-2} + 4)^{-10}$ **15.** $k(x) = \sqrt[3]{x^5 - 3x^2}$

16. $k(x) = x^3(5x^2 - 1)^2$ **17.** $k(x) = (2x^2 + 1)^{11}(x^3 - 1)^9$ **18.** $k(x) = [(3 - 2x + x^4)(x^2 - 4)]^7$

19. $k(x) = [(4\sqrt{x} - 3x^{-2})(3x^{4/3} + 10)]^{-3}$ **20.** $k(x) = \sqrt{(x^2 + 1)(x^2 - 3)}$

21. $k(x) = \sqrt{(x^3 + 3)^3(2x - 3)^{2/3}}$ **22.** $k(x) = \sqrt{x^2 - 1}\,\sqrt[3]{x^3 - 1}$

23. $k(x) = \left(\frac{x + 2}{x - 2}\right)^4$ **24.** $k(x) = \left(\frac{2x + 5}{3 - 4x}\right)^{-2}$

25. $k(x) = x^3\sqrt{4x + 3}$ **26.** $k(x) = 2x\sqrt[3]{x^2 + 8}$

27. $k(x) = \frac{3x + 1}{\sqrt{x^2 + 3}}$ **28.** $k(x) = \left(\frac{x^{-1} + \sqrt{x}}{x^{-1} - \sqrt{x}}\right)^3$

29. Use the second formulation of the chain rule to find dy/dt if $y = x^2$ and $x = 4 - 2t$. Check your answer by expressing y as a function of t and finding dy/dt directly.

30. Repeat Exercise 29 if $y = 2x^2 - 3x + 1$ and $x = 5t + 1$.

31. Use the second formulation of the chain rule to find du/dv if $u = \sqrt{x + 1}$ and $x = v^2 - 1$. Check your answer by expressing u as a function of v and finding du/dv directly.

32. Use the second formulation of the chain rule to find dy/dt if
(a) $y = \sqrt{x^2 + 1}$ and $x = t^2$ (b) $y = x^2$ and $x = t^2 + 1$.

33. Use the second formulation of the chain rule to find dy/dt if
(a) $y = \dfrac{1}{x}$ and $x = 4t - 5$ (b) $y = (x^3 + 3x - 1)^5$, $x = u^2$, and $u = 5t$.

34. Use the second formulation of the chain rule to find dz/dw if
(a) $z = 3\sqrt{v}$, $v = \dfrac{1}{q}$, and $q = w^2 + 1$ (b) $z = \dfrac{p - 1}{2 - p}$, $p = 5s + 1$, and $s = w^{-3}$.

Management

35. A firm's revenue function is given by

$$R(x) = x\left(1 - \frac{x^2}{100}\right)^2$$

and its cost function by

$$C(x) = \frac{x^2}{x + 1} + 1,$$

both valid for $x \geq 1$. Find the firm's marginal revenue and marginal profit when $x = 1$, $x = 10$, and $x = 20$, and interpret your results.

36. Suppose a firm's cost and revenue functions are given by

$$C(x) = \sqrt{x + 100} \qquad \text{and} \qquad R(x) = \sqrt{x^2 + 2},$$

respectively, for $x \geq 1$. Find the firm's marginal cost, marginal revenue, and marginal profit when $x = 10$, and interpret your results.

Life Science

37. A rare species of seal was placed on the endangered species list in 1984. It is estimated that the size of the population of this species in year t is given by

$$y = \sqrt{\frac{8t + 2}{2t + 9}},$$

where $t = 0$ represents 1984 and y is in thousands. Find the rate at which the population was changing in 1986 and 1987. Also find and interpret $\text{Lim}_{t \to +\infty} y$.

Social Science

38. Suppose the proportion of eligible voters who participated in presidential elections in the United States since World War II is y, where

$$y = \frac{1}{\sqrt{0.5t + 2}}$$

and $t = 0$ represents 1948, $t = 1$ represents 1952, and so on. Find the rate at which voter participation was changing in the election of 1980 and in the election of 1984.

3.3 RELATIVE MAXIMA AND MINIMA: THE FIRST DERIVATIVE TEST

Suppose a function has the graph shown in Figure 3–2. Notice that the graph comes to a peak at the point $(-1, 10)$, forms a trough at the point $(2, 1)$, and "flattens out" at the point $(4, 5)$. The tangent lines at these points have been drawn in to emphasize these facts. The point $(-1, 10)$ is a **relative maximum point,** and the point $(2, 1)$ is a **relative minimum point;** we call the point $(4, 5)$ a **saddle point.** (The term *saddle point* is not standard terminology, but it is convenient to use it to refer to a point where the graph flattens out. In Section 3.6 we will introduce the notion of a **point of inflection;** a saddle point is a particular kind of point of inflection.) When graphing and analyzing functions, we must find their relative maxima, relative minima, and saddle points. In this section we show how to use the first derivative of a function to help find these points.

Figure 3–3 shows the graph of a function f that has a relative maximum at $(a, f(a))$, a relative minimum at $(b, f(b))$, and a saddle point at $(c, f(c))$. Consider the point $(a, f(a))$. From the figure it is apparent that for all values of x near a, on either side of a, the inequality $f(x) \leq f(a)$ holds. The point $(a, f(a))$ is a relative maximum point exactly *because* this inequality holds. Similar inequalities define the relative minimum point $(b, f(b))$ and the saddle point $(c, f(c))$.

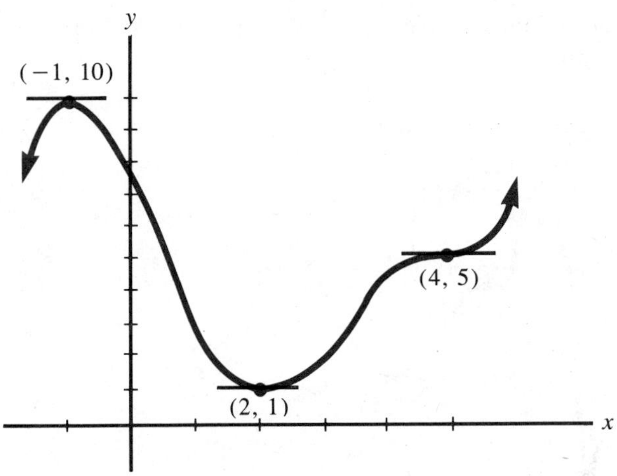

FIGURE 3–2

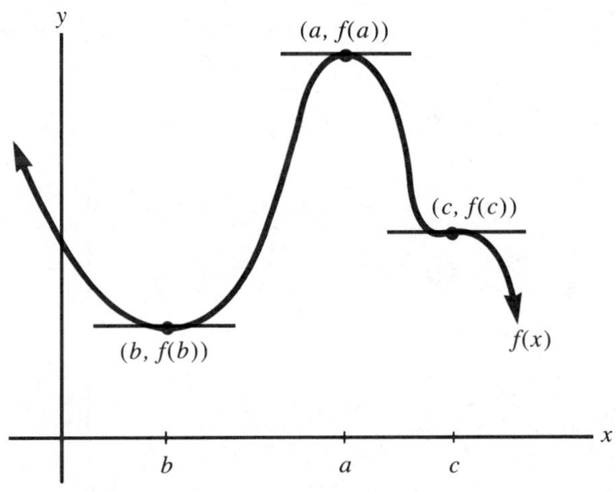

FIGURE 3–3

Relative Maxima, Relative Minima, and Saddle Points

Let f be a function.

1. The point $(a, f(a))$ is a **relative maximum point** for f if
$$f(x) \leq f(a)$$
for all values of x near a on either side of a.

2. The point $(b, f(b))$ is a **relative minimum point** for f if
$$f(x) \geq f(b)$$
for all values of x near b on either side of b.

3. The point $(c, f(c))$ is a **saddle point** for f if
$$f(x) > f(c)$$
for all values of x near c and on one side of c, and
$$f(x) < f(c)$$
for all values of x near c and on the other side of c.

EXAMPLE 1 The function whose graph is shown in Figure 3–4 has relative maxima at the points (4, 8) and (9, 7), relative minima at the points (3, 2) and (7, 4), and a saddle point at (5, 6).

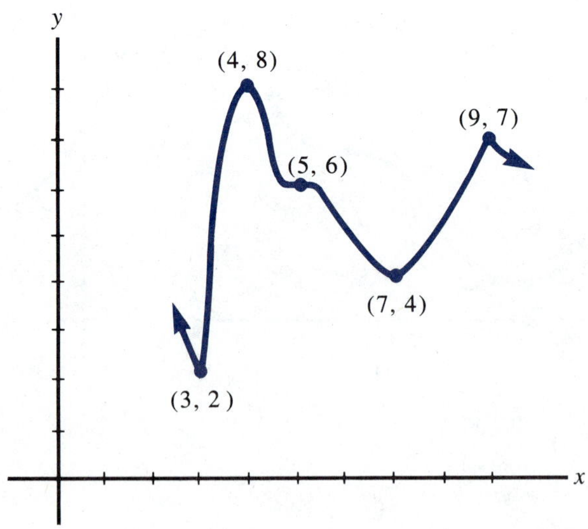

FIGURE 3–4

How can we find the relative maxima, relative minima, and saddle points of a function f? As Figure 3–3 indicates, if $(a, f(a))$ is a relative maximum point for f *and* there is a unique tangent line at $(a, f(a))$, then the tangent line must have a slope of zero: in other words, if $f'(a)$ exists, then $f'(a) = 0$. The same reasoning holds for relative mimima and saddle points. Therefore we can find at least some of the relative maxima, relative minima, and saddle points for f by finding the values of x for which $f'(x) = 0$. We might not find all relative maxima and minima for f by doing this, because f might have relative maxima or minima at points where $f'(x)$ does not exist. As an example of such behavior, the function of Figure 3–4 has a relative maximum at (9, 7) and a relative minimum at (3, 2), but $f'(9)$ and $f'(3)$ do not exist. (Why not?)

Our conclusion, then, is that if we are looking for the relative maxima, relative minima, and saddle points of a function f, we must find all values of x for which $f'(x) = 0$ and also all values of x for which $f'(x)$ does not exist. Such values are called **critical values** of the function.

Critical Values

If f is a function, the value $x = a$ is a **critical value** for f if either $f'(a) = 0$ or $f'(a)$ does not exist.

EXAMPLE 2 Find the critical values for the functions f and g defined by

$$f(x) = 2x^3 - 12x^2 + 18x + 1 \qquad \text{and} \qquad g(x) = x^{2/3}.$$

Since

$$f'(x) = 6x^2 - 24x + 18 = 6(x - 1)(x - 3)$$

exists for all x, the only critical values for f are those for which $f'(x) = 0$, namely, $x = 1$ and $x = 3$. Since

$$g'(x) = \frac{2}{3x^{1/3}}$$

is never zero (since its numerator is 2), and since $g'(x)$ does not exist when $x = 0$, the only critical value for g is $x = 0$.

Now we consider the sign of the derivative and what it tells us about the function. Since $f'(x)$ is the slope of the line tangent to f at $(x, f(x))$, it follows that if $f'(x) > 0$ for all x in an interval (c, d), all the tangent lines to f over the interval have positive slope, as in Figure 3–5; but as Figure 3–5 shows, this means that the function f must be increasing over the interval. Similarly, if $f'(x) < 0$ for all x in an interval, then, as Figure 3–6 demonstrates, f must be decreasing over the interval.

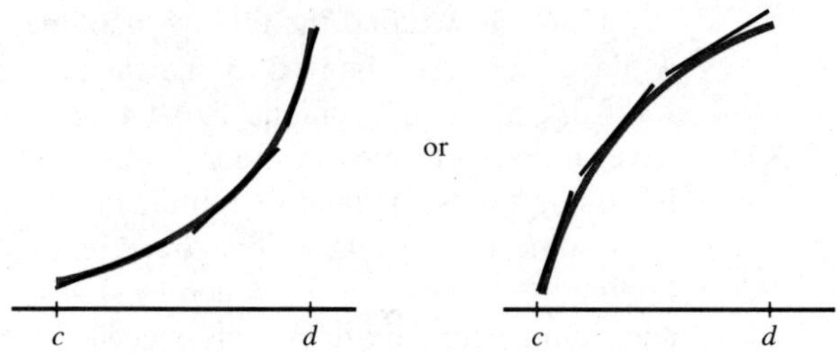

FIGURE 3–5 $f'(x) > 0$ for x in (c, d)

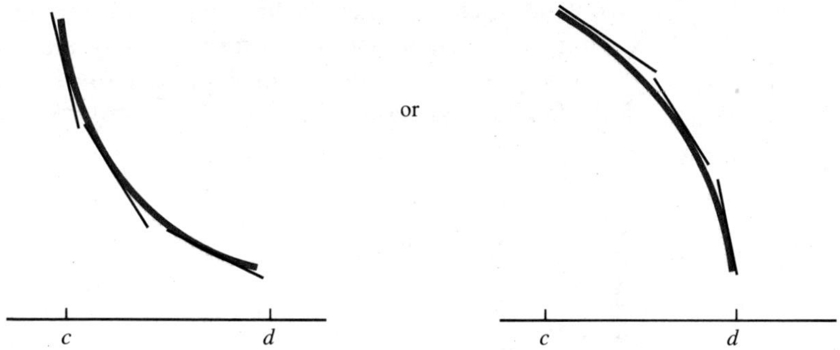

FIGURE 3–6 $f'(x) < 0$ for x in (c, d)

> **The Sign of the Derivative**
>
> **1.** If $f'(x) > 0$ for all x in an interval (c, d), then f is increasing over the interval.
>
> **2.** If $f'(x) < 0$ for all x in an interval (c, d), then f is decreasing over the interval.

EXAMPLE 3 Let $f(x) = 2x^3 - 12x^2 + 18x + 1$. Then

$$f'(x) = 6x^2 - 24x + 18 = 6(x - 1)(x - 3).$$

If $x < 1$, then

$$x - 1 < 0 \qquad \text{and} \qquad x - 3 < 0,$$

so

$$f'(x) = 6(x - 1)(x - 3) > 0;$$

therefore f is increasing over the interval $(-\infty, 1)$. If $1 < x < 3$, then

$$x - 1 > 0 \qquad \text{and} \qquad x - 3 < 0,$$

so

$$f'(x) = 6(x - 1)(x - 3) < 0;$$

therefore f is decreasing over the interval (1, 3). Finally, if $x > 3$, then

$$x - 1 > 0 \qquad \text{and} \qquad x - 3 > 0,$$

so

$$f'(x) = 6(x - 1)(x - 3) > 0;$$

therefore f is decreasing over the interval $(3, +\infty)$. Figure 3–7 shows the graph of f: note that f is increasing over the intervals $(-\infty, 1)$ and $(3, +\infty)$ and decreasing over the interval (1, 3).

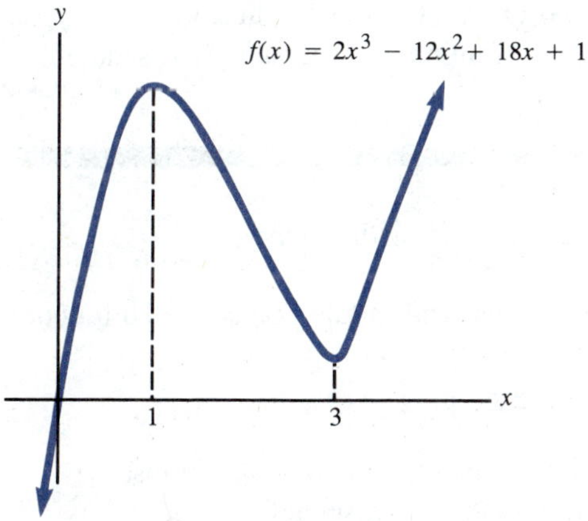

FIGURE 3–7

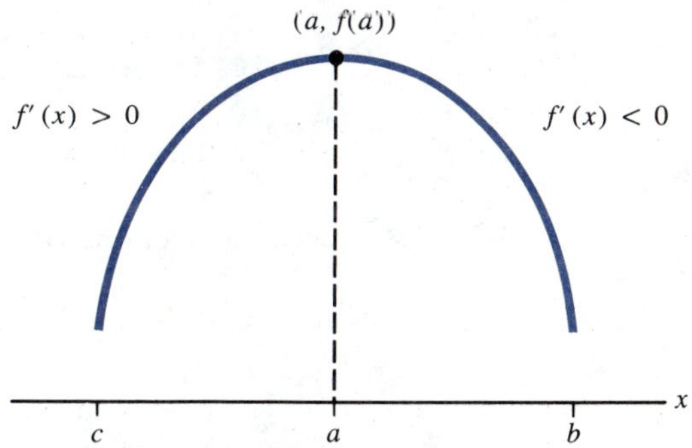

FIGURE 3–8

Now we return to the topic of critical values. Once we have found a critical value for a particular function, we must decide whether it yields a relative maximum, a relative minimum, or a saddle point. But we can do this by examining the sign of the derivative. For instance, suppose that $x = a$ is a critical value for f, and that $f'(x)$ is positive for all x in some interval (c, a) and negative for all x in some interval (a, b); then f must be increasing to the left of a and decreasing to the right of a (Figure 3–8). Thus the point $(a, f(a))$ must be a relative maximum. Similar arguments apply to relative minima and saddle points. Hence we can test each critical value to see whether it gives a relative maximum, a relative minimum, or a saddle point by examining the sign of the derivative on either side of the critical value.

The procedure we have outlined for finding relative maxima, relative minima, and saddle points is called the **first derivative test.**

The First Derivative Test

To find the relative maxima, relative minima, and saddle points of a function f, proceed as follows:

1. Find the critical values of f. These are the values of x for which $f'(x) = 0$ or for which $f'(x)$ does not exist.
2. Plot the critical values on the x-axis, thus dividing the x-axis into intervals.
3. Examine the sign of f' on each of the intervals obtained in step 2.
   a. If $f'(x) > 0$ for all x in an interval, f is increasing over the interval. We symbolize this with an upward-slanting arrow ↗.
   b. If $f'(x) < 0$ for all x in an interval, f is decreasing over the interval. We symbolize this with a downward-slanting arrow ↘.
4. The diagram obtained in step 3 shows which critical values yield relative maxima, which yield relative minima, and which yield saddle points.

We illustrate the use of the first derivative test with examples.

EXAMPLE 4 Let us find all relative maxima, relative minima, and saddle points for the function f defined by

$$f(x) = 2x^3 - 12x^2 + 18x + 1.$$

1. We saw in Example 2 that the only critical points of f were those obtained by setting $f'(x) = 0$ and solving for x, namely, 1 and 3.
2. We plot $x = 1$ and $x = 3$ on the x-axis:

3. We saw in Example 3 that f was increasing on $(-\infty, 1)$, decreasing on $(1, 3)$, and increasing on $(3, +\infty)$. Therefore we have the diagram

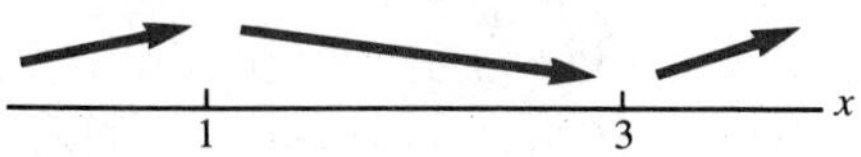

4. The diagram of step 3 shows that there is a relative maximum at $(1, f(1)) = (1, 9)$ and a relative minimum at $(3, f(3)) = (3, 1)$. There are no saddle points. The graph of this function is depicted in Figure 3–9.

EXAMPLE 5 We find the relative maxima, relative minima, and saddle points for the function g defined by $g(x) = x^{2/3}$.

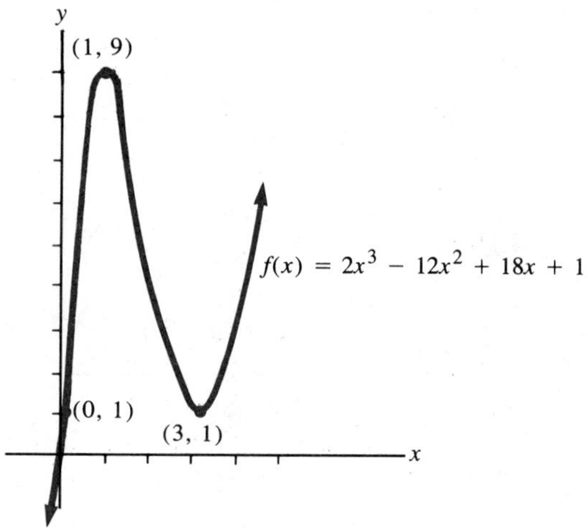

FIGURE 3–9

1. From Example 2 the only critical point of g is the point $x = 0$, for which the derivative $g'(x)$ does not exist.
2. We plot the critical point on the x-axis:

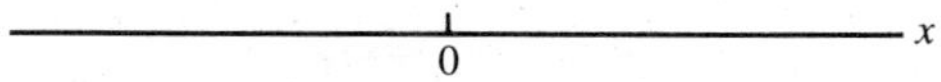

3. We determine the sign of

$$g'(x) = \frac{2}{3x^{1/3}}$$

on the intervals $(-\infty, 0)$ and $(0, +\infty)$. Since $x^{1/3} = \sqrt[3]{x}$ is negative when $x < 0$ and positive when $x > 0$, we have $g'(x) < 0$ when $x < 0$, and $g'(x) > 0$ when $x > 0$. Hence g is decreasing over the interval $(-\infty, 0)$ and increasing over the interval $(0, +\infty)$:

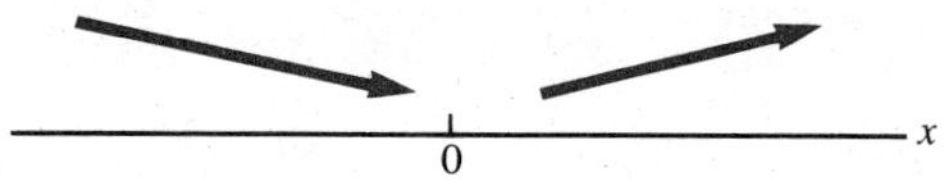

4. The diagram of step 3 shows that $(0, g(0)) = (0, 0)$ is a relative minimum point. There are no relative maxima or saddle points. The graph of $g(x)$ is shown in Figure 3–10.

Step 3 of the first derivative test requires that we determine the sign of $f'(x)$ on each of the intervals between critical points of f. If f' is continuous on such an interval, we can determine its sign by choosing a number from the interval and substituting it for x in $f'(x)$. This fact can often be used to simplify the work in step 3.

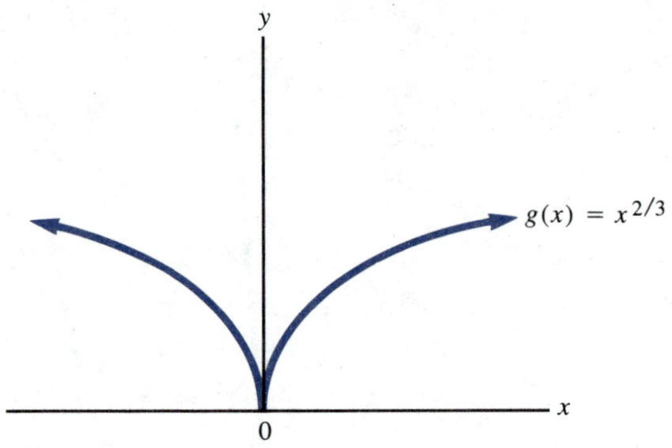

FIGURE 3–10

EXAMPLE 6 Let us find the relative maxima, relative minima, and saddle points for the function h given by

$$h(x) = -x^3 + 6x^2 + 15x.$$

1. Since

$$h'(x) = -3x^2 + 12x + 15 = -3(x + 1)(x - 5),$$

setting $h'(x) = 0$ and solving for x yields $x = -1$ and $x = 5$. There are no values of x for which $h'(x)$ does not exist.

2. We have

3. Since h' is a polynomial, it is continuous on the intervals $(-\infty, -1)$, $(-1, 5)$, and $(5, +\infty)$. Therefore we can determine its sign on each of these intervals by selecting a number from the interval and substituting it into $h'(x)$. For instance, if we select -2 from $(-\infty, -1)$, 0 from $(-1, 5)$, and 6 from $(5, +\infty)$, we have

$$h'(-2) = -3(-2 + 1)(-2 - 5) < 0,$$

$$h'(0) = -3(0 + 1)(0 - 5) > 0,$$

$$h'(6) = -3(6 + 1)(6 - 5) < 0.$$

Thus $h'(x)$ is negative on $(-\infty, -1)$ and $(5, +\infty)$ and positive on $(-1, 5)$. Hence

4. The diagram of step 3 shows that $(-1, h(-1)) = (-1, -8)$ is a relative minimum for h and that $(5, h(5)) = (5, 100)$ is a relative maximum for h. Figure 3–11 shows the graph of h.

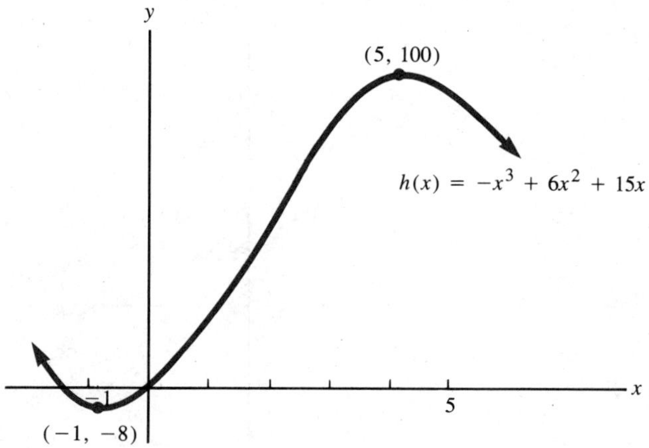

FIGURE 3–11

EXAMPLE 7 We find the relative maxima, relative minima, and saddle points of the function k defined by

$$k(x) = -x^4 + 8x^3 - 18x^2.$$

1. Since

$$k'(x) = -4x^3 + 24x^2 - 36x = -4x(x-3)^2,$$

setting $k'(x) = 0$ and solving for x yields $x = 0$ and $x = 3$. There are no values for which $k'(x)$ does not exist.

2. We have

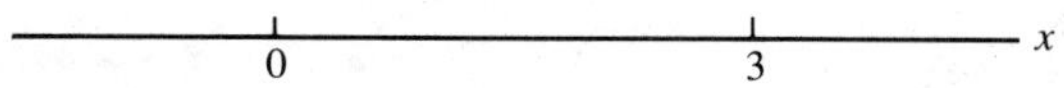

3. Note that k' is continuous on $(-\infty, 0)$, $(0, 3)$, and $(3, +\infty)$. (Why?) If we choose $x = -1$ from $(-\infty, 0)$, $x = 1$ from $(0, 3)$, and $x = 4$ from $(3, +\infty)$, we find that $k'(-1) = 64$ and $k'(1) = k'(4) = -16$. Therefore k is increasing over $(-\infty, 0)$ and decreasing over $(0, 3)$ and $(3, +\infty)$:

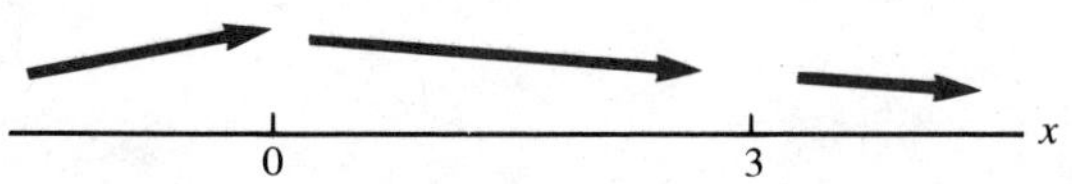

4. The diagram of step 3 shows that there is a relative maximum at $(0, k(0)) = (0, 0)$ and a saddle point at $(3, k(3)) = (3, -27)$. The graph of the function is shown in Figure 3–12.

Computer Manual: Program DERTEST1.

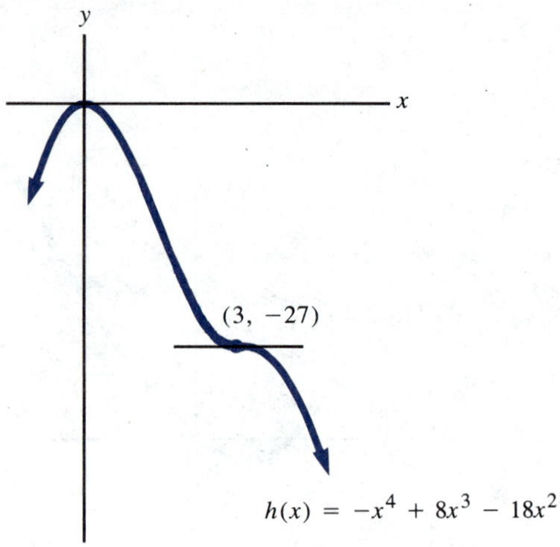

FIGURE 3–12

3.3 EXERCISES

In Exercises 1 through 4 find all relative maxima, relative minima, and saddle points for the function whose graph is shown.

1.

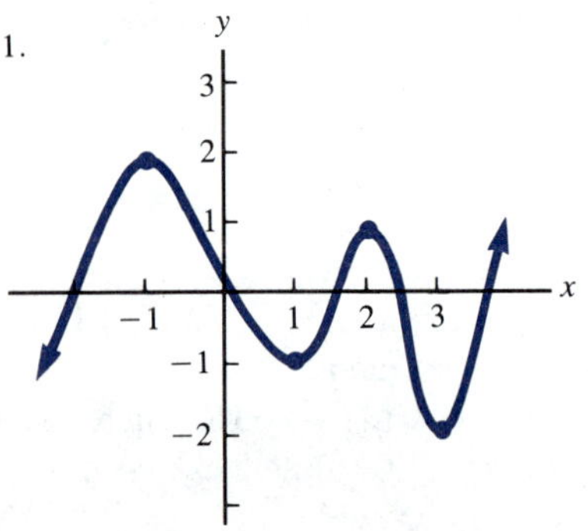

2.

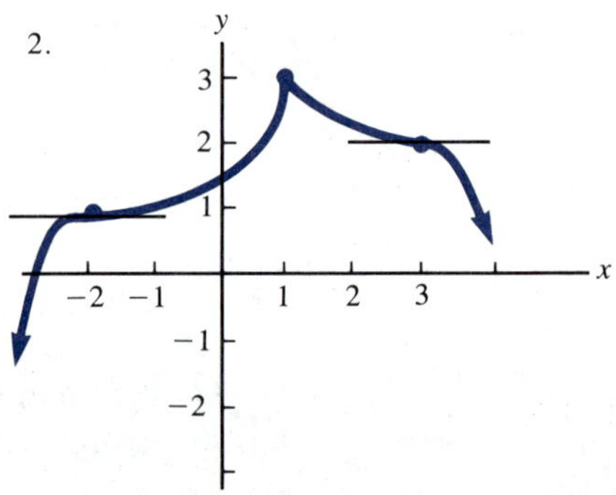

3.

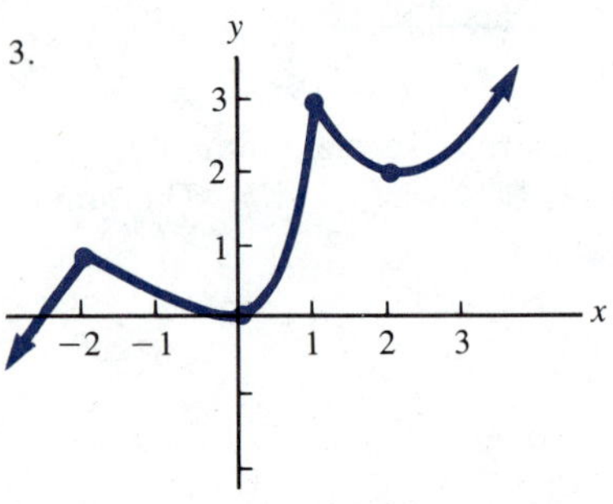

4.

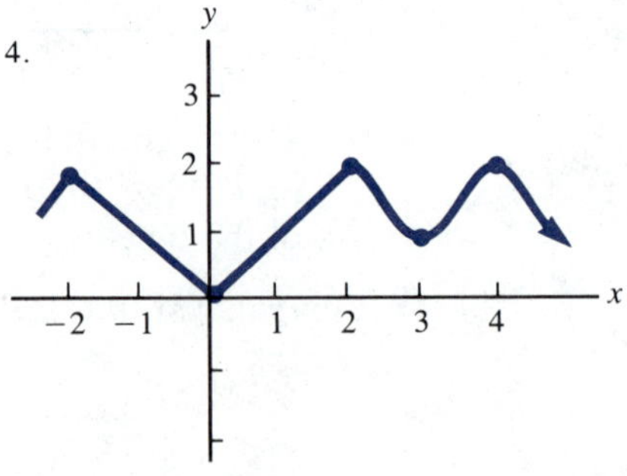

In Exercises 5 through 22 use the first derivative test to find all relative maxima, relative minima, and saddle points for the given function.

5. $f(x) = x^2 - 1$

6. $f(x) = x^2 - 2x + 1$

7. $f(x) = x^2 - 4x + 3$

8. $f(x) = -x^2 - 4x + 12$

9. $f(x) = -0.5x^2 + 3x - 5$

10. $f(x) = x^3$

11. $f(x) = x^3 - 2x^2$

12. $f(x) = -2x^3 + 15x^2 - 24x + 8$

13. $f(x) = -x^3 + 12x$

14. $f(x) = 2x^3 + 9x^2 - 108x + 20$

15. $f(x) = x^3 - 6x^2 + 12x + 2$

16. $f(x) = x^4$

17. $f(x) = -3x^4 + 4x^3 + 6x^2 - 12x$ [*Hint:* $f'(x) = -12(x - 1)(x^2 - 1)$.]

18. $f(x) = 3x^5 - 10x^3 + 15x + 1$ [*Hint:* $f'(x) = 15(x^2 - 1)^2$.]

19. $f(x) = x^{4/3}$

20. $f(x) = \dfrac{2x - 1}{x + 2}$

21. $f(x) = \dfrac{x + 1}{-3 - x^2}$

22. $f(x) = x + \dfrac{1}{x}$

Management

23. Amity Company's average cost function is given by

$$A(x) = 10x + \frac{4000}{x},$$

where x is the number of units produced, $x \geq 1$. If Amity's minimum average cost per unit occurs at a relative minimum of this function, find the minimum average cost per unit.

24. If Beatific Corporation spends x thousand dollars on advertising, its sales will be

$$f(x) = 2x^3 - 210x^2 + 6000x + 100{,}000$$

dollars. Because Beatific can spend at most \$40,000 on advertising, its maximum sales occur at a relative maximum of f. Find the firm's maximum sales and the amount that must be spent on advertising to attain maximum sales.

25. A worker's productivity is given by

$$f(t) = -0.05t^2 + 1.8t + 150,$$

where t is the number of years the worker has been on the job and $f(t)$ is the dollar value of the worker's average daily production in year t. The worker's maximum productivity occurs at a relative maximum of this function. Find the worker's maximum productivity and the year in which it occurs.

26. A radio station's share of the audience during 1986 was given by

$$s = \frac{t - 2}{t^2 + 12} + \frac{3}{10},$$

where t is in months, $1 \leq t \leq 12$, with $t = 1$ representing January 1986. If the station's maximum audience share occurred at a relative maximum of this function, find the maximum audience share and the time during the year when it occurred.

Life Science

27. The number of birds in a wildlife sanctuary in month t of the year is estimated to be

$$y = -t^3 + 5.25t^2 + 33t + 1000.$$

The maximum bird population occurs at a relative maximum of this function. Find the maximum bird population and the time during the year when it occurs.

28. During an epidemic the number of people inoculated during month t was given by

$$y = -24t^5 + 165t^4 + 800t^3, \qquad 0 \le t \le 10.$$

If the maximum number inoculated during a month occurred at a relative maximum of this function, find this maximum and when it occurred.

Social Science

29. Enrollment in the higher education system of a nation for the period 1990–2010 is forecast to be

$$y = 0.20t^5 - 7.5t^4 + 99t^3 - 594t^2 + 1620t + 5000$$

thousand students in year t, where $t = 0$ represents 1990. The maximum and minimum enrollments will occur at relative maxima and relative minima of this function. Find the maximum and minimum enrollments and the years in which they will occur. [*Hint:* $\frac{dy}{dt} = (t-3)(t-6)^2(t-15)$.]

30. Political advisors for a certain candidate estimate that if x thousand people vote in the next election, their candidate will receive y thousand votes, where

$$y = 10\sqrt{x} - 0.5x - 5.$$

If the maximum number of votes the candidate can receive occurs at a relative maximum of this function, find this maximum number of votes.

3.4 RELATIVE MAXIMA AND MINIMA: THE SECOND DERIVATIVE TEST

In this section we present another method for testing the critical values of a function to find out whether they yield relative maxima or relative minima. The test we will develop, which is called the **second derivative test,** involves substituting the critical values into the second derivative of the function.

In the previous section we noted that if f' is positive on an interval, then f is increasing over the interval. Since this is true for any function and its derivative, let us apply it to the function f' and its derivative f'': if $f''(x) > 0$ for all x in an interval, then f' must be increasing over the interval. But if f' is increasing over the interval, then the slopes of the tangent lines to f are increasing over the interval, as shown in Figure 3–13, for instance. Therefore in order for the graph of f to fit the tangent lines, it must bend upward when $f''(x) > 0$.

Similarly, if the derivative of a function is negative over an interval, then the function is decreasing over the interval. Thus if $f''(x) < 0$ for all x in an interval, then f' is decreasing over the interval. But then the slopes of the tangent lines to f are decreasing, as in Figure 3–14. Thus, in order to fit the tangent lines, the graph of f must bend downward over the interval.

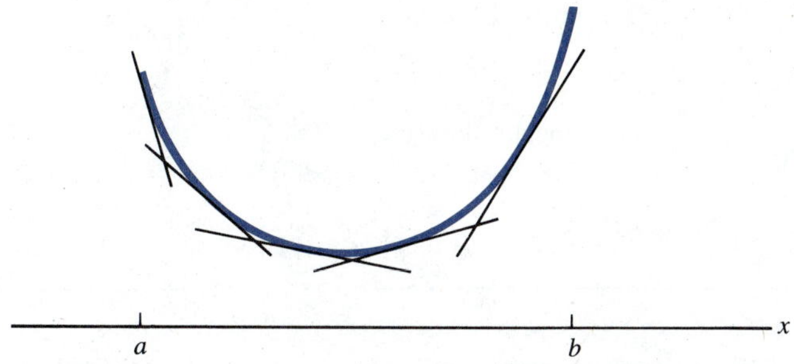

FIGURE 3–13 Slopes of Tangent Lines Increasing on *(a, b)*

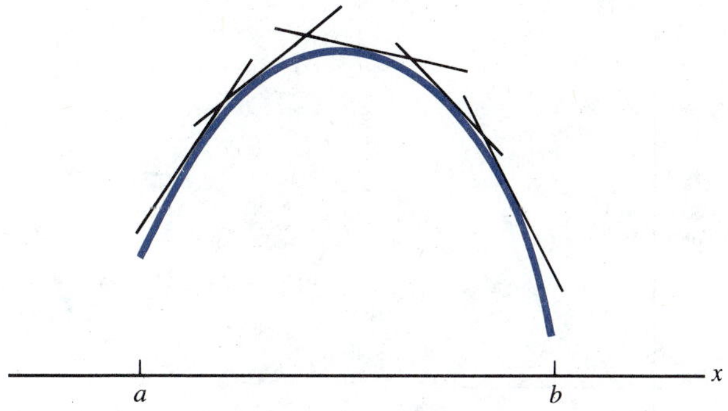

FIGURE 3–14 Slopes of Tangent Lines Decreasing on *(a, b)*

Concavity

1. If $f''(x) > 0$ for all x in an interval, the graph of f bends upward over the interval, as shown in Figure 3–15, and f is said to be **concave upward** over the interval.
2. If $f''(x) < 0$ over an interval, the graph of f bends downward over the interval, as shown in Figure 3–16, and f is said to be **concave downward** over the interval.

EXAMPLE 1 If $f(x) = x^2$, then $f''(x) = 2$. Hence $f''(x)$ is positive for all x, and therefore f is concave upward for all x. See Figure 3–17.

EXAMPLE 2 If $g(x) = \frac{1}{3}x^3 - x^2 - 3x + 2$, then $g''(x) = 2(x - 1)$. Note that if $x < 1$, then $g''(x) < 0$, while if $x > 1$, then $g''(x) > 0$. Thus $g''(x) < 0$ over the interval $(-\infty, 1)$, and $g''(x) > 0$ over the interval $(1, +\infty)$. Therefore g is concave downward over $(-\infty, 1)$ and concave upward over $(1, +\infty)$. See Figure 3–18.

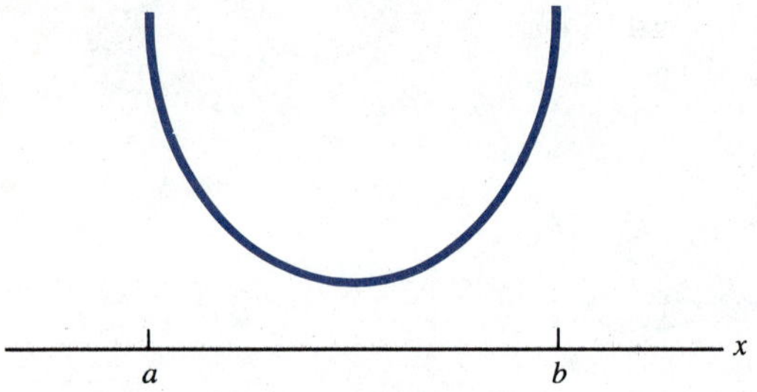

FIGURE 3–15 *f* Concave Upward over *(a, b)*

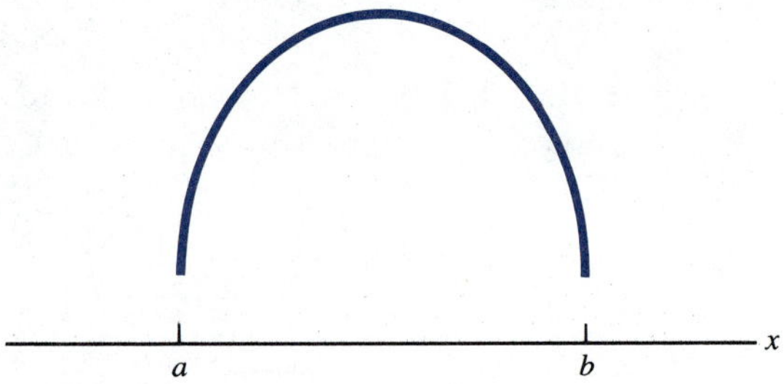

FIGURE 3–16 *f* Concave Downward over *(a, b)*

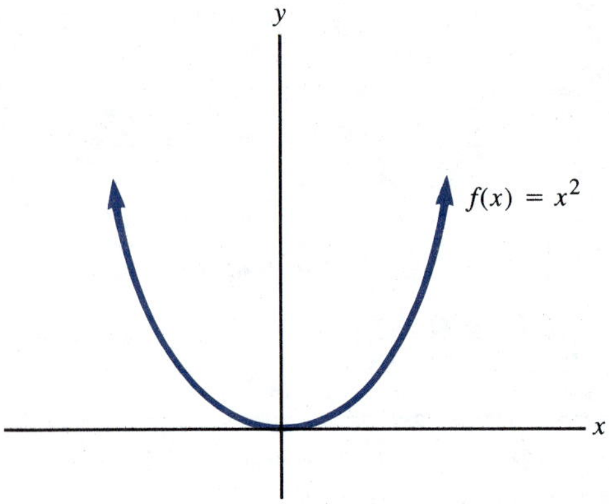

FIGURE 3–17

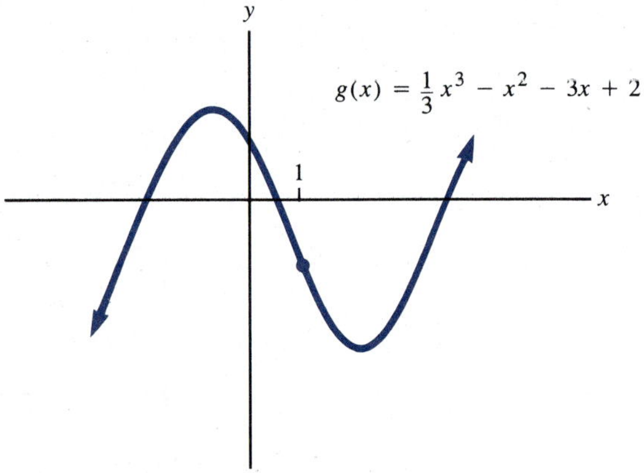

FIGURE 3–18

Now suppose that $f'(a) = 0$, so that $(a, f(a))$ is either a relative maximum, a relative minimum, or a saddle point for $f(x)$. Figure 3–19 depicts these three possibilities. But if $f''(a) > 0$, then $f(x)$ must be concave upward at $(a, f(a))$, so the only one of the three possibilities that can occur is the one in which $(a, f(a))$ is a relative minimum. Similarly, if $f''(a) < 0$, then $(a, f(a))$ must be a relative maximum. This shows that we can sometimes test critical points by substituting them into the second derivative.

The Second Derivative Test

1. If $f'(a) = 0$ and $f''(a) > 0$, then $(a, f(a))$ is a relative minimum point for f.
2. If $f'(a) = 0$ and $f''(a) < 0$, then $(a, f(a))$ is a relative maximum point for f.

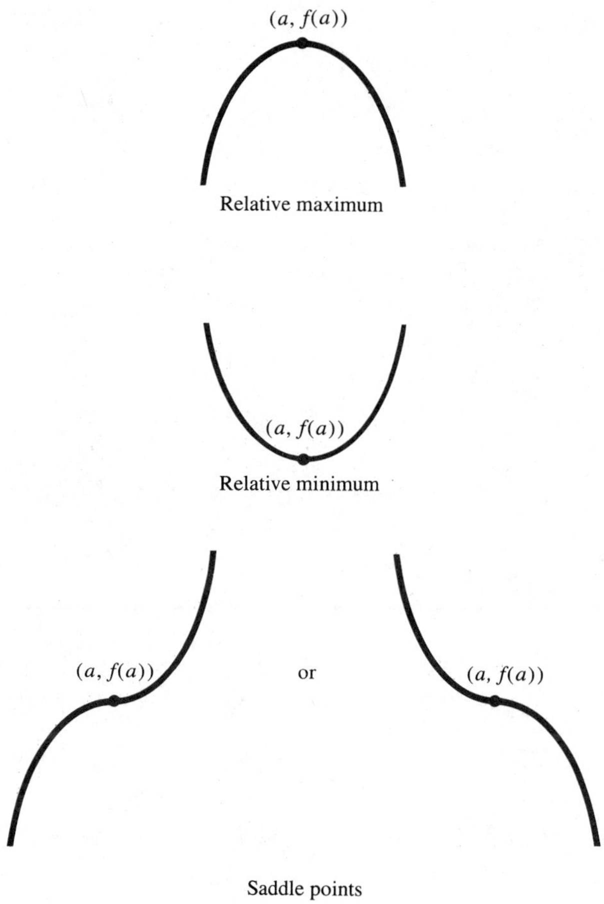

FIGURE 3–19

EXAMPLE 3 Let $f(x) = -2x^3 - 3x^2 + 36x + 5$. We have

$$f'(x) = -6x^2 - 6x + 36 = -6(x + 3)(x - 2)$$

and

$$f''(x) = -12x - 6 = -6(2x + 1).$$

Since $f'(-3) = 0$ and $f''(-3) = 30 > 0$, the point $(-3, f(-3)) = (-3, -76)$ is a relative minimum for f. Since $f'(2) = 0$ and $f''(2) = -30 < 0$, the point $(2, f(2)) = (2, 49)$ is a relative maximum for f.

The second derivative test is often easier to apply than the first derivative test, but it is not as general as the latter. If $f'(a) = 0$ and also $f''(a) = 0$, or if $f'(a) = 0$ and $f''(a)$ does not exist, then the second derivative test tells us nothing about whether $(a, f(a))$ is a relative maximum or a relative minimum. In such cases we must use the first derivative test.

EXAMPLE 4 Let $f(x) = -x^3 + 6x^2 - 12x + 8$. Then

$$f'(x) = -3(x-2)^2 \qquad \text{and} \qquad f''(x) = -6(x-2),$$

so $f'(2) = 0$ and $f''(2) = 0$. Thus the second derivative test cannot be used to classify the point $(2, f(2)) = (2, 0)$ as a relative maximum or a relative minimum for f. The first derivative test shows that $(2, 0)$ is a saddle point for f.

EXAMPLE 5 Let $g(x) = x^{4/3}$. We have

$$g'(x) = \frac{4}{3}x^{1/3} \qquad \text{and} \qquad g''(x) = \frac{4}{9}x^{-2/3}.$$

Therefore $g'(0) = 0$, but $g''(0)$ does not exist, and again the second derivative test cannot be used. The first derivative test shows that the point $(0, g(0)) = (0, 0)$ is a relative minimum.

Computer Manual: Program DERTEST2.

3.4 EXERCISES

In Exercises 1 through 16 use the second derivative test to find all relative maxima and relative minima for the given function. If the second derivative test cannot be applied, use the first derivative test.

1. $f(x) = x^2 - 8x + 2$
2. $f(x) = -x^2 + 6x + 5$
3. $f(x) = 3x^2 - 30x + 1$
4. $f(x) = x^2 - 10x + 25$
5. $f(x) = -2x^3 + 15x^2$
6. $f(x) = x^3 + 12x^2 + 48x + 3$
7. $f(x) = 2x^3 - 15x^2 - 216x + 25$
8. $f(x) = -4x^3 + 6x^2 - 3x + 5$
9. $f(x) = -x^4 + 8x^3 - 22x^2 + 24x$ [*Hint:* $f'(x) = -4(x-1)(x^2 - 5x + 6)$]
10. $f(x) = 3x^4 - 40x^3 + 150x^2 - 10$
11. $f(x) = x^5 - 15x^3$
12. $f(x) = -6x^5 + 45x^4 - 110x^3 + 90x^2 - 50$ [*Hint:* $f'(x) = -30x(x-1)(x^2 - 5x + 6)$]
13. $f(x) = 4x^5 + 15x^4 - 40x^2$ [*Hint:* $f'(x) = 20x(x-1)(x^2 + 4x + 4)$]
14. $f(x) = x^{8/5}$
15. $f(x) = x^{4/5}$
16. $f(x) = x^{5/3}$

Management

17. Caritas, Inc., knows that the demand function for its product is given by

$$p = 10{,}000 - 8x.$$

The company's maximum revenue occurs at a relative maximum of its revenue function. Use the second derivative test to find the maximum revenue.

18. Dismal Company's cost function is given by $C(x) = x^2 + 12x + 900$. The firm's minimum average cost per unit occurs at a relative minimum of its average cost function. Use the second derivative test to find the minimum average cost per unit.

19. Show that average cost per unit is minimized at the value of x for which average cost per unit equals marginal cost. You may assume that the cost function is concave upward for all $x \geq 0$.

Life Science

20. Suppose that t minutes after the injection of a temperature-reducing drug a patient's temperature is $y°$ F, where

$$y = -0.006t^2 + 0.36t + 102.8.$$

The maximum temperature occurs at a relative maximum of this function. How long after the injection of the drug does the maximum temperature occur? What is the maximum temperature?

Social Science

21. Police records for a large city suggest that the city's crime rate may be described by the equation

$$y = 12.5\sqrt{t} + \frac{400}{t}.$$

Here y is the number of crimes per thousand population in year t, and $t = 1$ represents 1970. The minimum crime rate occurs at a relative minimum of this function. Find the minimum crime rate and the year in which it occurred.

3.5 ABSOLUTE MAXIMA AND MINIMA

The largest value a function can assume over an interval is called its **absolute maximum** over the interval; the smallest value it can assume is its **absolute minimum** over the interval. For instance, Figure 3–20 shows the graph of a function that describes the concentration in parts per million of carbon monoxide in the air of a large city during 1986. The absolute maximum of this function is 15 ppm; it occurred on August 1. The absolute minimum of the function is 4 ppm, which occurred on December 31. It is often necessary to find the absolute maximum

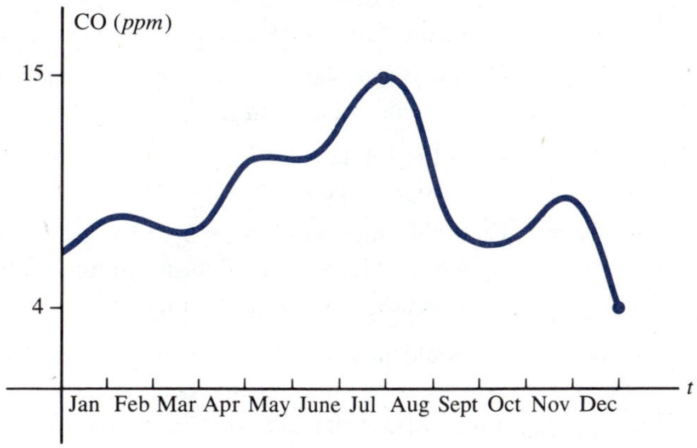

FIGURE 3–20

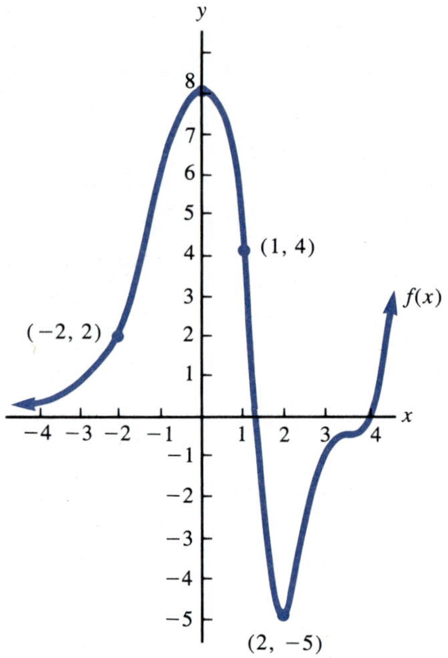

FIGURE 3–21

or absolute minimum of a function. In this section we show how to use the first derivative test to do this.

The existence and value of the absolute maximum and absolute minimum of a function depend not only on the function itself but also on the interval over which it is considered. For instance, consider the function f whose graph is pictured in Figure 3–21:

- Over the entire x-axis f has an absolute minimum of -5 at $x = 2$; however, f has no absolute maximum over the entire x-axis because as x approaches $+\infty$, the functional values $f(x)$ increase without bound.
- Over the interval $[-1, 4]$, f has an absolute minimum of -5 at $x = 2$ and an absolute maximum of 8 at $x = 0$.
- Over the interval $[-2, 1]$, f has an absolute minimum of 2 at $x = -2$ and an absolute maximum of 8 at $x = 0$.
- Over the interval $(-\infty, -2]$, f has an absolute maximum of 2 at $x = -2$; however, f has no absolute minimum over $(-\infty, -2]$ because as x approaches $-\infty$, the functional values $f(x)$ become arbitrarily small.

It should now be clear that the absolute maximum or absolute minimum of a function over an interval may or may not exist; if it does exist, it may occur at either a relative maximum or relative minimum within the interval or at an endpoint of the interval. Therefore in order to find the absolute maximum or absolute minimum of a function over an interval, we must examine its relative maxima or minima and the values of the function at the endpoints of the interval.

Finding the Absolute Maximum and Absolute Minimum

Let f be a function defined on an interval $[a, b]$.

1. The absolute maximum of f over $[a, b]$ is the largest of $f(a)$, $f(b)$, and all the relative maxima that lie within $[a, b]$.
2. The absolute minimum of f over $[a, b]$ is the smallest of $f(a)$, $f(b)$, and all the relative minima that lie within $[a, b]$.

The first derivative test is very useful when we must find absolute maxima and absolute minima, because the diagram of step 3 of the test tells us where to look for them. We illustrate with examples.

EXAMPLE 1 Let

$$f(x) = x^3 - 3x^2 - 45x + 1.$$

Then

$$f'(x) = 3x^2 - 6x - 45 = 3(x + 3)(x - 5),$$

and the diagram of step 3 of the first derivative test looks like this:

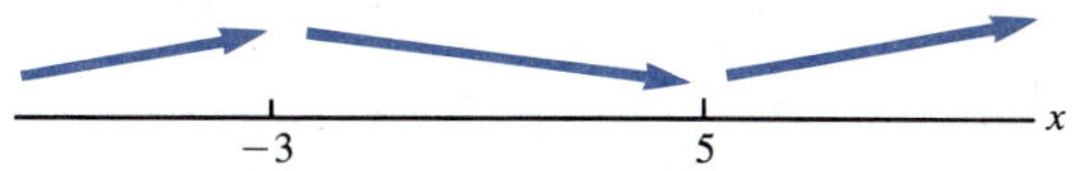

(Check this.) Suppose we want to find the absolute maximum and absolute minimum of f over the interval $[0, 8]$. Then we can only consider the function over this interval, so our diagram becomes

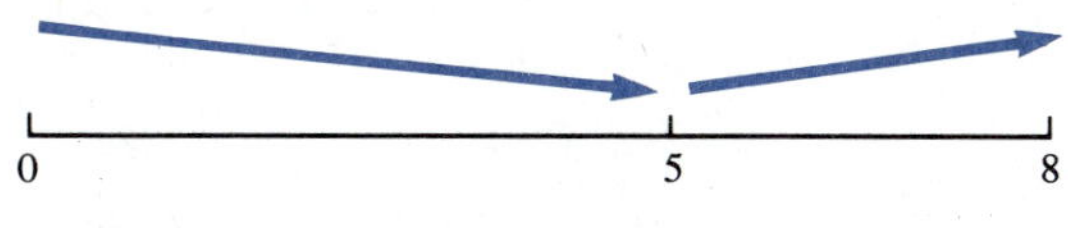

The diagram shows that the absolute minimum must occur at $x = 5$; therefore the absolute minimum is equal to $f(5) = -174$. It also shows that the absolute maximum must occur at either $x = 0$ or $x = 8$. Since $f(0) = 1$ and $f(8) = -39$, the absolute maximum occurs at $x = 0$ and is equal to 1.

EXAMPLE 2 Let f be as in Example 1, but this time let us find its absolute maximum and absolute minimum over the interval $[-5, 6]$. The diagram of step 3 of the first derivative test then becomes

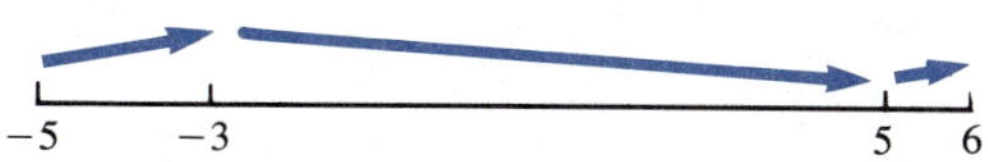

Therefore the absolute maximum of f over the interval $[-5, 6]$ must occur at either $x = -3$ or $x = 6$. Since $f(-3) = 82$ and $f(6) = -161$, the absolute maximum is 82 at $x = 3$. Similarly, the absolute minimum must occur at either $x = -5$ or $x = 5$. Since $f(-5) = 26$ and $f(5) = -174$, the absolute minimum is -174 at $x = 5$.

EXAMPLE 3 Again let f be as in Example 1.

a. Over the interval $[0, +\infty)$ the diagram becomes

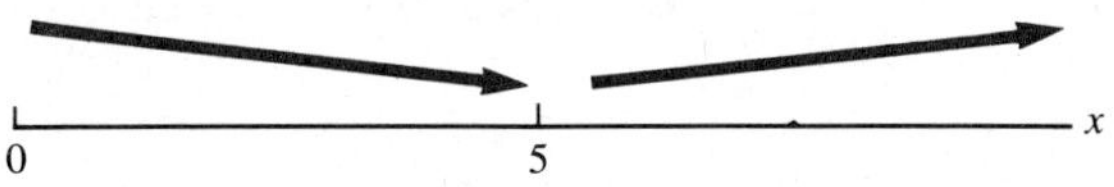

Thus over $[0, +\infty)$ the function has an absolute minimum of -174 at $x = 5$. It has no absolute maximum, since its functional values increase without bound as $x \to +\infty$.

b. Over the entire x-axis the diagram becomes

so over the x-axis the function has neither an absolute maximum nor an absolute minimum. (Why not?)

Now let us turn to some practical problems involving the absolute maximum and absolute minimum. Our first problem concerns profit maximization.

EXAMPLE 4 The profit function of Altruism, Inc., is given by

$$P(x) = -0.01x^3 + 300x - 10{,}000.$$

Let us find Altruism's maximum daily profit if the firm can make at most 400 units of product per day.

We seek to maximize

$$P(x) = -0.01x^3 + 300x - 10{,}000$$

for $0 \le x \le 400$, that is, over the interval $[0, 400]$.

Setting $P'(x) = 0$, we have

$$P'(x) = -0.03x^2 + 300 = 0,$$

which has solutions $x = 100$ and $x = -100$. Since -100 is not in our interval, we discard it, and our diagram is

Therefore the maximum profit is $P(100) = \$10{,}000$. Note that if we had said that Altruism could make at most 50 units per day, then the maximum profit would have been $P(50) = \$3750$. (Why?)

An **inventory problem** is one concerned with minimizing the total cost of purchasing, ordering, and carrying a product in inventory. Total inventory cost is thus the sum of purchasing cost, ordering cost, and carrying cost, where

purchasing cost = (number of units purchased) (price per unit),

ordering cost = (number of orders placed) (cost of placing an order),

and

carrying cost = (average number of units in inventory) times (cost of carrying a unit in inventory).

We solve an inventory problem by finding the **optimal reorder quantity,** which is the number of units that should be ordered each time to minimize total inventory cost.

EXAMPLE 5 Suppose the Shoe Shop sells 2000 pairs of Lazy Loafers each year. Each pair of Lazy Loafers is purchased for \$10, and it costs \$100 to place an order for them. Also, it costs \$1.60 to carry a pair of shoes in inventory for 1 year. Let us find the optimal reorder quantity for this problem.

Let x denote the number of pairs of Lazy Loafers that should be ordered each time an order is made. Since 2000 pairs of Lazy Loafers are required during the course of a year, they must be ordered $2000/x$ times per year. Suppose we assume that pairs of Lazy Loafers are sold at a constant rate and that they sell out just as a new order arrives; then since each order consists of x pairs, the average number of pairs in inventory at any time will be $x/2$. Therefore

$$\text{annual purchasing cost} = (2000 \text{ pairs})(\$10 \text{ per pair}) = \$20{,}000,$$

$$\text{annual ordering cost} = \left(\frac{2000}{x} \text{ orders}\right)(\$100 \text{ per order}) = \frac{\$200{,}000}{x},$$

$$\text{annual carrying cost} = \left(\frac{x}{2} \text{ pairs}\right)(\$1.60 \text{ per pair}) = \$0.80x.$$

Thus if $I(x)$ is the total annual inventory cost that results from ordering x pairs of Lazy Loafers per order, then

$$I(x) = 20{,}000 + \frac{200{,}000}{x} + 0.80x,$$

where $x > 0$. We must minimize the inventory cost function I over the interval $(0, +\infty)$.

Since $$I'(x) = -\frac{200{,}000}{x^2} + 0.80,$$

setting $I'(x) = 0$ and solving for x yields $x = \pm 500$. (Check this.) Because x must be positive, we may discard the critical value -500, and the diagram of the first derivative test looks like this:

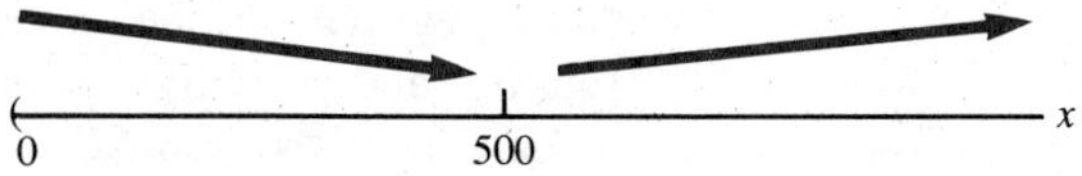

(Check this.) Note that we have placed a parenthesis "(" at $x = 0$ to denote that 0 is not included in our interval. It is clear that the minimum inventory cost occurs at $x = 500$. Therefore the optimal reorder quantity is 500 pairs of Lazy Loafers per order. This reorder quantity will result in 2000/500 = 4 orders per year, and the store's minimum annual inventory cost will be

$$I(500) = \$20{,}000 + \frac{\$200{,}000}{500} + \$0.80(500) = \$20{,}800.$$

Restrictions on the values a variable can assume, such as $0 \leq x \leq 400$ in Example 4 or $x > 0$ in Example 5, are called **constraints.** Many practical problems exhibit constraints that arise from the nature of the problem; this was the case in Examples 4 and 5, for instance. Another type of constraint arises when a problem is modeled using two variables: this constraint relates the variables and allows us to solve for one of them in terms of the other. We illustrate with an example.

EXAMPLE 6 We wish to construct a box with a square bottom and top and rectangular sides. The material for the sides costs \$2.00 per square foot and that for the top and bottom costs \$0.25 per square foot. We intend to spend \$24.00 on materials. Let us find the dimensions that will give us the box of maximum volume.

The box will look like this:

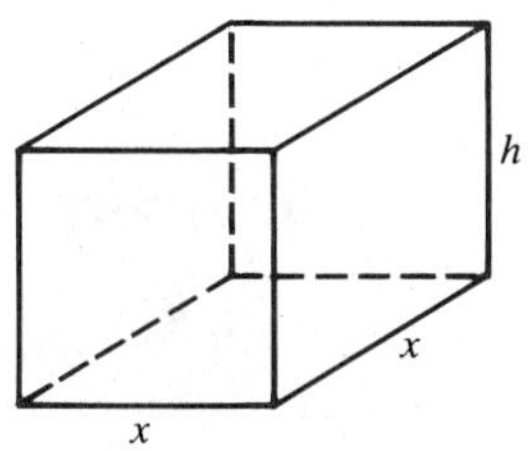

Here x and h are in feet. The volume V of the box is its length times its width times its height, or

$$V = x^2h$$

cubic feet. The cost of the material used to make the box is

$$\begin{aligned}&\$0.25\text{ (area of top + area of bottom)} + \$2.00\text{(area of four sides)}\\ &\quad = \$0.25(2x^2) + \$2.00(4xh)\\ &\quad = \$0.50x^2 + \$8.00xh,\end{aligned}$$

and this must equal \$24.00. Thus we seek to maximize

$$V = x^2h$$

subject to the constraint

$$0.5x^2 + 8xh = 24.$$

In order to apply our methods of maximization to the function V, we must express it in terms of a single variable. The constraint equation allows us to do this, for we may solve

$$0.5x^2 + 8xh = 24$$

for h to obtain

$$h = \frac{24 - 0.5x^2}{8x} = \frac{3}{x} - \frac{x}{16}$$

and then substitute this into the expression for V to get

$$V = x^2\left(\frac{3}{x} - \frac{x}{16}\right) = 3x - \frac{x^3}{16}.$$

Thus we seek to maximize

$$V = 3x - \frac{x^3}{16}$$

where $x > 0$. But

$$\frac{dV}{dx} = 3 - \frac{3x^2}{16} = 0$$

implies that

$$x^2 = 16,$$

so $x = \pm 4$. We may discard the negative critical value. To test the value $x = 4$, we use the second derivative test:

$$\frac{d^2V}{dx^2} = -\frac{6x}{16},$$

so

$$\left.\frac{d^2V}{dx^2}\right|_{x=4} = -\frac{24}{16} < 0,$$

and hence $x = 4$ yields a maximum. Therefore the dimensions that give the box of maximum volume are $x = 4$ feet and

$$h = \frac{3}{4} - \frac{4}{16} = \frac{1}{2}$$

foot, and the maximum volume is

$$V = 4^2\left(\frac{1}{2}\right) = 8$$

cubic feet.

Computer Manual: Program ABSOPT.

3.5 EXERCISES

1. Let f have the following graph:

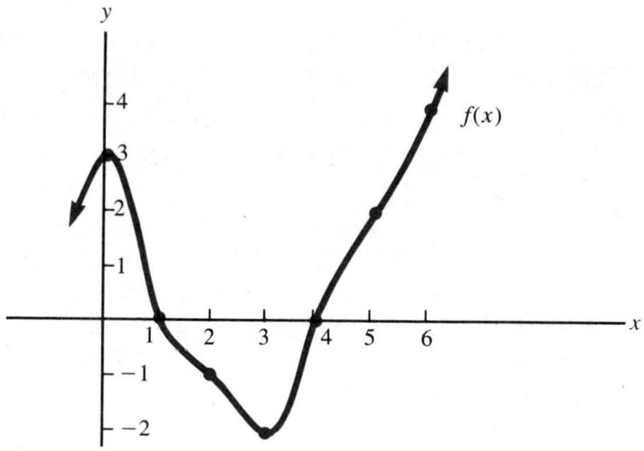

Find the absolute maximum and absolute minimum of f, if they exist, over

(a) the interval $[0, 4]$ **(b)** the interval $[2, 6]$
(c) the interval $(-\infty, 0]$ **(d)** the interval $[3, +\infty)$

2. Let g have the following graph:

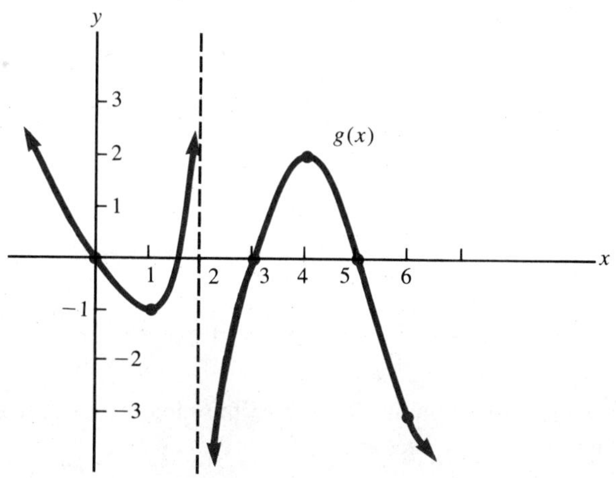

Find the absolute maximum and absolute minimum of g, if they exist, over

(a) the interval $[0, 2)$ **(b)** the interval $(2, 6]$
(c) the interval $[3, 6]$ **(d)** the x-axis

3. Find the absolute maximum and absolute minimum, if they exist, of the function f defined by

$$f(x) = 3x^2 - 18x - 5$$

over

(a) the interval $[0, 5]$ **(b)** the interval $[6, 8]$
(c) the interval $[0, +\infty)$ **(d)** the entire x-axis

4. Find the absolute maximum and absolute minimum, if they exist, of the function g defined by

$$g(x) = 2x^3 + 9x^2 - 60x + 1$$

over the interval

(a) $[-6, 6]$ **(b)** $[-6, 3]$ **(c)** $[1, +\infty)$ **(d)** $(-\infty, 0]$

Management

5. A firm's profit function is given by

$$P(x) = -2x^3 + 270x^2 - 4800x - 6000,$$

where x is in thousands of units and profit is in dollars. Find the firm's maximum profit if it can produce at most 100,000 units; if it can produce at most 50,000 units.

6. Derma Corporation's cost function is given by

$$C(x) = 0.01x^2 + 10x + 40{,}000.$$

Find Derma's minimum average cost per unit if it can produce at most 5000 units; if it can produce at most 1500 units.

7. The demand function for a company's product is defined by

$$p = 1000 - \frac{x}{200}.$$

Find the company's maximum revenue and the price per unit that the company must charge to attain it if the firm can produce at most 200,000 units; if it can produce at most 80,000 units.

8. A chain of furniture stores sells 10,000 sofas each year. Each sofa costs the chain \$500, and it costs \$200 to order sofas from the supplier. It also costs the chain \$100 to carry a sofa in inventory for a year. Assuming an average inventory equal to one half the optimal reorder quantity, find the optimal reorder quantity, the number of orders the chain must make each year, and the minimum annual inventory cost.

9. The Drive N' Shop supermarket sells 100 boxes of soap each week. Each box costs the store \$0.50, each order of soap costs \$2.00, and it costs \$0.01 to carry a box of soap in inventory for one week. Assuming an average inventory equal to one half the optimal reorder quantity, find the optimal reorder quantity, the number of orders required each week, and the minimum weekly inventory cost.

10. Octopus Industries makes radios. The annual demand for its radios is 50,000 units. It costs Octopus Industries \$70 to make a radio, \$2000 to set up a production run of radios, and \$2 to carry a radio in inventory for a year. Assuming an average inventory equal to one half the number of radios produced per production run, find the number of radios that Octopus should produce in each production run, the number of production runs required each year, and the minimum inventory cost for the radios. (*Hint:* Let x denote the number of production runs the company should make each year.)

11. Suppose that the demand for a product is Q units per time period, that it costs K dollars to place each order for the product, and that it costs A dollars to carry 1 unit of the product in inventory for the time period. Assuming that the average inventory is one half the optimal reorder quantity, show that the optimal reorder quantity is

$$x = \sqrt{\frac{2QK}{A}}.$$

12. A promoter can sell 20,000 tickets to a rock concert at \$10 each. For every dollar the price of a ticket is raised, it is estimated that 200 fewer tickets will be sold. Find the promoter's maximum revenue.

Life Science

13. An egg ranch has 120 chickens, each of which produces 250 eggs per year. If more chickens are squeezed into the chicken coop, the resulting overcrowding will reduce egg production by 1 egg per chicken per year for each additional chicken squeezed in. Find the maximum egg production and the number of chickens that will yield this maximum production.

14. Do Exercise 13 if each additional chicken squeezed in reduces egg production by 3 eggs per chicken per year.

15. A lake was stocked with trout in 1965, and t years later there were

$$y = -\frac{1}{3}t^3 + 12.5t^2 - 100t + 1000$$

trout in the lake. Find the maximum and minimum numbers of trout in the lake from 1965 through 1985.

16. If a farmer sows x seeds per square foot, the resulting yield will be y bushels of grain, where

$$y = -2x^3 - 39x^2 + 4620x.$$

Find the farmer's maximum yield if there are enough seeds on hand to sow 30 seeds per square foot; if there are enough to sow 20 seeds per square foot.

17. Poiseuille's law states that (if the units of measurement are properly chosen) the velocity of blood x units from the central axis of an artery of radius r is

$$V(x) = r^2 - x^2.$$

Find the position within the artery where blood flows the fastest.

18. If a blood vessel contracts, the velocity at which blood flows through it is affected. If the units of measurement are properly chosen, the average velocity of the blood flowing through an artery of radius x, where x is less than or equal to the normal radius r, is

$$V(x) = x^2(r - x).$$

Find the amount of contraction that maximizes the average velocity of the blood.

Social Science

19. In a test of spatial perception, a person is told to take a piece of string 20 centimeters long, cut it into two pieces, and bend one of the pieces into a circle and the other into a square in such a way as to minimize the total area of the two figures. Where should the string be cut in order to accomplish this?

20. Do Exercise 19 if the object of the test is to maximize the total area of the figures.

21. Minimize the sum of two positive numbers whose product is 36.

22. Maximize the product of two positive numbers whose sum is 36.

In Exercises 23 through 32 we suggest that you begin by drawing a picture of what is to be maximized or minimized.

23. What are the dimensions of the rectangular field of maximum area that can be fenced in with exactly 800 feet of fencing?

24. Prove that the rectangle of maximum area having a fixed perimeter P is actually a square with side $P/4$.

25. What are the dimensions of the rectangular field whose area is 10,000 square meters if the fencing around the field is to be a minimum?

26. Prove that the rectangle of minimum perimeter having a fixed area A is actually a square with side $\sqrt{A}$.

27. You wish to fence in a rectangular field. One side of the field will be bordered by a stone wall and therefore will not require any fencing. The area of the field must be 12,800 square feet, and fencing costs \$2 per foot. What should the dimensions of the field be in order to minimize the cost of the fencing if the stone wall is 200 feet long?

28. Do Exercise 27 if the stone wall is 100 feet long.

29. The base of a toy chest is to be a rectangle twice as long as it is wide. The material for the top and bottom of the chest costs \$5 per square foot, and that for the sides costs \$3 per square foot. The materials budget is \$240. Find the dimensions that maximize the volume of the toy chest.

30. A box with an open top is to have a square base and a volume of 500 cubic centimeters. If the material for the box costs \$0.50 per square centimeter, find the dimensions of the box that minimize its cost.

31. An open box is to be constructed from a 1-foot square sheet of tin by cutting out squares at the corners and folding the sides up, as shown in the figure:

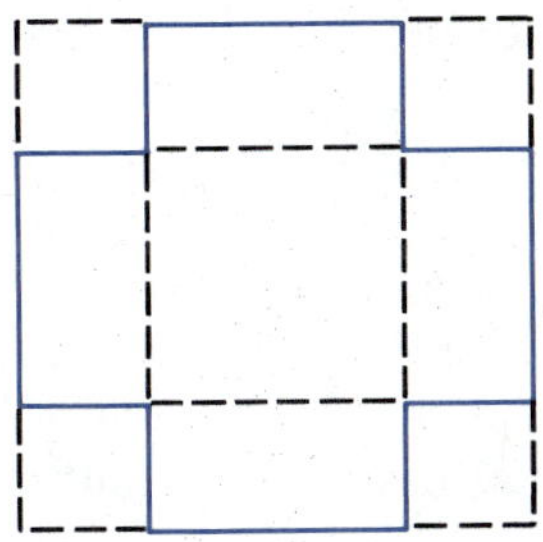

Find the maximum volume of the box.

32. A cylindrical container with a circular base and lid is to have a volume of 54 cubic inches. Find the dimensions that minimize the amount of material from which the container is made.

Management

33. Slicendice Company has \$120,000 to spend on advertising. Each radio ad will cost \$3000, and each newspaper ad \$6000. If Slicendice runs x radio ads and y newspaper ads, it will obtain

$$E = 10{,}000x - 30{,}000xy + 20{,}000y + 6{,}000{,}000$$

effective exposures to its ads. How many of each type of ad should the company run in order to maximize effective exposures?

34. Gruel World Corporation makes regular and deluxe model globes. Each regular globe costs \$4 to make, and each deluxe one costs \$8 to make. The corporation has budgeted \$800 for daily production costs. If it makes x regular and y deluxe globes per day, its daily profit is $P = -x^2 + 3xy - 400$ dollars. How many of each type of globe must the company produce in order to maximize daily profit?

3.6 CURVE SKETCHING

Sometimes a knowledge of the maxima and minima of a function, relative or absolute, is all we need to know in order to answer a question. However, if we want to analyze a function in the manner described earlier, we need to know

- The intervals over which the function is increasing and those over which it is decreasing.
- The location of its maxima, minima, and saddle points.
- The intervals over which the function is concave upward and those over which it is concave downward.

It is also useful to know the intercepts of the function and its asymptotes, if any. Given all this information, we can then make an accurate analysis of the behavior of the function. In this section we collect our results of Sections 3.3 and 3.4 in a procedure that will allow us to analyze functions with relative ease.

Recall that the first derivative test locates relative maxima, minima, and saddle points. The diagram used in the test also tells us the intervals over which the function is increasing and those over which it is decreasing. The second derivative tells us about the concavity of the function: if the second derivative is positive over an interval, the function is concave upward there; if the second derivative is negative, the function is concave downward over the interval. A point where the function changes its concavity is called a **point of inflection.** The function f whose graph is shown in Figure 3–22 has points of inflection at $(a, f(a))$ and $(b, f(b))$.

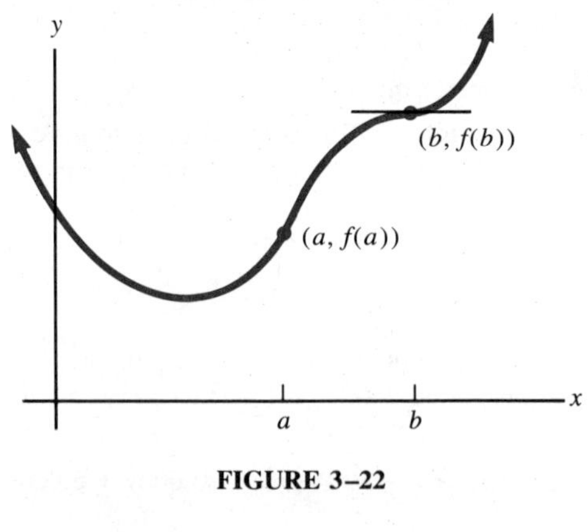

FIGURE 3–22

Note that a saddle point, such as $(b, f(b))$ in Figure 3–22, is a special kind of point of inflection. We can find points of inflection by setting the second derivative equal to zero and solving the resulting equation.

Suppose we use the first and second derivatives of a function to find its relative maxima, its relative minima, its saddle points and other points of inflection, and to determine where the function is increasing, where it is decreasing, where it is concave upward, and where it is concave downward. Suppose we also find the intercepts and asymptotes of the function. All this information will enable us to sketch an accurate graph of the function. Let us outline a procedure for this.

The Curve-Sketching Procedure

To sketch the graph of a function f, proceed as follows:

1. Use the first derivative test to locate all relative maxima, relative minima, and saddle points of the function, and to find the intervals over which it is increasing and those over which it is decreasing.

2. a. Find all solutions of the equation $f''(x) = 0$ and also all values of x for which $f''(x)$ does not exist.

b. Plot the values of x obtained in step 2(a) on the x-axis, thus dividing the x-axis into intervals.

c. Determine the sign of f'' on each interval of part (b). If $f''(x) > 0$ for all x in an interval, f is concave upward over the interval; if $f''(x) < 0$ for all x in an interval, f is concave downward over the interval.

d. Locate all points where the function changes its concavity: these are the points of inflection for f.

3. Find the intercepts of f.

a. If $f(0)$ is defined, the y-intercept is $(0, f(0))$.

b. To find the x-intercepts, set $f(x) = 0$ and solve for x. (In practice, this step is often omitted when the equation $f(x) = 0$ is not easily solved.)

4. Find the asymptotes for f.

a. If the denominator of $f(x)$ is zero and the numerator is nonzero when $x = c$, then the vertical line $x = c$ is an asymptote for f.

b. If

$$\lim_{x \to +\infty} f(x) = K \qquad \text{or} \qquad \lim_{x \to -\infty} f(x) = K,$$

then $y = K$ is an asymptote for f as $x \to +\infty$ or $x \to -\infty$, respectively. (Here K can be either a number or an expression involving x.)

5. Sketch the graph of f using the information obtained in steps 1 through 4.

EXAMPLE 1 Let us use the curve-sketching procedure to draw the graph of the quadratic function defined by

$$f(x) = x^2 - 8x + 15.$$

1. Since $f'(x) = 2x - 8$, the first derivative test yields the following diagram:

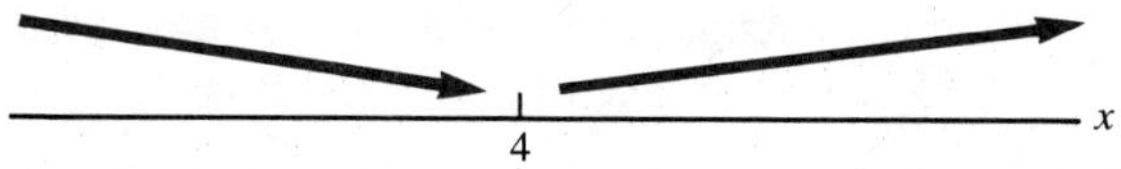

Hence the function has a relative minimum at $(4, f(4)) = (4, -1)$ and is decreasing over the interval $(-\infty, 4)$ and increasing over the interval $(4, +\infty)$.

2. **a.** Since $f''(x) = 2$, the equation $f''(x) = 0$ has no solution. Also, $f''(x)$ exists for all x.
   b. Therefore the diagram of this step is

   c. Since $f''(x) = 2$ is positive for all x, the function is concave upward for all x:

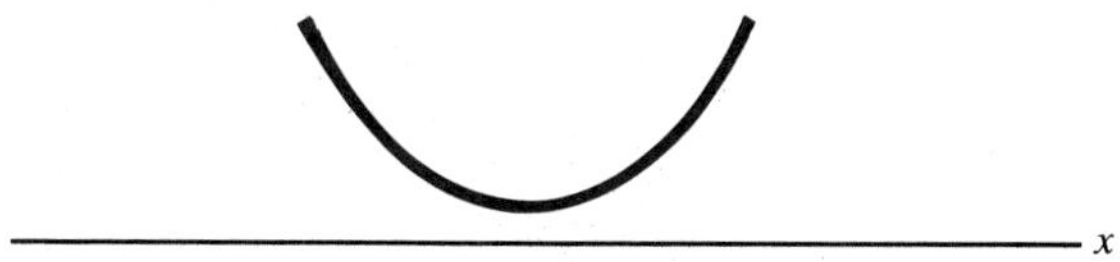

   d. The function never changes concavity, so it has no points of inflection.

3. **a.** The y-intercept is $(0, f(0)) = (0, 15)$.
   b. The x-intercepts occur at the solutions of

$$x^2 - 8x + 15 = 0,$$

or

$$(x - 3)(x - 5) = 0,$$

and thus are $(3, 0)$ and $(5, 0)$.

4. The function is a polynomial, and polynomials have no asymptotes.
5. The graph of the function is shown in Figure 3–23. Notice how each piece of information determined in steps 1 through 4 has been used to sketch the graph.

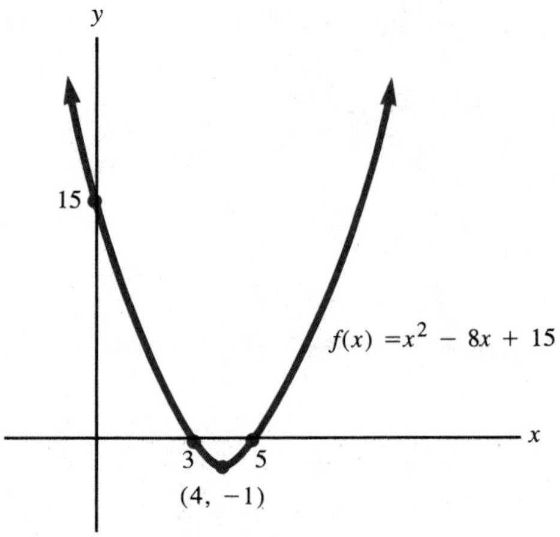

FIGURE 3–23

EXAMPLE 2 Let us sketch the graph of the function defined by

$$f(x) = x^3 - 3x^2 - 24x + 3.$$

1. Since

$$f'(x) = 3x^2 - 6x - 24 = 3(x + 2)(x - 4),$$

the diagram of the first derivative test is

Hence the function

has a relative maximum at $(-2, f(-2)) = (-2, 31)$,

has a relative minimum at $(4, f(4)) = (4, -77)$,

is increasing on the intervals $(-\infty, -2)$ and $(4, +\infty)$,

is decreasing on the interval $(-2, 4)$.

2. a. Since

$$f''(x) = 6x - 6 = 6(x - 1),$$

setting $f''(x) = 0$ and solving for x yields $x = 1$. The second derivative exists for all x.

b. Hence we have the diagram

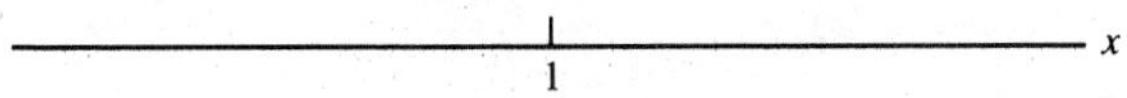

c. Since $f''(x) < 0$ if $x < 1$, f is concave downward on $(-\infty, 1)$. Since $f''(x) > 0$ if $x > 1$, f is concave upward on $(1, +\infty)$. Thus we obtain the diagram

d. Since f changes concavity at $(1, f(1)) = (1, -23)$, this is a point of inflection for the function.

3. **a.** The y-intercept is $(0, f(0)) = (0, 3)$.
b. The x-intercepts occur at the values of x for which

$$x^3 - 3x^2 - 24x + 3 = 0.$$

Since this equation is not easily solved, we omit this step.

4. The function is a polynomial and hence has no asymptotes.

5. The function is graphed in Figure 3–24.

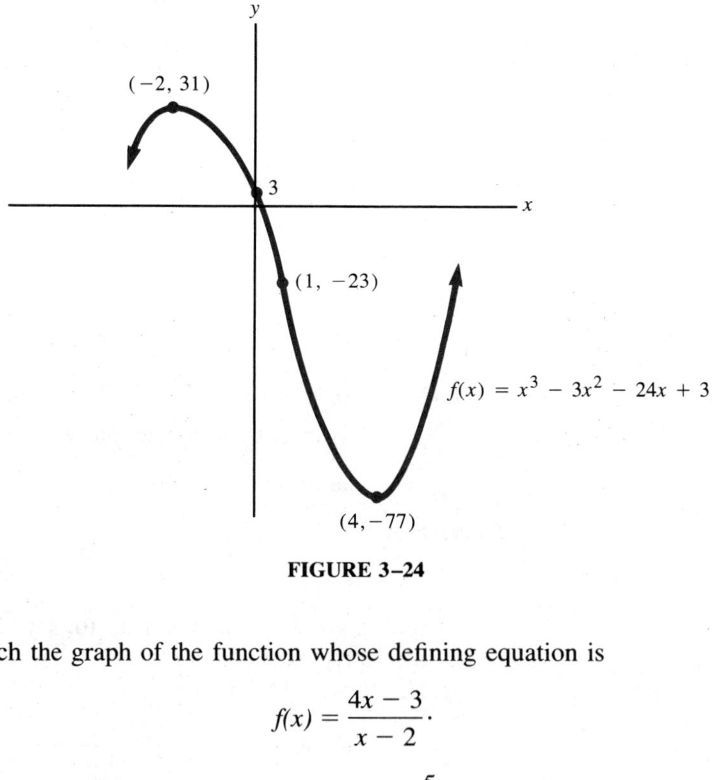

FIGURE 3–24

EXAMPLE 3 We sketch the graph of the function whose defining equation is

$$f(x) = \frac{4x - 3}{x - 2}.$$

1. Since
$$f'(x) = -\frac{5}{(x-2)^2},$$

the equation $f'(x) = 0$ has no solution. Since $f'(2)$ does not exist, the only critical value is $x = 2$, and we have the diagram

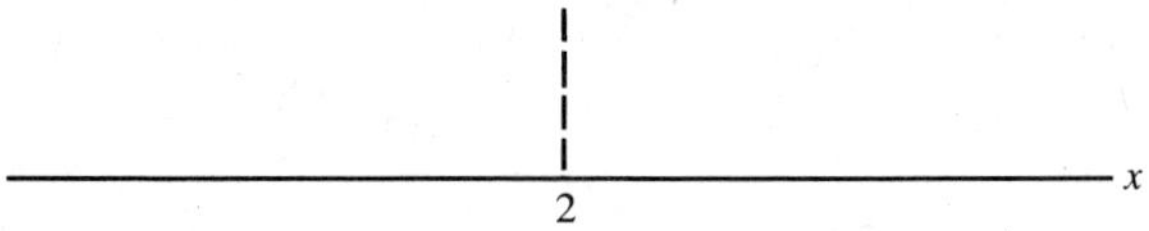

Note the vertical dotted line at $x = 2$: this reminds us that $f(2)$ is not defined, so the graph cannot intersect the line $x = 2$. Since $f'(x) < 0$ for all values of x, the function is decreasing on the intervals $(-\infty, 2)$ and $(2, +\infty)$:

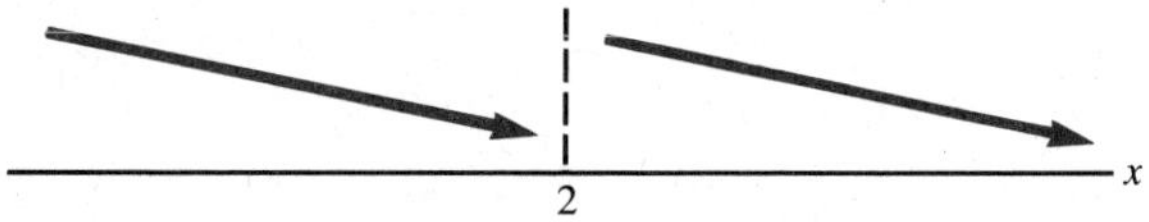

2. a. Since
$$f''(x) = \frac{10}{(x-2)^3},$$
the equation $f''(x) = 0$ has no solution. The second derivative does not exist at $x = 2$.

b. Our diagram for this step is

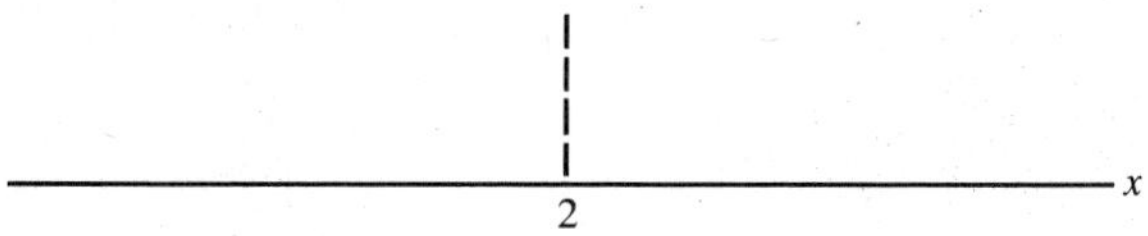

Again, note the dotted line at $x = 2$.

c. Since $f''(x) < 0$ if $x < 2$ and $f''(x) > 0$ if $x > 2$, the function is concave downward on the interval $(-\infty, 2)$ and concave upward on the interval $(2, +\infty)$:

d. The function has no points of inflection. (Why not?)

3. a. The y-intercept is $(0, f(0)) = (0, 3/2)$.

b. The x-intercepts occur at the values of x for which
$$\frac{4x-3}{x-2} = 0.$$
The only solution of this equation is $x = 3/4$. Thus the function's single x-intercept is $(3/4, 0)$.

4. **a.** When $x = 2$, the denominator of $f(x)$ is zero but the numerator is not zero. Therefore the line $x = 2$ is a vertical asymptote for f.

   b. Since

$$\lim_{x\to\pm\infty} \frac{4x-3}{x-2} = \lim_{x\to\pm\infty} \frac{4-3(1/x)}{1-2(1/x)} = \frac{4-3(0)}{1+2(0)} = 4,$$

   the line $y = 4$ is a horizontal asymptote for f as $x \to +\infty$ and as $x \to -\infty$.

5. The graph of the function is shown in Figure 3–25.

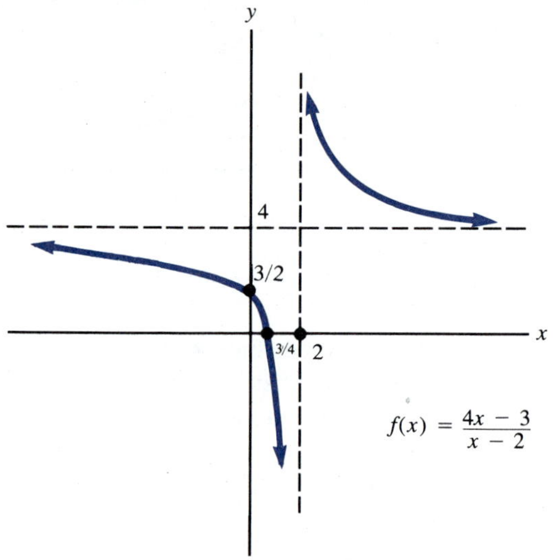

FIGURE 3–25

EXAMPLE 4 Suppose that Acme Company's profit function is given by

$$P(x) = -x^3 + 36x^2 - 240x - 100,$$

where x is in thousands of units and profit is in thousands of dollars. Let us analyze Acme's profits. We begin by sketching the graph of the profit function. Since x must be nonnegative, we can restrict our diagrams to the interval $[0, +\infty)$. We will leave out some of the details of the curve-sketching procedure: you should check these details.

1. The diagram of the first-derivative test is

   The relative maximum occurs at (20, 1500), and the relative minimum occurs at (4, −548).

2. The concavity diagram is

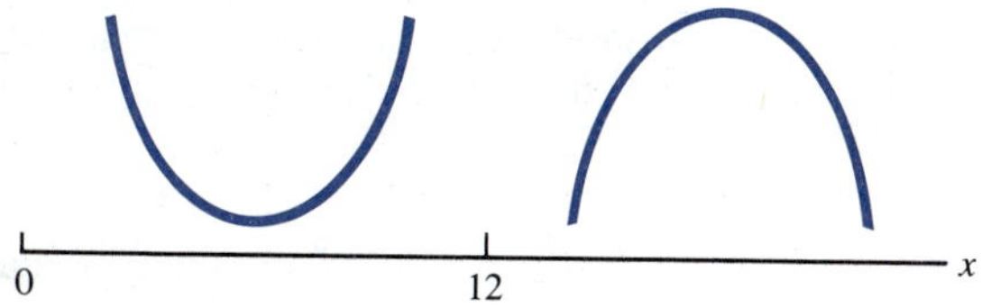

The point of inflection is (12, 476).

3. The y-intercept is (0, −100).
4. There are no asymptotes.
5. The graph of Acme's profit function is shown in Figure 3–26.

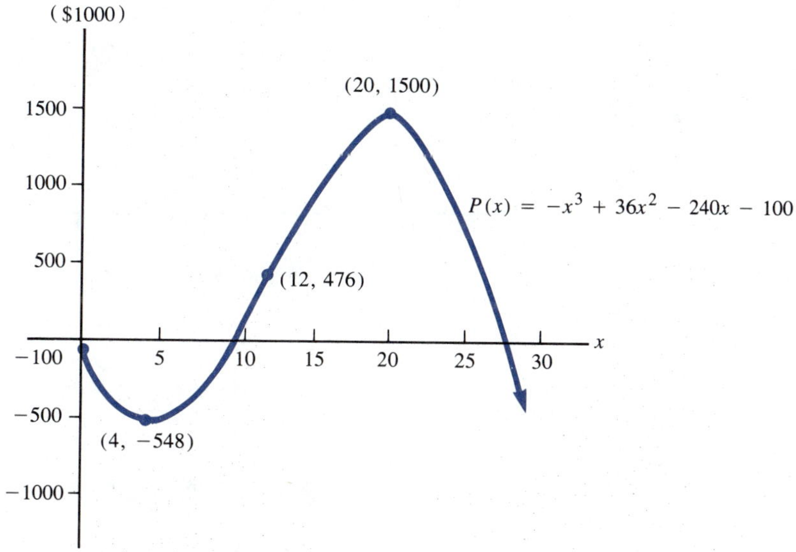

FIGURE 3–26

Now let us use the graph to analyze Acme's profit. Starting at $x = 0$ and moving to the right along the x-axis, we observe the following:

a. Since $P(0) = -100$, when Acme produces 0 units of its product it suffers a loss of $100,000. Thus Acme's fixed cost is $100,000.

b. As production increases from 0 units to 4000 units, Acme's profit decreases, but at a slower and slower rate, until it reaches a minimum of −$548,000 at 4000 units.

c. As production increases from 4000 units to 12,000 units, profit increases, and it does so at a faster and faster rate as the point of inflection is approached.

d. At a production level of 12,000 units, Acme's profit is $476,000. As production increases from 12,000 units to 20,000 units, profit continues to increase, but at a slower and slower rate, until it reaches a maximum of \$1.5 million at 20,000 units.

e. If production increases beyond 20,000 units, profit will decrease, and it will do so at a faster and faster rate as production goes up; eventually losses will begin to accrue at an ever-increasing rate.

Note also that careful drawing of the graph shows us that Acme will break even somewhere between 9000 and 10,000 units, and again somewhere near 27,000 units. The exact break-even quantities are the solutions of the equation

$$-x^3 + 36x^2 - 240x - 100 = 0.$$

The Newton–Raphson method of Section 2.6 applied to this equation yields approximate solutions $x = 9.4327$ and $x = 26.9605$. Thus the break-even quantities to the nearest unit are 9433 and 26,961.

Computer Manual: Program CURSKECH.

3.6 EXERCISES

In Exercises 1 through 21 use the curve-sketching procedure to sketch the graph of the given function.

1. $f(x) = x^2 - 2x - 15$

2. $f(x) = -2x^2 + 18x - 28$

3. $f(x) = 2x^3 + 3x^2 - 12x + 2$

4. $f(x) = -2x^3 + 27x^2 - 108x + 12$

5. $f(x) = \frac{1}{3}x^3 + 4x^2 + 16x + 2$

6. $f(x) = -x^3 + 12x^2 - 45x$

7. $f(x) = x^4 - 32x$

8. $f(x) = x^4 - 8x^2 + 1$

9. $f(x) = 6x^5 + 15x^4 + 10x^3$

10. $f(x) = \frac{1}{5}x^5 - \frac{8}{3}x^3 + 7x$

11. $f(x) = x^{2/3}$

12. $f(x) = x^{4/3}$

13. $f(x) = \frac{1}{x}$

14. $f(x) = \frac{2}{x^2}$

15. $f(x) = \frac{3}{x^3}$

16. $f(x) = \frac{1}{x - 1}$

17. $f(x) = \frac{2}{(x + 1)^2}$

18. $f(x) = \frac{x - 3}{x + 2}$

19. $f(x) = \frac{2x - 1}{3x + 2}$

20. $f(x) = \frac{x}{x^2 - 1}$

21. $f(x) = \frac{x^2 - 1}{x}$

22. Sketch the graph of the quadratic function defined by

$$y = ax^2 + bx + c$$

when $a > 0$ and when $a < 0$.

Management

23. Brawny Corporation's profit function is given by

$$P(x) = -x^3 + 90x^2 - 1500x - 9,$$

where x is in thousands of units and profit is in thousands of dollars. Sketch the graph of the profit function and analyze Brawny's profit in the manner of Example 4.

24. The average cost function of Gormless, Inc., is given by

$$A(x) = 2x + \frac{50}{x}.$$

Here x is in hundreds of units and $A(x)$ is in dollars. Draw the graph of A and discuss Gormless's average cost per hundred units produced.

25. Housing starts in a city are described by the equation

$$y = -3t^4 + 80t^3 - 750t^2 + 3000t + 500,$$

where t is in years, with $t = 0$ representing 1974, and y is the number of new houses started in the city in year t. Draw the graph and analyze housing starts in the city from 1974 through 1986. (*Hint:* $t - 10$ and $(t - 5)^2$ are factors of the derivative of y with respect to t.)

26. An economist estimates that if inflation is q percent per year, then families will save approximately s percent of their income, where

$$s = -0.001q^3 + 0.06q^2 - 1.2q + 12.$$

Draw the graph and analyze the effect of inflation on savings.

Life Science

27. The number of deer in a region is

$$y = -t^3 + 18t^2 - 96t + 2000,$$

where $t = 0$ represents 1976, t in years. Sketch the graph and discuss the changes in the deer population from 1976 through 1986. At one time deer hunting was banned in the region for a 4-year period. Can you tell when this occurred?

Social Science

28. The number of children attending public school in a large city in year t is

$$y = -\frac{t^4}{4} + \frac{14t^3}{3} - 20t^2 + 200,$$

where $t = 0$ represents 1985, and y is in thousands. Draw the graph and analyze enrollment from 1985 through 1997. Do you think it would make sense to sell school buildings in 1988 because of declining enrollment?

29. A developing nation intends to mount a campaign against illiteracy among its citizens. The plan is that t years after the campaign starts the proportion of the population that is literate should be L, where

$$L = \frac{t + 1}{t + 5}.$$

Draw the graph and discuss the campaign.

3.7 DIFFERENTIALS

In Section 2.7 we discussed the use of differential notation dy/dx for the derivative and stated that for the time being we would regard this expression not as a fraction but merely as another symbol for the derivative. Now it is time to demonstrate how the symbol dy/dx can indeed be regarded as the ratio of the **differentials** dy and dx. This view of the derivative as a ratio will be extremely useful to us in later chapters when we discuss the process of integration. We also show in this section how differentials can be used to make approximations.

The Differentials *dy* and *dx*

Let f be a function and let $(x, f(x))$ be a point on its graph. Let Δx be a number added to x. (The symbol Δ is the Greek letter capital delta, and Δx is read "delta x.") The number Δx is called an **increment** and is referred to as the **change in** $\boldsymbol{x}$; Δx may be positive or negative. Figure 3–27 depicts the situation if Δx is positive.

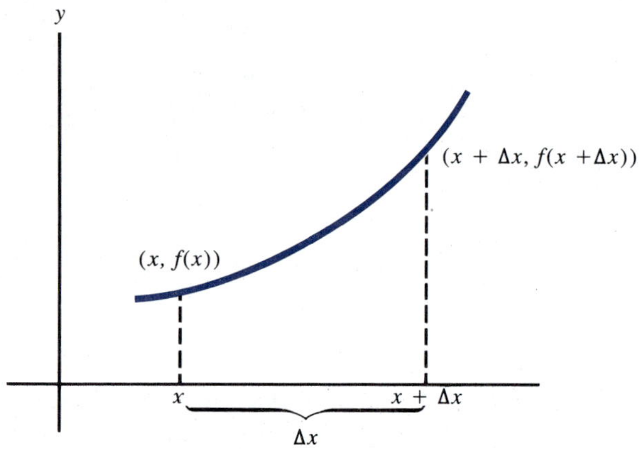

FIGURE 3–27

Now suppose we set $dx = \Delta x$ and dy equal to the vertical distance from the point $(x + \Delta x, f(x))$ to the tangent line to f at $(x, f(x))$. The number dx is the **differential of** $\boldsymbol{x}$, and dy is the **differential of** $\boldsymbol{y}$. See Figure 3–28. Since dy/dx is the slope of the tangent line, and since this slope is the derivative $f'(x)$, we have

$$\frac{dy}{dx} = f'(x).$$

This shows how the derivative may be regarded as a ratio of the differentials dy and dx.

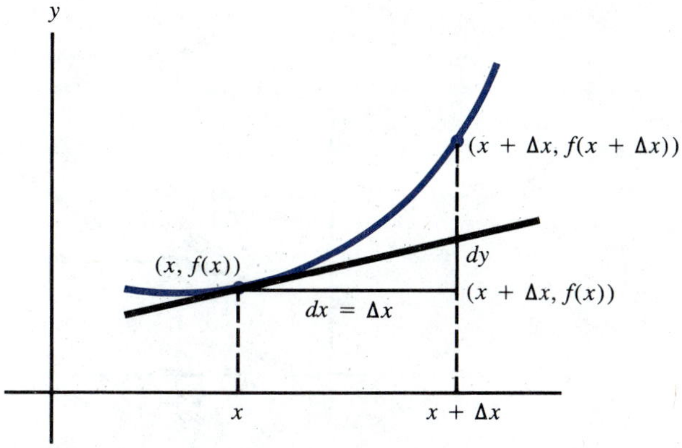

FIGURE 3–28

Now that we can regard dy and dx as individual quantities, we can solve the equation

$$\frac{dy}{dx} = f'(x)$$

for dy to obtain the differential formula:

> **The Differential Formula**
>
> If f is differentiable at x, then
>
> $$dy = f'(x)\,dx.$$

EXAMPLE 1 If $f(x) = x^2 - 5x + 3$, then $dy = (2x - 5)\,dx$.

Approximation by Differentials

As before, we let Δx be an increment added to x. The point $(x + \Delta x, f(x + \Delta x))$ is on the graph of f, and the quantity

$$\Delta f = f(x + \Delta x) - f(x)$$

is the **change in f** brought about by the change Δx in x. See Figure 3–29.

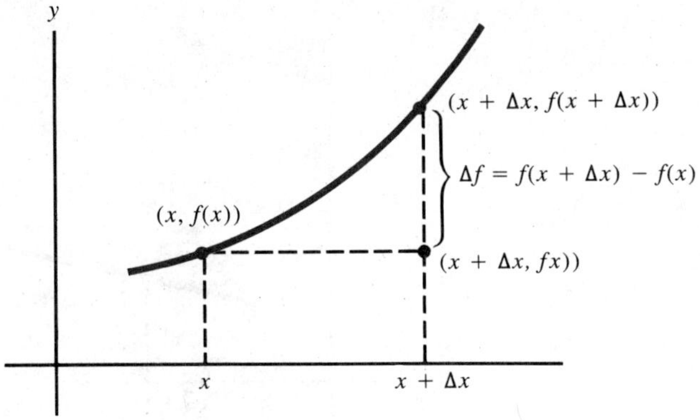

FIGURE 3–29

EXAMPLE 2 Let $f(x) = x^2$, $x = 1$, and take $\Delta x = 0.1$. Then

$$x + \Delta x = 1.1,$$

and thus

$$\Delta f = f(1.1) - f(1) = (1.1)^2 - 1^2 = 0.21.$$

See Figure 3–30.

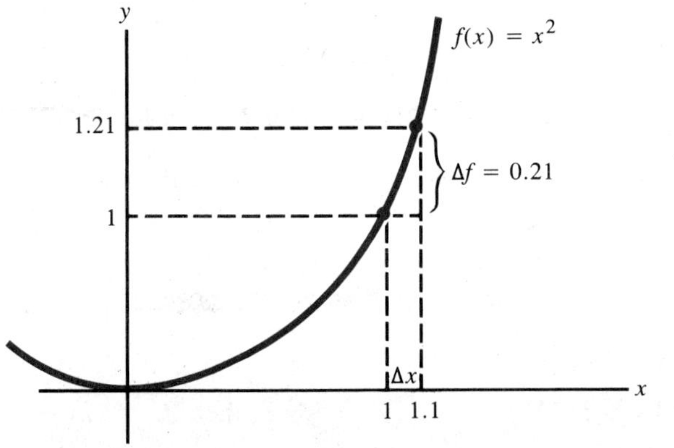

FIGURE 3–30

What happens if the increment Δx is allowed to approach zero? Figure 3–31 shows that as Δx approaches 0, the point

$$(x + \Delta x, f(x + \Delta x)) = (x + \Delta x, f(x) + \Delta f)$$

"slides" down the graph toward the point (x, y) and hence Δf approaches dy. Therefore for values of Δx that are close to (but not equal to) zero, we have

$$\Delta f \approx dy.$$

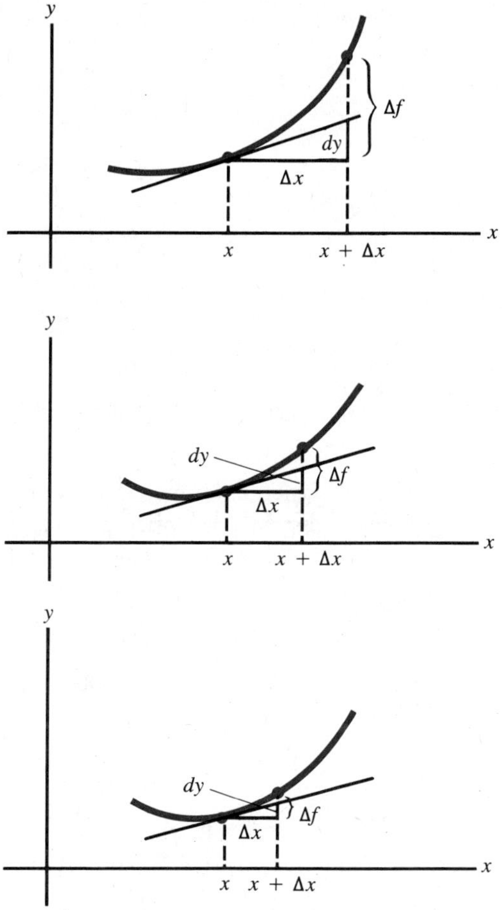

FIGURE 3–31 As $\Delta x \to 0$, $\Delta f \to dy$

This fact can be used to approximate functional values. If $f(x)$ is known and $\Delta x = dx$ is small and nonzero, then

$$f(x + \Delta x) = f(x) + \Delta f \approx f(x) + dy = f(x) + f'(x)\,dx = f(x) + f'(x)\,\Delta x.$$

Approximation by Differentials

If f is differentiable at x and Δx is close to but not equal to zero, then

$$f(x + \Delta x) \approx f(x) + f'(x)\,\Delta x.$$

EXAMPLE 3 Let us use differentials to approximate $\sqrt[4]{16.5}$. We let

$$f(x) = \sqrt[4]{x} = x^{1/4},$$

and set $x = 16$ and $\Delta x = 0.5$. Then

$$f(16) = \sqrt[4]{16} = 2,$$

and since $f'(x) = x^{-3/4}/4$

$$f'(16) = \frac{1}{4}(16)^{-3/4} = \frac{1}{4}(\sqrt[4]{16})^{-3} = \frac{1}{4}(2)^{-3} = \frac{1}{32}.$$

Thus

$$\sqrt[4]{16.5} = f(16 + 0.5) \approx f(16) + f'(16)(0.5) = 2 + \frac{1}{32}(0.5) = 2.015625.$$

EXAMPLE 4 The supply function for a commodity is given by $p = 3q^{2/3}$. Let us use differentials to find the approximate price per unit when the quantity demanded is 990 units.

Let $p = f(q) = 3q^{2/3}$ and set $q = 1000$ and $\Delta q = -10$. Then $f'(q) = 2q^{-1/3}$ and

$$\begin{aligned} p = f(990) &= f(1000 + (-10)) \approx f(1000) + f'(1000)(-10) \\ &= 3(1000)^{2/3} + 2(1000)^{-1/3}(-10) = 3(10)^2 + 2(10)^{-1}(-10) \\ &= 300 - 2 = 298 \end{aligned}$$

dollars.

The Proportional Change in $f(x)$

As Figure 3–29 shows, when x changes by an amount Δx, the functional value $f(x)$ changes by an amount Δf. The **proportional change in $f(x)$** is thus defined to be the ratio

$$\frac{\Delta f}{f(x)}.$$

Again using the fact that $\Delta f \approx dy$ when Δx is close to zero, we have

The Proportional Change in $f(x)$

If f is differentiable at x and Δx is close to but not equal to zero, then

$$\frac{\Delta f}{f(x)} \approx \frac{f'(x)\,\Delta x}{f(x)}.$$

If we multiply the proportional change in $f(x)$ by 100, we obtain the percentage change in $f(x)$.

EXAMPLE 5 One side of a square plot of land is measured to be 10 meters long, with a possible measurement error of ± 0.1 meter. Find the proportional and percentage errors that might occur in the value for the area of the plot.

The area of the plot is given by $A(x) = x^2$, where x is the length of one of its sides. The proportional change in area that could be introduced by a change of ± 0.1 in the value for the length of a side when $x = 10$ is thus

$$\frac{\Delta A}{A(x)} \approx \frac{A'(10)(\pm 0.1)}{A(10)} = \frac{2(10)(\pm 0.1)}{100} = \pm 0.02.$$

Hence the proportional error is ± 0.02, and the percentage error is ± 2 percent.

3.7 EXERCISES

The Differentials dy and dx

In Exercises 1 through 6 find the differential dy.

1. $f(x) = 3x^2 - 2x + 5$

2. $f(x) = (x^2 - 3x + 12)^5$

3. $y = x^3 - 4x^2 - 4x^{-3/4}$

4. $y = x + \frac{1}{x}$

5. $y = (3x^2 - 2x + 1)(x^2 + 4)^4$

6. $f(x) = \frac{x^2 - 1}{x^3 + 2x}$

Approximation by Differentials

In Exercises 7 through 12 use differentials to approximate the given quantity.

7. $\sqrt{15}$

8. $\sqrt{36.8}$

9. $\sqrt[3]{28}$

10. $\sqrt[3]{8.1}$

11. $\sqrt[5]{32.4}$

12. $\sqrt[5]{245}$

Management

13. The demand function for a product is given by $p = 5000q^{-1/2}$. Use differentials to find the approximate price per unit when quantity demanded is

(a) 2510 units;
(b) 3580 units.

14. The supply function for a commodity is given by $q = 200(p - 1)^{4/3}$. Use differentials to find the approximate quantity supplied when the price per unit is

(a) \$8.50 **(b)** \$28.25.

15. A firm's cost function is given by

$$C(x) = 30(x + 50)^{3/2} + 200.$$

The firm is currently making 350 units per day. Find the approximate additional cost if it makes 365 units per day. [*Hint:* Use the approximation by differentials formula in the form $f(x + \Delta x) - f(x) \approx f'(x)\,\Delta x$.]

16. A firm's weekly output of goods is worth $y = 20x^{1/5}$ thousand dollars, where x is the amount (in thousands of dollars) that the firm spends on labor during the week. If the firm is currently spending \$32,000 per week on labor, find the approximate change in the weekly output if new hires increase labor costs to \$35,000 per week.

Life Science

17. The number of bacteria in a culture t hours after an experiment has begun is given by

$$f(t) = 200t^{3/2} + 4000.$$

Use differentials to find the approximate number of bacteria in the culture 245 minutes after the beginning of the experiment.

18. The concentration of pollutants in a reservoir as of year t is given by

$$y = 10 + 2t - 0.5\sqrt{t + 3}.$$

Here y is in parts per million and $t = 0$ corresponds to January 1, 1986. Find the approximate change in the concentration of pollutants during the first three months of 1987.

Social Science

19. A rumor is started, and t days later $f(t)$ people have heard it, where

$$f(t) = 200\sqrt{t} + 1.$$

Use differentials to find the approximate number of people who have heard the rumor 9.5 days after it was started.

20. Referring to Exercise 19, find the approximate number of people who hear the rumor during the fifth day after it is started.

The Proportional Change in $f(x)$

21. The top of a square table is measured, and its length is 2 meters, with a possible error in the measurement of ± 1 centimeter (± 0.01 meter). Find the approximate proportional and percentage errors in both the area and the perimeter of the tabletop due to measurement error.

22. The radius of a circle is measured to be 10 inches, with a possible error of ± 0.2 inches. Find the approximate proportional and percentage errors in both the circumference and the area of the circle due to measurement error.

23. The radius of a sphere is measured to be 2.5 meters, with a possible error of ± 1 centimeter. Find the approximate proportional and percentage errors in both the volume ($4\pi r^3/3$) and the surface area ($4\pi r^2$) of the sphere due to measurement error.

24. Find the approximate proportional and percentage changes in the area of a square if the length of each of its sides is increased by 1 percent.

25. Find the approximate proportional and percentage changes in the volume of a cube if the length of each of its sides is increased by 1 percent.

26. Find the approximate proportional and percentage changes in the area of a circle if the length of its radius is decreased by 2 percent.

27. Find the approximate proportional and percentage changes in the volume of a sphere if the length of its radius is decreased by 1 percent.

Management

28. A firm's cost function is given by

$$C(x) = 0.5x^2 + 5x + 2000$$

and its revenue function by

$$R(x) = x^2 + 4x.$$

Find the proportional and percentage changes in cost and in revenue of making and selling 1 additional unit if the firm is currently producing 25 units.

29. A firm's profit function is given by

$$P(x) = -x^2 + 80x - 200.$$

If the firm is currently producing and selling 20 units, find the proportional and percentage changes in profit if 2 more units are produced and sold; if 1 fewer unit is produced and sold.

3.8 IMPLICIT DIFFERENTIATION

A defining equation of the form $y = f(x)$ is said to define y **explicitly** as a function of x. For instance, the equation $y = x^2 + 3x - 5$ defines y explicitly as a function of x. Sometimes, however, an equation involving x and y will not be solved for y in terms of x; examples are

$$4x + 2y = 12 \qquad \text{and} \qquad x^2 + y^2 = 1.$$

Such equations are said to define y **implicitly** as a function of x. It is sometimes necessary to be able to find the derivative of y with respect to x when y is defined implicitly as a function of x. In this section we show how to use **implicit differentiation** to accomplish this.

Equations that define y implicitly as a function of x can sometimes be solved to yield one or more explicitly defined functions. For example, the equation

$$4x + 2y = 12$$

can be solved for y to obtain the explicitly defined function

$$y = -2x + 6.$$

Similarly, the equation

$$x^2 + y^2 = 1$$

can be solved for y to yield the two explicitly defined functions

$$y = +\sqrt{1 - x^2} \qquad \text{and} \qquad y = -\sqrt{1 - x^2}.$$

However, not all implicitly defined functions can be converted to explicit ones, and even when such conversion is possible, it is often far too tedious to carry out. Therefore we determine the derivatives of implicitly defined functions by the method of **implicit differentiation,** which requires that we apply the rules of differentiation to all terms of the defining equation while keeping in mind the fact that y is actually a function of x. We illustrate with examples.

EXAMPLE 1 Let us find the derivative dy/dx if y is defined implicitly as a function of x by the equation

$$xy + y^2 = 1.$$

To remind ourselves that y is indeed a function of x, we will replace y by $f(x)$ in the equation:

$$xf(x) + [f(x)]^2 = 1.$$

Now we differentiate each term of this equation, using the product rule to differentiate the term $xf(x)$ and the chain rule to differentiate the term $[f(x)]^2$. We have

$$\frac{d}{dx}(xf(x)) + \frac{d}{dx}([f(x)]^2) = \frac{d}{dx}(1),$$

or

$$x\frac{d}{dx}(f(x)) + \frac{d}{dx}(x) \cdot f(x) + 2[f(x)]\frac{d}{dx}(f(x)) = 0,$$

or

$$x\frac{d}{dx}(f(x)) + 1 \cdot f(x) + 2f(x)\frac{d}{dx}(f(x)) = 0,$$

or

$$x\frac{dy}{dx} + y + 2y\frac{dy}{dx} = 0,$$

or

$$(x + 2y)\frac{dy}{dx} = -y.$$

Solving the last equation for dy/dx, we find that

$$\frac{dy}{dx} = -\frac{y}{x + 2y}.$$

EXAMPLE 2 Let us find dy/dx if

$$\frac{2x^2}{y} = x + y.$$

This time we will not replace y by $f(x)$, but we will simply remember as we differentiate that y is defined implicitly as a function of x. It will also be convenient to use prime notation in this example. We have

$$\left(\frac{2x^2}{y}\right)' = (x + y)'.$$

Now we differentiate the left side of the equation by the quotient rule and the right side by the sum rule to obtain

$$\frac{y(2x^2)' - 2x^2y'}{y^2} = (x)' + (y)',$$

or

$$\frac{4xy - 2x^2y'}{y^2} = 1 + y',$$

or

$$4xy - 2x^2y' = y^2 + y^2y',$$

so

$$(y^2 + 2x^2)y' = 4xy - y^2,$$

and thus

$$\frac{dy}{dx} = y' = \frac{4xy - y^2}{2x^2 + y^2}.$$

EXAMPLE 3 We find the equation of the line tangent to the circle

$$x^2 + y^2 = 5$$

at the point (1, 2). We have

$$\frac{d}{dx}(x^2) + \frac{d}{dx}(y^2) = \frac{d}{dx}(5),$$

or

$$2x + 2y\frac{dy}{dx} = 0.$$

Hence

$$\frac{dy}{dx} = -\frac{x}{y},$$

so at (1, 2) we have

$$\frac{dy}{dx} = -\frac{1}{2}.$$

Thus the slope of the line tangent to the circle at (1, 2) is $-\frac{1}{2}$, and the equation of the tangent line is therefore

$$y = -\frac{1}{2}x + b.$$

Since (1, 2) is on the line, $b = \frac{5}{2}$, and the equation is

$$y = -\frac{1}{2}x + \frac{5}{2}.$$

If y is defined implicitly as a function of x, and x is a function of a third variable t, then y is a function of t, and we can use implicit differentiation and the chain rule to find dy/dt.

EXAMPLE 4

Suppose
$$xy + y + x = 1$$
and that x is a function of t. Let us find dy/dt.

Differentiating with respect to t, we have
$$\frac{d}{dt}(xy) + \frac{d}{dt}(y) + \frac{d}{dt}(x) = \frac{d}{dt}(1),$$
or
$$x\frac{dy}{dt} + y\frac{dx}{dt} + \frac{dy}{dt} + \frac{dx}{dt} = 0,$$
or
$$(1 + x)\frac{dy}{dt} + (1 + y)\frac{dx}{dt} = 0.$$
Thus
$$\frac{dy}{dt} = -\frac{1 + y}{1 + x}\frac{dx}{dt}.$$

3.8 EXERCISES

In Exercises 1 through 12 use implicit differentiation to find dy/dx.

1. $2x + y = 2$ **2.** $5x - 3y = 0$ **3.** $xy = 1$

4. $3x - 2xy = -4$ **5.** $x^2 - 2y^2 = 1$ **6.** $y^2 + 4x^3 - 2y = 0$

7. $x^2y = 3$ **8.** $x^3y^2 = 4$ **9.** $\frac{5x}{y} = 1$

10. $\frac{x}{y} - \frac{y}{x} = 1$ **11.** $\sqrt{xy} + y = 0$ **12.** $\sqrt{\frac{x}{y}} + 2x + y^{-2} = 6$

In Exercises 13 through 20 find the equation of the line tangent to the graph of the function at the given point.

13. $x^2 + y^2 = 10, \quad (1, 3)$ **14.** $x^3 - y^3 = 0, \quad (1, 1)$

15. $2xy = -8, \quad (4, -1)$ **16.** $x^2y = 5, \quad \left(2, \frac{5}{4}\right)$

17. $2x^3y^2 - 8x = 0, \quad (1, -2)$ **18.** $xy^2 - yx^2 + 6 = 0, \quad (-1, -3)$

19. $\frac{x}{y} + x = 0, \quad (1, -1)$ **20.** $\frac{2y}{x} + y^2 = 0, \quad \left(-\frac{2}{3}, 3\right)$

In Exercises 21 through 26, x and y are functions of t. Find dy/dt.

21. $x^2 + y^2 = 4$ **22.** $2x + 3y + xy = 0$

23. $xy^2 + yx^2 = 1$ **24.** $y^2 + \frac{2x}{y} = 12$

25. $2xy^2 - 3(y + 2)^2 = 1$ **26.** $\frac{x}{y + 1} - \frac{3}{y^2} = 2x$

3.9 RELATED RATES

There is a common type of problem in which information about the rate of change of one variable with respect to another is known and it is required to find the rate of change of the second variable with respect to a third one. Such a problem is known as a **related rates** problem. Some examples of related rates problems are

- Finding the rate at which the demand for a product is changing if the rate at which its price is changing is known.
- Finding the rate at which the area of a circle is changing if the rate at which its radius is changing is known.
- Finding the rate at which the top of a ladder is sliding down a wall if the rate at which the bottom of the ladder is being pulled away from the base of the wall is known.

Related rates problems are solved with the aid of the chain rule. The key to solving related rates problems is to find an equation that relates the variable whose rate of change we want to the one whose rate of change we know; we can then differentiate this equation with the aid of the chain rule and solve for the unknown rate of change. Sometimes implicit differentiation is helpful. We will illustrate with examples.

EXAMPLE 1 Suppose the demand function for a product is given by

$$q = 10{,}000 - 400p,$$

where q is quantity demanded and p is the unit price of the product. Suppose also that the price is given as a function of time by the equation

$$p = 5 + 2\sqrt{t},$$

where t is time in months, with $t = 0$ representing the present. Let us find the rate at which demand for the product will be changing with respect to time 4 months from now.

We want to find the rate of change of quantity demanded q with respect to time t when $t = 4$; that is, we want to find dq/dt when $t = 4$. We can easily find dp/dt from the equation $p = 5 + 2\sqrt{t}$:

$$\frac{dp}{dt} = \frac{1}{\sqrt{t}}.$$

The equation that relates q and p is

$$q = 10{,}000 - 400p.$$

Using the chain rule, we differentiate this equation with respect to t and obtain

$$\frac{dq}{dt} = -400\frac{dp}{dt} = -\frac{400}{\sqrt{t}}.$$

Therefore

$$\left.\frac{dq}{dt}\right|_{t=4} = -\frac{400}{\sqrt{4}} = -200.$$

Thus 4 months from now demand for the product will be decreasing at the rate of 200 units per month.

EXAMPLE 2 A stone is dropped into a pool and the ripples created move outward in a circular pattern from the point where the stone entered the water. See Figure 3–32. The radius of the outermost ripple increases at a rate of 0.2 meters per second. Find the rate at which the area covered by the ripples is increasing when the radius of the outermost ripple is 1 meter.

If A denotes the area covered by the ripples, and r denotes the radius of the outermost ripple, then A and r are functions of time t. We know that the rate of change of r with respect to t is 0.2 meters per second; that is,

$$\frac{dr}{dt} = 0.2.$$

We must find dA/dt when $r = 1$. The equation that relates A and r is the formula for the area of a circle of radius r, namely,

$$A = \pi r^2.$$

If we differentiate this equation with respect to t, we obtain

$$\frac{dA}{dt} = 2\pi r\frac{dr}{dt}.$$

Therefore when $r = 1$,

$$\frac{dA}{dt} = 2\pi(1)(0.2) = 0.4\pi,$$

so the area covered by the ripples is increasing at a rate of 0.4π square meters per second.

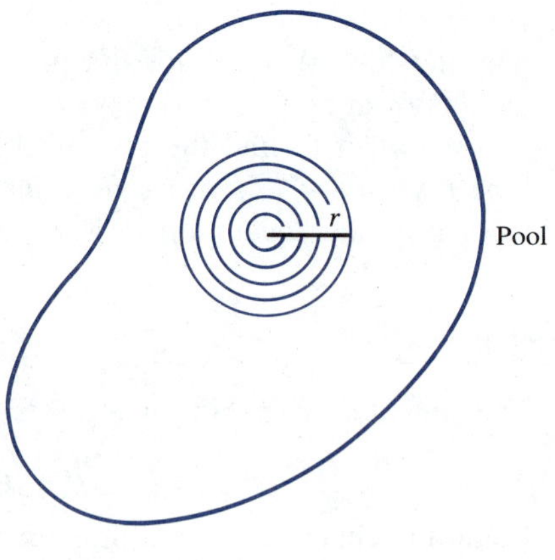

FIGURE 3–32

EXAMPLE 3 A 10-foot ladder is leaning against a wall. The bottom of the ladder is pulled away from the wall at a rate of 2 feet per second. Find the rate at which the top of the ladder is sliding down the wall when the bottom is 6 feet from the base of the wall.

We begin by sketching a picture of the situation. See Figure 3–33. Here x is the distance from the base of the wall to the bottom of the ladder, and y is the distance from the base of the wall to the top of the ladder. Both x and y change with time, so they are functions of time t. We must find dy/dt when x equals 6 feet. We know that $dx/dt = 2$ feet per second. We need an equation relating x and y that we can differentiate to find dy/dt. By the Theorem of Pythagoras we have

$$x^2 + y^2 = 10^2,$$

or

$$x^2 + y^2 = 100.$$

Differentiating this last equation implicitly with respect to t, we find that

$$2x\frac{dx}{dt} + 2y\frac{dy}{dt} = 0,$$

or

$$\frac{dy}{dt} = -\frac{x}{y}\frac{dx}{dt}.$$

Substituting 6 for x in the equation $x^2 + y^2 = 100$, we obtain

$$6^2 + y^2 = 100,$$

which yields

$$y = \sqrt{64} = 8.$$

Hence when $x = 6$,

$$\frac{dy}{dt} = -\frac{6}{8}(2) = -1.5$$

feet per second.

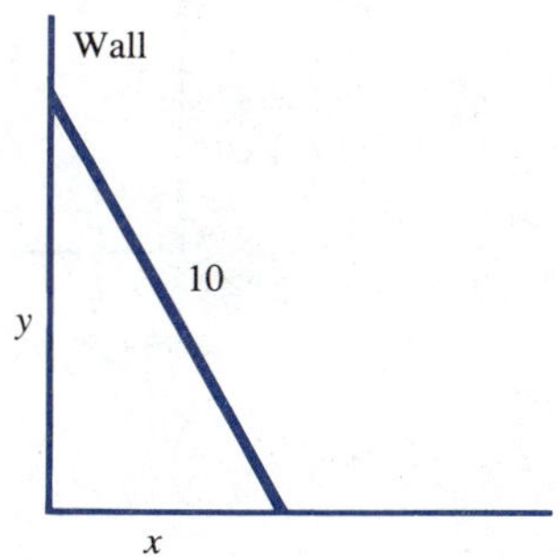

FIGURE 3–33

EXAMPLE 4 A person who is 5 feet tall is walking toward a building at a rate of 3 feet per second. The building is 100 feet high and has a light on its top. At what rate is the length of the person's shadow changing?

Figure 3–34 depicts the situation: x is the distance from the building to the person, and s is the length of the person's shadow. Both x and s are functions of time t. We must find ds/dt. We know that $dx/dt = -3$ feet per second. Note that dx/dt is *negative:* this is so because the distance x is *decreasing* as the person walks toward the building, and the derivative of a decreasing function must be negative.

We need to find an equation relating s and x that we can differentiate to obtain ds/dt. From Figure 3–34 we see that the triangles ABE and CDE are similar, so their corresponding sides are in proportion. Thus

$$\frac{100}{x+s} = \frac{5}{s},$$

which yields

$$s = \frac{x}{19}.$$

Therefore

$$\frac{ds}{dt} = \frac{1}{19}\frac{dx}{dt} = -\frac{3}{19}$$

feet per second.

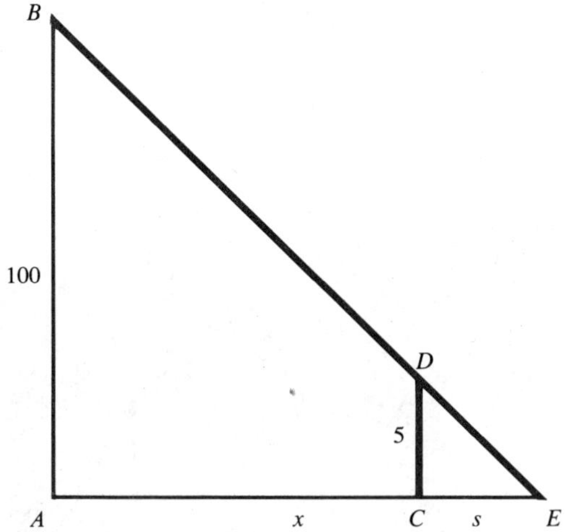

FIGURE 3–34

3.9 EXERCISES

1. If you blow up a balloon so that its radius increases at a rate of 2 centimeters per second, find the rate at which its surface area is changing 4 seconds after you begin to blow it up. (The surface area of a sphere of radius r is $4\pi r^2$.)
2. Referring to Exercise 1, find the rate at which the volume of the balloon is changing 4 seconds after you begin to blow it up. (The volume of a sphere of radius r is $\frac{4}{3}\pi r^3$.)
3. A person who is 6 feet tall walks toward an 18-foot-high street light at a rate of 2 feet per second. How fast is the length of the person's shadow changing?
4. A person who is 6 feet tall walks away from a 30-foot-high street light at a rate of 5 feet per second. How fast is the length of the person's shadow changing?
5. Two airplanes take off from the same airport at the same time. One flies due south at 300 miles per hour, the other due west at 400 miles per hour. At what rate is the distance between the planes changing 3 hours after takeoff?
6. Two cars are approaching an intersection, one from due north and the other from due east. The car from the north has a constant speed of 30 miles per hour. When the cars are each 1 mile from the intersection, the distance between them is decreasing at a rate of 40 miles per hour. How fast is the car coming from the east traveling at the instant when both cars are 1 mile from the intersection?
7. A water tank has the shape of a cone 30 meters tall with a circular base whose radius is 3 meters. Water is being drawn out of the tank at a rate of 5 cubic meters per minute. How fast is the water level falling when the water is 20 meters deep? (The volume of a cone of height h whose base is a circle of radius r is $\frac{1}{3}\pi r^2 h$.)
8. An observer on the shore is watching a ship steam through a canal. The ship heads down the middle of the canal, which is one-quarter mile wide, at a rate of 20 mph. Find the rate at which the distance between the ship and the observer is changing when the ship is 1 mile from the observer.

Management

9. The supply function for a product is given by
$$q = p^2 + 10p,$$
where q is quantity supplied and p is unit price. The price as of month t is given by
$$p = 25 - 0.5t,$$
where $t = 0$ represents the present. Find the rate at which quantity supplied will be changing 3 months from now.
10. A firm plans to build a new manufacturing plant. It is estimated that if the plant produces x units per month, its contribution to the firm's profits will be P dollars, where
$$P = 0.5x^2 - 3x - 50{,}000.$$

It is also estimated that t months after it opens, the plant will produce

$$x = 200\sqrt{t} + 50$$

units per month. Find the rate at which the plant's contribution to profits will be changing 16 months after it opens.

11. A store's monthly sales depend on the number of newspaper ads it runs, according to the equation

$$S = 20\sqrt{x + 100}.$$

Here S is monthly sales in thousands of dollars, and x is the number of ads run during a month. If the store is increasing the number of ads it runs at the rate of 5 per month, find the rate at which sales are increasing when 96 ads are run during a month.

Life Science

12. A circular oil slick is created as the result of the blowout of an offshore oil well. The radius of the slick increases at the rate of 400 feet per hour. Find the rate at which the area polluted by the slick is increasing 12 hours after the blowout.

13. The daily amount of smog produced by cars coming into a city is given by

$$y = 2 + 0.01x^{3/2},$$

where y is smog concentration in parts per million and x is the number of cars in thousands. The number of cars coming into the city is increasing at the rate of 250 per day. Find the rate at which the smog concentration will be changing when 225,000 cars are coming into the city.

14. Cases of a new disease are increasing at a rate of 10,000 per year, and the number of deaths y depends on the number of cases x according to the equation $y = 0.1\sqrt{x}$. Find an expression that gives the annual death rate due to the disease.

Social Science

15. If the population of a state is x million people, the state will need y thousand social workers in its department of social services, where

$$y = 3\sqrt{x} + 20.$$

Suppose the population of the state is increasing at a rate of 50,000 people per year. Find the annual rate at which the number of social workers needed is changing when the state's population is 5 million people.

3 REVIEW EXERCISES

Answers to all review exercises are at the back of the book.

In Exercises 1 through 5 find $f'(x)$.

1. $f(x) = (8x^3 - 5x^2 + 2x)(x^2 + 3x + 1)$

2. $f(x) = \dfrac{3x + 7}{5x - 3}$

3. $f(x) = \dfrac{2x^2 - x}{4x + 6}$

4. $f(x) = (2\sqrt{x} + 3)(x^3 - 1)$

5. $f(x) = \dfrac{(5x + 2)(x^2 + 3)}{2x^2 + 1}$

In Exercises 6 through 12 find $f'(x)$.

6. $f(x) = (3x^2 + 5x + 1)^{12}$

7. $f(x) = \sqrt{6x^3 - 2x + 3}$

8. $f(x) = (2x + 3)^6(x^2 - 5x)^8$

9. $f(x) = [(2x - 2)^4(x^2 + 1)]^5$

10. $f(x) = \left(\dfrac{3x - 1}{x^2 + 2}\right)^7$

11. $f(x) = \dfrac{(3x + 4)^5}{(5x^2 - 1)^4}$

12. $f(x) = \left(\dfrac{x^2 + 1}{x^2 - 1}\right)^{1/2}$

13. If $y = x^2 - 5x + 2$ and $x = 8t - 3$, find $\dfrac{dy}{dt}$.

14. If $v = 7w - 3\sqrt{w}$ and $u = (v^4 - 2v + 11)^{11}$, find $\dfrac{du}{dw}$.

15. The supply function for a commodity is given by

$$q = 5000(0.5p^2 - 2\sqrt{p})^{3/4}.$$

Find the rate of change of quantity supplied q when $p = \$16$.

16. Here is the graph of a function:

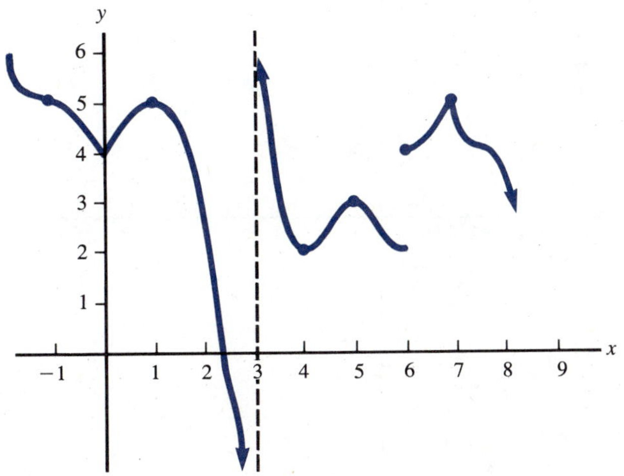

Find all relative maxima, relative minima, and saddle points for the function.

In Exercises 17 through 19 find all critical values, and use the first derivative test to locate all relative maxima, relative minima, and saddle points.

17. $f(x) = x^3 + 6x^2 - 180x - 1200$
18. $f(x) = x^4 + 4x^3$

19. $f(x) = \dfrac{x^2 - 8}{2x + 1}$

20. A firm's average cost per unit is given by

$$A(x) = 0.5x^2 - 54\sqrt{x} + 200.$$

Find the firm's minimum average cost per unit.

21. A new disease that affects birch trees is introduced into a locality, and t years later the proportion of birches that have the disease is

$$y = \frac{t^2 + 8t + 2}{4t^2 + 24}.$$

What was the maximum proportion of birches that contracted the disease?

In Exercises 22 and 23 use the second derivative test to find all relative maxima and relative minima.

22. $f(x) = x^3 + 6x^2 - 15x$

23. $f(x) = \dfrac{1}{5}x^5 - 4x^2 + 1$

24. The demand function for a firm's product is given by

$$p = 6000 - 2\sqrt{x}.$$

Use the second derivative test to find the firm's maximum revenue.

25. Here is the graph of a function:

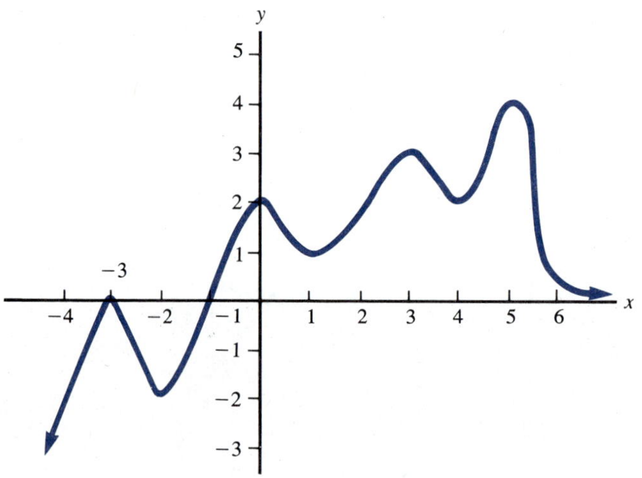

Find the absolute maximum and absolute minimum of the function over the interval

(a) $[0, 5]$ **(b)** $[0, 3]$ **(c)** $[-2, 5]$ **(d)** $[-2, 3]$
(e) $[0, +\infty)$ **(f)** $[5, +\infty)$ **(g)** $(-\infty, -2]$ **(h)** $(-\infty, 0]$

26. Find the absolute maximum and absolute minimum of the function defined by

$$f(x) = x^4 - 32x^2$$

over

(a) the x-axis (b) the interval $[-1, 1]$
(c) the interval $[-5, 6]$ (d) the interval $[1, 6]$

27. A firm's profit function is given by

$$P(x) = -x^3 + 150x^2 - 4800x.$$

Find its maximum profit if it can produce and sell at most 50 units; an unlimited number of units.

28. A car dealership sells 400 new cars per year. Each new car costs \$12,000, it costs the dealership \$1000 to order cars, and it costs \$2000 to carry a car in inventory for a year. Assuming that the average inventory of cars is one half the number ordered, find the optimal reorder quantity and the minimum annual inventory cost for the dealership.

29. Find the dimensions of a cylindrical container with a circular base and top if its volume is to be maximized and its surface area is to be 6 square feet. ($V = \pi r^2 h$ and $S = 2\pi r^2 + 2\pi rh$.)

In Exercises 30 through 32 use the curve-sketching procedure to sketch the graph of the function.

30. $f(x) = -x^3 + 6x^2 + 96x$

31. $f(x) = \dfrac{x^2 - 4}{x^2 - 9}$

32. $f(x) = \dfrac{x^3 + 1}{2x^2}$

33. A pollster finds that t months after inauguration y percent of the voters approve of the job that a governor is doing, where

$$y = 0.01\left(-\frac{t^3}{3} + 21t^2 - 360t + 6000\right), \qquad 0 \le t \le 46.$$

Sketch the graph of this function and analyze the governor's popularity.

34. Use differentials to approximate $\sqrt[3]{65}$.

35. A demand function is given by

$$p = \frac{80,000}{\sqrt{10q}},$$

where p is unit price and q is quantity demanded. If quantity demanded is currently 1000 units, find the proportional and percentage changes in price if 100 more units would be demanded.

36. A firm's revenue function is given by

$$R(x) = 25(3x + 1)^{4/5} - 25.$$

The firm is currently selling 341 units per week. What would be the approximate increase in revenue if they could sell 350 units per week?

37. Find $\frac{dy}{dx}$ if $8x^2y^2 - \frac{2x}{y^2} = 10$.

38. If x and y are functions of u and $3xy^2 + \frac{2y}{x} = 1$, find $\frac{dy}{du}$.

39. If price p and quantity demanded q for a certain commodity are related by the equation

$$q = -p^2 - 50p + 5000$$

and the price is increasing at a rate of \$0.50 per month, find the rate at which quantity demanded is changing with respect to time when the price is \$25.

40. If one person drives due north from point A at 55 miles per hour and another person leaves point A 2 hours after the first and drives due east at 50 miles per hour, find the rate at which the distance separating them is changing 4 hours after the second person left.

3 SUPPLEMENT: CONSUMPTION AND THE MULTIPLIER

It is not uncommon to find news items such as this: ''Today XYZ Corporation announced that it will build a new plant in Smallville. The plant will provide employment for 250 workers at an operating cost of some \$1 million per year. It is estimated that the presence of the plant will add \$4 million per year to the Smallville economy.''

How was the figure of \$4 million per year determined? It must have taken into account the fact that some of the income generated by an investment (such as the \$1 million per year that XYZ is investing in Smallville) will be saved and invested by those who receive it; this new round of investment will in turn produce more income, some of which will again be saved and invested to produce still more income, and so on. This ''chain reaction'' of income generation is called the **multiplier effect.** Let us see if we can quantify the multiplier effect.

Evidently the strength of the multiplier effect depends on the proportion of income saved and invested at each stage of the process. Therefore we need an expression that relates income to savings and investment. For simplicity we assume that all income that is saved is invested, so savings equals investment. Thus we need a relationship between income and savings.

Economists divide income into the portion that is saved and the portion that is not saved; the latter is called **consumption.** Thus,

$$\text{income} = \text{savings} + \text{consumption},$$

or

$$\text{consumption} = \text{income} - \text{savings}.$$

If we denote income by Y, consumption by C, and savings by S, then consumption is given as a function of income by a **consumption function**

$$C = Y - S.$$

EXAMPLE 1 Let

$$C = 5 + 8\sqrt{Y}$$

be a consumption function, with Y and C in millions of dollars. Thus when Y = \$100 million dollars, consumption is

$$C(100) = 5 + 8\sqrt{100} = 85$$

million dollars. The remaining \$100 − \$85 = \$15 million dollars of income is saved.

The derivative of C with respect to Y is the rate of change of consumption with respect to income, which is known as the **marginal propensity to consume.** The marginal propensity to consume is the proportion of an additional \$1 of income that would be spent rather than saved.

EXAMPLE 2 Let

$$C = 5 + 8\sqrt{Y}$$

as in Example 1. Then the marginal propensity to consume is

$$\frac{dC}{dY} = \frac{4}{\sqrt{Y}}.$$

For instance, when Y = \$100 million, we have

$$\text{marginal propensity to consume} = \frac{4}{\sqrt{100}} = \$0.40.$$

Hence at an income level of \$100 million, 40 cents of each additional dollar of income would be spent, and the remaining 60 cents would be saved. Note that the marginal propensity to consume $dC/dY = 4/\sqrt{Y}$ is a decreasing function of Y. Therefore in this example, as income increases a smaller proportion of it is spent and a larger proportion is saved.

The equation $C = Y - S$ shows that savings S is a function of income Y also. As we remarked previously, the strength of the multiplier effect depends on the proportion of income invested at each stage of the income-generation process. Thus, since we are assuming that savings equals investment, we can say that the multiplier effect is measured by the rate at which *income* changes with respect to *savings*, or dY/dS. The derivative dY/dS is called the **multiplier.**

From $C = Y - S$ we have

$$\frac{dC}{dS} = \frac{dY}{dS} - \frac{dS}{dS}.$$

Applying the chain rule to the left side gives

$$\frac{dC}{dY}\frac{dY}{dS} = \frac{dY}{dS} - 1.$$

Solving this last equation for dY/dS, we obtain

$$\frac{dY}{dS} = \frac{1}{1 - dC/dY} = \frac{1}{1 - \text{marginal propensity to consume}}.$$

EXAMPLE 3 Again let $C = 5 + 8\sqrt{Y}$. Then the multiplier is

$$\frac{dY}{dS} = \frac{1}{1 - 4/\sqrt{Y}} = \frac{\sqrt{Y}}{\sqrt{Y} - 4}.$$

Thus, at an income level of \$100 million,

$$\frac{dY}{dS} = \frac{\sqrt{100}}{\sqrt{100} - 4} = \frac{5}{3}.$$

Hence each additional \$1 of savings generates an additional \$1.67 in income.

EXAMPLE 4 Suppose the consumption function for Smallville is

$$C = 20 + 0.75Y,$$

where Y and C are in millions of dollars. The multiplier is then

$$\frac{1}{1 - dC/dY} = \frac{1}{1 - 0.75} = 4,$$

so if XYZ Corporation invests \$1 million per year in Smallville it will generate approximately \$4 million in additional income for the town.

The multiplier effect is what is behind the idea of ''pump priming'' in the economy: if there are many opportunities for investment, the multiplier is large; therefore a relatively small investment (the ''priming of the pump'') can lead to a large increase in income. Unfortunately, the reverse is also true: a relatively small decrease in investment can lead to a large decrease in income.

SUPPLEMENTARY EXERCISES

1. Let $C = 20 + 0.6Y$ be a consumption function, with Y and C in millions of dollars. Find and interpret the marginal propensity to consume at every income level.

2. Let $C = 50 + 10\sqrt{Y - 10}$ be a consumption function, with Y and C in billions of dollars. Find and interpret the marginal propensity to consume when the income level is

(a) \$410 billion **(b)** \$2510 billion

3. Find and interpret the multiplier for the consumption function of Exercise 1.

4. Find and interpret the multiplier for the consumption function of Exercise 2 when the income level is

   (a) \$410 billion **(b)** \$2510 billion

5. Since $C = Y - S$, we have $S = Y - C$. The derivative dS/dY is called the **marginal propensity to save.** Find and interpret the marginal propensity to save for

   (a) the consumption function of Exercise 1;

   (b) the consumption function of Exercise 2 when $Y =$ \$410 billion and when $Y =$ \$2510 billion.

4

Exponential and Logarithmic Functions

In this chapter we introduce **exponential functions** and **logarithmic functions.** These functions are important because they can be used to model many situations of growth and decline. For instance, they are often used in the study of population growth, investment growth, radioactive decay, the spread of epidemics, and similar topics. We discuss some of the applications of exponential and logarithmic functions in detail in the last two sections of this chapter. In the first three sections we define exponential and logarithmic functions, describe some of their elementary properties, and show how to differentiate them.

4.1 EXPONENTIAL FUNCTIONS

In this section we define exponential functions, discuss their characteristics, show how to graph them, and demonstrate how to use them to answer questions about the growth of money invested at compound interest. We repeatedly use the material on exponents and roots in Section 1.1.

Exponential Functions with Base b

Let x be a real number. If x is an integer or a rational number, then we can calculate 2^x using the results of Section 1.1. For instance,

$$2^0 = 1, \qquad 2^1 = 2, \qquad 2^2 = 2 \cdot 2 = 4,$$

$$2^{-1} = \frac{1}{2^1} = \frac{1}{2}, \qquad 2^{-2} = \frac{1}{2^2} = \frac{1}{4}, \qquad 2^{1/2} = \sqrt{2},$$

$$2^{2/3} = (\sqrt[3]{2})^2, \qquad 2^{-1/2} = \frac{1}{\sqrt{2}}, \qquad 2^{-2/3} = \frac{1}{(\sqrt[3]{2})^2}.$$

If x is an irrational number, we can approximate x to any desired degree of accuracy by a rational number, and this allows us to approximate 2^x to any desired degree of accuracy. For example, if $x = \sqrt{3}$, successively better approximations of x are

$$\sqrt{3} \approx 1.73 \qquad \text{and} \qquad \sqrt{3} \approx 1.732,$$

so successively better approximations of $2^{\sqrt{3}}$ are

$$2^{1.73} = 2^{173/100} = 3.317278 \qquad \text{and} \qquad 2^{1.732} = 2^{1732/1000} = 3.321880.$$

In this way we can calculate $2^{\sqrt{3}}$ to any required degree of accuracy. Similar reasoning allows us to calculate 2^x when x is any irrational number.

Now that we know how to calculate 2^x for any number x, suppose we define a function f by setting

$$f(x) = 2^x$$

for all x. This function is called an **exponential function with base 2.** Its graph is shown in Figure 4–1. Note that

- The domain of the function is the set of all real numbers.
- For every real number $y > 0$, there is exactly one real number x such that $2^x = y$.
- The y-intercept of the function is $(0, f(0)) = (0, 2^0) = (0, 1)$.
- The x-axis is an asymptote for the function as $x \to -\infty$; that is, $\text{Lim}_{x \to -\infty} 2^x = 0$.
- The function is continuous at every value of x.

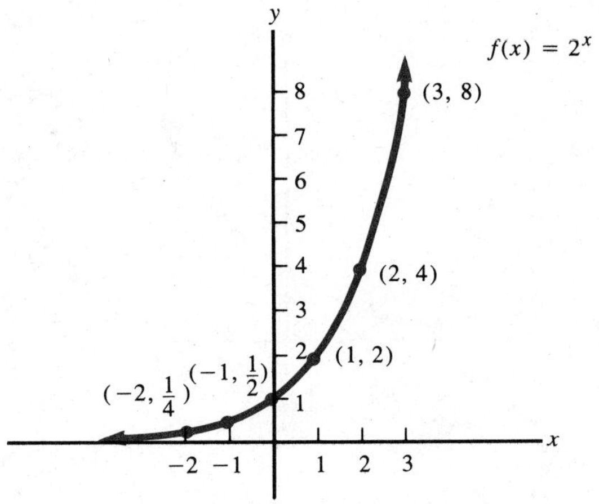

FIGURE 4–1

We can perform a similar analysis using any positive number b in place of 2. Therefore it is possible to consider the exponential function defined by the equation

$$f(x) = b^x$$

for any positive number b. Because $1^x = 1$ for all x, we omit the case $b = 1$.

> **Exponential Functions**
>
> Let b be any positive number, $b \neq 1$. The function f defined by
>
> $$f(x) = b^x$$
>
> is called an **exponential function with base *b*.**

Figure 4–2 shows the graph of $y = b^x$ when $b > 1$ and when $0 < b < 1$.

Functions of the form $f(x) = cb^{rx}$, where c and r are numbers, $c \neq 0$, $r \neq 0$, are also often referred to as exponential functions with base b. We graph exponential functions by substituting numbers for x in their defining equations and plotting the resulting points.

EXAMPLE 1

a. The functions f and g defined by $f(x) = 3^{2x}$ and $g(x) = 2(3^{2x})$ are both exponential with base 3. Their graphs are shown in Figure 4–3.

b. The functions f and g defined by $f(x) = 2^{-x/2}$ and $g(x) = (1/2)^{-x/2}$ are both exponential with base 2. Their graphs are shown in Figure 4–4.

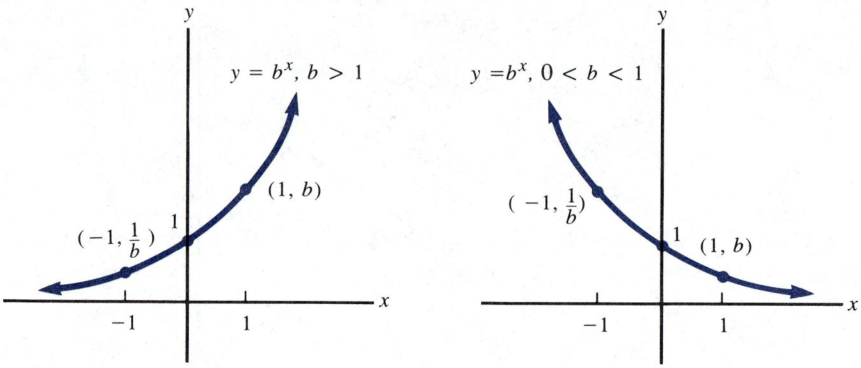

FIGURE 4–2

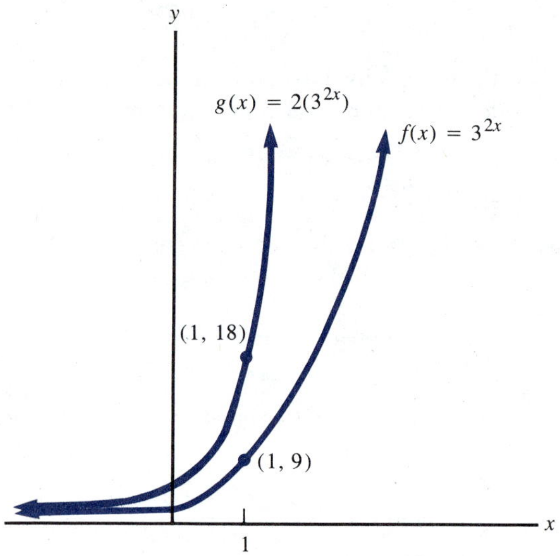

FIGURE 4–3

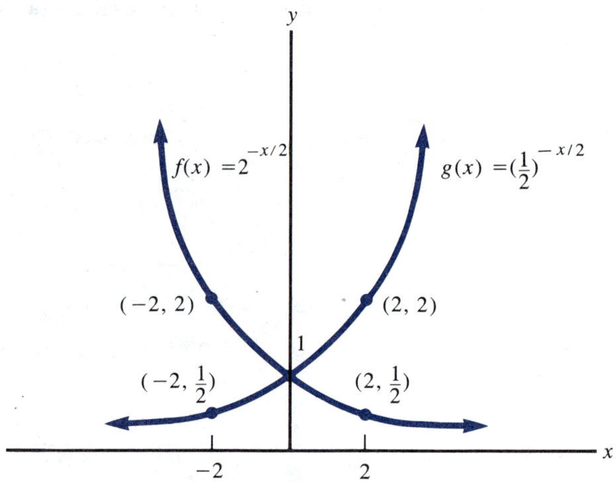

FIGURE 4–4

Compound Interest

As an application of exponential functions, let us consider the topic of **compound interest.** If an amount of money is invested, the investor receives a fee for allowing the borrower to use the money. The amount invested is the **principal,** and the fee is the **interest earned** on the principal. If the interest earned during a period depends not only on the principal but also on interest previously earned, we say that the interest is **compounded.** For instance, if \$10,000 is invested at an interest rate of 8 percent per year compounded annually, then at the end of each year the interest earned for that year will amount to 8 percent of the sum of the principal and any previously accumulated interest—that is, to 8 percent of the worth of the investment at the end of the previous year. Therefore

- At the end of one year the investment will be worth

$$\$10,000 + 0.08(\$10,000) = \$10,800.$$

- At the end of two years the investment will be worth

$$\$10,800 + 0.08(\$10,800) = \$11,664.$$

- At the end of three years the investment will be worth

$$\$11,664 + 0.08(\$11,664) = \$12,597.12.$$

In the preceding illustration the interest was compounded annually (at the end of every year), but interest may be compounded at the end of any time period: for example, semiannually (at the end of every six months), quarterly (at the end of every three months), monthly, daily, and so on. Suppose \$10,000 is invested at an interest rate of 8 percent per year compounded quarterly. The stated rate, here 8 percent per year, is an annual rate; if interest is to be compounded quarterly, then it will be compounded four times per year, so the interest rate for each three-month compounding period will be (8 percent)/4 = 2 percent. Therefore

- At the end of the first quarter the investment will be worth

$$\$10,000 + 0.02(\$10,000) = \$10,200.$$

- At the end of the second quarter the investment will be worth

$$\$10,200 + 0.02(\$10,200) = \$10,404.$$

- At the end of the third quarter the investment will be worth

$$\$10,404 + 0.02(\$10,404) = \$10,612.08.$$

- At the end of the fourth quarter the investment will be worth

$$\$10,612.08 + 0.02(\$10,612.08) = \$10,824.32.$$

Let us develop a formula for the worth of an investment of P dollars at the end of t years if the money earns interest at i percent per year compounded m times per year. The interest rate for each compounding period will then be (i percent)/m, or, in decimal form, $i/100m$. Let $r = i/100m$. Then

- At the end of the first compounding period the investment will be worth

$$P + rP = P(1 + r)$$

dollars.

- At the end of the second compounding period the investment will be worth

$$P(1 + r) + rP(1 + r) = P(1 + r)(1 + r) = P(1 + r)^2$$

dollars.

- At the end of the third compounding period the investment will be worth

$$P(1 + r)^2 + rP(1 + r)^2 = P(1 + r)^3$$

dollars.

Thus we can see that at the end of k compounding periods the investment will be worth

$$P(1 + r)^k$$

dollars. Therefore at the end of t years, when the money will have earned interest for mt compounding periods, the investment will be worth

$$P(1 + r)^{mt}$$

dollars. This is the compound interest formula.

The Compound Interest Formula

Let principal P be invested at an interest rate of i percent per year compounded m times per year. Then $r = i/100m$ is the interest rate per compounding period in decimal form. At the end of t years the investment will be worth A dollars, where

$$A = P(1 + r)^{mt}.$$

The amount A is called the **future value** of the principal. Note that A is an exponential function of t with base $1 + r$.

EXAMPLE 2 Suppose \$1000 is invested at 6 percent per year compounded semiannually. How much will the investment be worth at the end of 3 years?

Here P = \$1000, i percent = 6 percent, $m = 2$, and $t = 3$. Therefore

$$r = \frac{6}{100(2)} = 0.03,$$

and

$$A = \$1000(1 + 0.03)^{2 \cdot 3} = \$1000(1.03)^6 = \$1000(1.194052) = \$1194.05,$$

rounded to the nearest cent.

EXAMPLE 3 Suppose \$1000 is invested at 6 percent per year compounded quarterly. How much will the investment be worth at the end of 3 years?

Here P = \$1000, i percent = 6 percent, $m = 4$, and $t = 3$. Therefore

$$r = \frac{6}{100(4)} = 0.015,$$

and

$$A = \$1000(1 + 0.015)^{4 \cdot 3} = \$1000(1.015)^{12} = \$1000(1.195618) = \$1195.62,$$

rounded to the nearest cent.

Note that the only difference between the situation of Example 2 and that of Example 3 is the number of times interest is compounded each year, and that the more frequent compounding of Example 3 results in a greater worth for the investment at the end of the three-year period. This behavior holds in general: for a fixed annual interest rate of i percent, the more frequently interest is compounded, the faster it accumulates. Therefore for a fixed principal P, a fixed annual interest rate of i percent, and a fixed number of years t, the future value A will depend on the number of times m that interest is compounded each year: the larger m is, the larger A will be.

The Number *e*

Suppose we invest amount P at i percent per year for t years. To obtain the largest possible future value, we want the interest to be compounded as frequently as possible. This occurs when interest is compounded at every instant of time, and is known as **continuous compounding** of interest. In such a case m, the number of compounding periods per year, becomes infinite, and we have

$$A = \lim_{m \to +\infty} P(1+r)^{mt} = P \cdot \lim_{m \to +\infty} \left(1 + \frac{i/100}{m}\right)^{mt}.$$

For instance, if $P = \$1$, i percent $= 100$ percent, and $t = 1$ year, then

$$A = \lim_{m \to +\infty} \left(1 + \frac{1}{m}\right)^{m}.$$

The limit on the right side of this equation exists and is equal to a certain irrational number denoted by e:

$$\lim_{m \to +\infty} \left(1 + \frac{1}{m}\right)^{m} = e.$$

To eight significant digits, $e = 2.7182818$. More generally, it can be shown (though we shall not do so) that for any number a,

$$\lim_{m \to +\infty} \left(1 + \frac{a}{m}\right)^{mt} = e^{at}.$$

Therefore if we take $a = i/100$, we have

$$A = P \cdot \lim_{m \to +\infty} \left(1 + \frac{i/100}{m}\right)^{mt} = Pe^{(i/100)t},$$

and we have developed the formula for continuous compounding of interest.

The Formula for Continuous Compounding of Interest

If principal P is invested at i percent per year compounded continuously, then $r = i/100$ is the annual interest rate expressed in decimal form, and the future value A of the investment at the end of t years is

$$A = Pe^{rt}.$$

We can obtain the value of e^{rt} by using a calculator with an e^x-key, y^x-key, or INV LNx-key, or by looking it up in a table such as Table 1 in the back of this book.

EXAMPLE 4 If \$1000 is invested at 6 percent per year compounded continuously, then at the end of 3 years the investment will be worth

$$A = \$1000e^{(0.06)3} = \$1000e^{0.18}.$$

Using a calculator with an e^x key, we find that

$$e^{0.18} = 1.1972174.$$

Thus

$$A = \$1000(1.1972174) = \$1197.22,$$

to the nearest cent. Compare the results of this example with those of Examples 2 and 3.

Since the number e is positive, we may use it as the base for exponential functions: any function defined by an equation of the form $f(x) = ce^{rx}$ is an **exponential function with base *e*.** The graphs of the exponential functions defined by

$$f(x) = e^x \qquad \text{and} \qquad g(x) = e^{-x}$$

are shown in Figure 4–5.

Why would we want to use the number e as a base for exponential functions? There are two reasons. First, e often turns up when models are created for practical situations, just as it did in the case of continuously compounded interest. Second, the use of e as a base greatly simplifies the derivative formula for exponential functions, as we will see in Section 4.3.

Computer Manual: Programs POWERS, INTREST, and LIMIT.

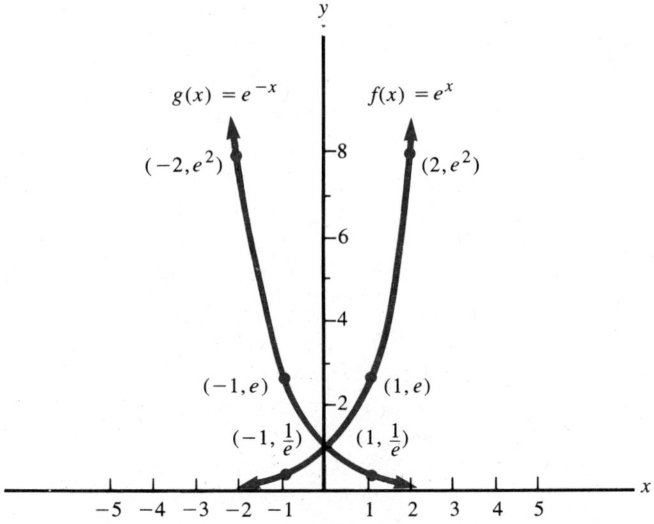

FIGURE 4–5

4.1 EXERCISES

Exponential Functions with Base *b*

In Exercises 1 through 8 graph the exponential function.

1. $f(x) = 3^x$ **2.** $f(x) = 3^{-x}$ **3.** $f(x) = 4(2^{3x})$ **4.** $f(x) = 4(2^{-3x})$

5. $y = \frac{1}{2}(5^{-x/2})$ **6.** $y = -(4^{-3x})$ **7.** $y = \left(\frac{1}{2}\right)^{2x}$ **8.** $y = \left(\frac{2}{3}\right)^{-x/4}$

Compound Interest

9. Find the worth at the end of 1 year of an investment of \$100,000 if it earns 6 percent per year compounded

(a) annually **(b)** semiannually

10. Find the worth at the end of 2 years of an investment of \$8000 if it earns 12 percent per year compounded

(a) annually **(b)** semiannually
(c) quarterly **(d)** monthly

11. If you invest \$10,000 at 10 percent per year compounded quarterly, how much will the investment be worth at the end of

(a) 3 months? **(b)** six months?

(c) 2 years? **(d)** $4\frac{3}{4}$ years?

12. If you borrow \$1000 at an annual interest rate of 6 percent per year, how much will you owe at the end of 5 years if interest is compounded
    (a) annually (b) quarterly

13. How much should you invest now at 10 percent per year compounded quarterly in order to have exactly \$10,000 at the end of 4 years?

14. How much must you invest now in order to have \$1000 at the end of 1 year if the interest rate you obtain is 12 percent per year compounded
    (a) annually? (b) semiannually?
    (c) quarterly? (d) monthly?

15. Show that a **nominal rate** of 8 percent per year compounded quarterly is equivalent to an **effective rate** of 8.24322 percent per year by computing the interest that an investment of \$1 will earn in 1 year at the nominal rate.

16. The concept of an effective interest rate (see Exercise 15) is important because it allows us to compare nominal interest rates that have different compounding periods. For instance, compare the nominal rates of 7.92 percent per year compounded monthly, 8 percent per year compounded quarterly, 8.10 percent per year compounded semiannually, and 8.25 percent per year compounded annually by finding their equivalent effective rates. Which of these rates would you prefer to obtain if you were investing money? Which would you prefer to pay if you were borrowing money?

The Number *e*

17. Find the value of an investment of \$100,000 at the end of 1 year if it earns 6 percent per year compounded continuously.

18. Find the value of an investment of \$10,000 at the end of
    (a) 5 years (b) 10 years
    (c) 20 years (d) 50 years
    if it earns 10 percent per year compounded continuously.

19. Find the value of an investment of \$1000 at the end of 1 year if it earns 12 percent per year compounded
    (a) semiannually (b) quarterly
    (c) monthly (d) continuously

20. How much would you have to invest at 8 percent per year compounded continuously in order to have exactly \$10,000 at the end of 5 years?

In Exercises 21 through 26 graph the exponential function.

21. $f(x) = e^{2x}$ 22. $f(x) = e^{-2x}$ 23. $f(x) = 3e^x$

24. $f(x) = \frac{1}{2}e^{-2x}$ 25. $y = 2e^{x/2}$ 26. $y = -4e^x$

27. By definition,

$$e = \lim_{m\to+\infty}\left(1 + \frac{1}{m}\right)^m.$$

Show that also

$$e = \lim_{x\to 0}(1 + x)^{1/x}.$$

4.2 LOGARITHMIC FUNCTIONS

In this section we introduce logarithms and logarithmic functions. Logarithmic functions are closely related to exponential functions and are important for that reason. We define the concept of the logarithm of a positive number, state the properties of logarithms, define logarithmic functions and examine some of their characteristics, and show how logarithms are used to solve exponential equations.

Logarithms

If b is a number, $b > 1$, and u is a positive number, the graph of the exponential function $y = b^x$ shows that there is one and only one number v such that $u = b^v$. See Figure 4–6. The number v is called the **logarithm to the base *b* of *u*** and is denoted by $\log_b u$:

Log$_b$ *u*

If $b > 1$ and u is positive, then

$$\log_b u = v \quad \text{means} \quad b^v = u.$$

Note that the logarithm of a number u is defined only if u is positive.

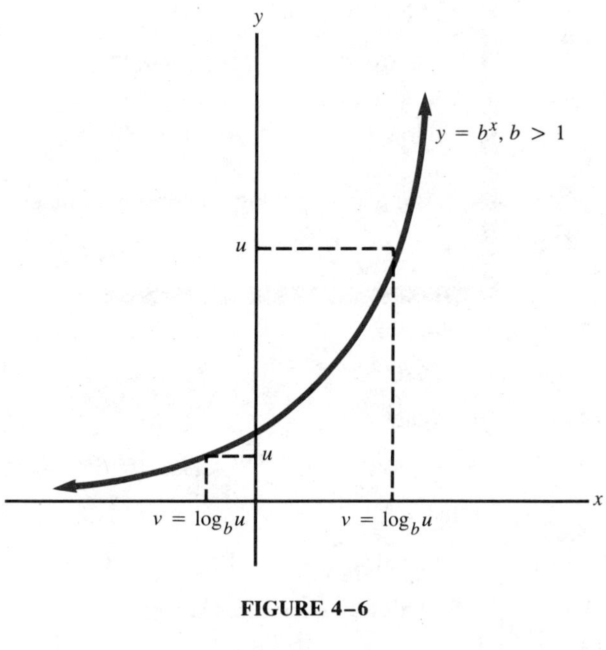

FIGURE 4–6

EXAMPLE 1 We have

$$\log_2 8 = 3 \quad \text{because} \quad 2^3 = 8;$$
$$\log_3 81 = 4 \quad \text{because} \quad 3^4 = 81;$$
$$\log_2 \frac{1}{2} = -1 \quad \text{because} \quad 2^{-1} = \frac{1}{2};$$
$$\log_3 \frac{1}{27} = -3 \quad \text{because} \quad 3^{-3} = \frac{1}{27}.$$

EXAMPLE 2 We have

$$\log_{10} 10 = 1, \qquad \log_{10} 100 = 2, \qquad \log_{10} 1000 = 3,$$
$$\log_{10} 0.1 = -1, \quad \log_{10} 0.01 = -2, \quad \log_{10} 0.001 = -3.$$

Note that if b is a number greater than 1, then

1. $\log_b 1 = 0$ because $b^0 = 1$.
2. $\log_b b = 1$ because $b^1 = b$.
3. $\log_b b^r = r$ because $b^r = b^r$.
4. $b^{\log_b u} = u$ because if $v = \log_b u$, then $b^v = u$ and thus $b^{\log_b u} = b^v = u$.

The preceding properties of logarithms are actually special cases of the following rules of logarithms, which themselves follow from the rules of exponents of Section 1.1. (See Exercises 17 and 18 at the end of this section.)

The Rules of Logarithms

Let b be a number, $b > 1$, and let A and B be positive numbers. Then

1. $\log_b(AB) = \log_b A + \log_b B$.
2. $\log_b \left(\frac{A}{B}\right) = \log_b A - \log_b B$.
3. $\log_b A^r = r \cdot \log_b A$ for any number r.
4. If $A = B$, then $\log_b A = \log_b B$.

EXAMPLE 3 Given that

$$10^{0.3010} = 2, \quad 10^{0.4771} = 3, \quad \text{and} \quad 10^{0.6990} = 5,$$

so that

$$\log_{10} 2 = 0.3010, \quad \log_{10} 3 = 0.4771, \quad \text{and} \quad \log_{10} 5 = 0.6990,$$

by the rules of logarithms we have

a. $\log_{10} 6 = \log_{10}(2 \cdot 3) = \log_{10} 2 + \log_{10} 3 = 0.3010 + 0.4771$
$= 0.7781.$

b. $\log_{10} 8 = \log_{10} 2^3 = 3 \cdot \log_{10} 2 = 3(0.3010) = 0.9030.$

c. $\log_{10}(2/3) = \log_{10} 2 - \log_{10} 3 = 0.3010 - 0.4771 = -0.1761.$

d. $\log_{10}(75/4) = \log_{10}(3 \cdot 5^2) - \log_{10} 2^2$
$= \log_{10} 3 + 2 \cdot \log_{10} 5 - 2 \cdot \log_{10} 2$
$= 0.4771 + 2(0.6990) - 2(0.3010)$
$= 1.2731.$

Just as exponential functions with base e are important, so the logarithm to the base e is important. The logarithm to the base e is called the **natural logarithm** and is symbolized by ln rather than by $\log_e$.

The Natural Logarithm

If u is a positive number, then

$$\ln u = \log_e u.$$

Thus,

$$v = \ln u \quad \text{means} \quad e^v = u.$$

We can find natural logarithms by using a calculator that has an ln-key or by looking them up in a table such as Table 2 at the back of the book.

EXAMPLE 4 Let us find ln 20. If we use a calculator with an ln-key, we find that $\ln 20 = 2.995732$. If we use Table 2 at the back of the book, we find that there is no entry for ln 20. There are entries for ln 2 and ln 10, however, and by the rules of logarithms,

$$\ln 20 = \ln(2 \cdot 10) = \ln 2 + \ln 10 = 0.6931 + 2.3026 = 2.9957.$$

Logarithmic Functions

A function defined by an equation of the form

$$f(x) = \log_b x,$$

where $x > 0$, is called a **logarithmic function with base *b*.** The graph of f is shown in Figure 4–7. Note that

- The domain of the function is the set of positive real numbers.
- For every number y there is one and only one positive number x such that $\log_b x = y$.
- The x-intercept of the function is the point $(1, 0)$.
- The negative y-axis is a vertical asymptote for the function.
- The function is continuous for every value of $x > 0$.

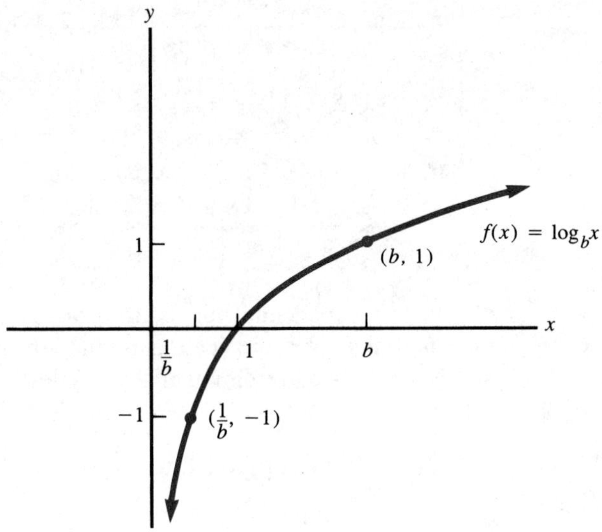

FIGURE 4–7

Furthermore, as Figure 4–8 indicates, the graph of the function defined by $f(x) = \log_b x$ is that of the function defined by $f(x) = b^x$ reflected about the line $y = x$.

In particular, the foregoing facts apply to the **natural logarithm function** defined by

$$f(x) = \ln x.$$

Figure 4–9 shows the graph of the natural logarithm function.

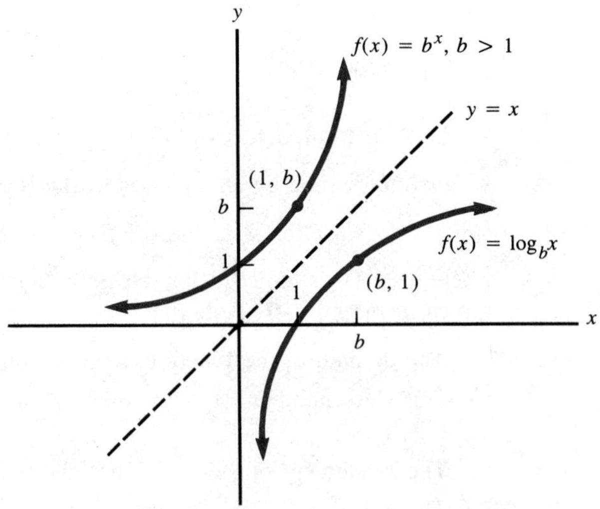

FIGURE 4–8

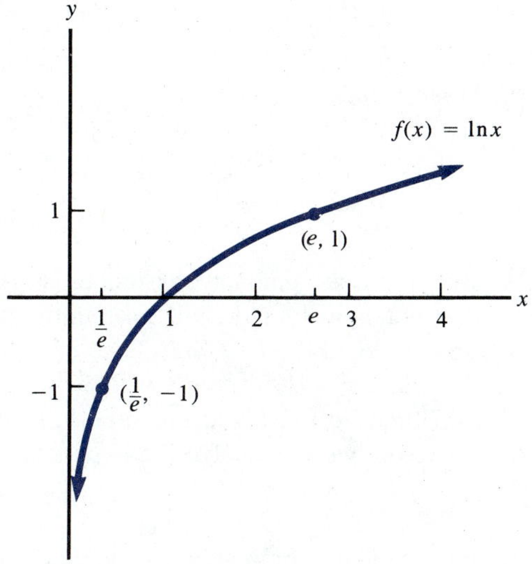

FIGURE 4–9

Composite functions which are given by equations of the form

$$f(x) = \log_b g(x) \qquad \text{or} \qquad f(x) = \ln g(x),$$

are also called logarithmic functions. Since the logarithm of a nonpositive number is not defined, it follows that $g(x)$ must be positive. Thus when we define a function by an equation such as

$$f(x) = \ln(x^2 - 2x + 3),$$

we understand that its domain consists of those values of x for which

$$x^2 - 2x + 3 > 0.$$

Solving Exponential Equations

Logarithms can be used to solve **exponential equations:** These are equations in which the quantity to be determined appears as an exponent. The method of solution requires that we take the logarithm of both sides of the equation. We could use logarithms to any base when doing this, but it will be convenient for us to use the natural logarithm. We illustrate with examples.

EXAMPLE 5 Let us solve the exponential equation

$$e^{2x} = 5.$$

Taking the natural logarithm of both sides of the equation, we obtain

$$\ln e^{2x} = \ln 5,$$

or

$$2x(\ln e) = \ln 5,$$

or $$2x(1) = 1.6094.$$

Hence $$x = \frac{1.6094}{2} = 0.8047.$$

EXAMPLE 6 The decay of a radioactive isotope is described by the equation

$$y = 200e^{-0.06t},$$

where y is the amount of isotope, in grams, present at time t, t in years. Let us find out how long it will take until only one half of the isotope remains; this time is called the **half-life** of the isotope.

Since $y = 200$ when $t = 0$ (why?), there are 200 grams of the isotope present initially. We must find out how long it will take until only 100 grams are left. Therefore we must solve the exponential equation

$$100 = 200e^{-0.06t}$$

for t. Dividing both sides of the equation by 200, we have

$$\frac{1}{2} = e^{-0.06t},$$

so that

$$\ln\frac{1}{2} = \ln e^{-0.06t} = (-0.06t)\ln e = -0.06t.$$

Hence

$$-0.6931 = -0.06t,$$

and solving for t yields

$$t = \frac{-0.6931}{-0.06} \approx 11.55.$$

Thus the half-life of the isotope is approximately 11.55 years. See Figure 4–10.

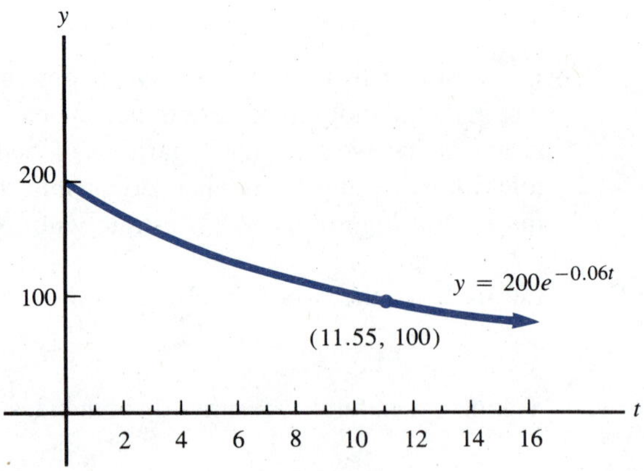

FIGURE 4–10

EXAMPLE 7 If \$10,000 is invested at 8 percent per year compounded quarterly, how long will it take until the investment is worth \$250,000?

We must solve the equation

$$250{,}000 = 10{,}000(1.02)^{4t}$$

for t. Dividing both sides by 10,000, taking ln of both sides, and simplifying yield

$$\ln 25 = 4t(\ln 1.02),$$

or

$$3.2189 = 4t(0.0198).$$

Therefore $t \approx 40.64$ years.

Computer Manual: Program LOGVALUE.

4.2 EXERCISES

Logarithms

In Exercises 1 through 14 evaluate the expression.

1. $\log_2 16$ **2.** $\log_2 32$ **3.** $\log_2 \frac{1}{16}$

4. $\log_2 \frac{1}{32}$ **5.** $\log_3 27$ **6.** $\log_5 1$

7. $\log_6 6$ **8.** $\log_5 \frac{1}{125}$ **9.** $\log_{10} 100{,}000$

10. $\log_{10} 0.0000001$ **11.** $\ln e^{2.85}$ **12.** $\ln e^{-7.8}$

13. $2^{\log_2 3}$ **14.** $e^{\ln 7}$

15. Using only the rules of logarithms and the facts that

$$\log_{10} 3 = 0.4771 \quad \text{and} \quad \log_{10} 5 = 0.6990,$$

find

(a) $\log_{10} 9$ **(b)** $\log_{10} 125$ **(c)** $\log_{10} 75$ **(d)** $\log_{10}\left(\frac{81}{25}\right)$

16. Find

(a) $\ln e$ **(b)** $\ln 1$ **(c)** $\ln 2.5$ **(d)** $\ln 12$ **(e)** $\ln \frac{5}{3}$

17. Show that the properties of logarithms (1)–(4), which were proved preceding the statement of the rules of logarithms, are actually special cases of the rules of logarithms.

18. Use the rules of exponents to prove the rules of logarithms.

Logarithmic Functions

In Exercises 19 through 24 graph the function.

19. $f(x) = \log_2 x$

20. $f(x) = \log_3 x$

21. $f(x) = \ln 3x$

22. $f(x) = 3(\ln x)$

23. $y = \ln x^3$

24. $y = \ln \frac{1}{x}$

Solving Exponential Equations

In Exercises 25 through 28 solve the equation for x.

25. $e^{2x} = 4$ **26.** $3e^{-3x} = 10$ **27.** $2^{5x} = 8$ **28.** $3^{-x/2} = 7$

29. If \$1000 is invested at 10 percent per year compounded **(a)** continuously and **(b)** quarterly, how long will it take until the investment is worth \$5000?

30. If \$1000 is invested at a certain annual interest rate, compounded continuously for 5 years, and at the end of that time has grown to \$2500, what was the interest rate?

Management

31. A firm's sales are given by the equation

$$y = 10{,}000e^{0.4t},$$

where y is annual sales, in dollars, in year t, and $t = 0$ represents 1986. How long will it take until the firm's annual sales reach \$100,000?

Life Science

32. A radioactive isotope is present in amount y grams at time t, where

$$y = 10e^{-0.025t}$$

and t is in years. Find the half-life of the isotope. How long will it take for 99 percent of the isotope to decay?

Social Science

33. The percentage of eligible voters who are actually registered to vote in a state is given by

$$y = 80e^{-0.0078t},$$

where y is the percentage of those eligible who are registered in year t, with $t = 0$ representing 1970. What percentage of eligible voters were registered in 1980? Find the year in which 60 percent of eligible voters will be registered.

4.3 DIFFERENTIATION OF LOGARITHMIC AND EXPONENTIAL FUNCTIONS

Now that we know what exponential and logarithmic functions are, we would like to be able to differentiate them, for as soon as we can do so we will be able to apply the procedures of calculus to the study of exponential and logarithmic models. In this section we develop differentiation formulas for logarithmic and exponential functions and demonstrate a technique of differentiation that uses logarithms.

Differentiation of Logarithmic Functions

Let $f(x) = \ln x$. To find $f'(x)$, we must take the limit as h approaches 0 of the difference quotient

$$\frac{f(x+h) - f(x)}{h} = \frac{\ln(x+h) - \ln x}{h}.$$

We rewrite the difference quotient using the rules of logarithms:

$$\begin{aligned}\frac{\ln(x+h) - \ln x}{h} &= \frac{1}{h}[\ln(x+h) - \ln x] \\ &= \frac{1}{h}\ln\frac{x+h}{x} \\ &= \ln\left(1 + \frac{h}{x}\right)^{1/h}.\end{aligned}$$

Now let $m = x/h$. Then

$$\frac{\ln(x+h) - \ln x}{h} = \ln\left(1 + \frac{1}{m}\right)^{m/x} = \ln\left[\left(1 + \frac{1}{m}\right)^m\right]^{1/x}.$$

But by definition,

$$\operatorname*{Lim}_{m \to +\infty}\left(1 + \frac{1}{m}\right)^m = e,$$

and it is possible to use the continuity of the natural logarithm function to show that

$$\operatorname*{Lim}_{m \to +\infty} \ln\left[\left(1 + \frac{1}{m}\right)^m\right]^{1/x} = \ln\left[\operatorname*{Lim}_{m \to +\infty}\left(1 + \frac{1}{m}\right)^m\right]^{1/x} = \ln e^{1/x}.$$

(The details of the argument are rather technical, and we omit them.) Thus, since $m = x/h \to +\infty$ as $h \to 0$, we have

$$f'(x) = \lim_{h \to 0} \frac{\ln(x+h) - \ln x}{h}$$

$$= \lim_{m \to +\infty} \ln\left[\left(1 + \frac{1}{m}\right)^m\right]^{1/x}$$

$$= \ln e^{1/x}$$

$$= \frac{1}{x} \ln e$$

$$= \frac{1}{x}.$$

The Derivative of ln *x*

If $f(x) = \ln x$, then

$$f'(x) = \frac{1}{x}.$$

EXAMPLE 1 If $f(x) = 2 \cdot \ln x$, then $f'(x) = 2(1/x) = 2/x$.

Now suppose that $u = g(x) > 0$ and consider $\frac{d}{dx}(\ln u)$. By the chain rule

$$\frac{d}{dx}(\ln u) = \frac{d}{du}(\ln u)\frac{du}{dx} = \frac{1}{u}\frac{du}{dx} = \frac{1}{g(x)}\frac{d}{dx}(g(x)).$$

Therefore we have

The Derivative of ln *g(x)*

If $f(x) = \ln g(x)$, then

$$f'(x) = \frac{g'(x)}{g(x)}.$$

EXAMPLE 2 If $f(x) = \ln(x^2 + 3x + 1)$, then

$$f'(x) = \frac{(x^2 + 3x + 1)'}{x^2 + 3x + 1} = \frac{2x + 3}{x^2 + 3x + 1}.$$

EXAMPLE 3 If $f(x) = x \cdot \ln x$, then by the product rule,

$$f'(x) = x(\ln x)' + (x)'\ln x = x\frac{1}{x} + 1 \cdot \ln x = 1 + \ln x.$$

EXAMPLE 4 Let us use the curve-sketching procedure to draw the graph of the function defined by

$$f(x) = x \ln x$$

over the interval $[e^{-2}, +\infty)$.

From Example 3, $f'(x) = 1 + \ln x$. Setting $f'(x) = 0$ and solving for x yield

$$1 + \ln x = 0,$$

or

$$\ln x = -1,$$

and hence

$$e^{\ln x} = e^{-1}.$$

Since $e^{\ln x} = x$, we have

$$x = e^{-1}.$$

Thus $x = e^{-1}$ is the only critical point. Since e^{-2} is in the interval $[e^{-2}, e^{-1})$, we may use it as a test value. Then

$$f'(e^{-2}) = 1 + \ln e^{-2} = 1 - 2 = -1$$

shows that $f'(x) < 0$ over the interval $[e^{-2}, e^{-1})$. Similarly, since e is in the interval $(e^{-1}, +\infty)$, we may use it as a test value. Then

$$f'(e) = 1 + \ln e = 1 + 1 = 2$$

implies that $f'(x) > 0$ over the interval $(e^{-1}, +\infty)$. Hence we have the diagram

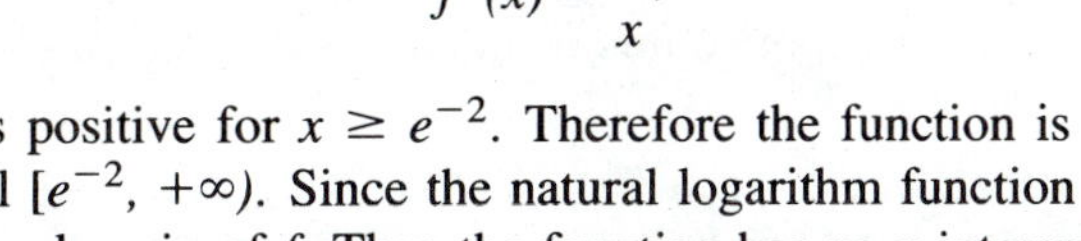

and there is a relative minimum at $(e^{-1}, f(e^{-1})) = (e^{-1}, -e^{-1})$.

The second derivative is

$$f''(x) = \frac{1}{x},$$

which is always positive for $x \geq e^{-2}$. Therefore the function is concave upward over the interval $[e^{-2}, +\infty)$. Since the natural logarithm function is not defined at 0, 0 is not in the domain of f. Thus the function has no y-intercept. If

$$x \ln x = 0,$$

then x 0 implies that $\qquad \ln x = 0,$

and hence that $\qquad e^{\ln x} = e^0 = 1.$

Thus $\qquad x = 1,$

and therefore the only x-intercept is the point $(1, 0)$. The function has no asymptotes. Its graph is shown in Figure 4–11.

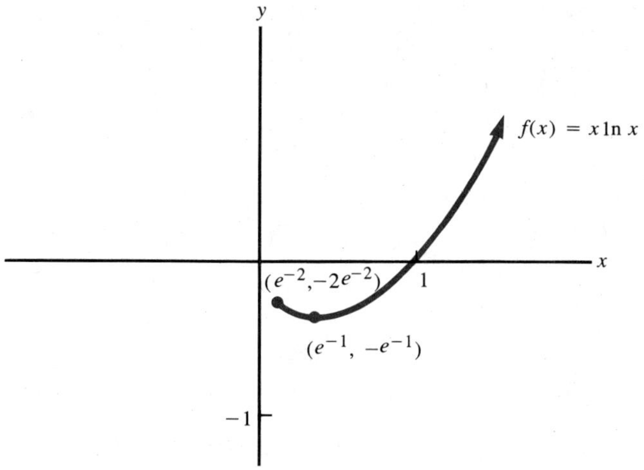

FIGURE 4–11

What about the derivative of a logarithmic function with base $b \neq e$? The rule for differentiating such a function is as follows:

The Derivative of $\log_b g(x)$

If $f(x) = \log_b g(x)$, then

$$f'(x) = \frac{1}{\ln b}\frac{g'(x)}{g(x)}.$$

A proof of this formula is requested in Exercise 15 at the end of this section.

EXAMPLE 5 If $f(x) = \log_2(x^2 + 1)$, then

$$f'(x) = \frac{1}{\ln 2}\frac{(x^2+1)'}{x^2+1} = \frac{2x}{(\ln 2)(x^2+1)}.$$

Differentiation of Exponential Functions

If we differentiate both sides of the identity

$$x = \ln e^x$$

using the rule for differentiating $\ln g(x)$, we find that

$$1 = \frac{(e^x)'}{e^x}.$$

Thus

$$(e^x)' = e^x.$$

The Derivative of e^x

If $f(x) = e^x$, then

$$f'(x) = e^x.$$

EXAMPLE 6 If $f(x) = 3e^x$, then $f'(x) = 3(e^x)' = 3e^x$.

If $f(x) = e^{g(x)}$, let $u = g(x)$. Then by the chain rule,

$$\frac{df}{dx} = \frac{d}{dx}(e^u) = \frac{d}{du}(e^u)\frac{du}{dx} = e^u\frac{du}{dx} = e^{g(x)}g'(x).$$

The Derivative of $e^{g(x)}$

If $f(x) = e^{g(x)}$, then

$$f'(x) = e^{g(x)}g'(x).$$

EXAMPLE 7 If $f(x) = e^{-3x}$, then

$$f'(x) = e^{-3x}(-3x)' = e^{-3x}(-3) = -3e^{-3x}.$$

EXAMPLE 8 If $f(x) = 2e^{x^2+x}$, then

$$f'(x) = 2e^{x^2+x}(x^2 + x)' = 2(2x + 1)e^{x^2+x}.$$

EXAMPLE 9 If $f(x) = xe^{-x}$, then by the product rule,

$$f'(x) = x(e^{-x})' + (x)'e^{-x} = xe^{-x}(-1) + 1 \cdot e^{-x} = e^{-x}(1 - x).$$

EXAMPLE 10 Let us sketch the graph of the function defined by

$$f(x) = xe^{-x}.$$

Setting $f'(x)$ equal to 0, we have

$$e^{-x}(1 - x) = 0.$$

Since e to any power is positive, this implies that $x = 1$. The diagram of the first-derivative test looks like this:

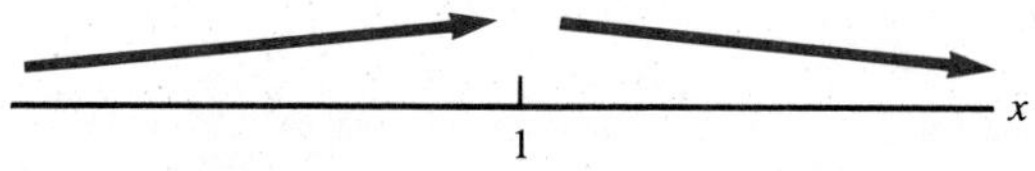

(Check this.) Therefore there is a relative maximum at $(1, f(1)) = (1, e^{-1})$.

The second derivative is

$$\begin{aligned} f''(x) &= e^{-x}(1 - x)' + (e^{-x})'(1 - x) \\ &= -e^{-x} - e^{-x}(1 - x) \\ &= e^{-x}(x - 2). \end{aligned}$$

Setting the second derivative equal to zero and solving for x yield $x = 2$. Since $f''(x) < 0$ when $x < 2$ and $f''(x) > 0$ when $x > 2$, the concavity diagram is

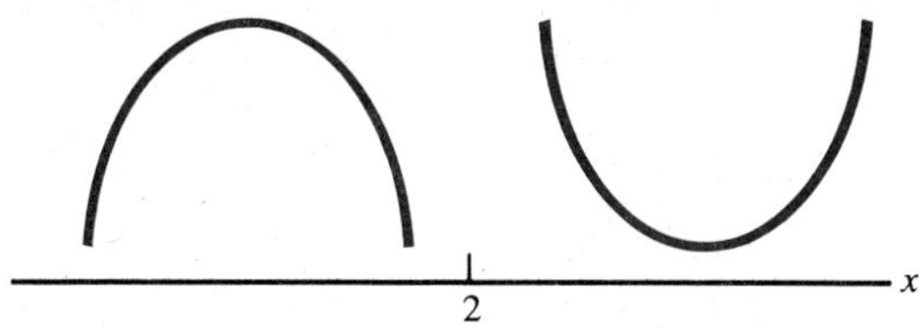

(Check this.) Hence there is a point of inflection at $(2, f(2)) = (2, 2e^{-2})$. The point (0, 0) is the only intercept. The function has no vertical asymptotes. However, as $x \to +\infty$, the value of $f(x) = xe^{-x}$ approaches zero. (Check this.) Therefore $\text{Lim}_{x\to+\infty} xe^{-x} = 0$, and the x-axis is an asymptote as $x \to +\infty$. The graph of the function is shown in Figure 4–12.

What about the derivative of an exponential function with base $b \neq e$? The rule for differentiating such a function is as follows:

The Derivative of $b^{g(x)}$

If $f(x) = b^{g(x)}$, then

$$f'(x) = (\ln b)b^{g(x)}g'(x).$$

A proof of this formula is requested in Exercise 34 at the end of this section.

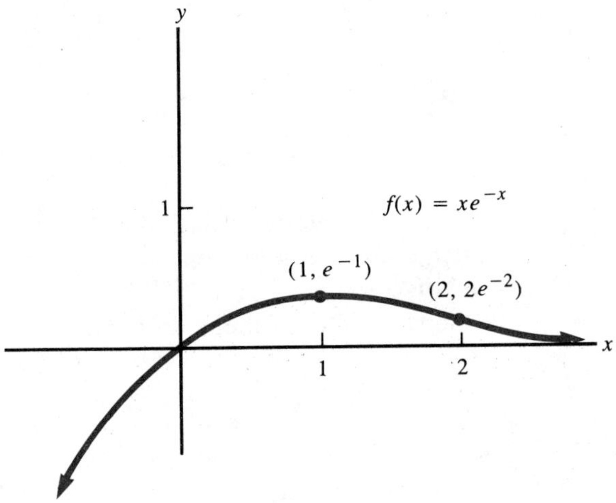

FIGURE 4–12

EXAMPLE 11 If $f(x) = 3^{2x}$, then

$$f'(x) = (\ln 3)3^{2x}(2x)' = 2(\ln 3)3^{2x}.$$

Note that the formula for the derivative of $e^{g(x)}$ is simpler than the formula for the derivative of $b^{g(x)}$. This is one reason why exponential functions with base e are usually preferred to those with other bases. Similar remarks apply to the preference for the natural logarithm as opposed to logarithms with bases other than e.

Logarithmic Differentiation

Sometimes it is convenient, or even necessary, to take the logarithm of both sides of an expression before we differentiate it. We illustrate with an example.

EXAMPLE 12 Let us differentiate $f(x) = x^2(x^2 + 1)^2(x^4 + 1)^4$. We could do this by repeatedly using the product rule and the chain rule, but we prefer to take the natural logarithm of both sides of the equation, differentiate the result, and solve for $f'(x)$. We have

$$\ln f(x) = \ln x^2(x^2 + 1)^2(x^4 + 1)^4,$$

or

$$\ln f(x) = \ln x^2 + \ln(x^2 + 1)^2 + \ln(x^4 + 1)^4,$$

which simplifies to

$$\ln f(x) = 2 \cdot \ln x + 2 \cdot \ln(x^2 + 1) + 4 \cdot \ln(x^4 + 1).$$

Now we differentiate both sides of the last equation to obtain

$$\frac{f'(x)}{f(x)} = \frac{2}{x} + 2\left(\frac{2x}{x^2 + 1}\right) + 4\left(\frac{4x^3}{x^4 + 1}\right),$$

so that
$$f'(x) = f(x)\left(\frac{2}{x} + \frac{4x}{x^2 + 1} + \frac{16x^3}{x^4 + 1}\right),$$

or
$$f'(x) = x^2(x^2 + 1)^2(x^4 + 1)^4\left(\frac{2}{x} + \frac{4x}{x^2 + 1} + \frac{16x^3}{x^4 + 1}\right).$$

The technique used in Example 12 is called **logarithmic differentiation.** It is often useful when the function to be differentiated is the product or quotient of several factors.

Sometimes logarithmic differentiation is the only way to proceed.

EXAMPLE 13 Let f be given by $f(x) = x^x$. None of our rules of differentiation apply to this function: it is neither of the form x^r, where r is a fixed real number, nor of the form b^x, where b is a fixed real number. In order to find $f'(x)$, we must use logarithmic differentiation:

$$\ln f(x) = \ln x^x,$$

so

$$\ln f(x) = x \ln x.$$

Differentiating both sides of this last equation, we obtain

$$\frac{f'(x)}{f(x)} = x\frac{1}{x} + 1 \cdot \ln x.$$

Therefore

$$f'(x) = f(x)[1 + \ln x] = x^x[1 + \ln x].$$

Computer Manual: Program DFQUOTNT.

4.3 EXERCISES

Differentiation of Logarithmic Functions

In Exercises 1 through 10 find $f'(x)$.

1. $f(x) = 3 \cdot \ln x$ **2.** $f(x) = \ln x^2$ **3.** $f(x) = \ln(7x + 2)$

4. $f(x) = \ln(x^2 + 5x - 3)$ **5.** $f(x) = x^2 \ln 2x$ **6.** $f(x) = \ln \dfrac{2x + 1}{3x + 2}$

7. $f(x) = (\ln x)^3$ **8.** $f(x) = \ln(\ln x)$ **9.** $f(x) = \ln(\ln(\ln x))$ **10.** $f(x) = \ln |x|$

In Exercises 11 through 14 use the curve-sketching procedure to draw the graph of the function.

11. $f(x) = (\ln x)^2$ **12.** $y = \dfrac{\ln x}{x}$, over $[e^{-2}, +\infty)$

13. $y = \dfrac{x}{\ln x}$, over $[e^{-2}, +\infty)$ **14.** $f(x) = \ln(ax + b)$, $a > 0, b > 0$

15. Prove the formula for the derivative of $\log_b g(x)$. (*Hint:* Show that

$$\log_b g(x) = \frac{\ln g(x)}{\ln b};$$

then differentiate.)

16. Find $f'(x)$ if

(a) $f(x) = \log_2 x$ **(b)** $f(x) = \log_3 5x$ **(c)** $f(x) = x^2 \log_{10} x^3$

Management

17. A firm's cost function is

$$C(x) = \ln \sqrt{x^2 + 2}$$

and its revenue function is

$$R(x) = \ln(100x + 1),$$

both valid for $x \geq 1$. Here cost and revenue are in millions of dollars. Find the firm's maximum profit.

18. If steel production in year t was y million tons, where

$$y = 96 \ln(t + 3)$$

and $t = 0$ represents 1945, find the rate at which steel production was changing in 1960, 1970, and 1980.

Life Science

19. The concentration of radioactive iodine in seawater increased during the period 1945–1965 due to aboveground testing of atomic weapons. Suppose the concentration of such iodine in year t, where $t = 0$ represents 1945, was

$$y = 3 \ln(5 + t)$$

parts per billion. Find the rate at which the iodine concentration was changing in 1955 and 1960.

Differentiation of Exponential Functions

In Exercises 20 through 29 find $f'(x)$.

20. $f(x) = e^{2x}$ **21.** $f(x) = 2e^{-x}$ **22.** $f(x) = -12e^{-x/2}$

23. $f(x) = 8e^{3x}$ **24.** $f(x) = 2e^{x^3}$ **25.** $f(x) = 5e^{x^2+3x+4}$

26. $f(x) = 2xe^{-4x}$ **27.** $f(x) = 3x^2e^{2x}$ **28.** $f(x) = e^{-x} \ln x$ **29.** $f(x) = \dfrac{e^x}{\ln x}$

In Exercises 30 through 33 use the curve-sketching procedure to draw the graph of the function.

30. $f(x) = xe^x$ **31.** $y = \dfrac{e^x}{x}$ **32.** $y = e^{-x^2}$

33. **(a)** $y = a - be^{-rx}, \quad a > b > 0, \quad r > 0$ **(b)** $y = a + be^{-rx}, \quad a > 0, \quad b > 0, \quad r > 0$

34. Prove the formula for the derivative of $b^{g(x)}$. (*Hint:* Show that

$$b^{g(x)} = e^{g(x)(\ln b)};$$

then differentiate.)

35. Find $f'(x)$ if

(a) $f(x) = 2^x$ **(b)** $f(x) = 3(5^{-2x})$ **(c)** $f(x) = 10^{e^x}$

Management

36. A firm's profit in year t is given by $P(t) = 100{,}000e^{-0.2t}$, where $t = 0$ represents 1985. Find the rate at which the firm's profit will be changing in 1990 and in 1995.

37. A company's average cost per unit is given by

$$f(x) = 2 + 3e^{-0.1x},$$

and its average revenue per unit by

$$g(x) = 5 - e^{-0.2x}.$$

Find the rate of change of average cost per unit, average revenue per unit, and average profit per unit when $x = 10$.

Social Science

38. A demographer predicts that the population (in thousands) of a city in year t will be

$$y = 700 + 100e^{-0.1t},$$

where $t = 0$ represents 1990. Find the rate at which the population of the city will be changing in the year 2000.

Logarithmic Differentiation

In Exercises 39 through 43 use logarithmic differentiation to find $f'(x)$.

39. $f(x) = x(x^3 + 1)^3(x^5 + 1)^5$

40. $f(x) = \dfrac{(x^2 + 3x + 5)^8(x^6 - 7x + 5)^4}{(x^3 - 8x)^2}$

41. $f(x) = x^{2x}$

42. $f(x) = x^{e^x}$

43. $f(x) = x^{x^x}$

4.4 APPLICATION: GROWTH AND DECLINE

If a quantity changes in such a way that its rate of change at time t depends directly on the amount of it present at time t, then its size can be modeled with an exponential function. We saw one example of this when we studied compound interest: the rate of growth of an investment at time t depends directly on the amount of the investment at time t, because the rate of growth is the interest earned, which is a percentage of the amount of the investment. It was because of this that we were able to model the size of the investment at time t by an exponential function and thus derive the compound interest formula. In this section we show how to construct exponential models for situations of growth and decline. We also briefly consider logarithmic models for growth.

Exponential Growth and Decline

Suppose we have a situation in which the rate of change of a quantity at time t depends directly on the amount of the quantity present at time t. Also suppose that the amount present is increasing with time. Then we may model the situation with an exponential function whose defining equation is of the form

$$y = ce^{rt}, \qquad c > 0, \quad r > 0.$$

Note that the rate of change of y does indeed depend directly on y:

$$\frac{dy}{dt} = rce^{rt} = ry.$$

The graph of

$$y = ce^{rt}, \qquad c > 0, \quad r > 0$$

is shown in Figure 4–13. Note that as t increases, y increases at an increasing rate. Situations that can be modeled with a function of this form are said to be situations of **exponential growth.**

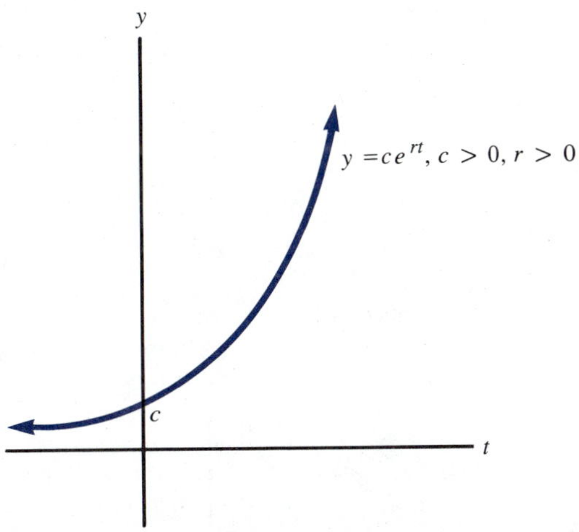

FIGURE 4–13

EXAMPLE 1 Suppose we have reason to believe that the rate of population growth for a caribou herd depends directly on the size of the herd. Suppose also that there were 10,000 caribou in the herd in 1985 and 11,000 in 1987, so that the size of the herd is increasing. We may then model the size of the herd by an exponential function whose defining equation is of the form

$$y = ce^{rt}, \qquad c > 0, \quad r > 0.$$

To complete the model, we must find numbers c and r such that the resulting function describes the size of the herd.

If we measure time in years, with $t = 0$ representing 1985, we must have

$$y = 10{,}000 \quad \text{when } t = 0 \qquad \text{and} \qquad y = 11{,}000 \quad \text{when } t = 2.$$

Substituting 0 for t and 10,000 for y in the model yields

$$10{,}000 = ce^{r(0)} = ce^0 = c \cdot 1 = c.$$

Therefore $c = 10{,}000$, so

$$y = 10{,}000e^{rt}.$$

To find r, we substitute 2 for t and 11,000 for y to obtain

$$11{,}000 = 10{,}000e^{r(2)} = 10{,}000e^{2r}.$$

Solving the exponential equation

$$11{,}000 = 10{,}000e^{2r}$$

for r, we have

$$\frac{11{,}000}{10{,}000} = e^{2r},$$

or

$$1.1 = e^{2r},$$

so

$$\ln 1.1 = \ln e^{2r} = 2r,$$

or

$$0.095310 = 2r.$$

Thus $r = 0.05$, rounded to two decimal places, and our model is

$$y = 10{,}000e^{0.05t}.$$

The graph of this function is shown in Figure 4–14.

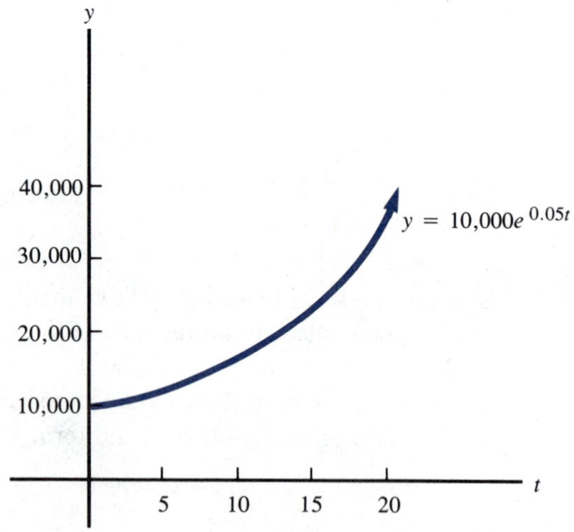

FIGURE 4–14

Now we can use the model for prediction. For instance, if we wish to predict the size of the herd in 1995, we need only substitute 10 for t in the model:

$$y = 10{,}000e^{0.05(10)} = 10{,}000e^{0.5} = 10{,}000(1.6487) = 16{,}487.$$

Similarly, if we wish to predict how long after 1985 it will take for the size of the herd to grow to 20,000 caribou, then we must substitute 20,000 for y and solve the exponential equation

$$20{,}000 = 10{,}000e^{0.05t}$$

for t. The solution is $t \approx 13.9$ years. (Check this.)

Again assume that we have a situation where the rate of change of some quantity at time t depends directly on the amount of it present at time t, but now suppose that the amount present is decreasing with time. Then we may model the situation with an exponential function whose defining equation is of the form

$$y = ce^{-rt}, \qquad c > 0, \quad r > 0.$$

Figure 4–15 shows the graph of such a function. Note that as t increases, y decreases at a decreasing rate. Situations that can be modeled by such a function are said to be situations of **exponential decline** or **exponential decay.**

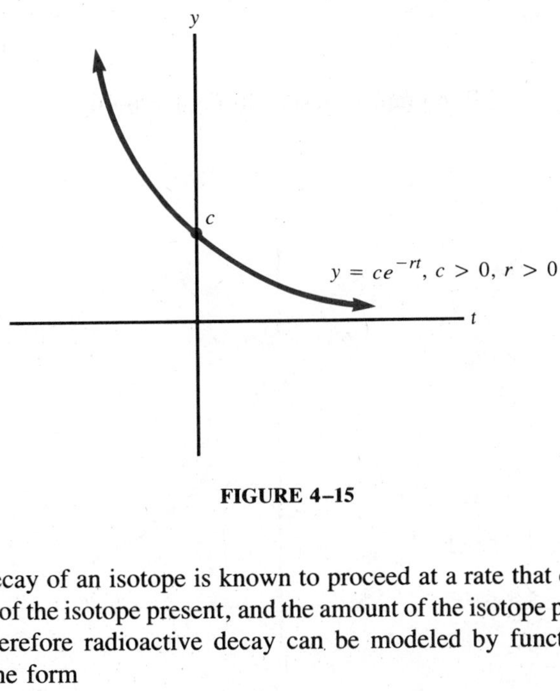

FIGURE 4–15

EXAMPLE 2 Radioactive decay of an isotope is known to proceed at a rate that depends directly on the amount of the isotope present, and the amount of the isotope present decreases with time. Therefore radioactive decay can be modeled by functions defined by equations of the form

$$y = ce^{-rt}, \qquad c > 0, \quad r > 0.$$

For instance, the radioactive isotope of uranium known as U^{235} has a half-life of 12 million years. (Thus any amount of U^{235} will decrease by one half in 12 million years.) Let us measure time in millions of years. When $t = 0$,

$$y = ce^{-r(0)} = ce^0 = c \cdot 1 = c,$$

so U^{235} is initially present in amount c; but since its half-life is 12 million years, when $t = 12$ we must have $y = \frac{1}{2}c$:

$$\frac{1}{2}c = ce^{-r(12)} = ce^{-12r}.$$

Solving this exponential equation for r, we find that

$$\frac{1}{2} = e^{-12r},$$

or

$$\ln\frac{1}{2} = \ln e^{-12r},$$

or

$$-0.693147 = -12r,$$

so $r = 0.06$, rounded to two decimal places. Therefore the model is

$$y = ce^{-0.06t}.$$

The graph is shown in Figure 4–16.

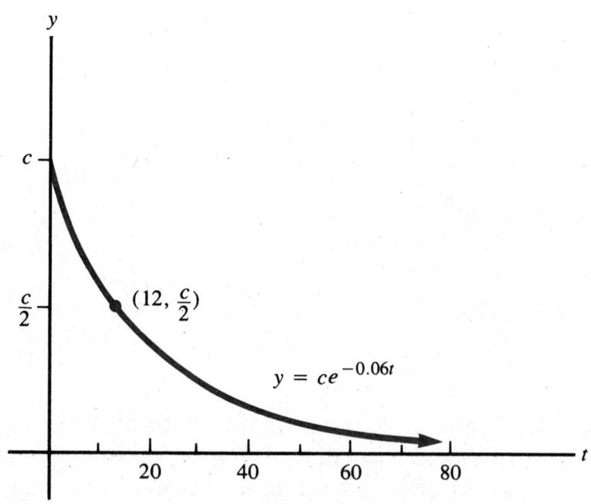

FIGURE 4–16

We see that at the end of 50 million years approximately 5 percent of the original amount of U^{235} will be left, because when $t = 50$ we have

$$y = ce^{-0.06(50)} = ce^{-3} = c(0.049787) \approx 0.05c.$$

How long will it be until only 1 percent of the original amount of U^{235} remains? One percent of the original amount is $0.01c$, and if we set $y = 0.01c$ in the model and solve

$$0.01c = ce^{-0.06t}$$

for t, we find that $t \approx 76.8$ million years. (Check this.)

Logarithmic Growth

Functions defined by equations of the form

$$y = \ln(at + b), \qquad a > 0, \quad b > 0,$$

are sometimes used to model situations of growth in which the rate of growth is decreasing. Figure 4–17 shows the graph of such a function when $b > 1$. Note that as t increases, y increases at a decreasing rate. A situation that can be modeled by such a function is said to be one of **logarithmic growth.**

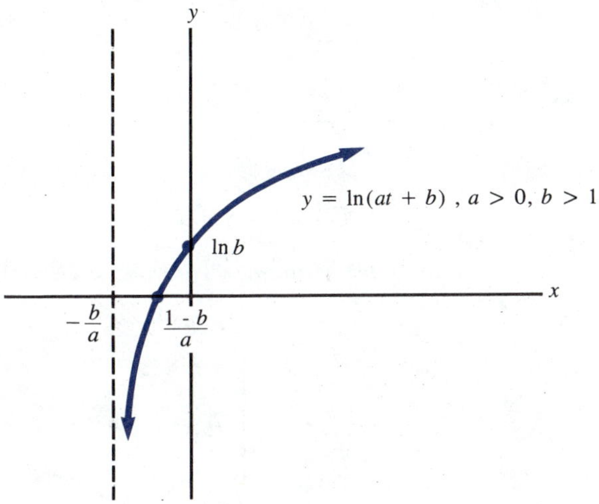

FIGURE 4–17

EXAMPLE 3 The amount of steel produced in a certain country has been increasing at a decreasing rate. The amount produced was 4 million tons in 1984 and 6 million tons in 1986. Let us describe the country's steel production with a model of the form

$$y = \ln(at + b), \qquad a > 0, \quad b > 0,$$

where y is in millions of tons of steel and t is in years, with $t = 0$ representing 1984.

When $t = 0$, $y = 4$, so

$$4 = \ln(a(0) + b) = \ln b.$$

But

$$4 = \ln b$$

implies that

$$e^4 = e^{\ln b} = b.$$

Similarly, since $y = 6$ when $t = 2$, we have

$$6 = \ln(2a + b) = \ln(2a + e^4),$$

so

$$e^6 = e^{\ln(2a+e^4)} = 2a + e^4,$$

which shows that

$$a = \frac{1}{2}(e^6 - e^4).$$

Converting a and b to decimal form, we have

$$a = \frac{1}{2}(e^6 - e^4) \approx 174.4, \qquad b = e^4 \approx 54.6.$$

Therefore our model is

$$y = \ln(174.4t + 54.6).$$

The graph is shown in Figure 4–18.

Computer Manual: Programs EXPGRWTH and LOGRWTH.

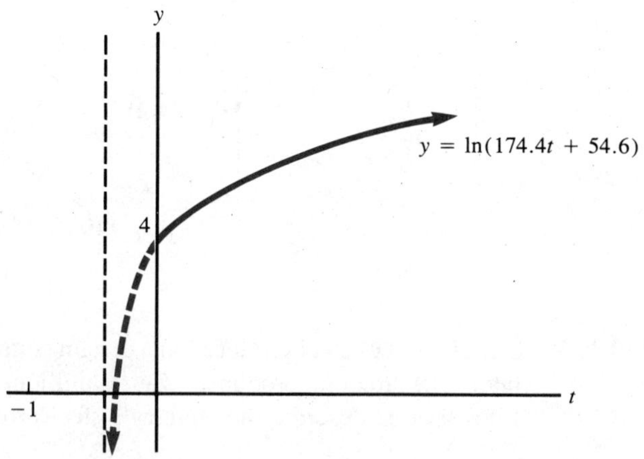

FIGURE 4–18

4.4 EXERCISES

Exponential Growth and Decline

In Exercises 1 through 14 assume that the rate of change of the quantity under discussion depends directly on the amount of it present.

Management

1. The annual sales of Ampersand, Inc. have been increasing. They were \$100,000 in 1980 and \$739,000 in 1985. Find

(a) a function that describes Ampersand's annual sales;
(b) the firm's sales in 1987;
(c) when Ampersand's sales will be \$15 million per year.

2. Battledore Company stopped advertising its product on January 1, and since then sales have been declining. January sales were 50,000 units, and February sales were 45,000 units. Find

(a) a function that describes Battledore's monthly sales;
(b) the company's sales in March;
(c) when sales will be 20,000 units per month.

3. The annual profits of Grimditch Associates were \$2.5 million in 1980 and \$1.4 million in 1983. Find

(a) a function that describes Grimditch's annual profits;
(b) the firm's profits in 1990;
(c) when Grimditch's annual profits will be \$250,000.

4. Dynamic Corporation's revenues have been doubling every 10 years and were \$1 million in 1985. Predict Dynamic's 1994 revenues. When will Dynamic's annual revenues attain the level of \$5.5 million per year?

5. The average cost per unit for Elwin Company was \$80 in 1980, and it has been declining by 50 percent every 20 years. When will Elwin's average cost per unit be \$30?

6. Under conditions of inflation the purchasing power of money declines. Suppose inflation persists at an annual rate of 10 percent per year: find a function that describes the purchasing power of a future dollar compared with today's dollar, and then predict the purchasing power of a dollar 10 years from today.

7. The demand for a commodity was 95,123 units when the price was \$1 per unit, and 90,484 units when the price was \$2 per unit. Find the demand function for the commodity and also find the quantity demanded when the price is \$5 per unit.

8. The supply of the commodity of Exercise 7 was 20,816 units when the price was \$1 per unit, and 23,470 units when the price was \$4 per unit. Find the supply function for the commodity and then find its equilibrium price.

Life Science

9. A radioactive isotope has a half-life of 16 days. Find

(a) a model that describes the amount of the isotope present as a function of time;
(b) the percentage of the original amount remaining at the end of 10 days;
(c) how long it will take for 90 percent of the isotope to decay.

10. The number of bacteria in a culture is growing. If there were 200 bacteria present to start with and 445 present 2 hours later, how long will it take until there are 10,000 bacteria?

11. Biologists are attempting to revive a lake that has suffered from acid rain. It is estimated that there are currently 80,000 fish in the lake, and that over the next 5 years this number will increase to 120,000 as the lake continues to revive. Find the number of fish the lake will support when the revivification effort is complete 12 years from now.

12. The number of cases of a new disease was 500 in 1983 and has been doubling every 18 months. Find a function that describes the number of cases of the disease as a function of time and use it to predict the number of cases in 1995.

Social Science

13. The population of the earth was approximately 4.85 billion in 1984 and 5 billion in 1986. Find a function that can be used to estimate the earth's population in the future. Estimate the earth's population in the year 2050. When will the population reach 100 billion?

14. A psychologist studying learning teaches a group of subjects a sequence of tasks, then tells them not to practice the tasks. After 10 days without practice, the typical subject can perform only one fifth of the tasks. Find a model that describes the proportion of tasks performed as a function of the time without practice. Predict the performance level of a subject who goes 2 weeks without practice.

Logarithmic Growth

In Exercises 15 through 18 use a logarithmic growth model.

Management

15. Annual aluminum production in a country was 7 million tons in 1950 and 10 million tons in 1980. Find a model that describes aluminum production in the country and predict
    (a) production in 1990;
    (b) the year in which production will reach 11 million tons.

16. When a firm spent \$200,000 on advertising, its sales were \$1.5 million; when it spent \$250,000 on advertising, its sales were \$1.6 million. Find a model that describes sales as a function of the amount spent on advertising. How much must the firm spend on advertising in order to have sales of \$2 million?

Life Science

17. If the concentration of radioactive iodine in seawater was 4.8 parts per million (ppm) in 1945 and 8.1 ppm in 1955, find a model that describes the concentration of such iodine as a function of time. The increase in radioactive iodine was largely a consequence of aboveground testing of atomic weapons. If damage to some forms of sea life would have occurred at a concentration of 9 ppm, when would such damage have begun had aboveground testing of atomic weapons not been halted in the early 1960s?

Social Science

18. If a state's aid to higher education was \$26.5 million in 1985 and \$27.2 million in 1986, predict the amount of such aid in 1990.

4.5 APPLICATION: LIMITED GROWTH AND DECLINE

The exponential models of growth and decline that we discussed in Section 4.4 assumed that the growth or decline was unlimited. In many circumstances this is unrealistic. For instance, the size of the caribou herd of Example 1, Section 4.4, certainly cannot grow without limitation, since, for one thing, there will be only a certain amount of food available for the caribou. Thus if we wish to make our models realistic, we must sometimes allow them to assume limited growth or decline. In this section we discuss such models and apply them to questions of population growth, learning, and product maturity.

Consider a function defined by an equation of the form

$$y = a - ce^{-rt}, \qquad a > c > 0, \quad r > 0$$

over the interval $[0, +\infty)$. Figure 4–19 shows the graph of such a function. Note the following:

1. As t increases, y grows.
2. As t increases, the rate at which y grows decreases.
3. As $t \to +\infty$, growth levels off and y approaches the limiting value a from below.

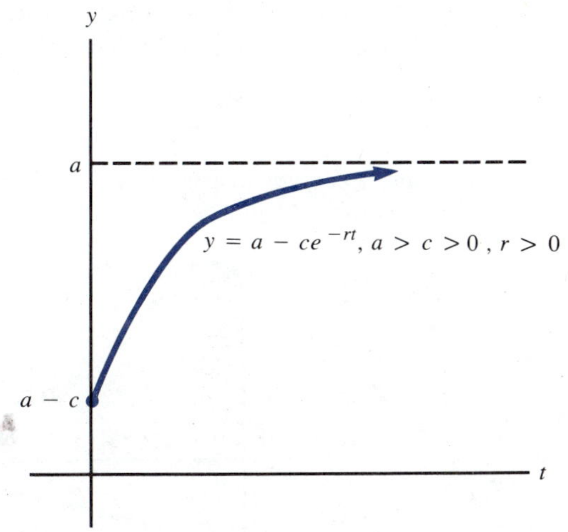

FIGURE 4–19

Functions of this type are often used to model situations of limited growth when the rate of growth decreases with time and some limiting value is approached from below.

EXAMPLE 1 Consider again the caribou herd of Example 1, Section 4.4. As before, suppose the size of the herd is 10,000 in 1985 and 11,000 in 1986, but now we will assume that as time goes on the growth of the herd will slow down and its size will be limited by the food supply to no more than 15,000. If we model the size of the herd by a function whose defining equation is of the form

$$y = a - ce^{-rt},$$

where $t = 0$ represents 1985, then the limiting value $a = 15{,}000$, so the model is

$$y = 15{,}000 - ce^{-rt}.$$

Since $y = 10{,}000$ when $t = 0$, we have

$$10{,}000 = 15{,}000 - ce^{-r(0)} = 15{,}000 - c.$$

Therefore $c = 5000$, and the model is

$$y = 15{,}000 - 5000e^{-rt}.$$

Finally, since $y = 11{,}000$ when $t = 2$, we have

$$11{,}000 = 15{,}000 - 5000e^{-r(2)},$$

or

$$-4000 = -5000e^{-2r},$$

or

$$0.80 = e^{-2r}.$$

Taking the natural logarithm of both sides of this last equation and solving for r give $r = 0.11$, rounded to two decimal places. (Check this.) Thus the final model is

$$y = 15{,}000 - 5000e^{-0.11t}.$$

The graph is shown in Figure 4–20. With this model our prediction for the size of the herd in 1995 is

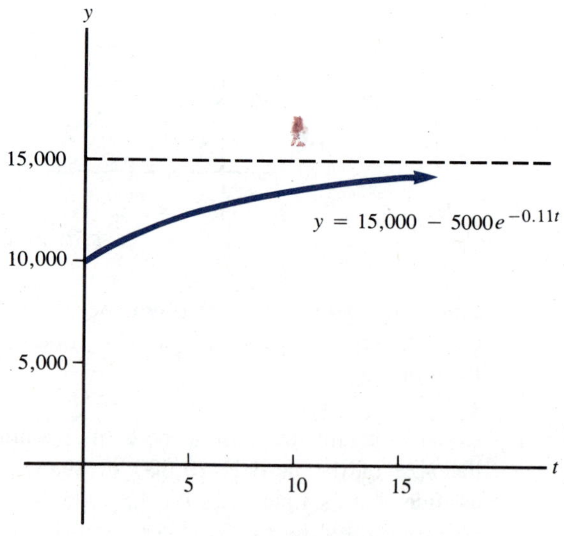

FIGURE 4–20

$$y = 15{,}000 - 5000e^{-0.11(10)} = 15{,}000 - 5000e^{-1.1} = 13{,}336.$$

An important application of limited growth is the **learning curve.** Learning curves relate productivity to experience. Typically, productivity increases as employees become more experienced, since with experience they become more efficient. Productivity cannot grow without limit, however, for even perfect efficiency would result in a certain finite productivity. Thus productivity increases toward some limit, so it can be described by a model of the form

$$y = a - ce^{-rt}.$$

EXAMPLE 2 Data collected at a plant that makes microwave carts show that a worker who has no experience can assemble carts at a rate of approximately 200 per week, whereas a worker who has 5 weeks of experience can assemble them at a rate of approximately 233 per week. It is also known that the most experienced and efficient workers can approach an assembly rate of 300 carts per week. Let us find the learning curve for a typical worker at the plant.

We use the model

$$y = a - ce^{-rt},$$

where t is worker experience, measured in weeks, and y is carts assembled per week. We may take the limiting value a to be 300, so our model is

$$y = 300 - ce^{-rt}.$$

We know that $y = 200$ when $t = 0$, and $y = 233$ when $t = 5$; hence

$$200 = 300 - ce^{-r(0)} = 300 - c,$$

so $c = 100$. But then

$$y = 300 - 100e^{-rt},$$

and the fact that $y = 233$ when $t = 5$ tells us that

$$233 = 300 - 100e^{-5r},$$

or

$$-67 = -100e^{-5r},$$

or

$$0.67 = e^{-5r}.$$

Taking the natural logarithm of both sides of this last equation and solving for r, we find that $r = 0.08$, to two decimal places. Hence the learning curve is

$$y = 300 - 100e^{-0.08t}.$$

Figure 4–21 shows the graph of this learning curve.

Now suppose that a worker must assemble at least 250 carts per week in order to cover the cost of his or her salary. Since $y \approx 247$ when $t = 8$ and $y \approx 251$ when $t = 9$, a worker cannot be expected to cover the cost of his or her salary until the ninth week on the job.

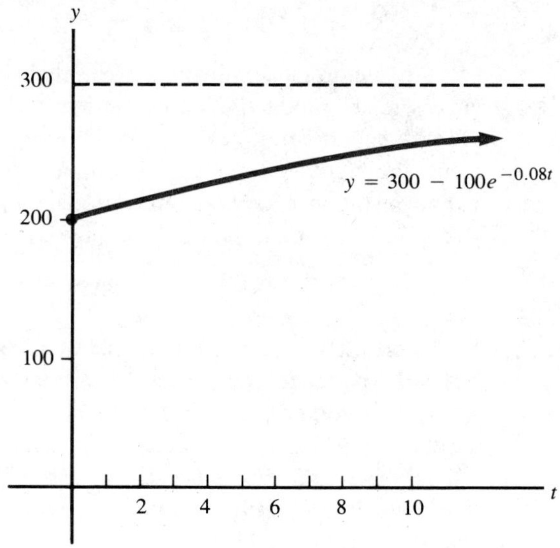

FIGURE 4–21

Over the interval $[0, +\infty)$ a function whose defining equation is of the form

$$y = a + ce^{-rt}, \qquad a > 0, \quad c > 0, \quad r > 0$$

has the graph shown in Figure 4–22. Note the following:

1. As t increases, y declines.
2. As t increases, the rate at which y declines decreases.
3. As $t \to +\infty$, the decline levels off and y approaches the limiting value a from above.

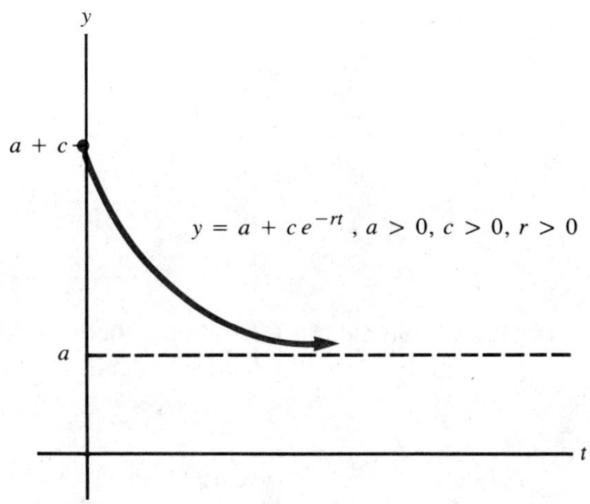

FIGURE 4–22

Therefore functions of this type are often used to model situations of limited decline when the rate of decline decreases with time and some limiting value is approached from above.

To illustrate, let us consider learning curves from a different viewpoint. In Example 2 we were interested in productivity as a function of experience, but sometimes it is more important to consider production time as a function of the number of units produced. What typically occurs is that each unit can be produced in slightly less time than the preceding one because production becomes more efficient with experience. Even with perfect efficiency, however, there must be some minimum time necessary to produce a unit. Hence the time required to produce a unit will approach some limiting value from above.

EXAMPLE 3 An airplane manufacturer about to begin production of a new plane estimates that it will take 5000 worker-hours to build the first plane, 4600 worker-hours to build the second one, and that it would take 2800 worker-hours to build a plane under conditions of perfect efficiency. Let us find the learning curve that describes the number of worker-hours needed to build the xth plane, $x \geq 1$.

Our model will be of the form

$$y = a + ce^{-rx}.$$

Since the limiting value is 2800 worker-hours, we may take $a = 2800$, so that

$$y = 2800 + ce^{-rx}.$$

When $x = 1$, $y = 5000$, and when $x = 2$, $y = 4600$. Therefore

$$5000 = 2800 + ce^{-r}$$

and

$$4600 = 2800 + ce^{-2r},$$

or

$$2200 = ce^{-r}$$

and

$$1800 = ce^{-2r}.$$

We divide the first equation by the second to obtain

$$\frac{2200}{1800} = \frac{ce^{-r}}{ce^{-2r}} = e^{r}.$$

Solving for r yields $r = 0.20$, and then

$$2200 = ce^{-0.20}$$

implies that $c = 2687$. Therefore our learning curve is

$$y = 2800 + 2687e^{-0.20x}.$$

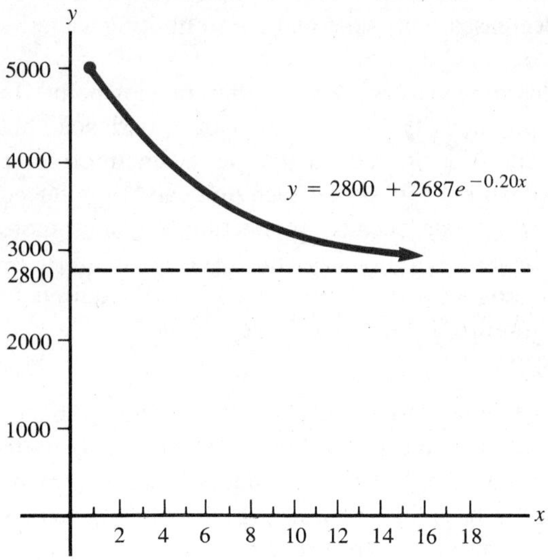

FIGURE 4–23

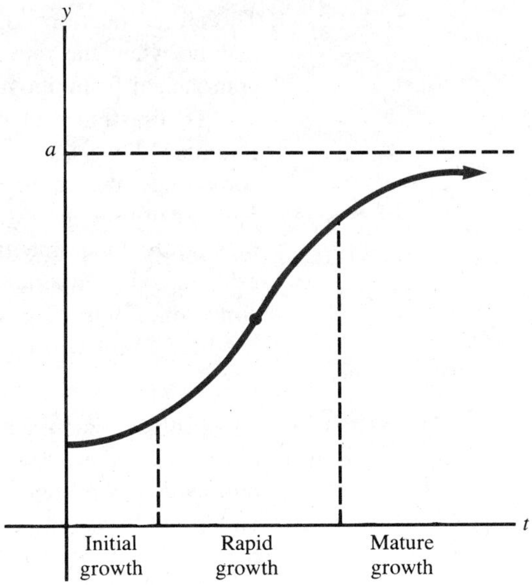

FIGURE 4–24

The graph is shown in Figure 4–23. If we wished, we could estimate the total number of worker-hours required to build the first 25 planes, say, by substituting $x = 1, 2, \ldots, 25$ into the model and summing the resulting y-values.

Another type of limited growth is that which can be modeled by an **S-curve,** such as is shown in Figure 4–24. Note that the *S*-curve can be divided into regions of initial growth, rapid growth, and mature growth. Note also that the rate of growth increases until the point of inflection is reached, but thereafter it decreases. The growths of individual companies, industries, and even entire economies often follow *S*-curves.

Several types of functions have *S*-curves as their graphs. One is the **logistic function** defined by

$$y = \frac{a}{1 + ce^{-rt}}, \qquad a > 0, \quad r > 0, \quad c > 1.$$

(See Exercise 1 at the end of this section.)

EXAMPLE 4 A new video game is placed in arcades, and t weeks later revenue from the game is running at a level of y thousand dollars per week, where

$$y = \frac{2}{1 + 9e^{-0.2t}}.$$

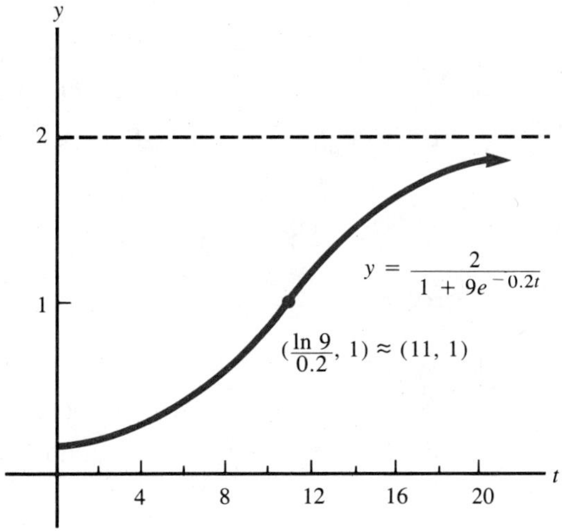

FIGURE 4–25

The graph of this function is shown in Figure 4–25. Note that revenue grows slowly for the first few weeks but then increases rapidly as the game gains popularity. After the 11th week revenue growth begins to slow, and ultimately levels off at approximately \$2000 per week.

We conclude this section by noting that in some instances of limited growth or decline the limiting value is not known beforehand. In such a case the limiting value must be found by substitution in the model.

EXAMPLE 5 The growth of bacteria in a culture is limited by the nutrients present. It is observed that at the start of an experiment a culture contains 1000 bacteria; 1 hour later it contains 5000 bacteria, and 2 hours later 7000 bacteria. If y denotes the number of bacteria in the culture at time t, t in hours, our limited growth model will be

$$y = a - ce^{-rt}.$$

We do not know the limiting value a for the number of bacteria in the culture and will have to find it from the data.

When $t = 0$, $y = 1000$, so

$$1000 = a - ce^{-r(0)} = a - c,$$

which implies that $a = 1000 + c$. But then

$$5000 = a - ce^{-r(1)} = 1000 + c - ce^{-r} = 1000 + c(1 - e^{-r})$$

and

$$7000 = a - ce^{-r(2)} = 1000 + c - ce^{-2r} = 1000 + c(1 - e^{-2r}),$$

or
$$4000 = c(1 - e^{-r})$$
and
$$6000 = c(1 - e^{-2r}).$$

If we divide the second equation by the first and use the fact that $1 - e^{-2r} = (1 - e^{-r})(1 + e^{-r})$, we obtain

$$\frac{6000}{4000} = \frac{c(1 - e^{-2r})}{c(1 - e^{-r})} = \frac{(1 - e^{-r})(1 + e^{-r})}{1 - e^{-r}} = 1 + e^{-r}.$$

Therefore
$$1.5 = 1 + e^{-r},$$
which yields
$$r = 0.693,$$
$$c = \frac{4000}{1 - e^{-r}} = \frac{4000}{0.5} = 8000,$$
$$a = 1000 + c = 9000.$$

Thus the limiting value for the number of bacteria in the culture is 9000, and the limited growth model is
$$y = 9000 - 8000e^{-0.693t}.$$

Computer Manual: Program LIMGRWTH.

4.5 EXERCISES

1. Use the curve-sketching procedure to graph the logistic function defined by
$$y = \frac{a}{1 + ce^{-rx}}, \qquad a > 0, \quad r > 0, \quad c > 1$$
over the interval $[0, +\infty)$.

Management

2. A worker with no experience can produce camshafts at a rate of 60 per week; with 1 week of experience he can produce them at a rate of 95 per week. The most experienced workers can approach a production rate of 200 camshafts per week. Find the learning curve and sketch its graph. How long will it be until a worker can produce camshafts at a rate of 140 per week?

3. Repeat Exercise 2 if a worker with 1 week of experience can produce camshafts at the rate of only 75 per week.

4. By the end of its first month of operation, a new plant fabricates units at a rate of 50,000 per year; by the end of its second month, at a rate of 60,000 units per year. The plant is designed to fabricate units at a maximum rate of 100,000 units per year. Find a model that describes the rate of production of the plant as a function of the time it has been in operation, and sketch its graph. How many units per year will the plant be fabricating at the end of 1 year of operation?

5. A computer manufacturer is introducing a new model of computer and estimates that it will take 2600 worker-hours to build the first unit and 2400 to build the second.

The manufacturer also estimates that it would take 2000 worker-hours to build a unit even with perfect efficiency. Find the manufacturer's learning curve and sketch its graph. How many worker-hours will be required to build the first 4 units?

6. An auto manufacturer knows that a newly hired worker who has no training will take an average of 4.0 minutes to install a steering wheel on a car, but that if the worker is given 1 hour of training, the average installation time falls to 3.4 minutes. The minimum possible time in which a steering wheel can be installed is 2.2 minutes. Find a model that describes average installation time as a function of training time and draw its graph. If the assembly line runs at such a speed that steering wheels must be installed in 3 minutes or less, how much training must new employees receive?

7. A manufacturing plant can fabricate units at the rate of 70,000 per year by the end of its first month of operation, 80,000 per year by the end of its second month of operation, and 88,000 per year by the end of its third month of operation. Find a model for the rate at which the plant fabricates units. What is the maximum fabrication rate of the plant?

8. Suppose the computer manufacturer of Exercise 5 does not know how many worker-hours would be required to build a unit with perfect efficiency, but estimates that it will take 4000 worker-hours to build the first unit, 3500 to build the second, and 3200 to build the third. Find the learning curve. What is the minimum time required to build a unit under conditions of perfect efficiency?

9. Suppose that sales of electronic calculators are described by the logistic function

$$y = \frac{100}{1 + 49e^{-0.4t}},$$

where y is annual sales of calculators in millions and t is time in years, with $t = 0$ representing 1970.

(a) What were calculator sales in 1970? in 1980?
(b) What will sales be in 1990?
(c) Sketch the graph of the function and analyze calculator sales.

10. The manufacturing capacity of a nation is given by

$$y = \frac{450}{1 + 300e^{-0.06t}},$$

where y is the volume of manufactured goods produced in the nation, measured in billions of 1986 dollars, and t is time in years, with $t = 0$ representing A.D. 1800. Graph this function, analyze the growth of the nation's manufacturing capacity, and predict its manufacturing capacity in the year 2000.

Life Science

11. The number of bacteria in a culture is limited to 40,000 by the amount of nutrients present. Suppose that 2000 bacteria are placed in the culture to begin with, and that 5 hours later there are 38,000 of them. Find a function that describes the number of bacteria in the culture and sketch its graph. How long does it take for the original number of bacteria to double?

12. A pond is stocked with 60 fish, and 1 year later contains 150 fish. The pond can support a maximum of 1200 fish. How many fish will the pond contain 10 years after it was stocked?

13. A farmer's crop is ready to be harvested. If it is harvested today, all of it will be saved, but if it rains and harvesting must be delayed, a smaller percentage will be saved. It is known that if harvesting is delayed 1 day, only 70 percent of the crop will be saved and that the minimum percentage salvageable under any conditions is 40 percent. Find a model that describes the percentage of the crop that will be saved as a function of the delay in harvesting. How much will be salvaged if harvesting is delayed for 4 days?

14. A patient is injected with a drug. At the moment of injection the concentration of the drug in the patient's blood is 0 ppm, 1 minute later it is 24 ppm, and 2 minutes after injection it is 40 ppm. Find a function that describes the concentration of the drug. What is the maximum concentration possible?

15. In a town of 12,000 people the spread of an influenza epidemic is described by the function

$$y = \frac{5000}{1 + 4999e^{-0.5t}},$$

where t is time in days since the first case of flu was diagnosed, and y is the number of flu cases contracted on or before day t. Graph this function and analyze the epidemic. What proportion of the town's inhabitants had the flu during the epidemic?

Social Science

16. The percentage of city voters who consider themselves independents was 22 percent in 1970 and 29 percent in 1980. Find a model that describes the percentage of independents as a function of time, if the maximum possible percentage who could consider themselves independents is 100 percent. If this model is valid, when will a majority of the voters consider themselves independents?

17. Two hundred students live in a college dormitory. One of them starts a rumor, and t hours later y of them have heard it, where

$$y = \frac{200}{1 + 199e^{-1.1t}}.$$

Graph this function and analyze the spread of the rumor. How long will it take from the time the rumor is started until everyone in the dormitory has heard it except the single resident who is its subject?

4 REVIEW EXERCISES

Answers to all review exercises are at the back of the book.

1. Graph the functions defined by
   (a) $f(x) = 5^{-2x}$ **(b)** $y = 2e^{3x}$ **(c)** $f(x) = 3e^{-0.8x}$ **(d)** $y = 3(4^{x/2})$
2. How much will you have at the end of 3 years if you invest \$5000 at 12 percent per year compounded
   (a) annually? **(b)** quarterly? **(c)** monthly? **(d)** continuously?
3. How much should you invest now in order to have \$20,000 two years from now if your investment will earn 6 percent per year compounded
   (a) annually? **(b)** quarterly? **(c)** monthly? **(d)** continuously?

4. Evaluate

(a) $\log_7 49$ (b) $\log_4 \frac{1}{64}$ (c) $\log_{10} 10$ (d) $\log_{11} 1$

(e) $\ln e^{1.2}$ (f) $e^{\ln 4}$ (g) $2 \cdot \ln 120$ (h) $\ln \frac{49}{64}$

5. Using only the facts that

$$\log_2 3 = 1.5850, \quad \log_2 5 = 2.3219, \quad \text{and} \quad \log_2 7 = 2.8074,$$

and the properties of logarithms, find

(a) $\log_2 4$ (b) $\log_2 49$ (c) $\log_2 \frac{1}{27}$ (d) $\log_2 35$ (e) $\log_2 \frac{25}{21}$

6. Graph

(a) $y = 2 \cdot \log_3 x$ (b) $y = 2 \cdot \ln 3x$ (c) $y = \ln \frac{x^2}{2}$

7. Solve for x:

(a) $2e^{-3x} = 10$ (b) $3^{x/3} = 10$

8. A radioactive isotope is present in amount y grams at time t, where

$$y = 10e^{-0.05t}$$

and t is in minutes. Find the half-life of the isotope.

9. If \$8000 is invested at 10 percent per year compounded semiannually, how long will it be until the investment is worth \$20,000?

In Exercises 10 through 19 find $f'(x)$.

10. $f(x) = \ln(x^3 + 3x - 2)$ 11. $f(x) = \ln(x^2 + 1)^2$ 12. $f(x) = \ln \frac{x^2}{2x - 1}$

13. $f(x) = \frac{\ln x^2}{\ln(2x - 1)}$ 14. $f(x) = 2e^{5x}$ 15. $f(x) = 4x^2e^{-x}$

16. $f(x) = e^{x^2} \ln x^2$ 17. $f(x) = \frac{e^x - 1}{e^x + 1}$ 18. $f(x) = e^{e^{-2x}}$ 19. $f(x) = x^{-x^2}$

20. Sketch the graph of $y = x^2e^{-x}$.

21. Sketch the graph of $y = x^2 \ln x$ on the interval $[e^{-2}, +\infty)$.

22. The money spent on social services by a state government in year t is

$$y = [\ln(3t + 2)]^{1/2}$$

million dollars. Here $t = 0$ represents 1984. Find the rate at which spending on social services was changing in 1986.

23. It is estimated that the concentration of pollutants in a lake t years from now will be

$$y = 200e^{-0.2t}$$

parts per million. Find the rate at which the concentration of pollutants in the lake will be changing 5 years from now.

24. An oil spill contaminates a pond; as a result the number of fish in the pond declines. The rate of decline depends directly on the number of fish present. If at the time of the spill there were 2000 fish in the pond and 4 days later there were 1340 of them, find a model that describes the number of fish as a function of time since the spill. How many fish will there be in the pond 10 days after the spill? How long will it take until only 10 percent of the original number remain?

25. The cost of raw materials used to make pallets has been doubling every 5 years. The cost was \$2 per pallet in 1980. Predict the cost of raw materials in 1987. When will the cost per pallet reach \$15?

26. A firm's profits are growing logarithmically. The profits were \$100,000 in 1975 and \$150,000 in 1985. Predict the firm's profits in 1990 and the year in which they will reach \$250,000 per year.

27. A newly hired employee can produce at the rate of 1600 units per month, whereas one with 1 month's experience can produce at the rate of 1675 units per month. The maximum production rate with perfect efficiency is 2000 units per month. Find and graph the learning curve. What will the employee's production rate be with 6 months of experience?

28. It takes 400 hours to produce the first unit of a new type of computer chip and 380 hours to produce the second unit. The minimum time necessary to produce a chip is estimated to be 250 hours. Find and graph the learning curve. How long will it take to produce the first 4 chips?

29. It takes a rat 350 seconds to negotiate a maze on its first attempt, 335 seconds on its second attempt, and 325 seconds on its third attempt. Find a model that describes the time needed to negotiate the maze as a function of the number of attempts. What is the theoretical minimum time required to negotiate the maze?

30. Sales of a new product are given by

$$y = \frac{12}{3 + 5e^{-0.4t}},$$

where t is time in months since the introduction of the product and y is monthly sales in thousands. Graph this function and analyze sales.

4 SUPPLEMENT: PRESENT VALUE AND ANNUITIES

In some of the exercises of Section 4.1 you were asked to find the amount P that would have to be invested now at compound interest in order to have a specified amount A at some future time: this amount P is called the **present value** of the future amount A. For instance, suppose money can be invested at 8 percent per year compounded annually; then the present value of \$1000 due two years from now is \$857.34, because \$857.34 invested now at 8 percent per year compounded annually will amount to \$1000 two years from now. (Use the compound interest formula to check this.) We can find the present value P of an amount A due t years in the future by solving the compound interest formula $A = P(1 + r)^{mt}$ for P.

The Present Value Formula

If money can be invested at i percent per year compounded m times per year, so that $r = i/100m$ is the interest rate per compounding period, then the present value of an amount A due t years in the future is

$$P = A(1 + r)^{-mt}.$$

EXAMPLE 1 If money is worth 12 percent per year compounded monthly (that is, if investments can be made at this interest rate), then the present value of \$100,000 due 10 years from now is

$$\begin{aligned} P &= \$100{,}000(1.01)^{-12(10)} \\ &= \$100{,}000(1.01)^{-120} \\ &= \$100{,}000(0.3029947) \\ &= \$30{,}299.47. \end{aligned}$$

Present values are important because they allow us to compare future returns with present costs. We illustrate with an example.

EXAMPLE 2 Suppose we have some money to invest and have decided that we will not accept a return on our investment of less than 8 percent per year compounded quarterly. If investment X will cost \$10,000 now and will pay us \$15,000 five years from now, whereas investment Y will cost \$9600 and will pay us \$18,000 eight years from now, which investment, if either, should we make, all other things being equal?

We must find the present value of each investment using our benchmark return of 8 percent per year compounded quarterly and then compare the present value of the investment to its cost:

$$\text{investment } X\text{:} \qquad \text{present value } P = \$15{,}000(1.02)^{-20} = \$10{,}094.57;$$

$$\text{investment } Y\text{:} \qquad \text{present value } P = \$18{,}000(1.02)^{-32} = \$\ \ 9551.40.$$

Note that for investment X,

$$\text{present value} - \text{cost} = \$10{,}094.57 - \$10{,}000 = \$94.57.$$

Thus, since its present value exceeds its cost, investment X returns more than 8 percent per year compounded quarterly. On the other hand, for investment Y,

$$\text{present value} - \text{cost} = \$9551.40 - \$9600 = -\$48.60;$$

since here the present value of the investment is less than its cost, investment Y does not return 8 percent per year compounded quarterly. Therefore we should select investment X.

Now we turn to the topic of **annuities.** An **annuity** is a sequence of regular payments that earn interest. For instance, suppose that at the end of each year we invest \$1000 at 8 percent per year compounded annually. If we continue this program for four years, we will have created a four-year annuity. Since we make each payment at the *end* of a year, when the four years are over

- The first \$1000 payment will have earned interest for three years.
- The second \$1000 payment will have earned interest for two years.
- The third \$1000 payment will have earned interest for one year.
- The fourth \$1000 payment will have earned no interest.

Using the compound interest formula of Section 4.1, we see that the value of the annuity at the end of four years will be

$$\$1000(1.08)^3 + \$1000(1.08)^2 + \$1000(1.08) + \$1000 = \$4506.11.$$

Of this \$4506.11, \$4000 is the total of our payments, and the remaining \$506.11 is the interest earned.

The sum that we evaluated in the preceding paragraph is an example of a **geometric sum:** a geometric sum is a sum of the form

$$S = a + aq + aq^2 + \cdots + aq^{n-1};$$

the number a is called the **first term** of S, and the number q is called the **ratio** of S. If we multiply both sides of the equation

$$S = a + aq + aq^2 + \cdots + aq^{n-1}$$

by q, we obtain

$$qS = aq + aq^2 + aq^3 + \cdots + aq^n.$$

Now if we subtract S from qS, all terms on the right side except a and aq^n subtract out, and we have

$$qS - S = aq^n - a,$$

or

$$(q - 1)S = a(q^n - 1).$$

Therefore if $q \neq 1$,

$$S = \frac{a(q^n - 1)}{q - 1}.$$

Formula for a Geometric Sum

If $q \neq 1$,

$$a + aq + aq^2 + \cdots + aq^{n-1} = \frac{a(q^n - 1)}{q - 1}.$$

EXAMPLE 3 The sum

$$S = 1000 + 1000(1.08) + 1000(1.08)^2 + 1000(1.08)^3$$

is geometric, with $a = 1000$, $q = 1.08$, and $n - 1 = 3$. Hence

$$S = \frac{1000[(1.08)^4 - 1]}{1.08 - 1} = \frac{1000(1.360489 - 1)}{0.08} = 4506.11.$$

We can use the formula for a geometric sum to develop a formula for the value of an annuity. Suppose we create an annuity by making a payment of P dollars at the *end* of each compounding period, and suppose each payment earns interest at i percent per year compounded m times per year. Then at the end of t years, interest on the first payment will have been compounded $mt - 1$ times, interest on the second payment $mt - 2$ times, . . . , interest on the next-to-last payment one time, and the last payment will earn no interest. Therefore the value of the annuity at the end of t years will be

$$P(1 + r)^{mt-1} + P(1 + r)^{mt-2} + \cdots + P(1 + r) + P,$$

where as usual, $r = i/100m$ is the interest rate per compounding period. But this is a geometric sum with first term P, ratio $1 + r$, and $n - 1 = mt - 1$. Therefore its value is

$$S = \frac{P[(1 + r)^{mt} - 1]}{(1 + r) - 1}.$$

The Annuity Formula

If an annuity has payment P made at the end of each compounding period and earns i percent interest per year compounded m times per year so that $r = i/100m$ is the interest rate per compounding period, then the value of the annuity at the end of t years will be

$$S = \frac{P[(1 + r)^{mt} - 1]}{r}.$$

EXAMPLE 4 Suppose that at the end of each month we deposit \$200 in an account that pays 9 percent per year compounded monthly. What will be the value of this annuity at the end of 25 years?

We have

$$P = \$200, \quad m = 12, \quad \text{and} \quad r = \frac{9}{100(12)} = 0.0075.$$

Therefore the value of the annuity at the end of 25 years will be

$$\begin{aligned} S &= \frac{\$200[(1.0075)^{12(25)} - 1]}{0.0075} \\ &= \frac{\$200[(1.0075)^{300} - 1]}{0.0075} \\ &= \frac{\$200[9.4084145 - 1]}{0.0075} \\ &= \$224{,}224.39, \end{aligned}$$

to the nearest cent.

We can apply the concept of present value to annuities. Suppose we are to receive an annuity in equal payments of P dollars at the end of each compounding period. If interest is compounded m times per year and the payments continue for t years, we must wait one compounding period for the first payment, two compounding periods for the second payment, . . . , mt compounding periods for the last payment. Hence by the formula for present value, the present value of the first payment is $P(1 + r)^{-1}$, the present value of the second payment is $P(1 + r)^{-2}$, . . . , the present value of the last payment is $P(1 + r)^{-mt}$. Thus the total present value V of the annuity is given by

$$\begin{aligned} V &= P(1+r)^{-1} + P(1+r)^{-2} + \cdots + P(1+r)^{-mt} \\ &= \frac{P}{1+r} + \frac{P}{1+r}\left(\frac{1}{1+r}\right) + \frac{P}{1+r}\left(\frac{1}{1+r}\right)^2 + \cdots \\ &\quad \cdots + \frac{P}{1+r}\left(\frac{1}{1+r}\right)^{mt-1}, \end{aligned}$$

which is a geometric sum with

$$a = \frac{P}{1+r}, \quad q = \frac{1}{1+r}, \quad \text{and} \quad n - 1 = mt - 1.$$

Therefore the present value of the annuity is

$$V = \frac{\dfrac{P}{1+r}\left[\left(\dfrac{1}{1+r}\right)^{mt} - 1\right]}{\dfrac{1}{1+r} - 1} = \frac{P[1 - (1+r)^{-mt}]}{r}.$$

The Present Value Formula for Annuities

If money is worth i percent per year compounded m times per year, so that $r = i/100m$ is the interest rate per compounding period, then an annuity with equal payments of amount P at the end of each compounding period for t years has present value

$$V = \frac{P[1 - (1+r)^{-mt}]}{r}.$$

EXAMPLE 5 If investments can be made at 10 percent per year compounded quarterly, let us find the present value of an annuity that pays \$5000 at the end of each quarter for 20 years.

We have $P = \$5000$, $m = 4$, $r = 10/100(4) = 0.025$, and $t = 20$, so

$$\begin{aligned} V &= \frac{\$5000[1 - (1.025)^{-4(20)}]}{0.025} \\ &= \frac{\$5000[1 - (1.025)^{-80}]}{0.025} \end{aligned}$$

$$= \frac{\$5000[1 - 0.1387045]}{0.025}$$

$$= \$172{,}259.09,$$

to the nearest cent.

EXAMPLE 6 Nocom Company is investigating the possibility of installing a new telephone system. Vendor A will sell Nocom an adequate system for $250,000, and vendor B will sell them a more sophisticated system for $325,000. Nocom estimates that installation of A's system will save them $8000 per month for the next 5 years, and B's will save them $7500 per month for the next 10 years. Nocom will not make any investment unless it returns at least 24 percent per year compounded monthly. All other things being equal, which system, if either, should Nocom purchase?

If we think of the savings from each of the systems as a monthly annuity, we must compare the present value of the annuities with the costs of the respective investments, using the benchmark rate of 24 percent per year compounded monthly. We have

$$\text{A's system:} \qquad \text{present value } V = \frac{\$8000[1 - (1.02)^{-60}]}{0.02} = \$278{,}087.09;$$

$$\text{B's system:} \qquad \text{present value } V = \frac{\$7500[1 - (1.02)^{-120}]}{0.02} = \$340{,}165.41.$$

Thus

$$\text{A's system:} \qquad \text{present value} - \text{cost} = \$278{,}087.09 - \$250{,}000 = \$28{,}087.09;$$

$$\text{B's system:} \qquad \text{present value} - \text{cost} = \$340{,}165.42 - \$325{,}000 = \$15{,}165.41.$$

Both systems return more than the required 24 percent per year compounded monthly, but A's system generates a higher rate of return than B's does, so Nocom should choose A's system.

Computer Manual: Programs PRESVALU and ANNUITY.

SUPPLEMENTARY EXERCISES

1. Find the present value of $25,000 due 10 years from now if money is worth 6 percent per year compounded quarterly.

2. If investments can be made at 9 percent per year compounded monthly, find the present value of $1000 due

(a) 1 year **(b)** 5 years **(c)** 10 years **(d)** 20 years

from today.

3. An investment in company X will cost $120,000 and will pay $225,000 six years from now. An investment in company Y will cost $160,000 and pay $325,000 eight years from now. All other things being equal, which investment (if either) would you make if you required a return of at least

(a) 6 percent per year compounded quarterly?
(b) 10 percent per year compounded annually?
(c) 12 percent per year compounded monthly?

4. Loans are often made on a **discounted** basis. This means that the borrower receives the present value of the loan and pays back the amount of the loan. Suppose you borrow \$5000 at 10 percent per year compounded quarterly for 2 years on a discounted basis. How much money will you actually receive from the lender? How much will you pay back at the end of the 2 years?

5. Find the value of an annuity that earns 10 percent per year compounded annually for 20 years if its payments of \$10,000 are made at the end of each year.

6. Find the value of an annuity that earns 8 percent per year compounded monthly if its payments of \$100 are made at the end of each month for

(a) 5 years **(b)** 10 years **(c)** 15 years **(d)** 20 years

7. Find the present value of an annuity that pays \$500 at the end of each quarter for 4 years if money is worth 6 percent compounded quarterly.

8. Find the present value of an annuity that pays \$2000 at the end of each month for 10 years if money is worth

(a) 6 percent compounded monthly.
(b) 9 percent compounded monthly.
(c) 12 percent compounded monthly.

9. Avaunt, Inc. will not undertake any investment unless it returns at least 18 percent per year compounded monthly. Avaunt is considering the development of 3 new products. The cost of developing each product and its estimated monthly profits once it is on the market are as follows:

	Development Costs	Monthly Profits
Product X:	\$500,000	\$20,000 for 6 years
Product Y:	600,000	10,000 for 12 years
Product Z:	700,000	40,000 for 3 years

If Avaunt has the resources to develop all 3 products, which, if any, should it proceed with, all other things being equal? What if Avaunt can afford to develop only 1 of the products?

10. The monthly payments on a mortgage loan form an annuity that is compounded monthly. The loan itself is a discounted loan (see Exercise 4), and thus the amount of the loan is the present value of the annuity formed by the payments. For instance, suppose a homeowner has a 25-year mortgage for \$150,000 at 12 percent per year compounded monthly: then \$150,000 is the present value of the annuity formed by the monthly mortgage payments over the 25-year period. For such a mortgage find the amount of the monthly payment and the total amount that the homeowner will pay over the life of the mortgage.

11. Suppose payments to an annuity are made at the *beginning* of each compounding period. Find a formula for the future value S of the annuity.

12. Use the result of Exercise 11 to find the future value of the annuities of Exercises 5 and 6 if the payments to these annuities are made at the beginning of each compounding period.

13. Suppose payments from an annuity are received at the *beginning* of each compounding period. Find a formula for the present value V of the annuity.

14. Use the result of Exercise 13 to find the present value of the annuities of Exercises 7 and 8 if the payments from these annuities are received at the beginning of each compounding period.

5

Integration

In Chapters 2 through 4 we were concerned with the problem of finding the rate of change of a function. We saw that the rate of change of a function is the derivative of the function, and we have exploited this fact in many interesting and important ways. Often, however, it is not the function that is known and its rate of change that must be found, but rather the rate of change that is known and the function that must be found. For instance,

- A biologist might know the rate of population growth for a species and wish to find a function that describes the size of its population.
- A physicist might know the velocity of a particle and want to find a function that describes its position.
- A manufacturer might know the marginal cost function for a product and wish to find its cost function.

In cases like these, instead of starting with a function f and differentiating it to find its rate of change f', we start with a function f that describes a rate of change and **antidifferentiate** it to find a function F such that $F' = f$. Such a function F is called an **antiderivative** of f.

We begin this chapter with two sections devoted to antiderivatives. In these sections we consider the set of all antiderivatives of a function: this set is known as the **indefinite integral** of the function. We state the rules of indefinite integrals and demonstrate some techniques for finding indefinite integrals. In the third section we introduce the **definite integral** of a function: the definite integral is a number that gives the total change of an antiderivative of the function over an interval. We conclude by showing how the definite integral may be used to find areas.

5.1 THE INDEFINITE INTEGRAL

We have discussed the motivation for antidifferentiation. In this section we consider the mechanics of the antidifferentiation process, introduce the concept of the indefinite integral of a function, and discuss some properties of the indefinite integral.

The Antiderivative of a Function

If f and F are functions such that

$$F'(x) = f(x)$$

for all x in the domain of f, then $F(x)$ is said to be an **antiderivative** of $f(x)$.

EXAMPLE 1 Let f be defined by $f(x) = 2x + 3$ and F by $F(x) = x^2 + 3x$. Then $F(x)$ is an antiderivative of $f(x)$, because

$$F'(x) = 2x + 3 = f(x)$$

for all x. More generally, if $G(x) = x^2 + 3x + c$, where c is any number, then

$$G'(x) = 2x + 3 + 0 = f(x)$$

for all x, and therefore $G(x)$ is an antiderivative for $f(x)$. Thus

$$f(x) = 2x + 3$$

has many antiderivatives: each choice of the number c in

$$G(x) = x^2 + 3x + c$$

results in a different antiderivative of $f(x)$.

As Example 1 suggests, if $f(x)$ has one antiderivative $F(x)$, then it has a whole *set* of antiderivatives, because if $F'(x) = f(x)$ for all x in the domain of f and c is any number, then

$$(F(x) + c)' = F'(x) + 0 = f(x)$$

for all x in the domain of f; hence $F(x) + c$ is an antiderivative for $f(x)$ also. On the other hand, if $F(x)$ and $G(x)$ are any two antiderivatives of $f(x)$, then there is some number c such that

$$G(x) = F(x) + c$$

for all x in the domain of f. To see this, set $H(x) = G(x) - F(x)$ for all x in the domain of f. Then

$$H'(x) = G'(x) - F'(x) = f(x) - f(x) = 0.$$

Hence the derivative of H is always zero, and therefore H has a horizontal tangent line at every point of its graph. But the only way this can happen is if the graph of H is itself a horizontal line. If this is so, then there is some number c such that $H(x) = c$ for every x in the domain of f. Thus $G(x) - F(x) = c$, or $G(x) = F(x) + c$ for all x in the domain of f.

Properties of the Antiderivative

If $F(x)$ is an antiderivative of $f(x)$, then

1. $F(x) + c$ is also an antiderivative of $f(x)$, for any number c.

2. Every antiderivative of $f(x)$ is of the form $F(x) + c$ for some number c.

Now we are ready to define the **indefinite integral** of a function f. This is just the set of all antiderivatives of $f(x)$.

The Indefinite Integral

The **indefinite integral** of a function f with respect to x, denoted by

$$\int f(x)\,dx,$$

is defined by

$$\int f(x)\,dx = F(x) + c,$$

where $F(x)$ is any antiderivative of $f(x)$ and c is a constant.

The symbol $\int$ is called an **integral sign,** and the expression $f(x)$ that follows the integral sign is called the **integrand.** The dx after the integrand is the differential of x discussed in Section 3.7; its presence tells us that the integral is taken with respect to the variable x. The constant c is referred to as the **constant of integration.** Note that to find the indefinite integral of f, we need only find *one* antiderivative $F(x)$ of $f(x)$.

EXAMPLE 2 Let us find

$$\int (2x + 3)\,dx.$$

Since $F(x) = x^2 + 3x$ is an antiderivative of $f(x) = 2x + 3$, we have

$$\int (2x + 3)\,dx = x^2 + 3x + c.$$

EXAMPLE 3 Let us find

$$\int (3x^2 + 2x + 1)\, dx.$$

Since $F(x) = x^3 + x^2 + x$ is an antiderivative of $f(x) = 3x^2 + 2x + 1$ (why?), it follows that

$$\int (3x^2 + 2x + 1)\, dx = x^3 + x^2 + x + c.$$

EXAMPLE 4 We claim that

$$\int (6t^5 - 6t^2 + 2t - 2)\, dt = t^6 - 2t^3 + t^2 - 2t + c.$$

The claim is true if the derivative of the right side with respect to t is the integrand. (Why?) Check that this is so.

Since the process of integration is the reverse of the process of differentiation, every rule of differentiation gives rise to a corresponding rule of integration. The rules are traditionally formulated in terms of a variable u. We can prove each rule by showing that the derivative with respect to u of the right side is the integrand.

The Rules of Integration

1. For any number k,

$$\int k\, du = ku + c.$$

2. For any number $r \neq -1$,

$$\int u^r\, du = \frac{u^{r+1}}{r+1} + c.$$

3. $\int \frac{1}{u}\, du = \ln |u| + c$. In other words,

if $u > 0$, $\quad \int \frac{1}{u}\, du = \ln u + c;$

if $u < 0$, $\quad \int \frac{1}{u}\, du = \ln(-u) + c.$

4. $\int e^u\, du = e^u + c.$

5. For any number k and function $f(u)$,

$$\int kf(u)\, du = k\int f(u)\, du.$$

6. For any functions $f_1(u)$ and $f_2(u)$,

$$\int [f_1(u) \pm f_2(u)]\, du = \int f_1(u)\, du \pm \int f_2(u)\, du.$$

Rule 6 remains true if the integrand is the sum or difference of more than two functions.

EXAMPLE 5

a. $$\int 2\,du = 2u + c.$$

$$\int -3\,dx = -3x + c.$$

$$\int dt = \int 1\,dt = t + c.$$

$$\int 0\,du = c.$$

b. $$\int u\,du = \int u^1\,du = \frac{u^{1+1}}{1+1} + c = \frac{u^2}{2} + c.$$

$$\int x^2\,dx = \frac{x^3}{3} + c.$$

$$\int \sqrt{x}\,dx = \int x^{1/2}\,dx = \frac{x^{3/2}}{3/2} + c = \frac{2}{3}x^{3/2} + c.$$

$$\int t^{-3}\,dt = \frac{t^{-3+1}}{-3+1} + c = -\frac{t^{-2}}{2} + c.$$

c. $$\int 4x^2\,dx = 4\int x^2\,dx = 4\frac{x^3}{3} + c = \frac{4}{3}x^3 + c.$$

$$\int \frac{4}{x}\,dx = 4\int \frac{1}{x}\,dx = 4\ln|x| + c.$$

$$\int -2e^t\,dt = -2\int e^t dt = -2e^t + c.$$

d. $$\int (u^2 + 3u)\,du = \int u^2\,du + \int 3u\,du$$
$$= \frac{u^3}{3} + 3\int u\,du$$
$$= \frac{u^3}{3} + \frac{3u^2}{2} + c.$$

e. $$\int (4x^4 - 3x^{-2} + 6)\,dx = 4\int x^4\,dx - 3\int x^{-2}\,dx + 6\int dx$$
$$= 4\frac{x^5}{5} - 3\frac{x^{-1}}{-1} + 6x + c$$
$$= \frac{4}{5}x^5 + \frac{3}{x} + 6x + c.$$

If $F'(x) = f(x)$ for all x in the domain of f, then $F(x)$ is an antiderivative of $f(x)$, so

$$\int f(x)\,dx = \int F'(x)\,dx = F(x) + c.$$

Thus if f is the derivative of a function F, we can find F by integrating $f(x)$ and eliminating the constant of integration. We illustrate with an example.

EXAMPLE 6 Suppose a firm's marginal cost function is given by

$$c(x) = 2x + 5.$$

Marginal cost is the derivative of cost, so we can find the firm's cost function by integrating $c(x)$ and eliminating the constant of integration. Thus

$$C(x) = \int c(x)\,dx = \int (2x + 5)\,dx = x^2 + 5x + c.$$

To eliminate the constant of integration, we must know the value of $C(x)$ for some value of x. For instance, suppose the firm's fixed cost is \$10,000. Then

$$10{,}000 = C(0) = 0^2 + 5(0) + c = c,$$

so $c = 10{,}000$ and the cost function is given by

$$C(x) = x^2 + 5x + 10{,}000.$$

Computer Manual: Program INTPOLY.

5.1 EXERCISES

In Exercises 1 through 26 find the indefinite integral.

1. $\int 3\,dx$
2. $\int \sqrt{2}\,dx$
3. $\int u^4\,du$
4. $\int x^{10}\,dx$
5. $\int t^{-4}\,dt$
6. $\int q^{-4/5}\,dq$
7. $\int 2x^3\,dx$
8. $\int \frac{5}{x^4}\,dx$
9. $\int -3\sqrt{y}\,dy$
10. $\int 2u^{1/3}\,du$
11. $\int -\frac{4}{x}\,dx$
12. $\int \frac{3}{t}\,dt, \quad t < 0$
13. $\int -3e^y\,dy$
14. $\int 12e^x\,dx$
15. $\int (x + 2)\,dx$
16. $\int (2v^2 - 3v + 1)\,dv$
17. $\int (x^3 + 3x^2 - 2x)\,dx$
18. $\int (-4x^3 + 3x^2 - 2x + 7)\,dx$
19. $\int (\pi x^3 - \sqrt[3]{3})\,dx$
20. $\int (-3x^{-5} + 4x^2 + 12)\,dx$
21. $\int \left(\frac{2}{t^2} + \frac{1}{t^3} + 3\right) dt$
22. $\int \left(\frac{2}{\sqrt{q}} - \frac{3}{q} + 1\right) dq, \quad q > 0$
23. $\int (100x^{99} + 99x^{98} + 98x^{97} + \cdots + 2x + 1)\,dx$
24. $\int -\left(\frac{1}{t^2} + \frac{2}{t^3} + \frac{3}{t^4} + \cdots + \frac{99}{t^{100}}\right) dt$

25. $\int (x + 5)(2x - 3)\, dx$

26. $\int (2x + 5)^2\, dx$

27. An airplane's velocity t hours after takeoff is $v = 300t$ kilometers per hour. If the plane flies in a straight line, find the distance it travels in the 2 hours after takeoff. (*Hint:* The distance the plane has traveled at takeoff is 0 kilometers.)

28. A rocket is launched straight up with a constant acceleration of 100 feet per second per second.
    (a) Find the rocket's velocity 10 seconds after launch (*Hint:* Its initial velocity is 0 feet per second.)
    (b) Find its altitude at the end of 10 seconds.

Management

29. A firm's marginal cost function is given by $c(x) = 2x + 1$, its fixed cost is \$5000, and its marginal revenue function is given by $r(x) = 15$. Find the firm's cost, revenue, and profit functions. (*Hint:* The revenue obtained from selling 0 units is \$0.)

30. A firm's marginal cost function is given by $c(x) = 5x + 20$, the cost of making 10 units is \$10,000, and its marginal revenue function is given by $r(x) = 40$. Find the firm's cost, revenue, and profit functions.

31. A firm's marginal profit function is given by $p(x) = x^2 - 100x + 1600$, and its profit from making and selling 3 units is \$4341. Find the firm's profit function.

Life Science

32. A nesting pair of blue herons is introduced into a bird sanctuary where blue herons did not previously exist. It is estimated that the population of blue herons in the sanctuary will increase at the rate of y individuals per year, where $y = \sqrt{t} + 4$ and t is time in years since the species was introduced into the sanctuary. Find a function that describes the size of the blue heron population in the sanctuary in year t.

Social Science

33. In 1976, 68 percent of the voters in a state were affiliated with a political party. Since then the percentage affiliated with a party has changed at the rate of $-0.6\sqrt{t}$ percent per year, where t is in years, with $t = 0$ representing 1976. Find a function that describes the percentage of voters affiliated with a political party in year t.

34. The property tax rate in a town is expressed in terms of dollars per thousand dollars of the assessed value of the property. The tax rate has been changing at a rate of $y = \dfrac{2}{t + 1}$ dollars per thousand dollars of assessed value, and was \$80 per thousand in 1984. Find a function that describes the property tax rate as a function of time.

5.2 FINDING INDEFINITE INTEGRALS

The rules of integration of Section 5.1 do not tell us how to integrate all functions of interest. For instance, none of the rules apply directly to the integrals

$$\int \sqrt{3x + 1}\, dx \qquad \text{and} \qquad \int x^2(x^3 + 1)^{10}\, dx.$$

In this section we discuss and illustrate two techniques for finding indefinite integrals: integration by guessing and integration by substitution.

Sometimes we can guess the general form that an integral must take, and then by differentiating this general form, find the exact expression for the integral. For instance, consider the integral

$$\int \sqrt{3x + 1}\, dx.$$

An antiderivative of $\sqrt{3x+1}$ must have the general form $(3x + 1)^{3/2}$. However, $(3x + 1)^{3/2}$ is not quite the antiderivative we seek, because its derivative is not quite equal to $\sqrt{3x + 1}$:

$$\frac{d}{dx}(3x + 1)^{3/2} = \frac{9}{2}(3x + 1)^{1/2} = \frac{9}{2}\sqrt{3x + 1}.$$

Thus we have an unwanted constant of 9/2. However, if we multiply $(3x + 1)^{3/2}$ by 2/9, the unwanted constant will cancel out:

$$\frac{d}{dx}\left[\frac{2}{9}(3x + 1)^{3/2}\right] = \frac{2}{9}\frac{d}{dx}(3x + 1)^{3/2} = \frac{2}{9}\frac{9}{2}\sqrt{3x + 1} = \sqrt{3x + 1}.$$

Thus

$$\frac{2}{9}(3x + 1)^{3/2}$$

is an antiderivative of $\sqrt{3x + 1}$, so

$$\int \sqrt{3x + 1}\, dx = \frac{2}{9}(3x + 1)^{3/2} + c.$$

EXAMPLE 1 Consider the integral

$$\int \frac{1}{3x + 2}\, dx, \qquad 3x + 2 > 0.$$

Since $3x + 2 > 0$, an antiderivative of $1/(3x + 2)$ must have the general form $\ln(3x + 2)$. But

$$\frac{d}{dx}(\ln(3x + 2)) = \frac{1}{3x + 2}\frac{d}{dx}(3x + 2) = \frac{3}{3x + 2}.$$

Therefore

$$\frac{d}{dx}\left(\frac{1}{3}\ln(3x + 2)\right) = \frac{1}{3}\frac{d}{dx}(\ln(3x + 2)) = \frac{1}{3}\frac{3}{3x + 2} = \frac{1}{3x + 2}.$$

Thus

$$\int \frac{1}{3x + 2}\, dx = \frac{1}{3}\ln(3x + 2) + c.$$

EXAMPLE 2 Let us find the indefinite integral

$$\int e^{-2x}\,dx.$$

Since

$$\frac{d}{dx}(e^{-2x}) = -2e^{-2x},$$

we have

$$\int e^{-2x}\,dx = -\frac{1}{2}e^{-2x} + c.$$

Guessing the general form of an integral does not always work, because often we cannot tell what the general form should look like. In any case, we would like to have a more systematic technique for performing integration. No technique of integration works for all integrals, but one widely used method is **integration by substitution,** which we now discuss.

Recall that in Section 3.7 we introduced the differentials dx and dy and showed that if $y = f(x)$ then $dy = f'(x)\,dx$. We shall use u rather than y as the dependent variable; thus if $u = f(x)$, then $du = f'(x)\,dx$. We can often use this relationship between du and dx to change the variable in an integral from x to u in such a way that the resulting integral in u is easy to find. The technique is called **integration by substitution,** because we substitute u for some portion of the integrand. We illustrate with an example.

EXAMPLE 3 Let us consider again the integral

$$\int \sqrt{3x + 1}\,dx.$$

Suppose we set $u = 3x + 1$. Then

$$du = (3x + 1)'\,dx = 3\,dx,$$

or

$$\frac{1}{3}\,du = dx.$$

If we substitute u for $3x + 1$ in the integrand and $\frac{1}{3}\,du$ for dx, we obtain

$$\int \sqrt{3x + 1}\,dx = \int \sqrt{u}\,\frac{1}{3}\,du = \frac{1}{3}\int u^{1/2}\,du$$

$$= \frac{1}{3}\frac{u^{3/2}}{3/2} + c = \frac{2}{9}u^{3/2} + c.$$

Back-substituting $3x + 1$ for u now yields

$$\int \sqrt{3x + 1}\,dx = \frac{2}{9}(3x + 1)^{3/2} + c.$$

This is the result we obtained previously by guessing.

How did we know to choose $u = 3x + 1$ in Example 3? The answer is that no other choice for u simplifies the integral. For instance, if we had set $u = 3x$, then $du = 3\,dx$ and substitution would give the integral

$$\int \sqrt{u+1}\,\frac{1}{3}\,du,$$

which is really no simpler than the one we started with. On the other hand, if we had let $u = \sqrt{3x+1}$, then we would have had

$$du = \frac{3}{2\sqrt{3x+1}}\,dx;$$

but with this expression for du, substitution does not even give an integral in the variable u. (Try it.)

EXAMPLE 4 Let us find

$$\int e^{-2x}\,dx,$$

using the technique of integration by substitution.

We set $u = -2x$. Then $du = -2\,dx$, so $-\frac{1}{2}\,du = dx$. Thus

$$\int e^{-2x}\,dx = \int e^{u}\left(-\frac{1}{2}\right)du = -\frac{1}{2}\int e^{u}\,du$$

$$= -\frac{1}{2}e^{u} + c = -\frac{1}{2}e^{-2x} + c,$$

which is the result we obtained in Example 2 by guessing.

Integration by substitution is often useful in dealing with more complicated integrals, such as

$$\int x^2(x^3+1)^{10}\,dx.$$

To illustrate, let $u = x^3 + 1$. Then $du = 3x^2\,dx$, or $\frac{1}{3}\,du = x^2\,dx$, so

$$\int x^2(x^3+1)^{10}\,dx = \int u^{10}\,\frac{1}{3}\,du = \frac{1}{3}\frac{u^{11}}{11} + c$$

$$= \frac{1}{33}(x^3+1)^{11} + c.$$

How did we know to let $u = x^3 + 1$? The answer is that we recognized part of the integrand, namely x^2, as the derivative (except for a constant factor) of another part, namely $x^3 + 1$.

> Whenever part B of an integrand is the derivative (except possibly for a constant factor) of part A of the integrand, the substitution u = part A will simplify the integral.

This is so because, under these conditions, part B of the integrand will be replaced by du when the substitution is performed.

EXAMPLE 5 Let us find $\int \frac{\ln x}{x}\, dx$.

We recognize $1/x$ as the derivative of $\ln x$. Therefore we make the substitution $u = \ln x$. Since $du = (1/x)\, dx$, we have

$$\int \frac{\ln x}{x}\, dx = \int (\ln x) \frac{1}{x}\, dx = \int u\, du = \frac{1}{2} u^2 + c = \frac{1}{2} (\ln x)^2 + c.$$

EXAMPLE 6 Find $\int xe^{-x^2} dx$.

Except for a constant factor of -2, x is the derivative of $-x^2$. Therefore we set $u = -x^2$, so $du = -2x\, dx$, $-\frac{1}{2}\, du = x\, dx$, and

$$\int xe^{-x^2}\, dx = \int e^{-x^2} x\, dx = \int -\frac{1}{2} e^u\, du = -\frac{1}{2} \int e^u\, du$$

$$= -\frac{1}{2} e^u + c = -\frac{1}{2} e^{-x^2} + c.$$

5.2 EXERCISES

In Exercises 1 through 26 find the integral.

1. $\int \sqrt{x+1}\, dx$ **2.** $\int \sqrt{4x-3}\, dx$ **3.** $\int (5t+2)^{3/2}\, dt$ **4.** $\int \frac{1}{(2x-1)^2}\, dx$

5. $\int (7x-3)^{12}\, dx$ **6.** $\int \frac{2}{1-2x}\, dx$ **7.** $\int \frac{-6}{5x+2}\, dx$ **8.** $\int e^{4x}\, dx$

9. $\int 3e^{-7x}\, dx$ **10.** $\int \frac{2}{\sqrt{5x+8}}\, dx$ **11.** $\int \frac{10}{(2x-3)^{2/3}}\, dx$ **12.** $\int 2x(x^2-2)^3\, dx$

13. $\int x(x^2+1)^4\, dx$ **14.** $\int -3x^2(x^3+1)^2\, dx$ **15.** $\int x^4(2x^5+10)\, dx$

16. $\int (2x+3)(x^2+3x+1)^4\, dx$ **17.** $\int (x-2)(2x^2-8x)^5\, dx$ **18.** $\int \frac{2x}{x^2+1}\, dx$

19. $\int \frac{-7t^2}{t^3+2}\,dt$ 20. $\int 3xe^{x^2}\,dx$ 21. $\int (2x+1)e^{2x^2+2x}\,dx$ 22. $\int \frac{2x+5}{x^2+5x-1}\,dx$

23. $\int \frac{2-3x^2}{x^3-2x}\,dx$ 24. $\int e^x e^{e^x}\,dx$ 25. $\int \frac{(\ln x)^2}{x}\,dx$ 26. $\int \frac{1}{x \ln x}\,dx$

Management

27. A firm's marginal cost and marginal revenue functions are given by

$$c(x) = 3x\sqrt{x^2+1} \qquad \text{and} \qquad r(x) = \frac{7}{2}x(x^2+1)^{3/4},$$

respectively. Find the firm's cost function and revenue function if the fixed cost is \$1000.

28. A firm's marginal profit function is given by

$$p(x) = \frac{x}{\sqrt{2x^2+9}}.$$

Find the firm's profit function if the profit from making and selling 6 units is \$52.50.

29. With no experience, a worker can assemble 100 units per week. With t weeks of experience, the worker can assemble

$$f(t) = 200e^{-2t}$$

units per week. Find a function that describes the worker's weekly production.

Life Science

30. It is estimated that the total world population of blue whales in 1978 was 250 and that the population since then has been changing at a rate of y individuals per year, where

$$y = \frac{20t}{(2t^2+1)^{3/2}}.$$

Here t is in years, with $t = 0$ representing 1978. Find a function that describes the blue whale population in year t.

31. The body produces antibodies as a reaction to vaccination. Suppose that t days after a vaccination, antibodies are being produced at a rate of

$$y = \frac{200}{5t+4}$$

per day, where y is in thousands of antibodies. Assuming there were no antibodies present before vaccination, find a function that describes antibody production t days after vaccination.

32. The concentration of smog produced by automobiles in a large city changes with the temperature at a rate given by

$$f(x) = \frac{0.006}{(0.02x+100)^{1/2}},$$

where x is temperature in degrees Celsius and the smog concentration is in parts per million (ppm). Find a function that describes the smog concentration if 6 ppm are produced at 0° C.

5.3 THE DEFINITE INTEGRAL

The concept of the **definite integral** of a function over an interval is one of the most important of calculus. As we shall see in the remainder of this chapter and in the next, definite integrals have many uses. In this section we will introduce the definite integral of a function in its aspect as the total change in any antiderivative of the function. We will also demonstrate how to apply the method of integration by substitution to definite integrals.

The Definite Integral as Total Change

The **definite integral** of a function over an interval is a number that expresses the total change in any antiderivative of the function over the interval. For instance, if

$$f(x) = 2x + 1,$$

any antiderivative of $f(x)$ is of the form

$$F(x) = x^2 + x + c.$$

The total change in F over the interval [1, 3] is thus

$$F(3) - F(1) = (3^2 + 3 + c) - (1^2 + 1 + c) = 12 - 2 = 10.$$

Hence we say that the definite integral of f over the interval [1, 3] is equal to 10. Note that when calculating a definite integral, we can use any antiderivative, since the constant of integration subtracts out. Therefore we omit the constant of integration when working with definite integrals.

Our formal definition of the definite integral is as follows:

The Definite Integral

If f is a function defined on the interval $[a, b]$, then the **definite integral of f over $[a, b]$,** symbolized by

$$\int_a^b f(x)\,dx,$$

is the total change in any antiderivative of $f(x)$ over the interval. Thus if $F(x)$ is any antiderivative of $f(x)$, then

$$\int_a^b f(x)\,dx = F(b) - F(a).$$

The numbers a and b are referred to as the **limits of integration.**

EXAMPLE 1 If f is defined by $f(x) = 4x - 5$, let us find the definite integral of f over the interval $[1, 2]$. Since $F(x) = 2x^2 - 5x$ is an antiderivative for $f(x)$, we have

$$\int_1^2 (4x - 5)\,dx = F(2) - F(1)$$

$$= (2(2)^2 - 5(2)) - (2(1)^2 - 5(1))$$

$$= 1.$$

If $F(x)$ is an antiderivative of $f(x)$, we usually write

$$\int_a^b f(x)\,dx = F(x)\Big|_a^b = F(b) - F(a)$$

when evaluating the definite integral. For instance, we could have carried out the integration of the preceding example as follows:

$$\int_1^2 (4x - 5)\,dx = (2x^2 - 5x)\Big|_1^2 = (2(2)^2 - 5(2)) - (2(1)^2 - 5(1)) = 1.$$

EXAMPLE 2 Let us find $\int_{-1}^3 (x^2 + 2x - 1)\,dx$.

We have

$$\int_{-1}^3 (x^2 + 2x - 1)\,dx = \left(\frac{x^3}{3} + x^2 - x\right)\Big|_{-1}^3$$

$$= \left(\frac{3^3}{3} + 3^2 - 3\right) - \left(\frac{(-1)^3}{3} + (-1)^2 - (-1)\right)$$

$$= 15 - \frac{5}{3}$$

$$= \frac{40}{3}.$$

EXAMPLE 3 Let us evaluate $\int_{-4}^4 2e^x\,dx$.

We have

$$\int_{-4}^4 2e^x\,dx = 2e^x\Big|_{-4}^4 = 2(e^4 - e^{-4}).$$

Note that the rules for indefinite integrals

$$\int kf(x)\,dx = k\int f(x)\,dx \qquad \text{for any number } k$$

and $$\int [f_1(x) \pm f_2(x)]\,dx = \int f_1(x)\,dx \pm \int f_2(x)\,dx$$

imply the corresponding rules for definite integrals:

$$\int_a^b kf(x)\,dx = k\int_a^b f(x)\,dx \qquad \text{for any number } k$$

and

$$\int_a^b [f_1(x) \pm f_2(x)]\,dx = \int_a^b f_1(x)\,dx \pm \int_a^b f_2(x)\,dx.$$

Now we demonstrate how the interpretation of the definite integral as total change can be exploited.

EXAMPLE 4 Suppose a city's population is growing at the rate of

$$f(t) = 12{,}000\sqrt{t}$$

persons per year, where t is in years with $t = 0$ representing 1985. Let us find the total change in population from 1985 to 1989.

Since $f(t)$ describes the rate of change of the population, its definite integral $\int_a^b f(t)\,dt$ gives the total change in the population from time $t = a$ to time $t = b$. Therefore we must evaluate

$$\int_0^4 12{,}000\sqrt{t}\,dt.$$

But

$$\int_0^4 12{,}000\sqrt{t}\,dt = 8000t^{3/2}\Big|_0^4 = 8000(4^{3/2} - 0^{3/2}) = 64{,}000.$$

Hence, over the period 1985–1989 the city will experience a net gain in population of 64,000 people.

EXAMPLE 5 A firm's marginal cost function is

$$c(x) = 2x^2 - 8x + 30.$$

The firm is currently producing 6 units per day. What will be the additional cost if the firm decides to produce 10 units per day?

Since marginal cost is the rate of change of cost, the total change in cost due to increasing production from $x = 6$ to $x = 10$ units is

$$\int_6^{10} (2x^2 - 8x + 30)\,dx = \frac{2}{3}x^3 - 4x^2 + 30x\Big|_6^{10}$$

$$= \left(\frac{2}{3}(10)^3 - 4(10)^2 + 30(10)\right) - \left(\frac{2}{3}(6)^3 - 4(6)^2 + 30(6)\right)$$

$$= \frac{1700}{3} - \frac{540}{3}$$

$$\approx \$386.67$$

per day.

EXAMPLE 6 An investment in a new computer will save a company money at the rate of

$$s(t) = 800 - 0.5t^2$$

dollars per month, where t is time in months since the installation of the computer. Figure 5–1 shows the graph of $s(t)$. Note that the rate of savings declines with time, reaching zero when $t = 40$.

The definite integral $\int_a^b s(t)\,dt$ gives the savings due to the computer from time $t = a$ to time $t = b$. Therefore to find the total savings due to the computer, we must evaluate the integral from the time the savings begin until the time they end. But the savings begin when $t = 0$ and end when $t = 40$. Thus the total savings due to the computer are

$$\int_0^{40} (800 - 0.5t^2)\,dt = 800t - \frac{t^3}{6}\Bigg|_0^{40}$$

$$= \left(800(40) - \frac{(40)^3}{6}\right) - \left(800(0) - \frac{0^3}{6}\right)$$

$$= \$21{,}333.33.$$

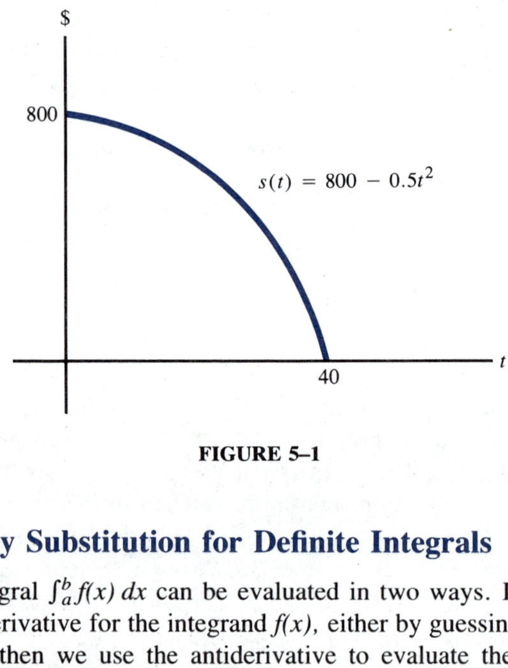

FIGURE 5–1

Integration by Substitution for Definite Integrals

The definite integral $\int_a^b f(x)\,dx$ can be evaluated in two ways. In the first method we find an antiderivative for the integrand $f(x)$, either by guessing or by integration by substitution; then we use the antiderivative to evaluate the definite integral. For instance, to evaluate

$$\int_0^1 x(x^2 + 1)^4\,dx$$

by substitution, we let $u = x^2 + 1$; then $du = 2\,dx$, and

$$\int x(x^2+1)^4\,dx = \frac{1}{2}\int u^4\,du = \frac{1}{10}u^5 + c = \frac{1}{10}(x^2+1)^5 + c.$$

Thus

$$\int_0^1 x(x^2+1)^4\,dx = \frac{1}{10}(x^2+1)^5\Big|_0^1 = \frac{1}{10}(2^5-1^5) = \frac{31}{10}.$$

Alternatively, we can apply the method of integration by substitution directly to the definite integral. If we do so, however, we must be careful to change the limits of integration when we change the variable from x to u in the integral. Thus, in our example, since $u = x^2 + 1$, it follows that

$$\text{when} \quad x = 0, \qquad u = 0^2 + 1 = 1$$

and

$$\text{when} \quad x = 1, \qquad u = 1^2 + 1 = 2.$$

Applying the method of integration by substitution directly to the definite integral, we now have

$$\int_0^1 x(x^2+1)^4\,dx = \frac{1}{2}\int_1^2 u^4\,du = \frac{1}{10}u^5\Big|_1^2 = \frac{1}{10}(2^5-1^5) = \frac{31}{10}.$$

EXAMPLE 7 Let us find $\int_1^2 \frac{x+1}{x^2+2x}\,dx$

by the method of substitution for definite integrals. If we let

$$u = x^2 + 2x,$$

then

$$du = 2(x+1)\,dx.$$

Thus,

$$\text{when} \quad x = 1, \qquad u = (1)^2 + 2(1) = 3,$$

$$\text{when} \quad x = 2, \qquad u = (2)^2 + 2(2) = 8.$$

Therefore

$$\int_1^2 \frac{x+1}{x^2+2x}\,dx = \frac{1}{2}\int_3^8 \frac{du}{u} = \frac{1}{2}\ln u\Big|_3^8$$

$$= \frac{1}{2}(\ln 8 - \ln 3) = \frac{1}{2}\ln\frac{8}{3}.$$

(Note that we did not use absolute values and write $\ln|u|$ when we integrated $1/u$ because we already knew that u was positive, since it was between 3 and 8.)

Computer Manual: Program DEFINT.

5.3 EXERCISES

The Definite Integral as Total Change

In Exercises 1 through 14 evaluate the definite integral.

1. $\int_0^4 x\,dx$
2. $\int_1^5 x^2\,dx$
3. $\int_0^3 (x^2 - 2x + 2)\,dx$
4. $\int_{-1}^2 (2x^2 - 3x + 2)\,dx$
5. $\int_{-1}^1 (2x^3 + 3x^2 - 2)\,dx$
6. $\int_{-2}^0 (x^4 - 5x^2 + 3x - 1)\,dx$
7. $\int_1^4 \sqrt{x}\,dx$
8. $\int_0^1 (x + \sqrt[3]{x})\,dx$
9. $\int_1^{64} (x^{5/3} - \sqrt{x})\,dx$
10. $\int_1^2 x^{-2}\,dx$
11. $\int_2^4 x^{-3}\,dx$
12. $\int_1^2 -\left(\frac{1}{t^2} + \frac{1}{t^3} + \frac{1}{t^4}\right)dt$
13. $\int_0^3 3e^x\,dx$
14. $\int_1^e \frac{1}{x}\,dx$

Management

15. Addlepate, Inc., has marginal cost function given by

$$c(x) = x^2 + 5x$$

and marginal revenue function given by

$$r(x) = 3x^2 + 8x,$$

where cost and revenue are in thousands of dollars and x is in thousands of units. The corporation is currently producing and selling 5000 units. What will be the additional cost and revenue if the firm increases production and sales to 8000 units?

16. Referring to Exercise 15, find the total cost and revenue generated by the first 5000 units Addlepate produces.

17. Apex Company has marginal revenue function given by

$$r(x) = 10x^{1/2},$$

where $x \geq 1$. If Apex currently sells 25 units per day, find the additional revenue if the firm could sell 100 units per day.

18. An investment in new equipment will save a firm $s(t)$ dollars per year, where

$$s(t) = 20{,}000 - 2000t$$

and t is measured in years since the installation of the equipment. Find the firm's total savings due to the new equipment.

Life Science

19. The pollutants in a pond are increasing at the rate of y grams per day, where

$$y = 10t + 3$$

and t is in days with $t = 0$ representing the present. Find the total amount of pollutants that will accumulate in the pond over the next 30 days.

20. A certain strain of influenza spreads in such a way that $f(t)$ persons per day contract the disease, where

$$f(t) = -5t^2 + 300t,$$

t in days. Find the number of persons who will contract the disease in the first 2 weeks of an epidemic that begins on day 0.

21. A fast-growing strain of bamboo grows $f(t)$ inches per day, where

$$f(t) = 6 + \frac{2}{\sqrt{t}},$$

t being the number of days since germination, and the formula is valid for $t \geq 10$. Find the amount the bamboo will grow in the 2 weeks from $t = 10$ to $t = 24$.

Social Science

22. In a certain metropolitan region the number of unemployed persons is increasing at a rate of y persons per year t, where

$$y = 60t^2 + 2000$$

and $t = 0$ is the present. Forecast the increase in unemployment in the region over the next 2 years.

Integration by Substitution for Definite Integrals

In Exercises 23 through 32 evaluate the definite integral by the method of integration by substitution for definite integrals.

23. $\int_{\ln 2}^{\ln 3} e^{3x}\,dx$

24. $\int_{-2}^{3} e^{-2x}\,dx$

25. $\int_{0}^{2} 2xe^{-x^2}\,dx$

26. $\int_{0}^{9} \frac{e^{\sqrt{x}}}{\sqrt{x}}\,dx$

27. $\int_{-1}^{e-2} \frac{6}{x+2}\,dx$

28. $\int_{1}^{2} \frac{4}{3x-2}\,dx$

29. $\int_{0}^{4} x^2(x^3-1)^2\,dx$

30. $\int_{0}^{1} \frac{2x}{x^2+5}\,dx$

31. $\int_{1}^{2} \frac{3x^2+x}{2x^3+x^2}\,dx$

32. $\int_{0}^{9} (\sqrt{x}+1)\sqrt{\frac{2}{3}x^{3/2}+x}\,dx$

Management

33. A firm intends to build a new plant. It is estimated that t years after the plant is built it will be producing at the rate of

$$f(t) = 100{,}000e^{0.01t}$$

units per year. Find the number of units the plant will produce in its first 5 years. How long must the plant operate if it is to produce 2 million units?

34. Able Associates has marginal profit function given by

$$p(x) = \frac{2000}{0.5x + 10}.$$

Find the firm's profit from making and selling its first 50 units.

Life Science

35. A patient receives a solution intravenously at a rate of

$$f(t) = 10.8e^{0.06t}$$

cubic centimeters per hour, t in hours. Find the amount of solution the patient receives during the first 24 hours of treatment.

Social Science

36. Research shows that people who learn a foreign language but never use it tend to forget their vocabulary at the rate of y percent per year, where

$$y = \frac{0.20}{t+1}.$$

Here t is time in years since the language was last used. How much of a language will be forgotten if it is not used for 10 years?

37. A politician's campaign manager estimates that voters who support the politician can be registered at a rate of y individuals per day t, where

$$y = 2e^{0.25t},$$

with $t = 0$ being the present. How many voters can be registered over the next 30 days?

5.4 THE DEFINITE INTEGRAL AND AREA

Let f be a function continuous and nonnegative over the interval $[a, b]$, as in Figure 5–2, and consider the shaded region bounded by the graph of f, the x-axis, and the lines $x = a$ and $x = b$. This region has an area, which we call the **area under *f* from *a* to *b*.** This area and the definite integral are related as follows: the area under f from a to b is the definite integral

$$\int_a^b f(x)\,dx.$$

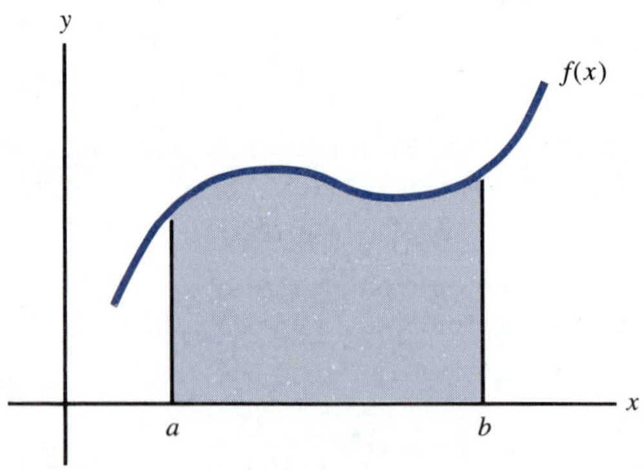

FIGURE 5–2

This remarkable fact, which is so important that it is known as the Fundamental Theorem of Calculus, is the subject of this section. We demonstrate the validity of the theorem and show how to use it to find the area of planar regions.

The Fundamental Theorem of Calculus

We state the theorem formally:

> **The Fundamental Theorem of Calculus**
>
> If f is continuous and nonnegative over the interval $[a, b]$, then the area under f from a to b is given by the definite integral
>
> $$\int_a^b f(x)\,dx.$$

Let us see why the theorem is true. If f is continuous and nonnegative over the interval $[a, b]$, as in Figure 5–2, then associated with f is an **area function** A, which is defined as follows: for each number x in the interval $[a, b]$,

$$A(x) = \text{the area under } f \text{ from } a \text{ to } x.$$

See Figure 5–3. Note that

$$A(b) = \text{the area under } f \text{ from } a \text{ to } b$$

and that

$$A(a) = \text{the area under } f \text{ from } a \text{ to } a = 0.$$

Now we will show that $A(x)$ is an antiderivative for $f(x)$. To do this, we must show that $A'(x) = f(x)$ for every number x in $[a, b]$. But

$$A'(x) = \lim_{h \to 0} \frac{A(x + h) - A(x)}{h},$$

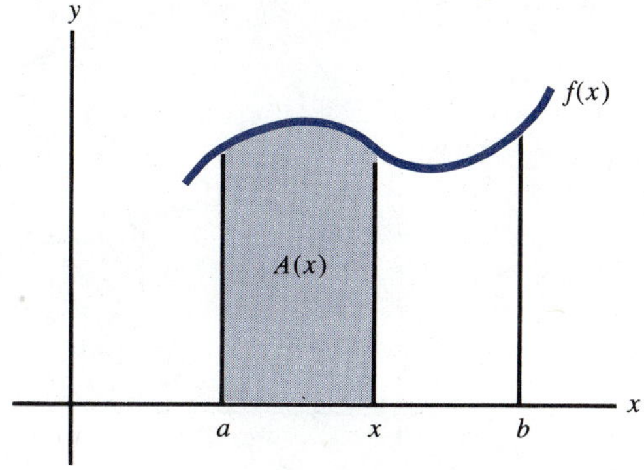

FIGURE 5–3

if the limit exists. However, if h is small enough so that $x + h$ is in the interval, then $A(x + h) - A(x)$ is just the area under f from x to $x + h$. See Figure 5–4. But as Figure 5–5 demonstrates, the area under f from x to $x + h$ can be approximated by the area of a rectangle whose height is $f(x + h)$ and whose base has length $x + h - x = h$; thus

$$A(x + h) - A(x) \approx hf(x + h).$$

Furthermore, it is clear that as $h \to 0$, the approximation becomes better and better. Therefore if we rewrite the approximation as

$$\frac{A(x + h) - A(x)}{h} \approx f(x + h)$$

and take the limit as $h \to 0$, we obtain

$$\lim_{h\to 0} \frac{A(x + h) - A(x)}{h} = \lim_{h\to 0} f(x + h).$$

But since f is continuous,

$$\lim_{h\to 0} f(x + h) = f(x)$$

by the second limit definition of continuity. Therefore

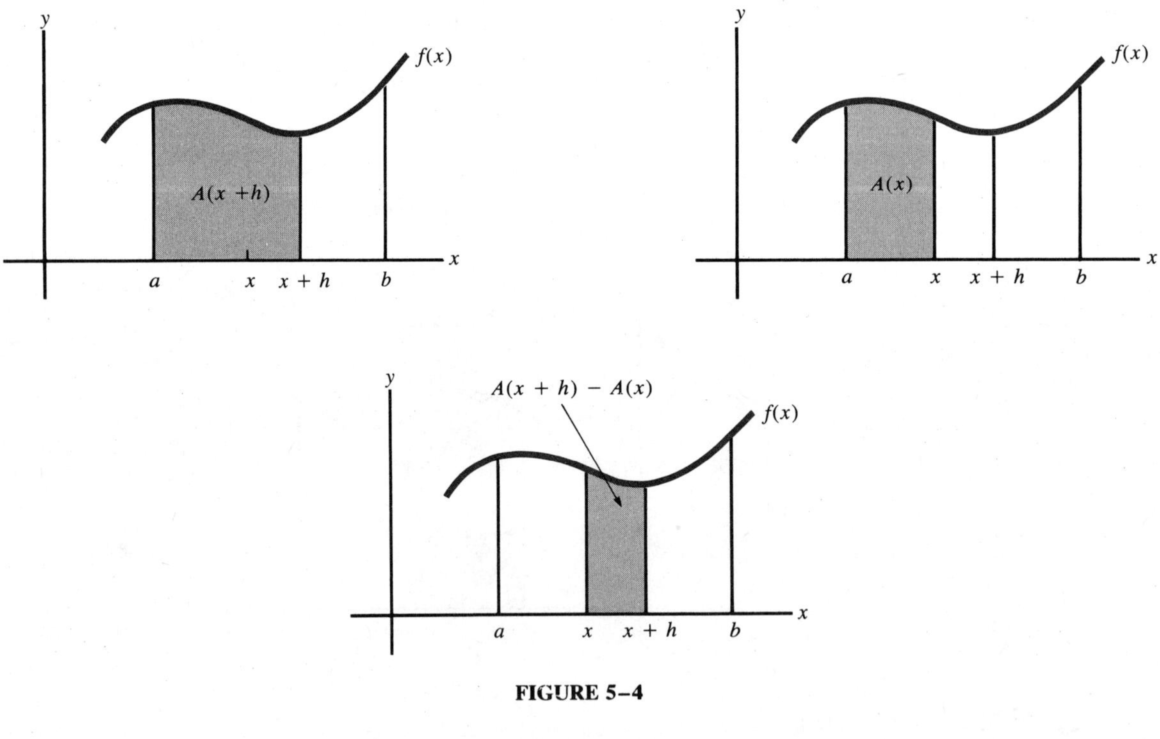

FIGURE 5–4

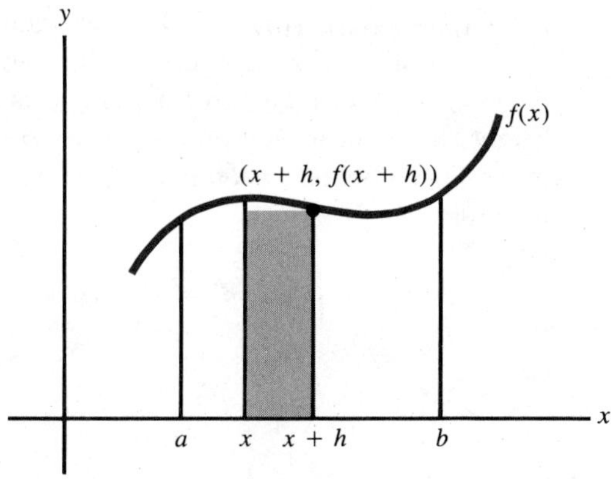

FIGURE 5–5

$$A'(x) = \lim_{h \to 0} \frac{A(x+h) - A(x)}{h} = \lim_{h \to 0} f(x+h) = f(x),$$

which shows that $A(x)$ is an antiderivative for $f(x)$.

Now suppose that $F(x)$ is any antiderivative for $f(x)$ on $[a, b]$. Then, as we saw in Section 5.1, there is some number c such that

$$F(x) = A(x) + c$$

for all x in $[a, b]$. But then

$$F(a) = A(a) + c = 0 + c = c,$$

so

$$F(x) = A(x) + F(a)$$

for all x in $[a, b]$. This in turn implies that when $x = b$,

$$A(b) = F(b) - F(a).$$

But $A(b)$ is the area under f from a to b, and

$$\int_a^b f(x)\,dx = F(b) - F(a).$$

Therefore the area under f from a to b is equal to the definite integral of f from a to b, and we have demonstrated the validity of the Fundamental Theorem of Calculus.

Finding Areas

The Fundamental Theorem of Calculus makes it easy to find the area of many regions in the xy-plane. We will illustrate with examples.

EXAMPLE 1 Let $f(x) = x$. We will find the area under f from 0 to 1. As Figure 5–6 shows, this is the area of a triangular region whose base has length 1 and whose height is also 1. Since the area of a triangle is one-half the length of its base times its height, the area in question must be ½. We obtain the same result with the aid of the Fundamental Theorem of Calculus, which tells us that the area is the definite integral

$$\int_0^1 f(x)\,dx = \int_0^1 x\,dx = \frac{1}{2}x^2\Big|_0^1 = \frac{1}{2}(1^2 - 0^2) = \frac{1}{2}.$$

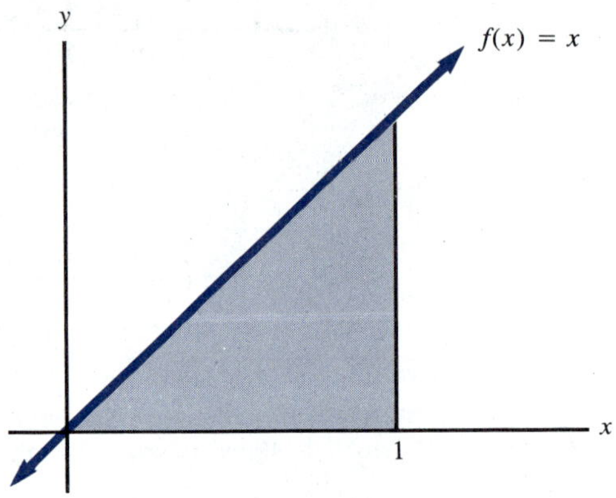

FIGURE 5–6

EXAMPLE 2 Let $f(x) = x^2$. We find the area of the region shaded in Figure 5–7, which is the area under f from -1 to 2, and thus is equal to

$$\int_{-1}^2 x^2\,dx = \frac{1}{3}x^3\Big|_{-1}^2 = \frac{1}{3}(2^3 - (-1)^3) = 3.$$

If the graph of a function f lies below the x-axis from $x = a$ to $x = b$, as in Figure 5–8, then the definite integral $\int_a^b f(x)\,dx$ will be a negative number; in such a case the area of the region which lies between the graph and the x-axis from a to b (the shaded region in Figure 5–8) is equal to the absolute value of the definite integral.

EXAMPLE 3 Suppose $f(x) = x^2 - 3x$ and we want to find the area of the region bounded by the graph of f and the x-axis from $x = 0$ to $x = 3$. This is the shaded region in Figure 5–9. Since the region lies below the x-axis, the definite integral is negative:

$$\int_0^3 f(x)\,dx = \int_0^3 (x^2 - 3x)\,dx$$

$$= \frac{1}{3}x^3 - \frac{3}{2}x^2\Big|_0^3$$

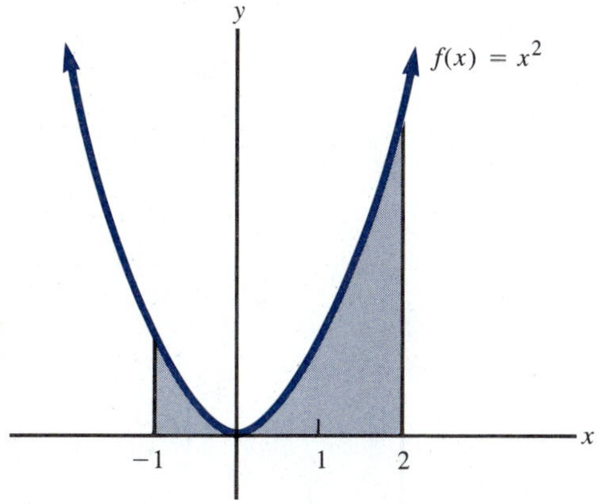

FIGURE 5–7

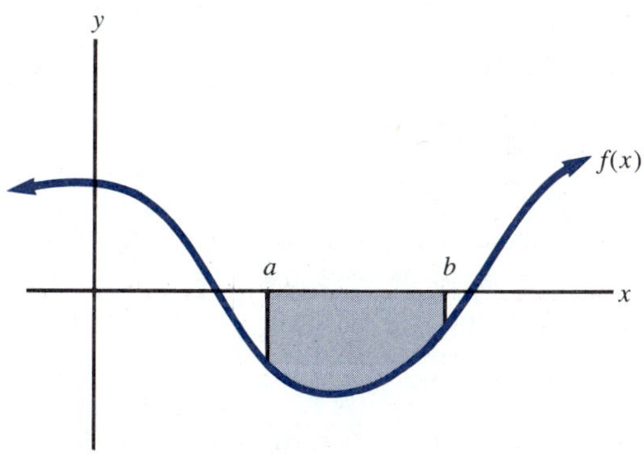

FIGURE 5–8

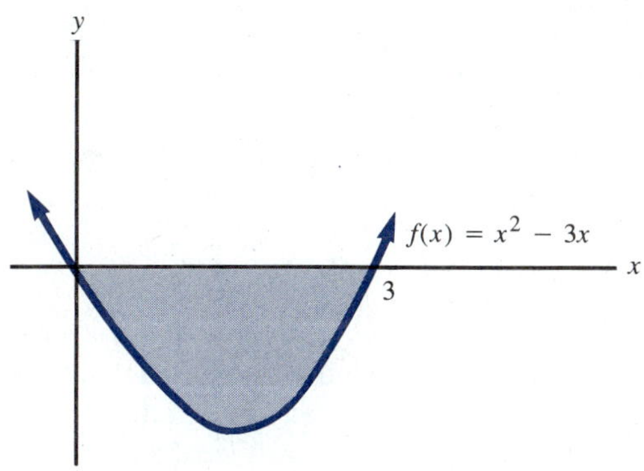

FIGURE 5–9

$$= \frac{1}{3}3^3 - \frac{3}{2}3^2$$

$$= -\frac{9}{2}.$$

Therefore the region has area equal to $|-9/2| = 9/2$.

If we need to find the area of a region part of which lies above the x-axis and part below, we can find the areas of the separate parts of the region and add them.

EXAMPLE 4 Let $f(x) = x^3 - 3x^2 + 2x$. Suppose we need to find the area of the shaded region in Figure 5–10. The area of the part of the region that is above the x-axis is

$$\int_0^1 (x^3 - 3x^2 + 2x)\,dx = \left(\frac{1}{4}x^4 - x^3 + x^2\right)\Big|_0^1 = \frac{1}{4}.$$

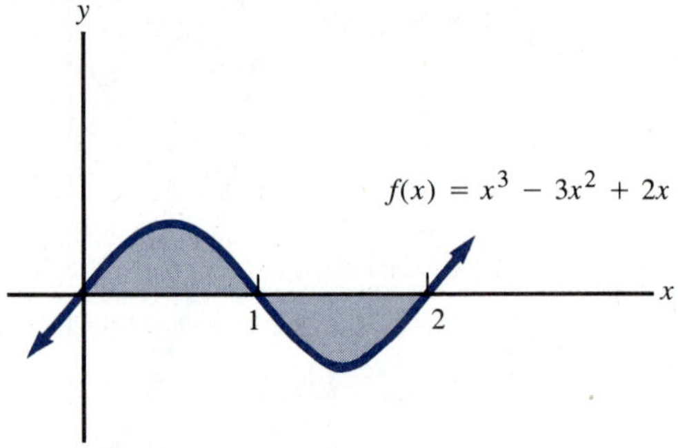

FIGURE 5–10

The area of the part of the region that is below the x-axis is the absolute value of

$$\int_1^2 (x^3 - 3x^2 + 2x)\,dx = \left(\frac{1}{4}x^4 - x^3 + x^2\right)\Big|_1^2 = -\frac{1}{4}.$$

Thus the area of this region is $|-\tfrac{1}{4}| = \tfrac{1}{4}$. The area of the entire region is therefore $\tfrac{1}{4} + \tfrac{1}{4} = \tfrac{1}{2}$.

The preceding examples show that when using the Fundamental Theorem of Calculus to find the area of a planar region, it is essential before doing the integration to determine whether the region lies above the x-axis, below it, or partly above it and partly below it. Thus careful curve sketching is required *before* any integration is performed.

Considering the definite integral as an area leads to many interesting consequences. For instance, it is true that for any number a,

$$\int_a^a f(x)\,dx = 0.$$

This is so because, as noted previously, the area under f from a to a must be zero. Similarly, if c is any number such that $a < c < b$, then

$$\int_a^b f(x)\,dx = \int_a^c f(x)\,dx + \int_c^b f(x)\,dx,$$

because, as Figure 5–11 shows, the area under f from a to b is equal to the area under f from a to c plus the area under f from c to b.

Another consequence of regarding the definite integral as an area is that total change can be represented by area. For instance, we have seen (Example 5, Section 5.3) that since marginal cost is the rate of change of cost, the definite integral of marginal cost from a to b is the total change in cost over the interval $[a, b]$; but this integral is also the area under the marginal cost curve from a to b. Hence, total change in cost is represented by the area of a region under the marginal cost curve. The same reasoning applied to any rate-of-change function leads to the following conclusion:

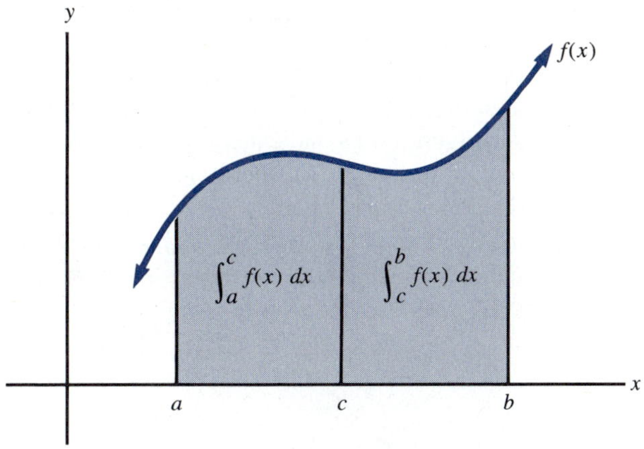

FIGURE 5–11

Total Change as Area

If f describes the rate of change of some quantity Q over the interval $[a, b]$, then the total change in Q over $[a, b]$ is represented by the area under f from a to b.

See Figure 5–12. This is an important fact to remember, because in business, economics, and the sciences, arguments concerning total change are often presented in terms of areas under some rate-of-change function.

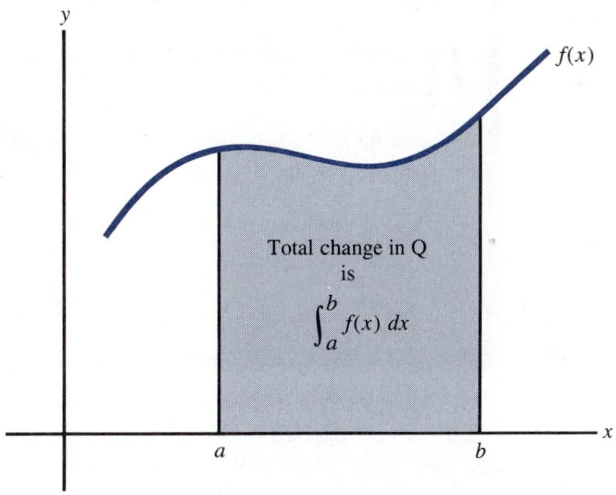

FIGURE 5–12 ***f* Describes Rate of Change of *Q***

The Area Bounded by Two Curves

Figure 5–13 shows a region bounded by the graphs of the functions f and g. The area of such a region is referred to as the **area bounded by the curves f and g.** As Figure 5–14 indicates, we can find the area bounded by the curves by first finding the points of intersection of the curves and then subtracting the area under the lower curve from the area under the upper one. Since the area under the lower curve in Figure 5–14 is $\int_a^b g(x)\,dx$ and that under the upper curve is $\int_a^b f(x)\,dx$, the area bounded by the curves is

$$\int_a^b f(x)\,dx - \int_a^b g(x)\,dx = \int_a^b [f(x) - g(x)]\,dx.$$

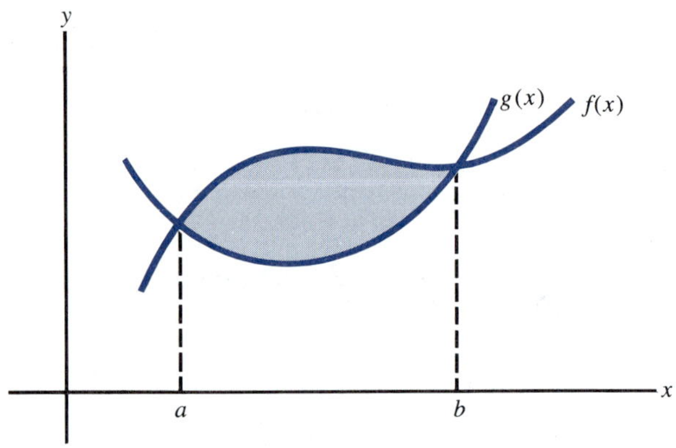

FIGURE 5–13

Finding the Area Bounded by Two Curves

To find the area bounded by two curves, where f is the upper curve and g is the lower curve:

1. Set $f(x) = g(x)$ and solve for x to find the values $x = a$ and $x = b$ at which the curves intersect.
2. The area bounded by the curves is then

$$\int_a^b [f(x) - g(x)]\,dx.$$

Figure 5–14 pictured the curves as nonnegative, but this is not necessary: the procedure for finding the area bounded by two curves works even if part or all of the region between them lies below the x-axis. However, it *is* necessary that in the integral of step 2, f must be the upper curve and g the lower one.

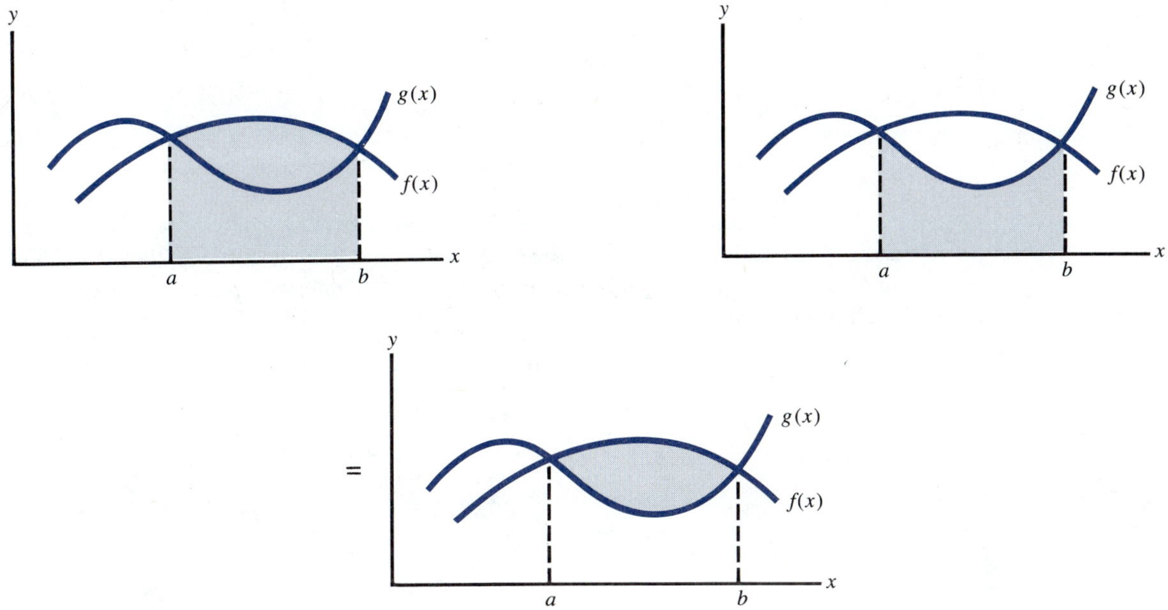

FIGURE 5–14 $\int_a^b f(x)\,dx - \int_a^b g(x)\,dx = \int_a^b [f(x) - g(x)]\,dx$

EXAMPLE 5 Let us find the area of the region bounded by the graphs of p and q, where

$$p(x) = 8x - x^2 \qquad \text{and} \qquad q(x) = x^2 - 4x + 10.$$

Since we must know which curve is the upper one and which the lower one, we draw the graphs of these functions. See Figure 5–15. To find the points where the curves intersect, we set $p(x) = q(x)$ and solve for x:

$$8x - x^2 = x^2 - 4x + 10,$$

so

$$2x^2 - 12x + 10 = 0,$$

or

$$2(x - 1)(x - 5) = 0.$$

Therefore the curves intersect at $(1, p(1)) = (1,7)$ and $(5, p(5)) = (5,15)$. Since p is the upper curve and q the lower one over the interval $[1, 5]$, it follows that the area of the region in question is equal to

$$\int_1^5 [p(x) - q(x)]\,dx = \int_1^5 [(8x - x^2) - (x^2 - 4x + 10)]\,dx$$

$$= \int_1^5 (-2x^2 + 12x - 10)\,dx$$

$$= -\frac{2}{3}x^3 + 6x^2 - 10x \Big|_1^5$$

$$= \frac{64}{3}.$$

Our procedure for finding the area bounded by two curves requires that we subtract the lower curve from the upper one. But which curve is the lower and which is the upper may change from interval to interval. Again, careful curve sketching is essential.

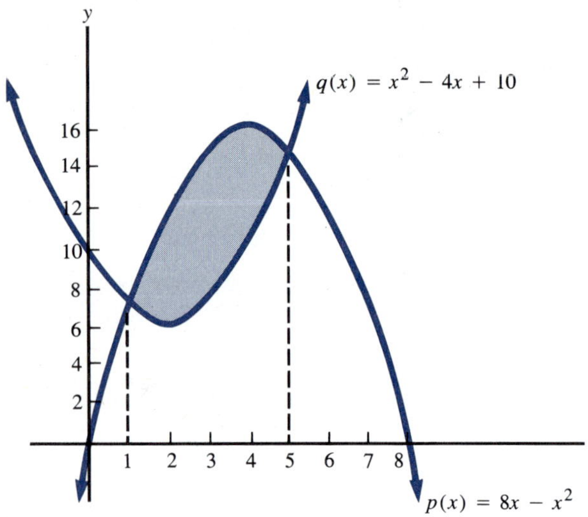

FIGURE 5–15

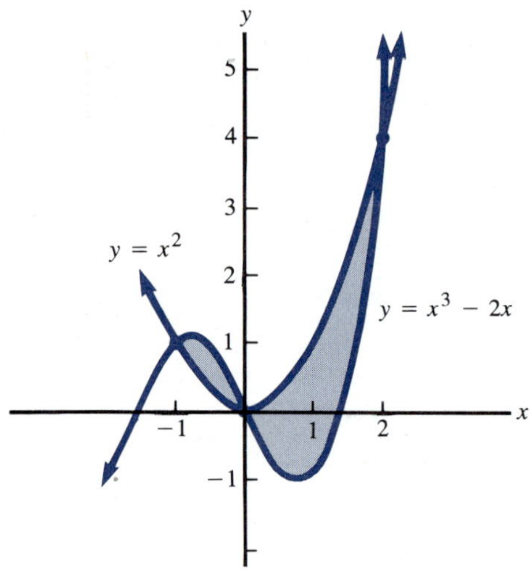

FIGURE 5–16

EXAMPLE 6 Find the area between the curves

$$y = x^2 \qquad \text{and} \qquad y = x^3 - 2x.$$

Setting the equations equal to each other and solving for x yields

$$x^2 = x^3 - 2x,$$

or

$$x^3 - x^2 - 2x = 0,$$

or

$$x(x + 1)(x - 2) = 0,$$

so the points of intersection are $(0, 0)$, $(-1, 1)$, and $(2, 4)$. See Figure 5–16. Note that over the interval $[-1, 0]$ the upper curve is $y = x^3 - 2x$, but over the interval $[0, 2]$ the upper curve is $y = x^2$. Therefore the area of the region bounded by the two curves is

$$\int_{-1}^{0} [(x^3 - 2x) - x^2]\,dx + \int_{0}^{2} [x^2 - (x^3 - 2x)]\,dx$$

$$= \left(\frac{1}{4}x^4 - \frac{1}{3}x^3 - x^2\right)\Big|_{-1}^{0} + \left(-\frac{1}{4}x^4 + \frac{1}{3}x^3 + x^2\right)\Big|_{0}^{2}$$

$$= 0 - \left(\frac{1}{4} + \frac{1}{3} + 1\right) + \left(-\frac{16}{4} + \frac{8}{3} + 4\right) - 0$$

$$= \frac{37}{12}.$$

Computer Manual: Program DEFINT.

5.4 EXERCISES

Finding Areas

In Exercises 1 through 10 find the area under f from a to b.

1. $f(x) = x^2 + 3x + 5, \quad a = 1, \quad b = 3$

2. $f(x) = x^2 + 3x + 5, \quad a = -1, \quad b = 1$

3. $f(x) = x^2 + 3x + 5, \quad a = -1, \quad b = 3$

4. $f(x) = x^3 - 3x^2 - 6x + 8, \quad a = -2, \quad b = 1$

5. $f(x) = x^3 - 3x^2 - 6x + 8, \quad a = 1, \quad b = 4$

6. $f(x) = x^3 - 3x^2 - 6x + 8, \quad a = -2, \quad b = 4$

7. $f(x) = \frac{1}{x}, \quad a = \frac{1}{2}, \quad b = e$

8. $f(x) = e^{2x}, \quad a = 0, \quad b = 4$

9. $f(x) = xe^{-x^2}, \quad a = 0, \quad b = 3$

10. $f(x) = \frac{2}{x+1}, \quad a = 0, \quad b = 3$

Management

11. The typical worker in a jewelry plant can produce $f(t)$ necklaces per hour, where

$$f(t) = -3t^2 + 36t,$$

t in hours.

(a) Graph this function.

(b) Interpret the number of necklaces an average worker can produce in an 8-hour day as an area, and find this number.

12. A firm's marginal cost and marginal revenue functions are defined by

$$c(x) = x^2 - 80x + 3600 \qquad \text{and} \qquad r(x) = -x^2 + 100x,$$

respectively.

(a) Graph these functions.

(b) Suppose the firm is currently making and selling 20 units per day. Interpret as areas the additional cost and revenue that would result from doubling production and sales, and then find the additional cost and revenue.

Life Science

13. A virus grows within the human body at a rate of y organisms per hour, where

$$y = 2000e^{0.02t}$$

and t is measured in hours since infection occurred.

(a) Graph this function.
(b) Interpret the number of viral organisms that are produced during the first 3 hours after infection as an area and find this number.
(c) If death results when the total number of viral organisms produced since infection reaches 1 million, how long will it be from infection until death if no treatment is given?

The Area Bounded by Two Curves

In Exercises 14 through 21 find the area bounded by f and g.

14. $f(x) = x,\ g(x) = x^2$
15. $f(x) = x^2,\ g(x) = x^3$
16. $f(x) = -x + 4,\ g(x) = -x^2 + 4x$
17. $f(x) = x^2 - 9x + 26,\ g(x) = -x^2 + 9x - 14$
18. $f(x) = x,\ g(x) = x^3$
19. $f(x) = \sqrt{x},\ g(x) = x$
20. $f(x) = \sqrt{x},\ g(x) = x^2$
21. $f(x) = |x|,\ g(x) = 0.5x + 3$
22. Find the area of the region bounded by the graphs of f, g, and h, where $f(x) = 4 + 2x$, $g(x) = 7 - x$, and $h(x) = -3 + 4x$.
23. Find the area of the region bounded by the graphs of f, g, and h, where $f(x) = 2x + 5$, $g(x) = 8 + x$, and $h(x) = 29 - 2x$.
24. Find the area of the region bounded by the graphs of $y = x^2 - 4x + 14$, $y = -x^2 + 6x - 9$, $x = 1$, and $x = 4$.
25. Find the area of the region bounded by the graphs of $y = e^x$, $y = e^{-x}$, $x = -1$, $x = 1$, and $y = 0$ (the x-axis).

Management

26. A firm's new drill press is saving s thousand dollars per month in raw materials cost, where $s = 150 - t^2$. Also, the cost of the labor needed to run the new press is increasing at the rate of $c = 0.5t^2$ thousand dollars per month. Here t is months since the press was installed. Graph these functions, interpret the firm's net savings as an area, and find the net savings.

Social Science

27. Emigrants are leaving a nation at the rate of $f(t)$ thousand persons per year, where $f(t) = 20e^{0.5t}$, and immigrants are entering the nation at the rate of $g(t)$ thousand persons per year, where $g(t) = 40e^{0.4t}$. Here $t = 0$ is 1985. Graph these functions, and then find the net population change due to the combination of emigration and immigration over the period

(a) 1985–1990 (b) 1985–1995.

5 REVIEW EXERCISES

Answers to all review exercises are at the back of the book.

1. Find

(a) $\int(-4x^5 + 3x^2 - 8x + 14)\,dx$ **(b)** $\int(-2x^{-2} + 3x^{4/5} + 4)\,dx$

(c) $\int\left(-2t^{-3} + \frac{2}{t^2} + 4\sqrt{t} + 1\right)dt$ **(d)** $\int 2e^x\,dx$

(e) $\int -\frac{3}{x}\,dx$ **(f)** $\int\left(\frac{2}{x} - 3e^x\right)dx$

2. A firm's marginal cost and marginal revenue functions are defined by

$$c(x) = 6x + 2 \qquad \text{and} \qquad r(x) = 20x,$$

respectively. Find the firm's cost and revenue functions if its fixed cost is \$5000.

3. During an epidemic the number of cases of influenza increases at a rate of $y = 300\sqrt{t}$ cases per day t. Find a function that describes the number of cases as a function of time if there were 2000 cases on day 4 of the epidemic.

In Exercises 4 through 10 find the indefinite integral.

4. $\int\sqrt{4x+1}\,dx$ **5.** $\int 3x(x^2+4)^3\,dx$ **6.** $\int 3x^2e^{x^3}\,dx$

7. $\int\frac{2x}{4x^2+1}\,dx$ **8.** $\int\frac{6x+2}{(3x^2+2x+1)^{4/3}}\,dx$ **9.** $\int 5xe^{-x^2}\,dx$

10. $\int\frac{1}{x(\ln x)^2}\,dx$

In Exercises 11 through 18 find the definite integral.

11. $\int_0^2 (x^2 - 3x + 1)\,dx$ **12.** $\int_{-2}^{-1}(4x^3 - x^2 + 1)\,dx$ **13.** $\int_0^{\ln 2} e^{3x}\,dx$

14. $\int_4^9(\sqrt{x} + x^{-2})\,dx$ **15.** $\int_{4/5}^{(e+3)/5}\frac{10}{5x-3}\,dx$ **16.** $\int_0^1 xe^{-x^2/2}\,dx$

17. $\int_0^2\frac{x^2}{x^3+1}\,dx$ **18.** $\int_e^{e^2}\frac{1}{x\ln x}\,dx$

19. The population of a town is growing at the rate of

$$f(t) = 15\sqrt{t} + 8$$

persons per year, where t is in years with $t = 0$ representing the present. What will be the change in the town's population size over the next 4 years?

20. A firm's marginal cost function is given by

$$c(x) = 3x + 10 + \frac{500}{x},$$

valid for $x \geq 1$. If the firm is currently producing 10 units, find the additional cost of producing 2 more units.

21. Let $f(x) = 4x^{-3}$. Find the area under f from 1 to 4.

22. Let $f(x) = \dfrac{x}{x^2 + 1}$. Find the area under f from 0 to 2.

23. Let $f(x) = -x^2 + 13x - 36$. Find the area of the region bounded by the x-axis and the graph of f.

24. Find the area bounded by the x-axis and the graph of $y = x^3 - 2x^2 - 15x$.

25. Find the area of the region bounded by p and q, where

$$p(x) = -x^2 + 6x + 2 \qquad \text{and} \qquad q(x) = x + 2.$$

26. It is estimated that the number of junk bonds on the market is increasing at the rate of

$$f(t) = \frac{50}{t + 2}$$

million bonds per year. Here $t = 0$ represents 1984.

(a) Find the area under f from 0 to b. What does this area represent?

(b) When will there be a total of 100 million junk bonds on the market?

5 SUPPLEMENT: THE DEVELOPMENT OF THE DEFINITE INTEGRAL

In this chapter we defined the definite integral of a function over an interval to be the total change in an antiderivative of the function over the interval. We then showed that the definite integral can be used to find areas. The historical development of the definite integral reversed these steps: first came the integral as area, then the realization that integrals could be evaluated with the aid of antiderivatives. In this supplement we give a brief account of the historical development of the definite integral.

We can consider integration as a type of limiting process in which areas are obtained as limits of sums of simpler areas; in this sense its beginnings go back to antiquity. The Greek mathematician Archimedes (287–212 B.C.) found the areas of many geometric figures by using a technique in which the figure under question was "filled up" with simpler figures whose areas could be easily determined. This method of integration was known as the **method of exhaustion.** We will demonstrate how the method of exhaustion may be employed to find the area P of a parabolic region such as that pictured in Figure 5–17.

We let D be the midpoint of the segment AB, and let AB serve as the base of a triangle ACB ($\triangle$ ACB), where C is a point on the parabola such that CD is parallel to the axis of symmetry of the parabola. See Figure 5–18, where we have chosen C so that CD lies on the axis of symmetry. As a first approximation, we have

$$P \approx \text{area of } \triangle \text{ ACB}.$$

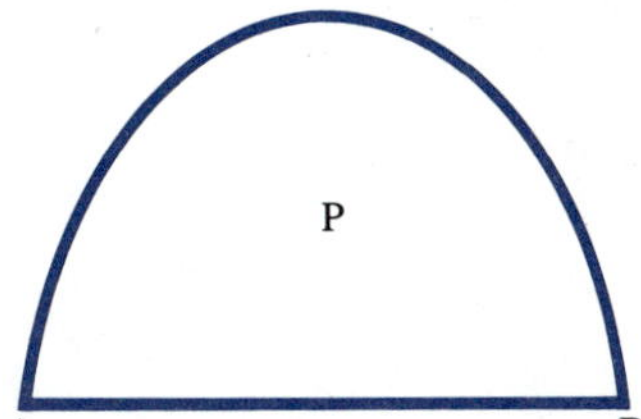

FIGURE 5–17

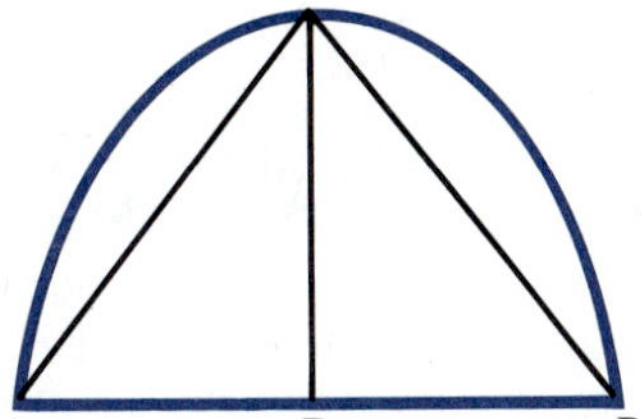

FIGURE 5–18

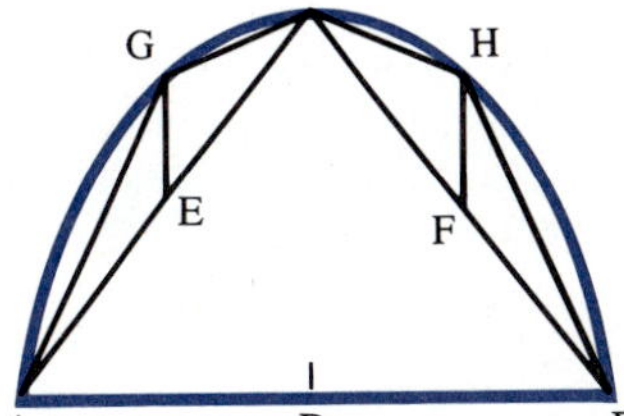

FIGURE 5–19

Now we repeat the construction with E and F the midpoints of AC and CB, respectively, and G and H the points on the parabola such that GE and HF are parallel to the axis of symmetry of the parabola. See Figure 5–19. We thus obtain the triangles AGC and CHB, and

$$P \approx \text{area of } \triangle \text{ ACB} + \text{area of } \triangle \text{ AGC} + \text{area of } \triangle \text{ CHB}.$$

Clearly, this approximation is better than the previous one. It is possible to show geometrically that

$$\text{area of } \triangle \text{ AGC} + \text{area of } \triangle \text{ CHB} = \frac{1}{4}(\text{area of } \triangle \text{ ACB}).$$

Therefore $P \approx \frac{5}{4}(\text{area of } \triangle \text{ ACB})$.

Now we repeat the construction with segments AG, GC, CH, and HB as the bases of four new triangles, to obtain a better approximation:

$$P \approx \frac{5}{4}(\text{area of } \triangle \text{ ACB}) + \frac{1}{4^2}(\text{area of } \triangle \text{ ACB})$$

$$= \frac{21}{16}(\text{area of } \triangle \text{ ACB}).$$

As the process continues, the triangles ''fill up'' or ''exhaust'' the parabolic region, and hence the sum of their areas approaches the value of P. It can be shown that after n repetitions of the construction, the sum of the areas of the triangles is

$$\frac{4}{3}\left(1 - \frac{1}{4^n}\right)(\text{area of } \triangle \text{ ACB}).$$

But since $\frac{1}{4^n}$ can be made arbitrarily small by taking n large enough, it follows that

$$P = \frac{4}{3}(\text{area of } \triangle \text{ ACB}).$$

In modern terminology we would say that

$$\begin{aligned}P &= \lim_{n\to+\infty} \text{(sum of the areas of the triangles)}\\ &= \lim_{n\to+\infty} \frac{4}{3}\left(1 - \frac{1}{4^n}\right) (\text{area of } \triangle \text{ ACB})\\ &= \frac{4}{3}(1 - 0)(\text{area of } \triangle \text{ ACB})\\ &= \frac{4}{3}(\text{area of } \triangle \text{ ACB}).\end{aligned}$$

We can use this result to determine the area of a parabolic segment by simply finding the length of the segments AB and CD of Figure 5–18. For instance, if AB has length 1 and CD has length ¼, then

$$\text{area of } \triangle \text{ ACB} = \frac{1}{2}(1)\left(\frac{1}{4}\right) = \frac{1}{8},$$

and

$$P = \frac{4}{3}\cdot\frac{1}{8} = \frac{1}{6}.$$

The method of exhaustion depends heavily on the geometric properties of the region under consideration and is difficult to apply in general situations. A better method for finding areas as the limits of sums of simpler areas became possible after the French mathematician René Descartes (1596–1650) introduced the cartesian coordinate system: This allowed mathematicians to regard curves as the graphs of equations and hence to consider regions bounded by curves as being defined by equations. This in turn made it possible to express the areas of such regions as limits of sums of areas of rectangles. We will illustrate how this is done for the area of the parabolic region of Figure 5–20, which can be described as the region bounded by the x-axis and the graph of the equation $y = x(1 - x)$.

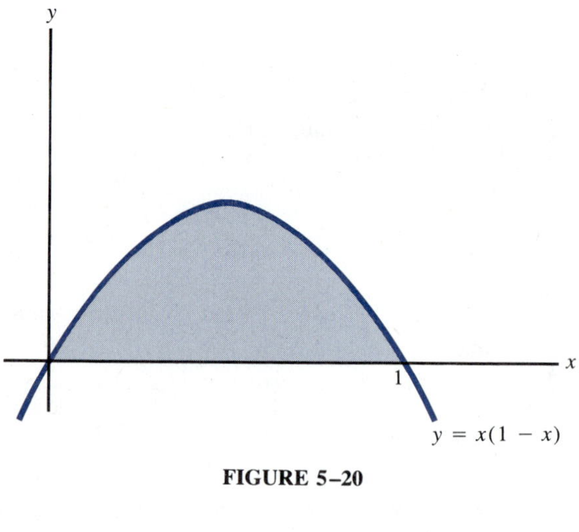

FIGURE 5–20

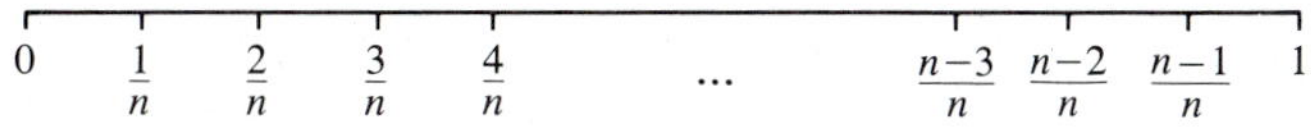

FIGURE 5–21

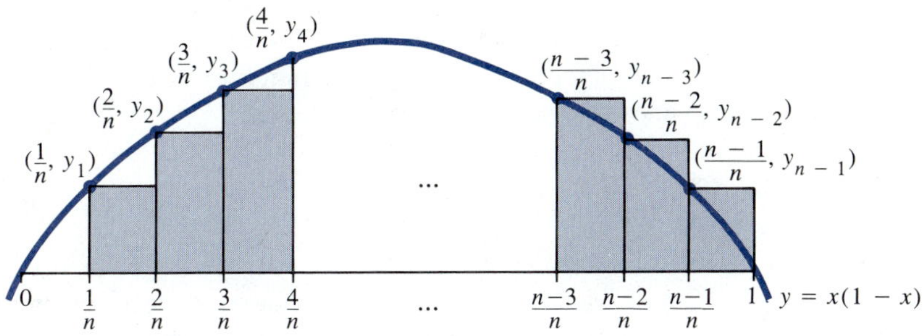

FIGURE 5–22

We begin by dividing the interval [0, 1] into n subintervals, each of width $1/n$ (Figure 5–21). On each subinterval $[k/n, (k + 1)/n]$ we construct a rectangle whose height is y_k, where

$$y_k = \frac{k}{n}\left(1 - \frac{k}{n}\right).$$

See Figure 5–22. Clearly the area P of the parabolic region is approximated by the sum of the areas of the rectangles:

$$\begin{aligned} P &\approx \frac{1}{n}y_1 + \frac{1}{n}y_2 + \cdots + \frac{1}{n}y_{n-1} \\ &= \frac{1}{n}\left[\left(\frac{1}{n} - \frac{1}{n^2}\right) + \left(\frac{2}{n} - \frac{2^2}{n^2}\right) + \cdots + \left(\frac{n-1}{n} - \frac{(n-1)^2}{n^2}\right)\right]. \end{aligned}$$

Furthermore, the more rectangles we use (that is, the larger we take n to be), the more closely the sum of their areas will approach the value of P. It can be shown that the sum

$$\frac{1}{n}\left[\left(\frac{1}{n} - \frac{1}{n^2}\right) + \left(\frac{2}{n} - \frac{2^2}{n^2}\right) + \cdots + \left(\frac{n-1}{n} - \frac{(n-1)^2}{n^2}\right)\right]$$

is equal to

$$\frac{1}{6}\left(1 - \frac{1}{n^2}\right).$$

Since $1/n^2$ can be made arbitrarily small by taking n large enough, it follows that $P = 1/6$. In other words, since

$$P = \lim_{n\to+\infty} \text{(sum of the areas of the rectangles)},$$

we have $P = \lim_{n \to +\infty} \frac{1}{6}\left(1 - \frac{1}{n^2}\right) = \frac{1}{6}(1 - 0) = \frac{1}{6}$.

This technique of expressing the area of a region as a limit of sums of areas of rectangles is important, and we shall encounter it again in Chapter 6. For our purposes here, it is enough to note that it constitutes a very general method for finding areas, and that it was known to mathematicians before the time of Newton and Leibniz. Thus if f is a function that is nonnegative over the interval $[a, b]$ and we *define* the definite integral

$$\int_a^b f(x)\,dx$$

to be the area under f from $x = a$ to $x = b$, then we can say that mathematicians possessed a technique for evaluating definite integrals (i.e., finding areas) before Newton and Leibniz. What was not appreciated was the fact that definite integrals could be evaluated by antidifferentiation. The realization that this is so is usually credited to the English mathematician Isaac Barrow (1630–1677), who was Newton's teacher; however, it was not until Newton and Leibniz put differentiation on a firm basis that this fact could be easily exploited.

Once the notion of the derivative of a function was formalized, mathematicians could easily show that definite integrals could be evaluated by finding antiderivatives for their integrands. The proof was as follows. By definition,

$$\int_a^x f(x)\,dx = \text{area under } f \text{ from } a \text{ to } x.$$

Set

$$A(x) = \int_a^x f(x)\,dx.$$

Then just as in Section 5.4, $A'(x) = f(x)$ and if F is an antiderivative for f,

$$\int_a^b f(x)\,dx = F(b) - F(a).$$

Since $A'(x) = f(x)$, the Fundamental Theorem of Calculus is often written as

$$\frac{d}{dx}\int_a^x f(x)\,dx = f(x).$$

In this form the Fundamental Theorem of Calculus deserves its name, for it makes explicit the relationship between the two great concepts of calculus—the derivative and the integral—by showing that they are inverse operations in the same sense that addition and subtraction are inverse operations: each can be used to undo the other. Thus the Fundamental Theorem is the cord that ties the two subjects of differential calculus and integral calculus together into one coherent theory.

Suggestion

If you would like to read more about the development of the integral, here are some books to consult:

Bell, E. T., *Men of Mathematics,* Simon and Schuster, New York, 1937.

Boyer, Carl B., *History of the Calculus and Its Conceptual Development,* Dover, New York, 1959.

Boyer, Carl B., *A History of Mathematics,* Wiley, New York, 1968.

Eves, Howard, *An Introduction to the History of Mathematics,* Holt, Rinehart, and Winston, New York, 1961.

Eves, Howard, *Great Moments in Mathematics before 1650,* Dolciani Mathematical Expositions, No. 5, The Mathematical Association of America, 1980.

Eves, Howard, *Great Moments in Mathematics after 1650,* Dolciani Mathematical Expositions, No. 7, The Mathematical Association of America, 1983.

6

More about Integration

In Chapter 5 we defined the definite integral $\int_a^b f(x)\,dx$ to be the total change in any antiderivative of $f(x)$ from $x = a$ to $x = b$. We also showed how the definite integral can be interpreted as an area and used to find the area of planar regions. Still another way to view the definite integral is as an **accumulator.** Section 5.1 discusses what this means and shows how this viewpoint allows us to use the definite integral to find the average value of a function. Section 5.2 continues with further examples of the integral as an accumulator, this time in economics and finance. We next consider the problem of evaluating the definite integral when we cannot find an antiderivative for the integrand. In the fourth section we expand the notion of the definite integral to encompass cases in which the interval of integration is infinite. We conclude with a presentation of the technique of integration known as integration by parts and a discussion of the use of a table of integrals.

6.1 RIEMANN SUMS AND THE INTEGRAL AS ACCUMULATOR

In this section we show how the definite integral may be used to find the total accumulation of a quantity over an interval. We then illustrate the use of the integral as an accumulator by using it to find the average value of a function over an interval.

Riemann Sums

Let n be any positive integer. Suppose we divide the interval $[a, b]$ into n equal subintervals. Then each of the subintervals will have length Δx, where

$$\Delta x = \frac{1}{n}(\text{length of } [a, b]) = \frac{1}{n}(b - a) = \frac{b - a}{n}.$$

See Figure 6–1. Note that the endpoints of the subintervals are

$$a = a_0, a_1, a_2, \ldots, a_{n-1}, a_n = b.$$

Thus the ith subinterval is $[a_{i-1}, a_i]$, and its length is Δx.

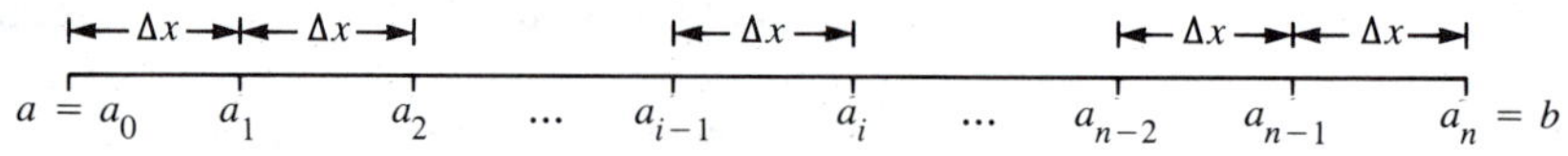

FIGURE 6–1

Now let f be a function defined on $[a, b]$. From each subinterval $[a_{i-1}, a_i]$ we choose a single number x_i, find the functional value $f(x_i)$ (see Figure 6–2), and form the sum

$$f(x_1)\,\Delta x + f(x_2)\,\Delta x + \cdots + f(x_n)\,\Delta x = [f(x_1) + f(x_2) + \cdots + f(x_n)]\,\Delta x.$$

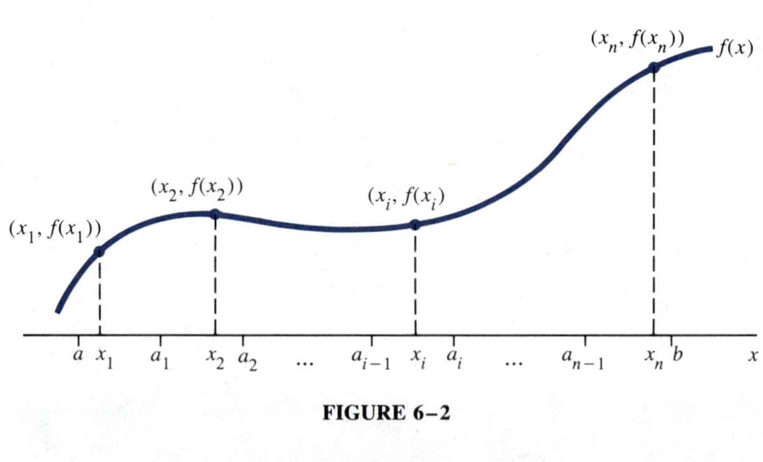

FIGURE 6–2

This sum is called a **Riemann sum for f over $[a, b]$.** If we construct a rectangle of height $f(x_i)$ on each subinterval $[a_{i-1}, a_i]$, the area of the rectangle will be $f(x_i)\,\Delta x$. See Figure 6–3. Thus the Riemann sum is the sum of the areas of all such rectangles. See Figure 6–4, where the Riemann sum is the sum of the areas of the rectangles.

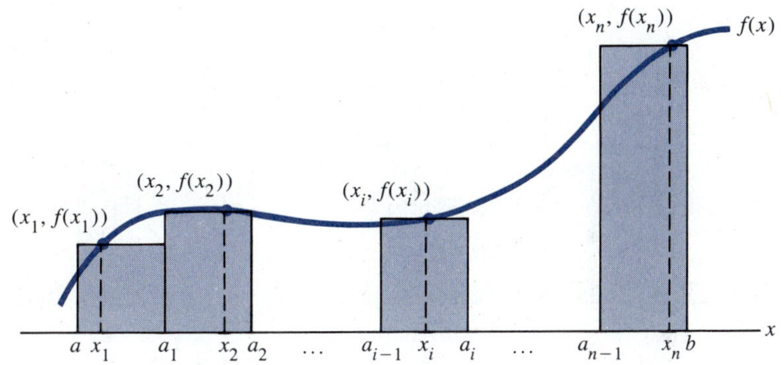

FIGURE 6–3

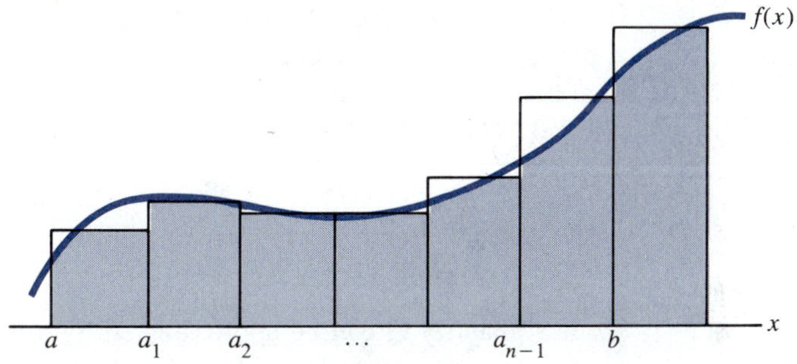

FIGURE 6–4

What happens if we let n become larger and larger? As $n \to +\infty$, $\Delta x \to 0$; in other words, as n becomes larger and larger, we subdivide the interval $[a, b]$ into more and more subintervals of shorter and shorter length. But as Figure 6–5 shows, as this occurs the total area of the rectangles approaches the area under f from a to b, which is the definite integral of f from $x = a$ to $x = b$. Therefore the limit of all Riemann sums for f over $[a, b]$ as $\Delta x \to 0$ is the definite integral.

The Integral as the Limit of Riemann Sums

$$\int_a^b f(x)\,dx = \lim_{\Delta x \to 0} [f(x_1) + f(x_2) + \cdots + f(x_n)]\,\Delta x.$$

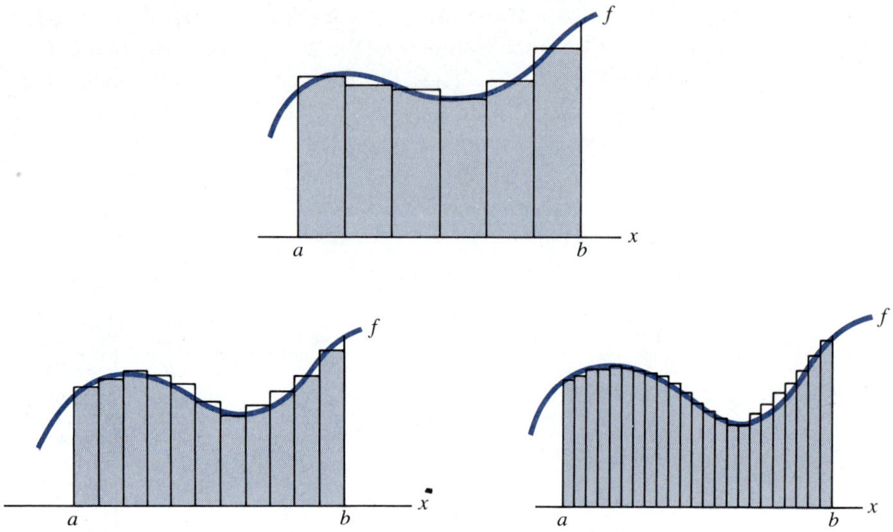

FIGURE 6–5

For our purposes the significance of this formula is as follows: if we can approximate a quantity Q by a Riemann sum in such a manner that the approximation becomes exact as Δx approaches 0, then Q will be equal to the definite integral. This is what we mean when we say that the integral can be used as an accumulator.

The Integral as an Accumulator

If a quantity Q can be approximated by a Riemann sum for some function f over some interval $[a, b]$, and if the approximation becomes exact as $\Delta x \to 0$, then

$$Q = \int_a^b f(x)\,dx.$$

The Average Value of a Function

We average a collection of numbers by summing them and dividing the sum by the number of them. For instance, the average of 2, −4, 12, 14, and 6 is

$$\frac{2 + (-4) + 12 + 14 + 6}{5} = \frac{30}{5} = 6.$$

Now suppose f is a function defined on $[a, b]$. We will show how to find the average value of f on $[a, b]$.

Let M denote the average value of f on $[a, b]$. Then M ought to be the sum of all functional values $f(x)$, for all x in $[a, b]$, divided by the number of such x's. We cannot find M directly because there are infinitely many such x's, but we can approximate M by choosing n equally spaced numbers $x_1, x_2, \ldots, x_n$ from $[a, b]$; then the average of these n functional values is

$$\frac{f(x_1) + f(x_2) + \cdots + f(x_n)}{n}.$$

Thus

$$M \approx \frac{f(x_1) + f(x_2) + \cdots + f(x_n)}{n}.$$

As we choose more and more functional values, that is, as $n \to +\infty$, the preceding approximation becomes better and better. Let us rewrite the approximation as

$$M \approx \frac{1}{b-a}\left[f(x_1)\frac{b-a}{n} + f(x_2)\frac{b-a}{n} + \cdots + f(x_n)\frac{b-a}{n}\right],$$

or

$$(b-a)M \approx [f(x_1) + f(x_2) + \cdots + f(x_n)]\frac{b-a}{n},$$

or

$$(b-a)M \approx [f(x_1) + f(x_2) + \cdots + f(x_n)]\,\Delta x,$$

where $\Delta x = (b-a)/n$. Since the sum on the right is a Riemann sum for f over $[a, b]$, and since the approximation becomes exact as $\Delta x \to 0$ (that is, as $n \to +\infty$), we have

$$(b-a)M = \int_a^b f(x)\,dx,$$

or

$$M = \frac{1}{b-a}\int_a^b f(x)\,dx.$$

The Average Value of a Function

The average value of a function f over the interval $[a, b]$ is

$$M = \frac{1}{b-a}\int_a^b f(x)\,dx.$$

Note that the average value M of f over $[a, b]$ has the property that

$$(b-a)M = \int_a^b f(x)\,dx;$$

in other words, M is the value of $f(x)$ over $[a, b]$ such that a rectangle with base $[a, b]$ and height M has the same area as the region under the graph of f from $x = a$ to $x = b$. See Figure 6–6.

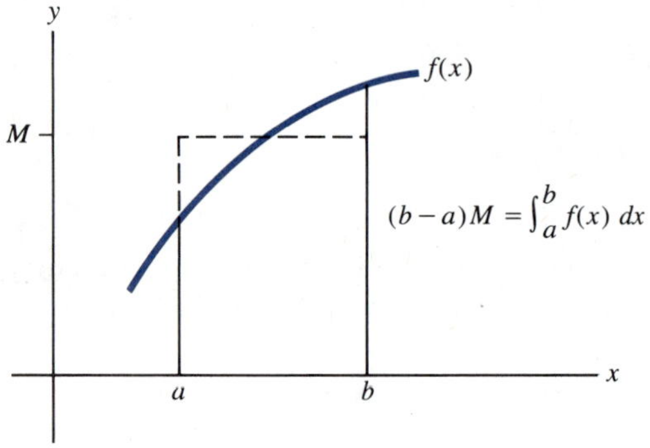

FIGURE 6–6

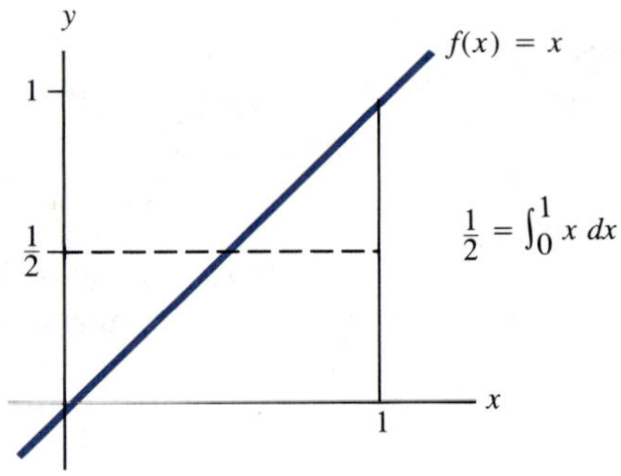

FIGURE 6–7

EXAMPLE 1 Let f be the function with defining equation $f(x) = x$. The average value of f over the interval $[0, 1]$ is

$$M = \frac{1}{1-0}\int_0^1 x\,dx = \frac{x^2}{2}\bigg|_0^1 = \frac{1}{2}.$$

Thus the rectangle of height ½ over the interval $[0, 1]$ has the same area as the region under f from $x = 0$ to $x = 1$. See Figure 6–7.

EXAMPLE 2 Over the course of each month (30 days) a retailer's inventory declines linearly from 300 units to 60 units. What is the retailer's average monthly inventory?

If we let y be the number of units in inventory on day t, $0 \le t \le 30$, then y is a linear function of t such that $y = 300$ when $t = 0$ and $y = 60$ when $t = 30$. Hence

$$y = \frac{60 - 300}{30 - 0}t + 300$$

or

$$y = -8t + 300,$$

$0 \le t \le 30$. Therefore the average number of units in inventory each month is

$$\frac{1}{30-0}\int_0^{30} (-8t + 300)\,dt = \frac{1}{30}(-4t^2 + 300t)\bigg|_0^{30} = \frac{1}{30}(5400) = 180.$$

Computer Manual: Program DEFINT.

6.1 EXERCISES

The Average Value of a Function

In Exercises 1 through 4 find the average value of the function over the given interval.

1. $f(x) = x^3$, $[0, 1]$
2. $f(x) = x^3 - 3x + 2$, $[1, 2]$
3. $f(x) = e^{-x}$, $[0, 2]$
4. $f(x) = \frac{1}{x}$, $[e, e^2]$
5. The velocity of a freely falling body t seconds after the start of its fall is $v = 32t$ feet per second. Find the average velocity of a freely falling body over the first 4 seconds of its fall.
6. If a rocket is fired straight up, its altitude t minutes after launch is

$$s = -t^2 + 40t$$

kilometers. Find the rocket's average altitude during the second 5 minutes of its flight.

7. Show that if f is the constant function defined by $f(x) = c$ for all x, then the average value of f over any interval $[a, b]$ is equal to c. Interpret this result geometrically.

Management

8. Dour Company's annual profits are given by the equation

$$P = 10e^{0.06t},$$

where P is in millions of dollars and t is in years, with $t = 0$ representing the present. Find Dour's average annual profit over the past 5 years and forecast their average annual profit over the next 5 years.

9. A new plant produces

$$y = 0.1t^2 + 0.5t + 400$$

units per month, where t is in months since the plant began operating. Find the plant's average monthly production during its first year of operation.

10. A worker can assemble compact disk players at the rate of $8\sqrt{t}$ per hour, where $0 \leq t \leq 8$. Find the worker's average hourly assembly rate over the 8-hour workday.
11. A retailer's inventory declines linearly from 250 units at the beginning of the month to 20 units at the end of the month. Find the retailer's average inventory for the month (30 days).
12. A manufacturer's inventory of a finished product over a 90-day period is y units, where

$$y = 100 - \frac{1}{90}t^2, \qquad 0 \leq t \leq 90.$$

Find the manufacturer's average inventory over the period.

Life Science

13. During a chemical experiment the temperature of a solution was

$$T = 20 + 2t - t^2$$

degrees Celsius t seconds after the experiment started. If the experiment lasted 5 seconds, find the average temperature of the solution during the experiment.

14. The concentration of sulfur dioxide in the air of a city at any time during the year is

$$f(t) = -t^3 + 16t^2 - 48t + 50$$

parts per million, where t is time in months since the beginning of the year. Find the average concentration of sulfur dioxide in the air during March.

Social Science

15. If the population of the world is y billion people in year t, where

$$y = 5e^{0.015t}$$

and $t = 0$ represents 1985, find the average population of the world from 1985 to 2035.

16. Polling shows that t months before an election y voters exhibited an interest in the election, where

$$y = -2000t^2 + 10{,}000t + 40{,}000$$

and t is in months before the election, $0 \leq t \leq 3$. Find the average number of voters who expressed an interest in the election over this 3-month period.

6.2 APPLICATIONS: THE INTEGRAL AS ACCUMULATOR

In this section we continue our use of the definite integral as an accumulator. We show how to use the integral to calculate the quantities known in economics as producers' and consumers' surpluses, and how to apply the integral to the study of streams of income. We also briefly discuss a general procedure for setting up the integral as an accumulator.

Producers' Surplus and Consumers' Surplus

Let the equation $p = s(q)$ define a supply function for some commodity: here q is the number of units of the commodity supplied to the market, and p is the resulting unit price. Similarly, let $p = d(q)$ define the demand function for the commodity. See Figure 6–8. Note that the market equilibrium quantity for the commodity is q_E, and its market equilibrium price is p_E.

In a free market the unit price of the commodity must stabilize at p_E. To see this, suppose the unit price is p_0. If $p_0 < p_E$, then $q_0 < q_E$ and there will be a shortage of the commodity, as shown in Figure 6–9. As consumers compete for the scarce commodity, they will bid up its price, and therefore the price will

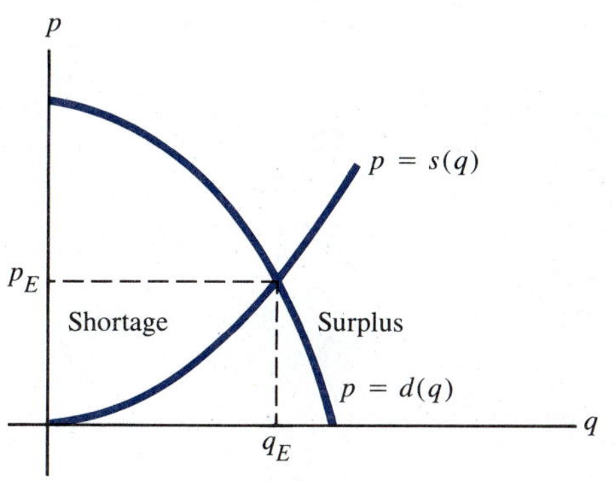

FIGURE 6–8

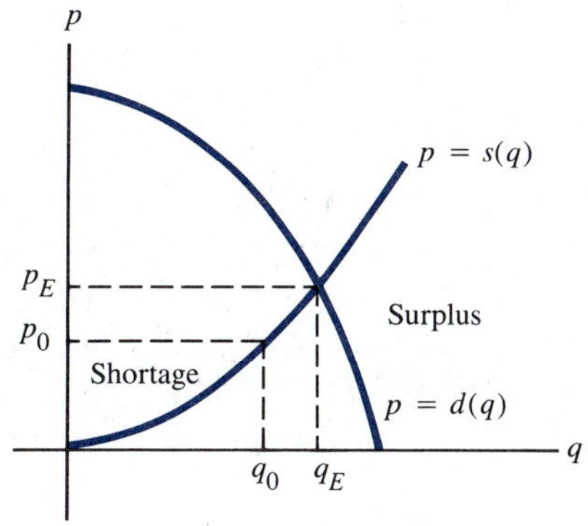

FIGURE 6–9

rise. On the other hand, if $p_0 > p_E$, then $q_0 > q_E$ and there will be a surplus of the commodity, as in Figure 6–10. In this case, producers must reduce the price in order to sell the commodity, so the price will fall. Thus the only stable price in the long run is the equilibrium price p_E.

For the moment, let us consider only the supply curve. Some producers would be willing to supply the commodity at prices lower than the equilibrium price p_E; these producers gain when the price is at equilibrium. For instance, Figure 6–11 indicates that some producers are willing to supply quantity q_1 of the commodity at price $p_1 < p_E$; when the price is at equilibrium, these producers gain an amount equal to $(p_E - p_1)q_1$. The total amount that producers as a group gain in this manner when the price is at equilibrium is called **producers' surplus.** We will use the integral as an accumulator to find producers' surplus.

Let us divide the interval $[0, q_E]$ on the q-axis into n subintervals, each of length Δq. For each i we select an arbitrary number q_i in the ith subinterval (see Figure 6–12): on this subinterval Δq units will be supplied at the approximate price of $s(q_i)$. Thus the total gained by producers over the ith subinterval (the area of the shaded region in Figure 6–12) is approximately $[p_E - s(q_i)]\,\Delta q$. Hence

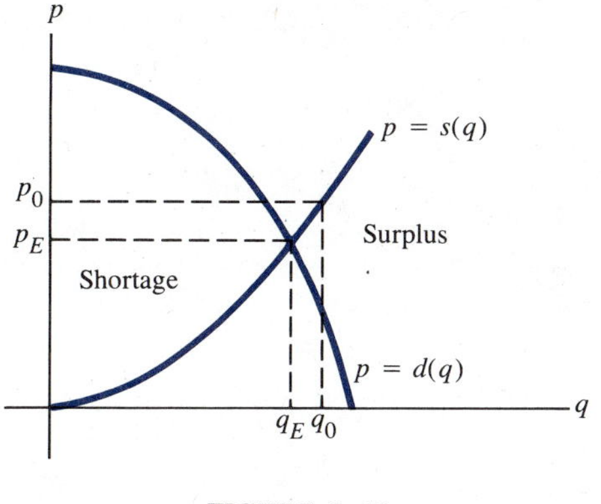

FIGURE 6–10

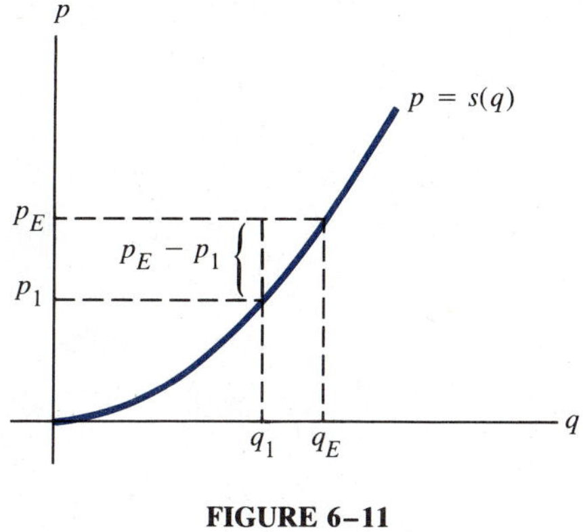

FIGURE 6–11

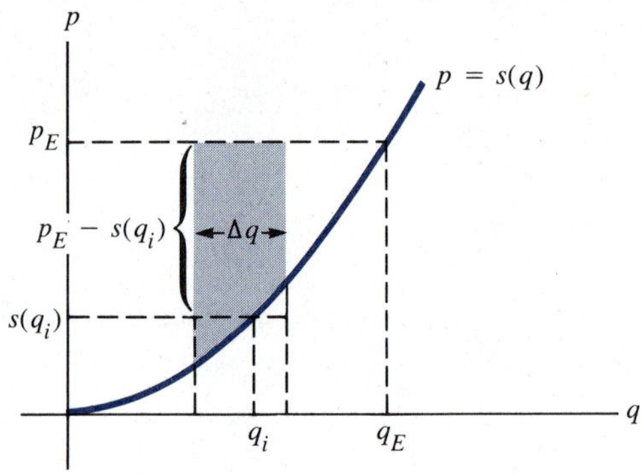

FIGURE 6–12

$$\text{producers' surplus} \approx [p_E - s(q_1)]\,\Delta q + \cdots + [p_E - s(q_n)]\,\Delta q.$$

Since this approximation becomes exact as Δq approaches zero, it follows that producers' surplus is the definite integral from $q = 0$ to $q = q_E$ of $p_E - s(q)$.

Producers' Surplus

If $p = s(q)$ defines a supply function for a commodity and (q_E, p_E) is the commodity's market equilibrium point, then producers' surplus for the commodity is

$$\int_0^{q_E} [p_E - s(q)]\,dq.$$

Producers' surplus is thus the area of the region shaded in Figure 6–13.

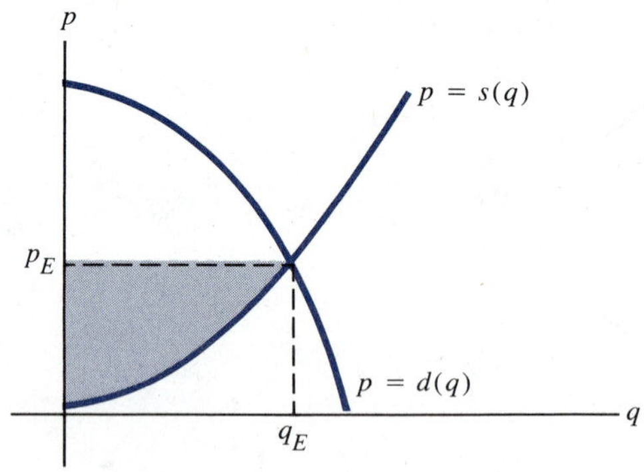

FIGURE 6–13 Producers' Surplus

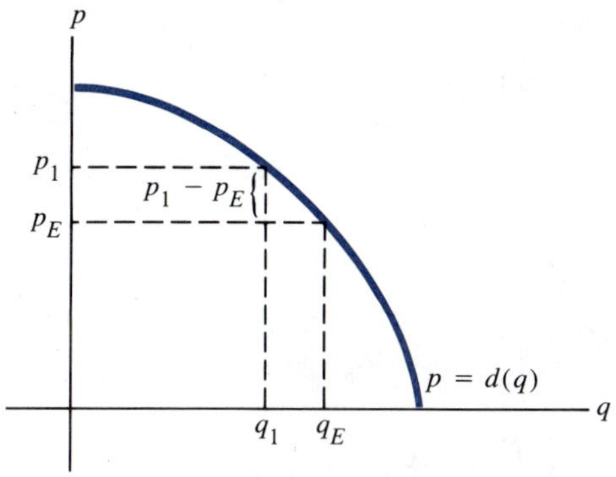

FIGURE 6–14

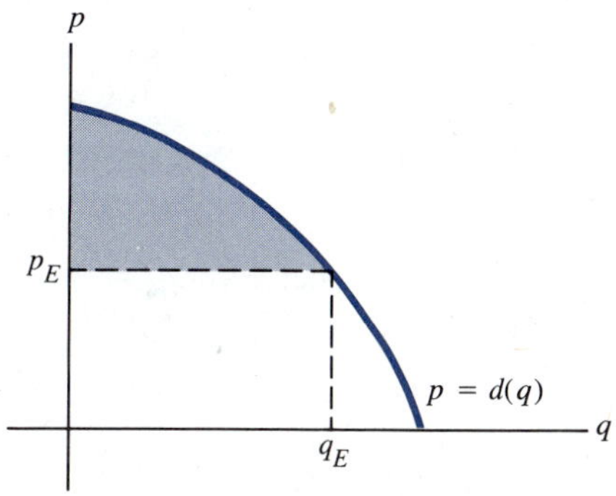

FIGURE 6–15 Consumers' Surplus

Analogous to producers' surplus is **consumer's surplus.** Figure 6–14 depicts the situation. Here the equation $p = d(q)$ defines the demand function. Consumers who would be willing to purchase the commodity at a higher price than p_E save when the price is at p_E. For example, if some consumers would have purchased q_1 units at price $p_1 > p_E$, as in the figure, then they save an amount equal to $(p_1 - p_E)q_1$ when the price is at equilibrium. **Consumers' surplus** is the total amount saved by consumers as a group when the price is at equilibrium. Just as we did for producers' surplus, we can approximate consumers' surplus by a Riemann sum

$$[d(q_1) - p_E]\,\Delta q + \cdots + [d(q_n) - p_E]\,\Delta q$$

(Exercise 11 at the end of this section asks for a proof of this) and thus obtain the following result.

Consumers' Surplus

If $p = d(q)$ defines a demand function for a commodity and (q_E, p_E) is the commodity's market equilibrium point, then consumers' surplus for the commodity is

$$\int_0^{q_E} [d(q) - p_E]\,dq.$$

Thus consumers' surplus is the area of the region shaded in Figure 6–15.

EXAMPLE 1 Suppose the supply function for a commodity is defined by

$$s(q) = 0.009q^2$$

and its demand function by

$$d(q) = 30 - 0.003q^2.$$

Solving the equation $s(q) = d(q)$ for q yields

$$0.009q^2 = 30 - 0.003q^2,$$

or

$$q^2 = \frac{30}{0.012} = 2500.$$

Therefore $q_E = 50$ and $p_E = s(50) = 22.50$. See Figure 6–16. Thus we have

$$\begin{aligned}\text{producers' surplus} &= \int_0^{q_E} [p_E - s(q)]\,dq \\ &= \int_0^{50} [22.50 - 0.009q^2]\,dq \\ &= \$750,\end{aligned}$$

and

$$\begin{aligned}\text{consumers' surplus} &= \int_0^{q_E} [d(q) - p_E]\,dq \\ &= \int_0^{50} [30 - 0.003q^2 - 22.50]\,dq \\ &= \$250.\end{aligned}$$

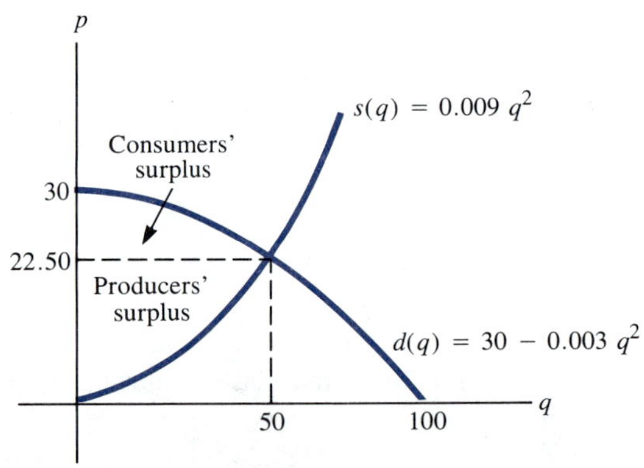

FIGURE 6–16

Income Streams

Any firm continually receives income from its business activities. This is referred to as the firm's **stream of income.** Most of the income that a firm receives is reinvested, either to keep the business going or in other ways. For purposes of analysis it is often convenient to assume that income flows into the firm continuously over the course of a year and that it is immediately reinvested at a certain interest rate that is compounded continuously. Thus we assume that at time t income is

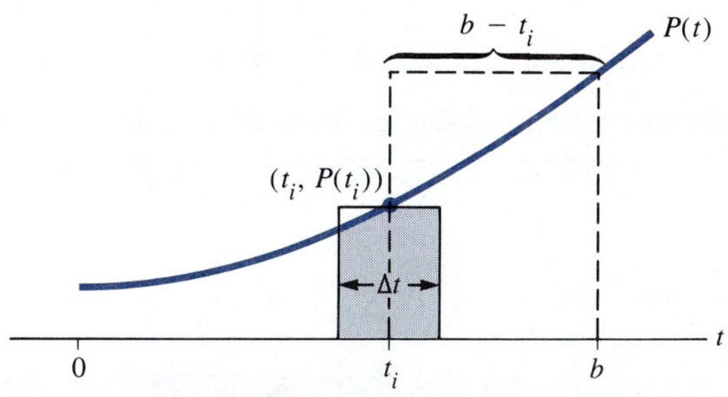

FIGURE 6–17

flowing into the firm at the rate of $P(t)$ dollars per year, where $P(t)$ is some continuous function, and that this income is immediately reinvested at $100r$ percent per year compounded continuously. We now ask the following question: if the firm continues to receive and reinvest income in this manner, what will be the value of its income stream over the next b years? To answer this question, we will again use the integral as an accumulator.

Suppose we divide the time interval $0 \le t \le b$ into n subintervals, each of length Δt, and select a number t_i in each subinterval. See Figure 6–17. Note that the rate at which income is flowing into the firm at time $t = t_i$ is $P(t_i)$, and that this money is reinvested at $100r$ percent per year compounded continuously for $b - t_i$ years. Hence by the formula for continuously compounded interest of Section 4.1, the money received by the firm at time $t = t_i$ will grow to

$$P(t_i)e^{r(b-t_i)}$$

dollars at the end of the b-year period. Since the total amount of income flowing into the firm over the ith subinterval is approximately

$$P(t_i)\,\Delta t$$

(the area of the rectangle over the ith subinterval), and since this amount will accumulate at interest for approximately $b - t_i$ years, the future value at the end of the b-year period of the income received over the ith subinterval is approximately

$$[P(t_i)\,\Delta t]e^{r(b-t_i)}.$$

This approximation will become exact as $\Delta t \to 0$. Thus the total future value of the income stream at the end of b years can be approximated by a Riemann sum

$$P(t_1)e^{r(b-t_1)}\,\Delta t + \cdots + P(t_n)e^{r(b-t_n)}\,\Delta t,$$

and the approximation will become exact as $\Delta t \to 0$. Therefore we have the following result for the future value of a stream of income.

The Future Value of a Stream of Income

If income flows in continuously at a rate of $P(t)$ dollars per year and is reinvested at $100r$ percent per year compounded continuously, then the future value of the income stream at the end of b years is

$$\int_0^b P(t)e^{r(b-t)}\,dt.$$

EXAMPLE 2 The earnings of Zymax, Inc. are currently flowing into the firm at a constant rate of \$100,000 per year and are increasing at the rate of 6 percent per year. If we assume that earnings flow into the company continuously over the course of a year and are immediately reinvested at 6 percent per year compounded continuously, then we have $P(t) = \$100{,}000$ and $r = 0.06$. Thus the value of Zymax's earnings stream over the next 4 years will be

$$\begin{aligned}\int_0^b P(t)e^{r(b-t)}\,dx &= \int_0^4 100{,}000e^{0.06(4-t)}\,dt \\ &= \int_0^4 100{,}000e^{0.24}e^{-0.06t}\,dt \\ &= 100{,}000e^{0.24}\int_0^4 e^{-0.06t}\,dt \\ &= 100{,}000e^{0.24}\left[\frac{e^{-0.06t}}{-0.06}\Bigg|_0^4\right] \\ &= 100{,}000e^{0.24}\left[\frac{e^{-0.24}-1}{-0.06}\right] \\ &= -\frac{100{,}000}{0.06}(1-e^{0.24}) \\ &= \$452{,}082.\end{aligned}$$

EXAMPLE 3 A new computer will save a company money at the rate of $P(t) = 2t$ thousand dollars per year, where t is in years, with $t = 0$ representing the present. Find the value of this stream of savings over the next 10 years if the company reinvests the savings at 10 percent per year compounded continuously. The answer is given by the integral

$$\begin{aligned}\int_0^{10} 2te^{0.10(10-t)}\,dt &= \int_0^{10} 2te^{1-0.10t}\,dt \\ &= 2e\int_0^{10} te^{-0.10t}\,dt.\end{aligned}$$

But, as we will show in Section 6.5,

$$\int ue^{au}\,du = \frac{ue^{au}}{a} - \frac{e^{au}}{a^2} + c.$$

(Check this by differentiating the right side with respect to u). Thus

$$\begin{aligned}
\int_0^{10} 2te^{0.10(10-t)}\,dt &= 2e\left[\frac{te^{-0.10t}}{-0.10} - \frac{e^{-0.10t}}{(-0.10)^2}\bigg|_0^{10}\right] \\
&= 2e\left[\left(-\frac{10e^{-1}}{0.10} - \frac{e^{-1}}{0.01}\right) - \left(0 - \frac{1}{0.01}\right)\right] \\
&= 2e[-100e^{-1} - 100e^{-1} + 100] \\
&= 200e - 400 \\
&= 143.656,
\end{aligned}$$

or \$143,656.

Using the Integral as an Accumulator

In this section we have shown how to use the integral as an accumulator to find three different quantities: producers' surplus, consumers' surplus, and the value of a stream of income. If we look back at these examples, we can see that it was not actually necessary to write out the Riemann sums in each case. The essential steps in setting up a definite integral to find the value of some quantity Q over an interval $[a, b]$ are as follows.

Procedure for Setting Up the Integral as an Accumulator

1. Divide the interval $[a, b]$ into subintervals, each of which has length Δx.
2. Approximate the value of Q over a typical subinterval by an expression of the form $f(x)\,\Delta x$ in such a way that the approximation will become exact as $\Delta x \to 0$.
3. Then the value of Q over $[a, b]$ is $\int_a^b f(x)\,dx$.

There are many situations other than those considered here in which this procedure can be profitably employed. Some of these are introduced in the exercises.

Computer Manual: Program DEFINT.

6.2 EXERCISES

Producers' Surplus and Consumers' Surplus

In Exercises 1 through 6 find producers' surplus and consumers' surplus using the given supply and demand functions.

1. supply: $p = 0.06q$
   demand: $p = -0.94q + 220$

2. supply: $p = 0.5q$
   demand: $p = 10 - 0.05q^2$

3. supply: $p = \sqrt{q}$
   demand: $p = -2q + 55$

4. supply: $p = 5e^{0.8q}$
   demand: $p = 20e^{-0.2q}$

5. supply: $p = \sqrt{q + 5}$
   demand: $p = \dfrac{27}{q + 5}$

6. supply: $p = \dfrac{4}{3}(q - 4)$
   demand: $p = \dfrac{400}{q + 1}$

Income Streams

7. A firm's earnings are flowing in continuously at a constant rate of \$1 million per year and are reinvested at 6 percent per year compounded continuously. Find the value of the firm's stream of earnings over the next 5 years.

8. Repeat Exercise 7 if the firm's earnings are flowing in at the rate of

$$P(t) = 1 + 0.2t$$

million dollars per year. Here t is time in years, with $t = 0$ representing the present. You may use the integration formula for ue^{au} that was stated in Example 3.

9. If a firm is generating profits at the rate of

$$P(t) = 20e^{0.01t}$$

million dollars per year, where t is in years and $t = 0$ is the present, find the value of the firm's stream of profits over the next 3 years if all profits are reinvested at 12 percent per year compounded continuously.

10. It is estimated that a company's investment in industrial robots will save it

$$P(t) = 200 + 12t$$

thousand dollars per year over the next 10 years, where t is time in years with $t = 0$ representing the present. Find the value of the stream of savings due to the robots if the company reinvests the savings at 10 percent per year compounded continuously.

Using the Integral as an Accumulator

11. Use the procedure for setting up the integral as an accumulator to show that consumers' surplus is equal to

$$\int_0^{q_E} [d(q) - p_E]\, dq.$$

Management

12. **Carrying Cost.** Suppose that it costs \$$p$ to carry one unit in inventory for one time period and that the number of units in inventory at time t is given by an inventory function $I(t)$. Then the total cost of carrying the inventory from time $t = 0$ to time $t = b$ is

$$\int_0^b pI(t)\,dt.$$

Use the procedure for setting up the integral as an accumulator to show that this is so.

13. Refer to Exercise 12. If it costs a store \$4 to carry one unit in inventory for one month and if the number of units in inventory at time t is given by

$$I(t) = 40{,}000 - 2000t,$$

where t is in months with $t = 0$ representing the present, find the store's total carrying cost for the next year.

14. Suppose it costs a manufacturer \$0.20 to carry a unit in inventory for one week and that the number of units in inventory as of week t is given by

$$I(t) = 12{,}000 - 150t^2,$$

where $t = 0$ represents the present. Find the manufacturer's total carrying cost over the next 4 weeks.

Life Science

15. **Survival–Renewal Functions.** Suppose that $p(t)$ is the size of a population at time t, $t \geq 0$. Let $f(t)$ be the proportion of the original population that survives at time t; the function f is called a **survival function.** Let $g(t)$ denote the number of individuals added to the population at time t; the function g is called a **renewal function.** Use the integral as an accumulator to show that the size of the population at time $t = b$ is

$$p(b) = p(0)f(b) + \int_0^b g(t)f(b - t)\,dt.$$

16. Use the result of Exercise 15 to find the size of a caribou herd 6 months from now if there are 2000 caribou in the herd now, the proportion of the 2000 who will survive t months from now is given by

$$f(t) = \frac{1}{1 + 0.02t},$$

and the number of caribou added to the herd is constant at 100 per month.

17. Use the result of Exercise 15 to find the number of rabbits you will have 1 year from now if you start with 100 rabbits, the proportion of the original population that survives as of month t is given by

$$f(t) = e^{-0.1t},$$

and the number of rabbits being born at time t is given by $g(t) = 20t$.

18. **Blood Flow.** Poiseuille's Law states that the velocity of blood in an artery depends on the distance of the blood from the center of the artery according to the equation

$$v(x) = c(b^2 - x^2).$$

Here x is the distance from the central axis of the artery, b is the radius of the artery, c is a constant, and $v(x)$ is the velocity of the blood in cubic centimeters per second. See the figure. Use the integral as accumulator to show that the rate of blood flow through the artery is

$$2\pi c \int_0^b (b^2x - x^3)\,dx = \frac{\pi c b^4}{2}$$

cubic centimeters per second. (*Hint:* Divide the artery into rings of width Δx and choose a number x_i in each ring; the approximate area of the ith ring is $2\pi x_i\,\Delta x$, and the approximate speed of blood in the ith ring is the area of the ring times $v(x_i)$.)

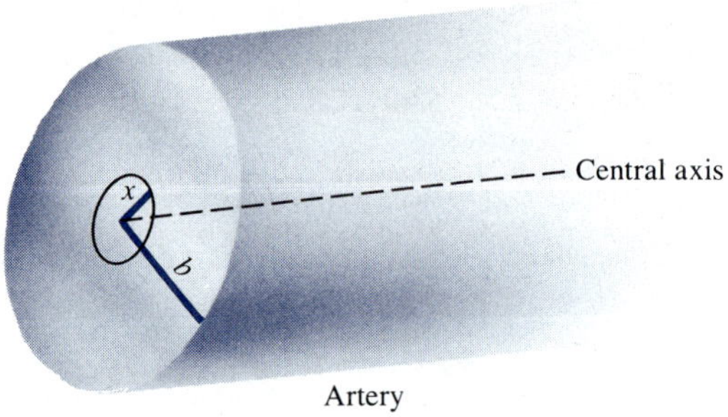

19. Use the result of Exercise 18 to find the rate at which blood flows through an artery of radius 1 centimeter; of radius 2 centimeters.

Social Science

20. A religious cult currently has 10,000 adherents. The proportion of them who will remain members of the cult t weeks from now is given by

$$f(t) = e^{-0.05t},$$

and the number of new members recruited during week t is given by

$$g(t) = 1000e^{-0.05t}.$$

Find the size of the cult 1 year (52 weeks) from now. (*Hint:* See Exercise 15.)

21. A prison has space for 3500 inmates and currently has a population of 3000 inmates. The proportion of inmates who will remain in the prison t months from now is given by

$$f(t) = \frac{1}{1 + 0.005t},$$

and new prisoners are arriving at the rate of 40 per month. Will the prison be overcrowded 20 months from now? 21 months from now?

22. **Population Density.** Suppose that the population x kilometers from the center of a region is $f(x)$ persons per square kilometer. Show that the total population within b miles of the center of the region is

$$2\pi \int_0^b xf(x)\,dx.$$

(*Hint:* Divide the region into circular rings as in Exercise 18.)

23. Refer to Exercise 22. Find the total population within 10 kilometers of the center of a city if there are $f(x) = 1000e^{-0.1x}$ persons per square kilometer x kilometers from the center.

6.3 NUMERICAL INTEGRATION

We have defined the definite integral of a function f over an interval $[a, b]$ to be $F(b) - F(a)$, where $F(x)$ is an antiderivative of $f(x)$. There are situations when it is necessary to evaluate $\int_a^b f(x)\,dx$ and it is not possible to find a simple expression for the antiderivative $F(x)$ of $f(x)$. For instance, there is no simple formula or defining equation for the antiderivative of

$$f(x) = e^{-x^2};$$

nevertheless, in statistics we must be able to evaluate integrals such as

$$\int_0^1 e^{-x^2}\,dx.$$

In a case like this we can use **numerical integration** to approximate the value of the integral. Numerical integration can be used to approximate a definite integral to any degree of accuracy. It is also useful when the integrand is defined by a collection of data points rather than by a functional expression.

Methods of numerical integration have become increasingly important with the advent of the computer. In this section we discuss and illustrate three of the most common methods: the midpoint rule, the trapezoidal rule, and Simpson's rule.

The Midpoint Rule

We saw in Section 6.1 that the definite integral $\int_a^b f(x)\,dx$ can be approximated by a sum of areas of rectangles. The **midpoint rule** requires that when we make such an approximation we choose the numbers x_i to be the midpoints of their respective subintervals. Figure 6–18 shows how this is done. Since

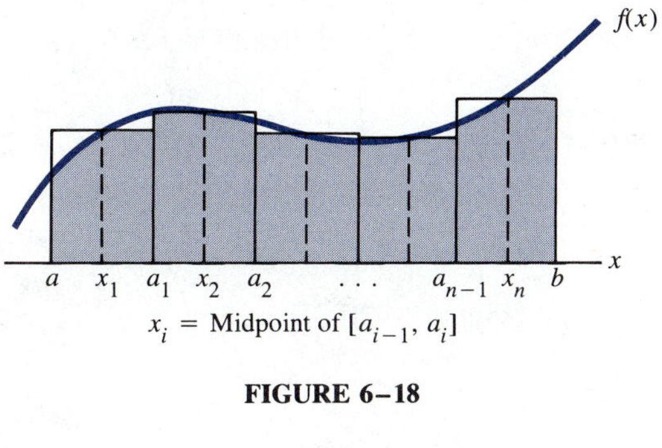

FIGURE 6–18

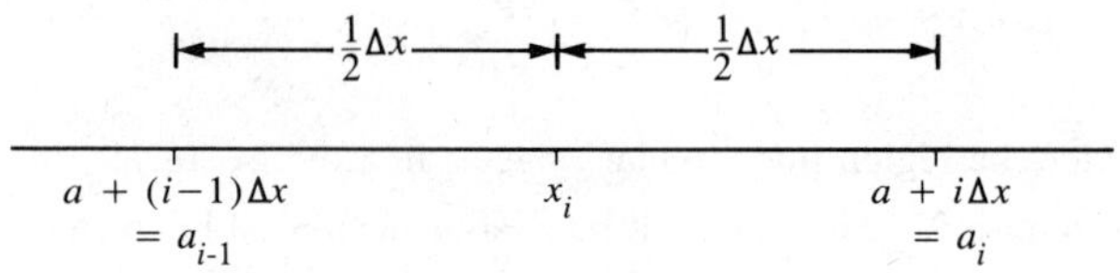

x_i = Midpoint of $[a_{i-1}, a_i]$

FIGURE 6–19

$$[a_{i-1}, a_i] = [a + (i - 1)\,\Delta x,\ a + i\,\Delta x]$$

for each i (Figure 6–19), the midpoint of the ith subinterval is

$$a + (i - 1)\,\Delta x + \frac{1}{2}\,\Delta x = a + \frac{2i - 1}{2}\,\Delta x.$$

Thus the sum of the areas of the rectangles in Figure 6–18 is

$$f\left(a + \frac{1}{2}\,\Delta x\right)\Delta x + f\left(a + \frac{3}{2}\,\Delta x\right)\Delta x + \cdots + f\left(a + \frac{2n - 1}{2}\,\Delta x\right)\Delta x,$$

which yields

The Midpoint Rule

If n is a positive integer and $\Delta x = \dfrac{b - a}{n}$, then

$$\int_a^b f(x)\,dx \approx \left[f\left(a + \frac{1}{2}\,\Delta x\right) + f\left(a + \frac{3}{2}\,\Delta x\right) + \cdots + f\left(a + \frac{2n - 1}{2}\,\Delta x\right)\right]\Delta x.$$

This approximation becomes exact as $\Delta x \to 0$, that is, as $n \to +\infty$.

EXAMPLE 1 Let us use the midpoint rule to approximate the definite integral

$$\int_1^2 x^2\,dx$$

using $n = 5$ rectangles. Then $a = 1$, $b = 2$, and $\Delta x = (2 - 1)/5 = 1/5$, so

$$a + \frac{1}{2}\,\Delta x = \frac{11}{10}, \quad a + \frac{3}{2}\,\Delta x = \frac{13}{10}, \quad a + \frac{5}{2}\,\Delta x = \frac{15}{10},$$

$$a + \frac{7}{2}\,\Delta x = \frac{17}{10}, \quad \text{and} \quad a + \frac{9}{2}\,\Delta x = \frac{19}{10}.$$

See Figure 6–20. Therefore

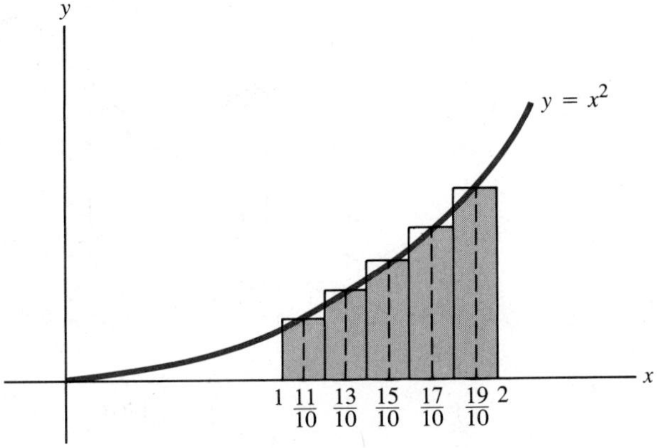

FIGURE 6–20

$$\int_1^2 x^2\,dx \approx \left[\left(\frac{11}{10}\right)^2 + \left(\frac{13}{10}\right)^2 + \left(\frac{15}{10}\right)^2 + \left(\frac{17}{10}\right)^2 + \left(\frac{19}{10}\right)^2\right]\frac{1}{5}$$

$$= \frac{1165}{100}\cdot\frac{1}{5} = 2.330.$$

Since

$$\int_1^2 x^2\,dx = \frac{1}{3}x^3\Big|_1^2 = \frac{7}{3} = 2.333333$$

to six decimal places, our approximation of 2.330 is accurate to two decimal places. We could obtain a better approximation to the integral by using more rectangles, that is, by using a larger value of n.

The virtue of the midpoint rule is its simplicity; however, it often does not produce a good approximation for a definite integral unless the value of n is very large. Therefore other methods of numerical integration are usually preferable.

The Trapezoidal Rule

The **trapezoidal rule** uses the areas of trapezoids to approximate the definite integral. Figure 6–21 shows how this is done. The area under the graph of f from a to b, which is the definite integral of f from $x = a$ to $x = b$, is approximated by the sum of the areas of n trapezoids. As before, we set $\Delta x = (b - a)/n$ and partition the interval $[a, b]$ into n subintervals $[a, a_1]$, $[a_1, a_2]$, . . . , $[a_{n-1}, b]$, where $a_i = a + i\,\Delta x$ for each i.

If a trapezoid has bases of lengths b_1 and b_2 and an altitude of length h, then its area is $(b_1 + b_2)h/2$. The ith trapezoid in Figure 6–21 has as its bases the line segments from $(a_{i-1}, 0)$ to $(a_{i-1}, f(a_{i-1}))$ and from $(a_i, 0)$ to $(a_i, f(a_i))$; its altitude is the interval $[a_{i-1}, a_i]$ on the x-axis. Therefore the area of the ith trapezoid is

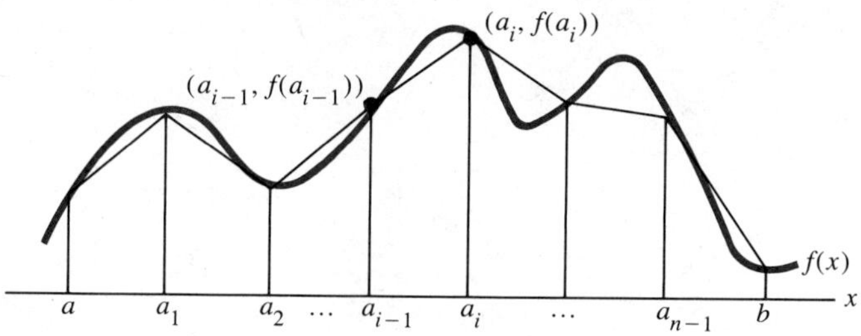

FIGURE 6–21

$$\frac{1}{2}(f(a_{i-1}) + f(a_i))\,\Delta x.$$

The sum of the areas of all n trapezoids is thus

$$\frac{1}{2}[f(a) + f(a_1)]\,\Delta x + \frac{1}{2}[f(a_1) + f(a_2)]\,\Delta x + \cdots$$

$$+\frac{1}{2}[f(a_{n-2}) + f(a_{n-1})]\,\Delta x + \frac{1}{2}[f(a_{n-1}) + f(b)]\,\Delta x,$$

which simplifies to

$$\left[\frac{1}{2}f(a) + f(a_1) + f(a_2) + \cdots + f(a_{n-1}) + \frac{1}{2}f(b)\right]\Delta x.$$

(Check this.) Therefore we have

The Trapezoidal Rule

If n is a positive integer, $\Delta x = \dfrac{b-a}{n}$, and $a_i = a + i\,\Delta x$ for each i, then

$$\int_a^b f(x)\,dx \approx \left[\frac{1}{2}f(a) + f(a_1) + f(a_2) + \cdots + f(a_{n-1}) + \frac{1}{2}f(b)\right]\Delta x.$$

This approximation becomes exact as $n \to +\infty$.

EXAMPLE 2 Use the trapezoidal rule to approximate

$$\int_1^2 \frac{1}{x}\,dx$$

with $n = 5$ trapezoids. Then $a = 1$, $b = 2$, and $\Delta x = (2 - 1)/5 = 1/5$, so

$$a = 1, \quad a_1 = a + \Delta x = \frac{6}{5}, \quad a_2 = a + 2\,\Delta x = \frac{7}{5},$$

$$a_3 = a + 3\,\Delta x = \frac{8}{5}, \quad a_4 = a + 4\,\Delta x = \frac{9}{5}, \quad \text{and} \quad b = 2.$$

See Figure 6–22. Therefore

$$\begin{aligned}\int_1^2 \frac{1}{x}\,dx &\approx \left[\frac{1}{2}\cdot\frac{1}{1} + \frac{1}{6/5} + \frac{1}{7/5} + \frac{1}{8/5} + \frac{1}{9/5} + \frac{1}{2}\cdot\frac{1}{2}\right]\frac{1}{5} \\ &= \left[\frac{1}{2} + \frac{5}{6} + \frac{5}{7} + \frac{5}{8} + \frac{5}{9} + \frac{1}{4}\right]\frac{1}{5} \\ &= 0.6956349.\end{aligned}$$

Since

$$\int_1^2 \frac{1}{x}\,dx = \ln x \Big|_1^2 = \ln 2 - \ln 1 = \ln 2 = 0.693147,$$

our approximation using the trapezoidal rule is accurate to two decimal places. We could obtain a more accurate approximation by using a larger value of n.

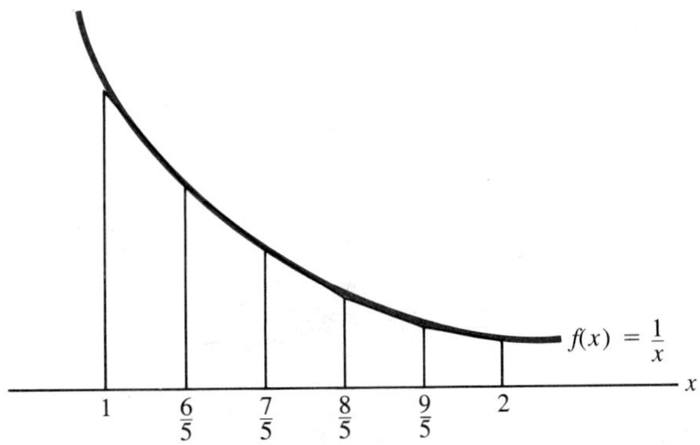

FIGURE 6–22

In Section 5.4 we remarked that the total change in a quantity Q can be thought of as an area under the graph of the function that describes the rate of change of Q. With the aid of numerical integration we can exploit this fact to estimate total change when the function f is unknown, provided some of its functional values have been determined. We illustrate with an example.

EXAMPLE 3 Suppose a firm wishes to find the total cost of increasing its daily production from 100 units per day to 125 units per day. This cost is equal to

$$\int_{100}^{125} c(x)\,dx,$$

where $c(x)$ denotes the marginal cost of x units. Suppose also that the firm does not know its marginal cost function, but only has information on its value at selected points, as follows:

Units Produced	Marginal Cost
100	36
105	28
110	26
115	30
120	36
125	44

See Figure 6–23. Then by the trapezoidal rule with $a = 100$, $b = 125$, and $\Delta x = 5$, we have

$$\int_{100}^{125} c(x)\,dx \approx \left[\frac{1}{2}c(100) + c(105) + c(110) + c(115) + c(120) + \frac{1}{2}c(125)\right]5$$

$$= \left[\frac{1}{2}(36) + 28 + 26 + 30 + 36 + \frac{1}{2}(44)\right]5$$

$$= 800$$

dollars.

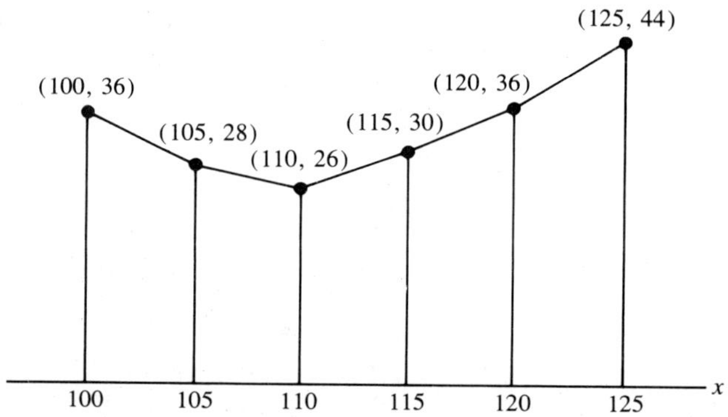

FIGURE 6–23

Simpson's Rule

Another widely used method of numerical integration is **Simpson's rule,** which approximates the definite integral by areas of parabolic regions. Figure 6–24 shows how this is done. As usual,

$$\Delta x = (b - a)/n \qquad \text{and} \qquad a_i = a + i\,\Delta x$$

for each i. Here n must be an even positive integer. It can be shown that the area of the region bounded by $x = a_{i-1}$, $x = a_{i+1}$, the x-axis, and the parabolic

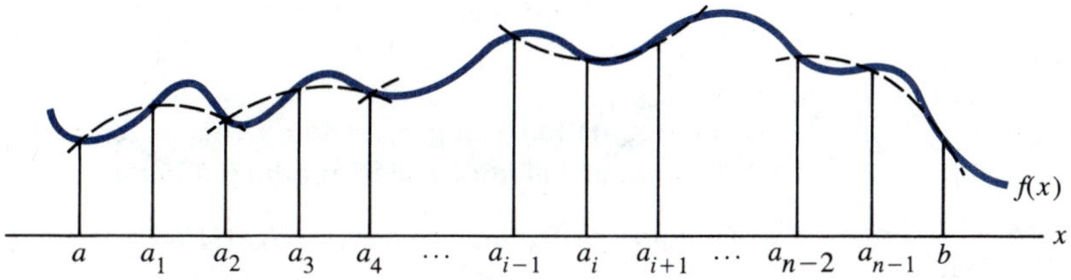

FIGURE 6–24

arc that passes through the points $(a_{i-1},f(a_{i-1}))$, $(a_i,f(a_i))$, and $(a_{i+1},f(a_{i+1}))$ on the graph of f is

$$[f(a_{i-1}) + 4f(a_i) + f(a_{i+1})]\,\frac{\Delta x}{3}\,.$$

The sum of all such terms simplifies to

$$[f(a) + 4f(a_1) + 2f(a_2) + 4f(a_3) + \cdots + 2f(a_{n-2}) + 4f(a_{n-1}) + f(b)]\,\frac{\Delta x}{3}\,.$$

Therefore Simpson's rule may be stated as follows:

Simpson's Rule

If n is an even positive integer, $\Delta x = \dfrac{b-a}{n}$, and $a_i = a + i\,\Delta x$ for each i, then

$$\int_a^b f(x)\,dx \approx [f(a) + 4f(a_1) + 2f(a_2) + \cdots + 2f(a_{n-2}) + 4f(a_{n-1}) + f(b)]\,\frac{\Delta x}{3}\,.$$

This approximation becomes exact as $n \to +\infty$.

EXAMPLE 4 Use Simpson's rule with $n = 4$ subintervals to approximate

$$\int_1^2 \frac{1}{x}\,dx.$$

Then $a = 1$, $b = 2$, and $\Delta x = 1/4$, so

$$a = 1, \quad a_1 = \frac{5}{4}, \quad a_2 = \frac{3}{2}, \quad a_3 = \frac{7}{4}, \quad \text{and} \quad b = 2.$$

Thus

$$\int_1^2 \frac{1}{x}\,dx \approx \left[\frac{1}{1} + 4\,\frac{1}{5/4} + 2\,\frac{1}{3/2} + 4\,\frac{1}{7/4} + \frac{1}{2}\right]\frac{1/4}{3} = 0.693254$$

to six decimal places. Since

$$\int_1^2 \frac{1}{x}\,dx = \ln x \Big|_1^2 = \ln 2 - \ln 1 = \ln 2 = 0.693147,$$

our approximation using Simpson's rule is accurate to three decimal places. As usual, a larger value of n will produce a better approximation.

Computer Manual: Program NUMERINT.

6.3 EXERCISES

The Midpoint Rule

In Exercises 1 through 4 use the midpoint rule with the given value of n to approximate the given definite integral. In each case compare your answer with the actual value of the integral.

1. $\int_1^2 x^2\,dx, \quad n = 6$ **2.** $\int_1^2 x^2\,dx, \quad n = 8$ **3.** $\int_2^4 \frac{1}{x}\,dx, \quad n = 5$ **4.** $\int_2^4 \frac{1}{x}\,dx, \quad n = 6$

The Trapezoidal Rule

In Exercises 5 through 8 use the trapezoidal rule with the given value of n to approximate the given definite integral. In each case compare your answer with the actual value of the integral.

5. $\int_0^1 x^2\,dx, \quad n = 4$ **6.** $\int_0^1 x^2\,dx, \quad n = 8$

7. $\int_2^5 (2x + 3)\,dx, \quad n = 4$ **8.** $\int_{-2}^2 (e^x - e^{-x})\,dx, \quad n = 8$

9. Use the trapezoidal rule with $n = 4$ to approximate

$$\int_0^1 \frac{1}{x^2 + 1}\,dx.$$

The actual value of this integral is $\pi/4$. How does your answer compare?

10. Use the trapezoidal rule with $n = 4$ to approximate

$$\int_0^1 e^{-x^2}\,dx.$$

Management

11. A firm estimates its marginal cost at various production levels to be as follows:

Production (units)	10	12	14	16	18	20
Marginal cost ($)	200	202	204	207	212	220

Use these data and the trapezoidal rule to approximate the firm's cost if it increases production from 10 to 20 units.

12. An efficiency expert sampled a new worker's daily production at various times during the worker's first 3 weeks on the job. The result was the following graph:

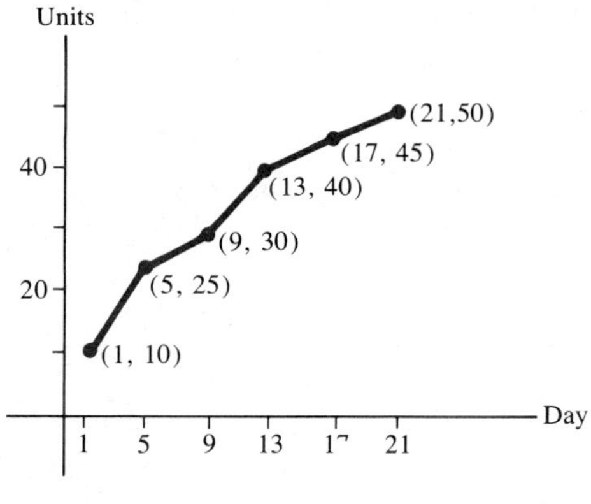

Use the trapezoidal rule to approximate the area under the graph from $x = 1$ to $x = 21$ and interpret your result.

Social Science

13. A psychologist measures the rate at which college freshmen can master new concepts in mathematics as a function of the number of weeks they have spent in college mathematics courses and finds the following:

Number of weeks of math completed	5	10	15	20	25
Number of new concepts mastered per week	2	6	10	13	14

Use the trapezoidal rule to approximate the total number of new concepts mastered from week 5 through week 25.

Simpson's Rule

In Exercises 14 through 17 use Simpson's rule with the given value of n to approximate the given definite integral. In each case compare your answer with the actual value of the integral.

14. $\int_0^1 (x^2 + 1)\,dx, \quad n = 4$

15. $\int_0^1 x^3\,dx, \quad n = 6$

16. $\int_{-2}^2 e^{2x}\,dx, \quad n = 8$

17. $\int_1^4 \frac{1}{x^2}\,dx, \quad n = 4$

18. Repeat Exercise 9 using Simpson's rule and compare your result with that of Exercise 9.

19. Repeat Exercise 10 using Simpson's rule and compare your result with that of Exercise 10.

Management

20. Acme Computer Company estimates that the number of worker-hours that will be required to build the xth unit of its new computer is as follows:

Unit number	1	4	7	10	13
Worker-hours	4000	3200	3000	2850	2750

Use Simpson's rule to approximate the total number of worker-hours that will be required to build units 1 through 13.

Life Science

21. The following graph shows the rate at which pollutants entered a lake at various times during one day of sampling:

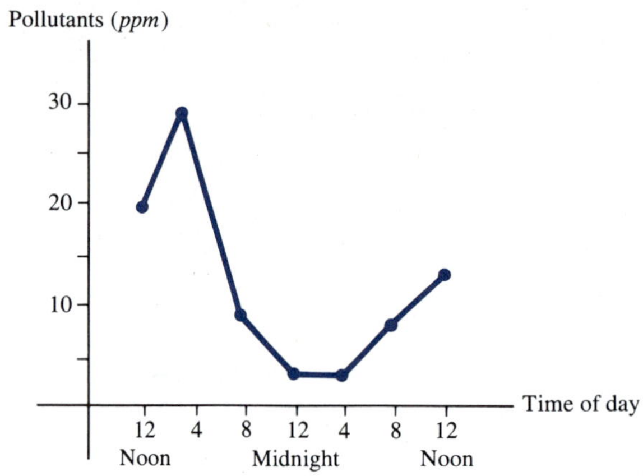

Using Simpson's rule to approximate the area under the curve and interpret your result.

6.4 IMPROPER INTEGRALS

Sometimes it is necessary to consider areas of regions that extend infinitely far to the left or right along the x-axis. For instance, the shaded region in Figure 6–25 is meant to extend infinitely far to the right along the x-axis. As we shall see, the area of such a region may or may not be finite. In this section we show how to determine if the area of such a region is finite, and how to find it if it is. In so doing, we use **improper integrals:** These are definite integrals that have at least one infinite limit of integration.

Consider the function f defined by $f(x) = e^{-x}$. The region under the graph of f above the x-axis and to the right of $x = 0$ (Figure 6–26) is important in probability and statistics. Let us see if this region has a finite area.

If t is any positive number, then the area under $f(x) = e^{-x}$ from 0 to t is certainly finite, and we know that it is equal to

$$\int_0^t e^{-x}\,dx.$$

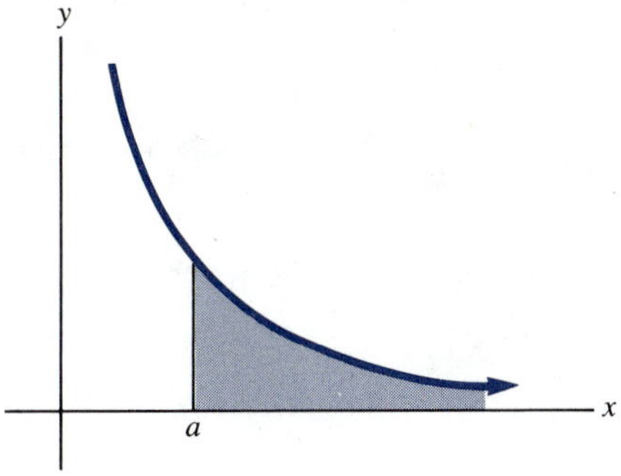

FIGURE 6–25

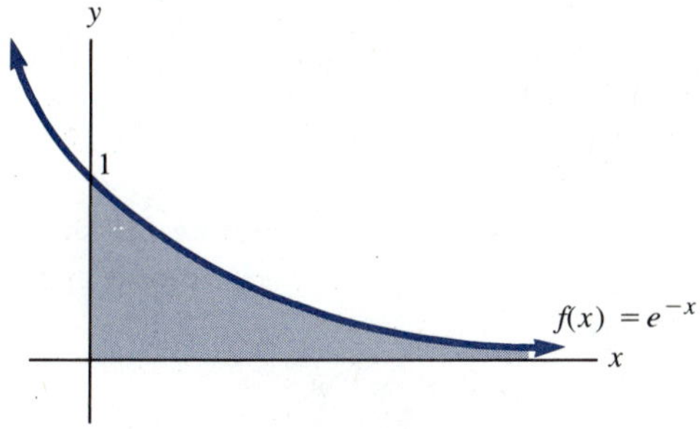

FIGURE 6–26

As $t \to +\infty$, the area under $f(x)$ from 0 to t will approach the area of the region in question. See Figure 6–27. Thus, if

$$\lim_{t \to +\infty} \int_0^t e^{-x}\, dx$$

exists, then the region in question has area equal to the value of the limit; if the limit does not exist, then the area of the region is not finite. Now,

$$\lim_{t \to +\infty} \int_0^t e^{-x}\, dx = \lim_{t \to +\infty} \left(-e^{-x} \right) \Big|_0^t = \lim_{t \to +\infty} (-e^{-t} + 1);$$

but, as Figure 6–26 shows,

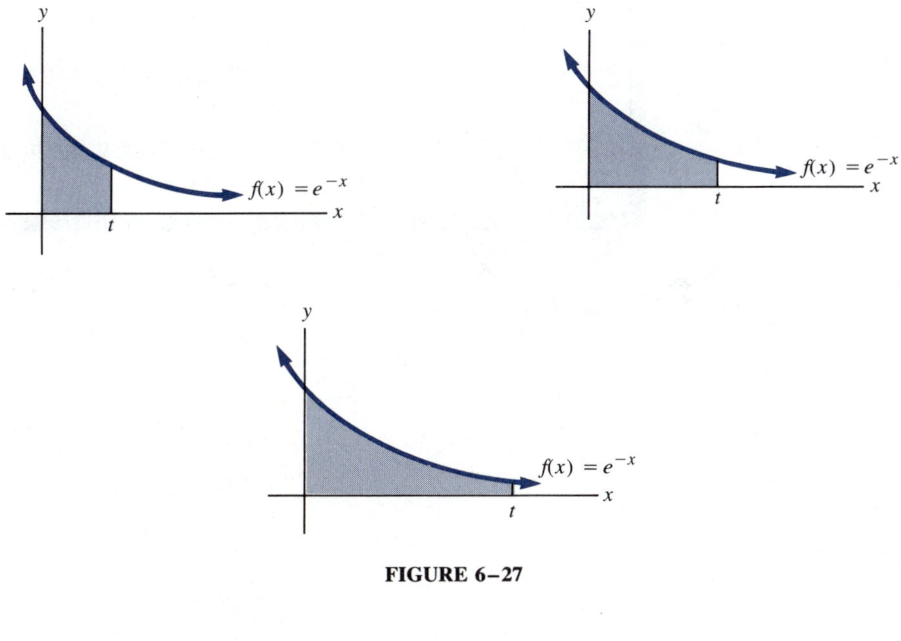

FIGURE 6–27

$$\lim_{t\to+\infty} e^{-t} = 0,$$

so
$$\lim_{t\to+\infty} \int_0^t e^{-x}\,dx = 0 + 1 = 1.$$

Therefore the shaded region of Figure 6–26 has finite area equal to 1.

It is customary to write a limit such as

$$\lim_{t\to+\infty} \int_0^t e^{-x}\,dx$$

as an integral:
$$\lim_{t\to+\infty} \int_0^t e^{-x}\,dx = \int_0^{+\infty} e^{-x}\,dx.$$

The integral on the right is called an **improper integral** because one of its limits of integration is infinite.

Improper Integrals

The **improper integrals**

$$\int_a^{+\infty} f(x)\,dx \qquad \text{and} \qquad \int_{-\infty}^b f(x)\,dx$$

are defined by

$$\int_a^{+\infty} f(x)\,dx = \lim_{t\to+\infty} \int_a^t f(x)\,dx$$

and

$$\int_{-\infty}^b f(x)\,dx = \lim_{t\to-\infty} \int_t^b f(x)\,dx.$$

The improper integrals are said to **converge** if the limits on the right side exist; otherwise they are said to **diverge.**

EXAMPLE 1 For the improper integral
$$\int_2^{+\infty} 2x^{-3}\,dx$$

we have
$$\int_2^{+\infty} 2x^{-3}\,dx = \lim_{t\to+\infty} \int_2^t 2x^{-3}\,dx$$

$$= \lim_{t\to+\infty} \left(-x^{-2} \right) \Big|_2^t$$

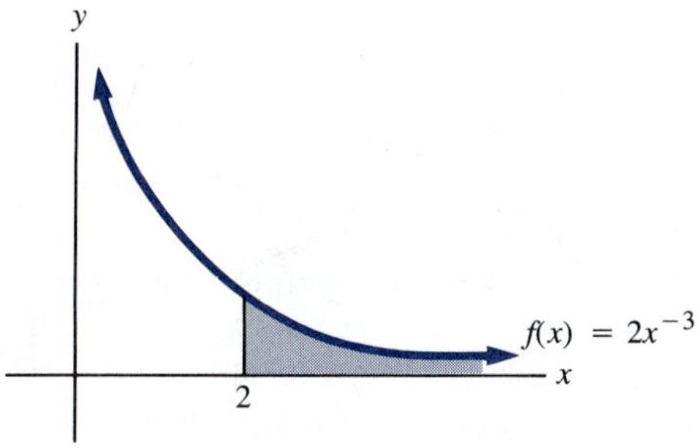

FIGURE 6–28

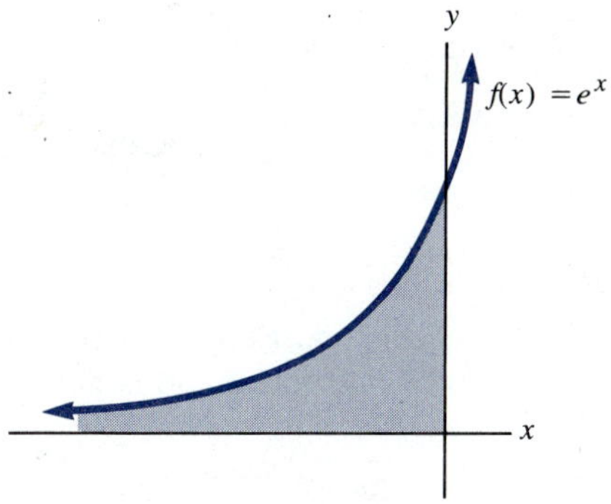

FIGURE 6–29

$$= \operatorname*{Lim}_{t \to +\infty} \left(-\frac{1}{t^2} + \frac{1}{4} \right)$$

$$= 0 + \frac{1}{4}$$

$$= \frac{1}{4}.$$

Therefore the improper integral

$$\int_2^{+\infty} 2x^{-3}\,dx$$

converges, and the shaded region of Figure 6–28 has an area of ¼.

EXAMPLE 2 Let us evaluate the improper integral

$$\int_{-\infty}^{0} e^x\,dx.$$

See Figure 6–29. We have

$$\int_{-\infty}^{0} e^x\,dx = \operatorname*{Lim}_{t \to -\infty} \int_t^0 e^x\,dx = \operatorname*{Lim}_{t \to -\infty} e^x \Big|_t^0 = \operatorname*{Lim}_{t \to -\infty} (1 - e^t).$$

But as Figure 6–29 shows, $\operatorname{Lim}_{t \to -\infty} e^t = 0$, so

$$\int_{-\infty}^{0} e^x\,dx = 1 - 0 = 1.$$

Hence the integral converges, and the shaded region in Figure 6–29 has an area of 1.

EXAMPLE 3 The improper integral

$$\int_1^{+\infty} \frac{1}{x}\, dx$$

diverges, because

$$\int_1^{+\infty} \frac{1}{x}\, dx = \lim_{t \to +\infty} \int_1^t \frac{1}{x}\, dx = \lim_{t \to +\infty} \ln x \Big|_1^t = \lim_{t \to +\infty} (\ln t),$$

and $\text{Lim}_{t \to +\infty} (\ln t)$ does not exist (see Figure 4–9). Therefore the shaded area in Figure 6–30 is not finite.

We may also define an improper integral having two infinite limits of integration by setting

$$\int_{-\infty}^{+\infty} f(x)\, dx = \int_{\infty}^{a} f(x)\, dx + \int_a^{+\infty} f(x)\, dx$$

for any number a. The integral on the left converges if and only if both of the integrals on the right converge.

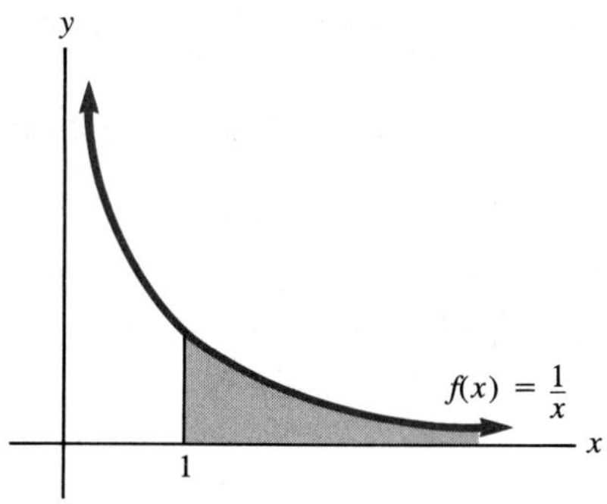

FIGURE 6–30

EXAMPLE 4 The improper integral

$$\int_{-\infty}^{+\infty} e^{-x}\, dx$$

diverges, because

$$\int_{-\infty}^{+\infty} e^{-x}\, dx = \int_{-\infty}^{0} e^{-x}\, dx + \int_0^{+\infty} e^{-x}\, dx,$$

and

$$\int_{-\infty}^{0} e^{-x}\, dx = \lim_{t \to -\infty} \int_t^0 e^{-x}\, dx = \lim_{t \to -\infty} (-1 + e^{-t})$$

diverges. (Why?)

6.4 EXERCISES

In Exercises 1 through 12 determine whether the given integral converges or diverges; if it converges, find its value and interpret your result as an area.

1. $\int_1^{+\infty} \frac{1}{x^2}\,dx$

2. $\int_{-\infty}^{1} \frac{1}{x^2}\,dx$

3. $\int_2^{+\infty} \frac{1}{x^3}\,dx$

4. $\int_{-\infty}^{-2} \frac{1}{x^3}\,dx$

5. $\int_4^{+\infty} \frac{1}{x}\,dx$

6. $\int_1^{+\infty} \frac{1}{x^{1/3}}\,dx$

7. $\int_1^{+\infty} \frac{x}{x^2+1}\,dx$

8. $\int_0^{+\infty} e^x\,dx$

9. $\int_{-\infty}^{0} e^x\,dx$

10. $\int_{-\infty}^{+\infty} e^x\,dx$

11. $\int_0^{+\infty} xe^{-x^2}\,dx$

12. $\int_{-\infty}^{+\infty} xe^{-x^2}\,dx$

13. If $0 \leq r \leq 1$, show that

$$\int_1^{+\infty} \frac{1}{x^r}\,dx$$

diverges.

14. If $r > 1$, show that

$$\int_1^{+\infty} \frac{1}{x^r}\,dx$$

converges.

Management

15. A firm's marginal cost function is given by

$$c(x) = \frac{100{,}000}{x^{5/4}}, \qquad x \geq 1.$$

Find the firm's total cost if it makes an unlimited number of units.

6.5 FURTHER TECHNIQUES OF INTEGRATION

To this point we have discussed only two methods for finding the indefinite integral of a function: integration by guessing an antiderivative and integration by substitution. No technique of integration is applicable to all integrals, and over the years many different techniques have been developed to handle different types of indefinite integrals. Most of these techniques of integration are beyond the scope of this book, but one of them, **integration by parts,** is so useful that we must present it. In this section we discuss integration by parts and demonstrate how to perform integration with the aid of a table of integrals.

Integration by Parts

If u and v are functions of x, the product rule for differentiation states that

$$\frac{d}{dx}(uv) = u\frac{dv}{dx} + v\frac{du}{dx}.$$

If we multiply both sides of this equation by the differential dx, we may write it in differential form as

$$d(uv) = u\,dv + v\,du,$$

or

$$u\,dv = d(uv) - v\,du.$$

Integrating both sides of this last equation, we obtain

$$\int u\,dv = \int d(uv) - \int v\,du.$$

But uv is an antiderivative for $d(uv)$, so we have

$$\int u\,dv = uv - \int v\,du.$$

This is known as the **integration by parts** formula.

Integration by Parts Formula

$$\int u\,dv = uv - \int v\,du.$$

EXAMPLE 1 Let us use the integration by parts formula to find

$$\int xe^{-x}\,dx.$$

We set $u = x$ and $dv = e^{-x}\,dx.$

Then $du = dx$ and $v = \int e^{-x}\,dx = -e^{-x},$

so

$$\begin{aligned}\int xe^{-x}\,dx &= \int u\,dv \\ &= uv - \int v\,du \\ &= x(-e^{-x}) - \int (-e^{-x})\,dx\end{aligned}$$

$$= -xe^{-x} + \int e^{-x}\,dx$$

$$= -xe^{-x} - e^{-x} + c.$$

(Note that when we found v we did not include the constant of integration and write

$$v = \int e^{-x}\,dx = -e^{-x} + c;$$

instead, we simply postponed adding the constant of integration until the final step.)

EXAMPLE 2 How did we know in Example 1 to let $u = dx$ and $dv = e^{-x}\,dx$? If we had tried the other possibility,

$$u = e^{-x} \qquad \text{and} \qquad dv = x\,dx,$$

then we would have obtained

$$du = -e^{-x}\,dx \qquad \text{and} \qquad v = \int x\,dx = \frac{x^2}{2},$$

and thus

$$\int xe^{-x}\,dx = \int u\,dv$$

$$= uv - \int v\,du$$

$$= \frac{x^2}{2}e^{-x} - \int\left(-\frac{x^2}{2}e^{-x}\right)dx.$$

But the integral on the right is more complicated than the one we started with. Therefore this choice of u and dv is not a useful one.

Integration by parts can often be used to convert an integral that cannot be handled into one that can, as in Example 1. Examples 1 and 2 together suggest that the key to integration by parts is to choose u and dv in such a way that $\int v\,du$ is easy to find. Note that when selecting u and dv, dv must include the differential dx of the integral.

EXAMPLE 3 Let us use integration by parts to find

$$\int \ln x\,dx.$$

If we choose

$$u = \ln x \qquad \text{and} \qquad dv = dx,$$

then

$$du = \frac{1}{x}\,dx \qquad \text{and} \qquad v = \int dx = x,$$

so
$$\int v\,du = \int x\frac{1}{x}\,dx = \int 1\,dx,$$

which is certainly easy to find. Hence this is a promising choice for u and dv. We have

$$\begin{aligned}\int \ln x\,dx &= \int u\,dv\\ &= uv - \int v\,du\\ &= x\ln x - \int 1\,dx\\ &= x\ln x - x + c.\end{aligned}$$

Sometimes it it necessary to use integration by parts more than once.

EXAMPLE 4 Let us find

$$\int x^2e^{-x}\,dx.$$

We let

$$u = x^2 \qquad \text{and} \qquad dv = e^{-x}\,dx.$$

Then

$$du = 2x\,dx \qquad \text{and} \qquad v = \int e^{-x}\,dx = -e^{-x}.$$

so

$$\int v\,du = \int(-2xe^{-x})\,dx = -2\int xe^{-x}\,dx;$$

but we know how to integrate xe^{-x} from Example 1. Therefore

$$\begin{aligned}\int x^2e^{-x}\,dx &= \int u\,dv\\ &= uv - \int v\,du\\ &= -x^2e^{-x} - \int(-2xe^{-x})\,dx\\ &= -x^2e^{-x} + 2\int xe^{-x}\,dx.\end{aligned}$$

But from Example 1,

$$\int xe^{-x}\,dx = -xe^{-x} - e^{-x} + c.$$

Thus

$$\begin{aligned}\int x^2e^{-x}\,dx &= -x^2e^{-x} + 2(-xe^{-x} - e^{-x}) + c\\ &= -e^{-x}(x^2 + 2x + 2) + c.\end{aligned}$$

When using integration by parts with definite integrals, the term uv must be evaluated at the limits of integration.

EXAMPLE 5 Let us evaluate

$$\int_0^2 xe^{-x}\,dx$$

using integration by parts. As in Example 1, we take

$$u = x \qquad \text{and} \qquad dv = e^{-x}\,dx,$$

so

$$du = dx \qquad \text{and} \qquad v = \int e^{-x}\,dx = -e^{-x}.$$

Thus

$$\begin{aligned}\int_0^2 xe^{-x}\,dx &= -xe^{-x}\Big|_0^2 + \int_0^2 e^{-x}\,dx \\ &= (-2e^{-2}) - (-0e^{-0}) - \left[e^{-x}\Big|_0^2\right] \\ &= -2e^{-2} - [e^{-2} - e^{-0}] \\ &= 1 - 3e^{-2}.\end{aligned}$$

Integral Tables

Many integrals cannot be found using only the techniques we have presented. For example, the integrals

$$\int \sqrt{x^2 - 4}\,dx \qquad \text{and} \qquad \int \frac{1}{4x^2 - 9}\,dx$$

cannot be determined by any method we have discussed so far. Fortunately there are tables of integrals that show how to integrate these and many other functions. Mathematical handbooks contain extensive tables of integrals; the integration formulas in the back endpapers of this book form an abbreviated version of such an integral table.

EXAMPLE 6 Let us use the integration formulas in the back endpapers to find

$$\int \sqrt{x^2 - 4}\,dx.$$

We search through the list of formulas until we find one in which the integrand has the form of our integrand $\sqrt{x^2 - 4}$. Formula 8, which tells us that

$$\int \sqrt{u^2 - a^2}\,du = \frac{1}{2}[u\sqrt{u^2 - a^2} - a^2 \ln |u + \sqrt{u^2 - a^2}|] + c,$$

has an integrand of the proper form if we take

$$u = x \qquad \text{and} \qquad a = 2.$$

Note that with the substitutions $u = x$ (so that $du = dx$) and $a = 2$ the integral of formula 8 exactly matches our integral. Thus,

$$\int \sqrt{x^2 - 4}\, dx = \int \sqrt{u^2 - a^2}\, du$$

$$= \frac{1}{2}[u\sqrt{u^2 - a^2} - a^2 \ln |u + \sqrt{u^2 - a^2}|] + c,$$

$$= \frac{1}{2}[x\sqrt{x^2 - 4} - 4 \ln |x + \sqrt{x^2 - 4}|] + c.$$

When we match an integral with a formula from a table of integrals, the entire integral, including the differential, must match the formula in the table exactly. This sometimes requires making a substitution or rewriting the integral.

EXAMPLE 7 Let us find

$$\int \frac{1}{4x^2 - 9}\, dx.$$

The integrand matches that of formula 11 in the endpapers if we take

$$u = 2x \qquad \text{and} \qquad a = 3.$$

But if $u = 2x$, then

$$du = 2\, dx \qquad \text{or} \qquad \frac{1}{2}\, du = dx.$$

Thus,

$$\int \frac{1}{4x^2 - 9}\, dx = \int \frac{1}{u^2 - 3^2} \frac{1}{2}\, du$$

$$= \frac{1}{2} \int \frac{1}{u^2 - 3^2}\, du$$

$$= \frac{1}{2}\left[\frac{1}{2 \cdot 3} \ln \left|\frac{u - 3}{u + 3}\right|\right] + c$$

$$= \frac{1}{12} \ln \left|\frac{2x - 3}{2x + 3}\right| + c.$$

EXAMPLE 8 Let us find

$$\int \frac{x}{4x^2 + 12x + 9}\, dx.$$

This integral does not appear to match any of those in the endpapers. However,

$$4x^2 + 12x + 9 = (2x + 3)^2,$$

and

$$\int \frac{x}{(2x + 3)^2}\, dx$$

matches formula 16 exactly if $u = x$ (so that $du = dx$), $a = 2$, and $b = 3$. Therefore

$$\int \frac{x}{4x^2 + 12x + 9}\,dx = \int \frac{x}{(2x+3)^2}\,dx$$
$$= \int \frac{u}{(2u+3)^2}\,du$$
$$= \frac{1}{4}\left[\ln|2u+3| + \frac{3}{2u+3}\right] + c$$
$$= \frac{1}{4}\left[\ln|2x+3| + \frac{3}{2x+3}\right] + c.$$

Computer Manual: Program DEFINT.

6.5 EXERCISES

Integration by Parts

In Exercises 1 through 12 integrate by parts.

1. $\int xe^x\,dx$ **2.** $\int x^2e^x\,dx$ **3.** $\int_0^1 x^3e^{-x}\,dx$ **4.** $\int x^3e^{-x^2}\,dx$

5. $\int x\ln x\,dx$ **6.** $\int_1^e x^2\ln x\,dx$ **7.** $\int \sqrt{x}\ln x\,dx$ **8.** $\int (\ln x)^2\,dx$

9. $\int (\ln x)^3\,dx$ **10.** $\int_e^{e^2} \frac{\ln x}{x^2}\,dx$ **11.** $\int_1^2 x(x+2)^5\,dx$ **12.** $\int x^2(x+1)^{3/2}\,dx$

Management

13. A firm's marginal cost function is given by

$$c(x) = x\sqrt{x+1}.$$

Find the firm's cost function if its fixed cost is \$0.

Integral Tables

In Exercises 14 through 25 use the integration formulas in the back endpapers to find the integral.

14. $\int \frac{1}{x^2 - 16}\,dx$ **15.** $\int \frac{1}{x\sqrt{81 - x^2}}\,dx$ **16.** $\int \frac{1}{3x\sqrt{4 + 9x^2}}\,dx$ **17.** $\int \frac{x}{3x+2}\,dx$

18. $\int \frac{1}{x\sqrt{25 - 4x^2}}\,dx$ **19.** $\int \frac{1}{8 - 2x^2}\,dx$ **20.** $\int \frac{1}{x^2 + 3x}\,dx$ **21.** $\int \frac{2}{\sqrt{9x^2 - \frac{1}{4}}}\,dx$

22. $\int \frac{3x}{36x^2 + 60x + 25}\,dx$ **23.** $\int 4\sqrt{x^2 + 5}\,dx$ **24.** $\int \sqrt{2x^2 - 1}\,dx$ **25.** $\int 2^{3x}\,dx$

6 REVIEW EXERCISES

Answers to all review exercises are in the back of the book.

1. Find the average value of the function f defined by $f(x) = xe^{-x^2}$ over the interval $[0,2]$.
2. A company's annual return on investment is given by
$$f(t) = 100e^{-0.05t},$$
where $f(t)$ is in thousands of dollars and $t = 0$ represents 1985. Forecast the firm's average annual return on investment over the period
(a) 1990–1995 **(b)** 1995–2005
3. A merchant's inventory over a 60-day period is given by
$$y = 1000e^{-0.02t}, \qquad 0 \le t \le 60.$$
Find the average inventory over the period.
4. There were y thousand riders each day on a city's transportation system, where
$$y = -0.1t^2 + 2.4t + 45.6$$
and $0 \le t \le 30$. Find the average number of riders over the 30-day period.
5. Find producers' surplus and consumers' surplus if the supply function is given by
$$p = 5q + 50$$
and the demand function by
$$p = -q^2 - 20q + 400.$$
6. A firm's earnings are flowing in at the rate of \$50 million per year and are compounding continuously at 15 percent per year. Find the value of the firm's earnings over the next 10 years.
7. Let f be a function which is positive over the interval $[a, b]$. If we revolve the graph of f about the x-axis, we obtain a solid known as a **solid of revolution:**

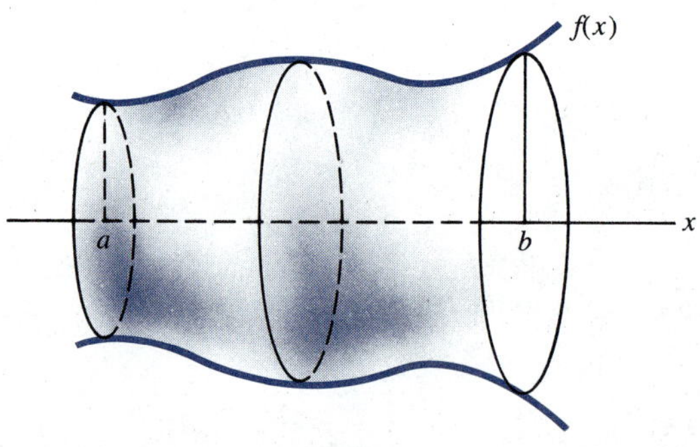

A small subinterval of $[a, b]$ of length Δx thus generates a portion of the solid of revolution whose volume can be approximated by that of a circular disk of radius $f(x)$ and width Δx:

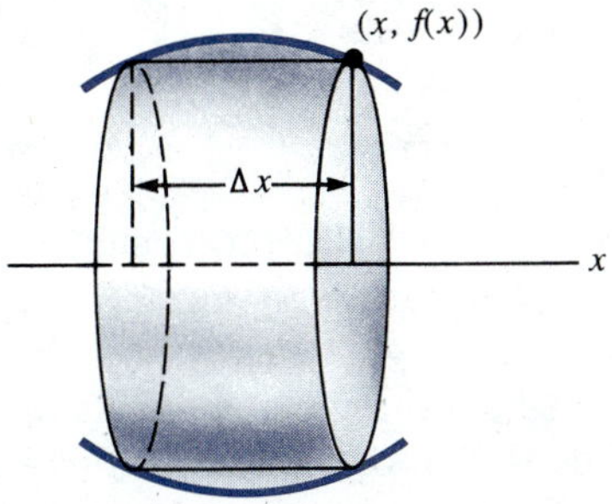

The volume of such a circular disk is $\pi[f(x)]^2\,\Delta x$. Use the integral as an accumulator to show that the volume of the solid of revolution is given by

$$\pi\int_a^b [f(x)]^2\,dx.$$

8. Use the result of Exercise 7 to find the volume of the solid of revolution obtained by revolving the given function about the x-axis over the given interval:

(a) $f(x) = x$, $[0, 1]$;
(b) $f(x) = x^2$, $[1, 2]$;
(c) $f(x) = e^{-x}$, $[0, 2]$.

9. Use the midpoint rule, the trapezoidal rule, and Simpson's rule with $n = 6$ to approximate the value of

$$\int_0^1 (4 - x^4)\,dx.$$

Compare your result in each case with the actual value of the integral.

10. A criminologist sampled the police records of a city and found the following:

Day	1	5	9	13	17	21	25	29
Crimes per day	125	128	136	132	130	124	118	116

Use the trapezoidal rule to estimate the total number of crimes reported over the 29-day period.

In Exercises 11 through 13 determine if the improper integral converges; if it does, find its value and interpret your result as an area.

11. $\displaystyle\int_3^{+\infty} \frac{2x^2}{x^3 + 1}\,dx$

12. $\displaystyle\int_{-\infty}^{-2} x^{-4}\,dx$

13. $\displaystyle\int_{-\infty}^{+\infty} f(x)\,dx$, where $f(x) = \begin{cases} e^{2x} & \text{for } x \le 0, \\ e^{-3x} & \text{for } x > 0. \end{cases}$

In Exercises 14 through 17 find the indefinite integral.

14. $\displaystyle\int 4xe^{-4x}\,dx$

15. $\displaystyle\int 3x(x + 5)^6\,dx$

16. $\displaystyle\int \frac{1}{2x^2 + 4x}\,dx$

17. $\displaystyle\int \frac{1}{16 - 4x^2}\,dx$

6 SUPPLEMENT: THE PRESENT VALUE OF A STREAM OF INCOME

In the supplement to Chapter 4 we discussed annuities and their present values. It is often useful to think of a stream of income, such as was considered in Section 6.2, as a **continuous annuity,** that is, as an annuity whose annual payment is spread continuously throughout the year. When this is done, we can apply the present value concept to the stream of income. In this supplement we show how to find the present value of a stream of income, and we illustrate some of the uses of this concept.

Suppose a stream of income flows into a firm continuously, arriving at the rate of $I(t)$ dollars per year at time t. Suppose also that money is worth $100s$ percent per year compounded continuously. (The rate $100s$ is called the **discount rate.**) We will find an expression for the present value of this income stream over the interval $[0, b]$.

If we divide the time interval $[0, b]$ into small subintervals, each of width Δt, then the amount of income that flows in during any particular subinterval is approximately $I(t)\,\Delta t$ dollars. From the interest formula

$$A = Pe^{rt}$$

of Section 4.1, we see that the present value P of an amount A due t years in the future when money is worth $100r$ percent per year compounded continuously is given by

$$P = Ae^{-rt}.$$

Therefore the present value of the amount $I(t)\,\Delta t$ that will arrive during the subinterval and is to be discounted at $100s$ percent per year is equal to

$$[I(t)\,\Delta t]e^{-st}$$

dollars. Using the integral as an accumulator, we thus find that the total present value of the stream of income over the time period $[0, b]$ is

$$\int_0^b I(t)e^{-st}\,dt.$$

The Present Value of a Stream of Income

If a stream of income flows in continuously, arriving at the rate of $I(t)$ dollars per year at time t, and if its future value is discounted at $100s$ percent per year compounded continuously, then the present value of the stream of income over the next b years is

$$\int_0^b I(t)e^{-st}\,dt.$$

EXAMPLE 1 Suppose that each year Blink, Inc. earns \$100,000 in revenue. What is the present value of Blink's revenue stream over the next 4 years if Blink discounts future revenues at 8 percent per year compounded continuously? The answer is

$$\begin{aligned}\int_0^4 100{,}000e^{-0.08t}\,dt &= -\frac{100{,}000}{0.08}\left(e^{-0.08t}\Big|_0^4\right)\\ &= -\frac{100{,}000}{0.08}(e^{-0.32}-1)\\ &= \$342{,}313.70.\end{aligned}$$

EXAMPLE 2 Suppose Blink, Inc. of the previous example expects its revenue to continue to be \$100,000 per year forever. Then the present value of Blink's future revenue for all time is

$$\begin{aligned}\int_0^{+\infty} 100{,}000e^{-0.08t}\,dx &= \lim_{b\to+\infty}\int_0^b 100{,}000e^{-0.08t}\,dt\\ &= -\frac{100{,}000}{0.08}\lim_{b\to+\infty}(e^{-0.08b}-1)\\ &= -1{,}250{,}000(0-1)\\ &= \$1{,}250{,}000.\end{aligned}$$

Another viewpoint of the present value of an income stream takes into account the fact that the money received can be invested. Thus, suppose that income is currently arriving continuously at the rate of P dollars per year, that it compounds continuously at $100r$ percent per year, and that it is discounted at $100s$ percent per year. Then at time t in the future the amount P will have grown to

$$A(t) = Pe^{rt}$$

dollars. But the present value of the amount $A(t)$ is equal to

$$A(t)e^{-st} = Pe^{rt}e^{-st} = Pe^{(r-s)t}.$$

Therefore by the usual argument using the integral as accumulator, the present value of the income stream over the next b years is

$$\int_0^b Pe^{(r-s)t}\,dt.$$

The Present Value of an Invested Stream of Income

If a stream of income is currently flowing in at a rate of P dollars per year, and if this income is invested at $100r$ percent per year compounded continuously and its future value is discounted at $100s$ percent per year compounded continuously, then the present value of the income received over the next b years is

$$\int_0^b Pe^{(r-s)t}\,dt.$$

EXAMPLE 3 Suppose that a firm's profits are currently flowing in at a rate of \$1 million per year and that it reinvests all profits. If the reinvested profits earn 8 percent per year compounded continuously and if the firm discounts future profits at 12 percent per year, then the present value of its profits over the next 5 years, in millions of dollars, will be

$$\begin{aligned}\int_0^5 1e^{(0.10-0.12)t}\,dt &= \int_0^5 e^{-0.02t}\,dt\\ &= -\frac{e^{-0.02t}}{0.02}\Bigg|_0^5\\ &= -\frac{e^{-0.1}-1}{0.02}\\ &\approx 4.758,\end{aligned}$$

or approximately \$4,758,000.

SUPPLEMENTARY EXERCISES

1. If profits are flowing into Aurora Corporation at a rate of 2 million dollars per year, and if Aurora discounts future profits at 10 percent per year compounded continuously, find the present value of Aurora's future profits
   (a) over the next 6 years **(b)** forever
2. Do Exercise 1 if profits are flowing in at a rate of $I(t) = 2 + 5t$ million dollars per year.
3. Suppose Borealis Company expects to be earning profits at the rate of
   $$I(t) = 1 + e^{0.05t}$$
   million dollars per year t years from now. If the company discounts future earnings at 8 percent per year compounded continuously, find the present value of its expected earnings
   (a) over the next 10 years **(b)** over the next 50 years
   (c) forever

4. Camran, Ltd. is currently earning profits at the rate of \$5 million per year. Camran reinvests its profits at 9 percent per year compounded continuously and discounts future profits at 10 percent per year compounded continuously. Find the present value of Camran's profits

(a) over the next 3 years **(b)** forever

5. Do Exercise 4 if Camran's discount rate per year is

(a) 20 percent **(b)** 9 percent

6. You are considering buying some rental property as an investment. You know that you can rent the property now for \$10,000 per year and that the rent will increase at 6 percent per year compounded continuously. If you discount future income at 10 percent per year compounded continuously, find the present value of the rent you will receive if you keep the property

(a) 10 years **(b)** 40 years **(c)** forever

7. Refer to Exercise 6. How long would you have to hold the property to receive rent having a present value of \$100,000?

8. Refer to Exercise 6. Suppose you intend to hold the property for 10 years and then sell it.

(a) If you could sell it for \$100,000, what is the maximum amount that you would be willing to pay for it now?

(b) If the property will cost you \$150,000, how much will you have to sell it for in order to earn more, on a present value basis, than it costs?

9. A certain stock pays a dividend of \$14 per share, and this is increasing at a rate of 8 percent per year compounded continuously. If future income is discounted at 16 percent per year compounded continuously, find the present value of all future dividends you will receive if you buy the stock and hold it for

(a) 5 years **(b)** 50 years **(c)** forever

10. Refer to Exercise 9.

(a) Would you be willing to pay \$100 per share for the stock now if you knew you could sell it 5 years from now for \$100? If you could sell it 5 years from now for \$50?

(b) Suppose the stock costs \$75 a share now. How much would you have to sell it for 5 years from now in order to earn more, on a present value basis, than it costs?

7

Functions of Several Variables

The functions we have studied up until now have been **functions of one variable,** that is, functions with just one independent variable. However, in many applications we must consider functions with more than one independent variable. For instance, suppose a firm produces two products. Its total cost will then depend on the number of units x of the first product *and* the number of units y of the second product that it makes. For instance, the firm's total cost C might be given by the equation

$$C = 20x + 30y + 50{,}000;$$

this equation defines C as a function of the two independent variables x and y. Similarly, a farmer's crop yield Y might depend on the amount q of fertilizer and the amount w of weedkiller used on the crop, the amount r of rainfall the crop receives, and the time t that the farmer devotes to cultivating the crop, according to the equation

$$Y = 0.5q + 2.2w + 2\sqrt{r} + 1.2t^2.$$

This equation defines crop yield Y as a function of the four independent variables q, w, r, and t. Functions such as these are called **functions of several variables.**

In this chapter we study functions of several variables. We begin by illustrating how functions of several variables are defined and interpreted, then introduce the three-dimensional cartesian coordinate system and show how functions of two variables may be graphed as surfaces in three-dimensional space. Next we discuss derivatives of functions of several variables, show how these derivatives may be interpreted, and demonstrate how to find relative maxima and minima for functions of two variables. We also present a method by means of which functions of several variables can be maximized or minimized subject to constraints. The final section of the chapter is devoted to integrating functions of two variables.

7.1 FUNCTIONS OF MORE THAN ONE VARIABLE

An equation such as

$$f(x,y) = x^2 + 2xy + 5y$$

defines f as a function of two independent variables x and y. The domain of such a function f is a set of ordered pairs (a, b), where a and b are numbers. We evaluate the function at the ordered pair (a, b) by substituting a for x and b for y in the defining equation.

EXAMPLE 1 Let

$$f(x, y) = x^2 + 2xy + 5y.$$

Then

$$\begin{aligned} f(0, 0) &= 0^2 + 2(0)(0) + 5(0) = 0, \\ f(2, 3) &= 2^2 + 2(2)(3) + 5(3) = 31, \\ f(-1, 2) &= (-1)^2 + 2(-1)(2) + 5(2) = 7, \\ f(4, y) &= 4^2 + 2(4)y + 5y = 16 + 13y, \\ f(x, 1) &= x^2 + 2x(1) + 5(1) = x^2 + 2x + 5, \\ f(x + 1, b) &= (x + 1)^2 + 2(x + 1)b + 5b \\ &= x^2 + 2x + 1 + 2xb + 2b + 5b \\ &= x^2 + 2(1 + b)x + 7b + 1, \end{aligned}$$

and so on. Since we may substitute any number for x and any number for y in the defining equation for f, the domain of f is the set of all ordered pairs (x, y).

Similarly, a function defined by an equation such as

$$f(x, y, z) = \frac{x - y}{y - z}$$

is a function of three variables. The domain of such a function is a set of ordered triples (a, b, c), where a, b, and c are numbers. We evaluate the function at the ordered triple (a, b, c) by substituting a for x, b for y, and c for z in its defining equation.

EXAMPLE 2 Let

$$f(x, y, z) = \frac{x - y}{y - z}.$$

Then

$$f(1, 1, 0) = \frac{1 - 1}{1 - 0} = 0,$$

and $$f(2, -1,3) = \frac{2-(-1)}{-1-3} = -\frac{3}{4}.$$

However, $f(1, 2, 2)$ is not defined. (Why not?) More generally, notice that

$$f(x, y, z) = \frac{x-y}{y-z}$$

is not defined if and only if $y = z$. Thus the domain of this function is $\{(x, y, z) \mid y \neq z\}$.

Functions of several variables can also be defined by equations that relate a dependent variable to two or more independent variables.

EXAMPLE 3 Let

$$z = xe^x + \ln y.$$

Then

$$\text{when } x = 0 \text{ and } y = 1, \qquad z = 0e^0 + \ln 1 = 0 + 0 = 0;$$

$$\text{when } x = 1 \text{ and } y = 1, \qquad z = 1e^1 + \ln 1 = e + 0 = e;$$

$$\text{when } x = -2 \text{ and } y = 2, \qquad z = -2e^{-2} + \ln 2;$$

and so on. Note that since $\ln y$ is not defined when $y \leq 0$, the function defined by this equation has domain $\{(x, y) \mid y > 0\}$.

Functions of four or more variables are defined and evaluated in a manner similar to that illustrated in the preceding examples.

As we mentioned at the beginning of this chapter, functions of several variables arise naturally in many situations. We illustrate with examples.

EXAMPLE 4 Suppose that a store sells two products, X and Y, and that the sales price of one product is not affected by the sales price of the other. If each unit of product X sells for \$15 and each unit of product Y for \$20, then the store's revenue from selling x units of X and y units of Y is

$$R(x, y) = 15x + 20y$$

dollars.

EXAMPLE 5 Tropicoid Corporation makes two models of its instant camera, the regular and the deluxe. The company's cost if it makes x units of the regular model and y units of the deluxe model is

$$C(x, y) = 40x + 60y + 0.01xy + 20{,}000$$

dollars. Note that the company's fixed cost is therefore

$$C(0, 0) = 20{,}000$$

dollars.

Now suppose Tropicoid makes a fixed number, say 1000 units, of the regular model; then its cost will be

$$C(1000, y) = 40(1000) + 60y + 0.01(1000)y + 20{,}000 = 60{,}000 + 70y$$

dollars. Therefore if production of the regular model is held fixed at 1000 units, each additional unit of the deluxe model made will cost \$70. Similarly, if production of the regular model is held fixed at 2000 units, then

$$C(2000, y) = 100{,}000 + 80y,$$

which tells us that each additional unit of the deluxe model produced will cost \$80. On the other hand, if production of the deluxe model is held fixed at 2000 units, then

$$C(x, 2000) = 140{,}000 + 60x$$

shows that each additional unit of the regular model produced will cost \$60.

EXAMPLE 6 Suppose a company sells both electric heaters and kerosene heaters. Let x be the unit price of the electric heaters and y the unit price of the kerosene heaters. The demand for either product may depend on both the price of that product and the price of the other; thus, if E is the demand for electric heaters, then E may depend on both x and y. For example, we might have

$$E = -100x + 500y + 10{,}000.$$

Then if electric heaters sell for \$40 each and kerosene heaters for \$60 each, the demand for electric heaters will be

$$E = -100(40) + 500(60) + 10{,}000 = 36{,}000$$

units.

Let us analyze the demand function

$$E = -100x + 500y + 10{,}000.$$

If the price y of kerosene heaters remains fixed at, say, $y = b$, then the demand for electric heaters is given by

$$E = -100x + 500b + 10{,}000.$$

Therefore each \$1 rise in the price x of electric heaters will result in a decrease of 100 in the quantity of electric heaters demanded by consumers. In other words, if the price of kerosene heaters remains fixed, a rise in the price of electric heaters leads to a decrease in the quantity of electric heaters demanded. On the other hand, if the price of electric heaters is fixed at, say, $x = a$, then

$$E = -100a + 500y + 10{,}000,$$

which shows that each increase of \$1 in the price y of kerosene heaters leads to an increase of 500 in the quantity of electric heaters demanded. Thus, if the price of electric heaters remains fixed while the price of kerosene heaters goes up, demand for electric heaters will increase.

Computer Manual: Program TWOVARS.

7.1 EXERCISES

In Exercises 1 through 10 find the domain of the function and the indicated functional values, if possible.

1. $f(x, y) = x^2 + 2xy + y^2$, $f(0, 0)$, $f(1, 3)$
2. $f(x, y) = \dfrac{2x}{y + 1}$, $f(-2, -3)$, $f(4, -1)$
3. $f(p, q) = \sqrt{p} + 2\sqrt{q}$, $f(0, 4)$, $f(9, 16)$
4. $f(x, y) = \dfrac{2x^2 - 3y^2}{x + y}$, $f(1, 1)$, $f(4, 0)$
5. $f(u, v) = \sqrt{2uv} + 1$, $f(0, 0)$, $f(3, -2)$
6. $f(x, y) = e^{-xy} + e^{x/y}$, $f(1, 0)$, $f(2, 2)$
7. $f(x, y, z) = x^2 + y^2 + z^2$, $f(1, 2, -1)$, $f(0, 2, 5)$
8. $f(r, s, t) = \dfrac{1}{r} + \dfrac{2}{s^2 - t^2}$, $f(1, 2, 3)$, $f(-2, 2, 2)$
9. $f(x, y, z) = \ln x + 2 \ln y + 3 \ln z$, $f(1, 1, 1)$, $f(e, e^2, e^3)$
10. $f(x, y, z, w) = \ln|x| + \ln|y| + \ln|z| + \ln|w|$, $f(-1, 1, 1, 1)$, $f(e, -e, -e, e)$
11. Let $z = 2x^2 + \dfrac{x}{y} + 2$. Find z when $x = 0$ and $y = 1$; when $x = 2$ and $y = 2$.
12. Let $w = e^{z-2x+3y}$. Find w when $x = -1$, $y = z = 2$.

Management

13. A refinery produces 2 grades of gasoline, regular and premium. If the cost of producing one grade does not affect the cost of producing the other, if it costs 40 cents to produce each gallon of regular and 50 cents to produce each gallon of premium, and if the refinery's fixed cost is \$100,000 per day, find
    (a) its cost function;
    (b) the cost of making 20,000 gallons of each grade of gasoline.
14. A firm makes two products, A and B. The profit contribution of one product is not affected by that of the other. If each unit of product A contributes \$60 toward profit, each unit of product B contributes \$75 toward profit, and the fixed cost is \$30,000, find the profit function and the firm's profit when it makes and sells 300 units of A and 500 of B.
15. A store sells white paint and blue paint. If the prices at which the 2 colors are sold do not affect each other, if x is the price per gallon of the white paint and y the price per gallon of the blue paint, and if the demand for white paint is

$$W = 400 - 6x + 2y.$$

gallons and that for blue paint is

$$B = 100 + 4x - 10y$$

gallons, find
    (a) the function that describes the store's revenue from paint sales;
    (b) the store's revenue if it sells the white paint for \$5 per gallon and the blue for \$7 per gallon.

16. Suppose the demand function for product A is given by

$$A(x, y) = 200 - 8x + 3y,$$

where x is the unit price of A and y the unit price of another product B.

(a) Find the demand for A if its price is \$10 per unit and the price of B is \$12 per unit.

(b) If the unit price of A is fixed at p dollars, what can you say about the demand for A?

(c) If the unit price of B is fixed at q dollars, what can you say about the demand for A?

17. A firm makes two products, X and Y. Its cost function is given by

$$C(x, y) = 2000 + 40x + 20y,$$

where x is the number of units of X produced and y the number of units of Y produced.

(a) What is the firm's fixed cost?

(b) What is the cost if 150 units of X and 100 units of Y are produced?

(c) What can you say about the cost if the number of units of X produced is held fixed? If the number of units of Y produced is held fixed?

(d) Find a function that describes the firm's average cost per unit, and find the average cost per unit when 150 units of X and 100 units of Y are produced.

(e) What can you say about the average cost per unit when production of X is held fixed at 150 units? When production of Y is held fixed at 100 units?

18. The revenue function for the firm of Exercise 17 is given by

$$R(x, y) = 12x\sqrt{y}.$$

(a) Find the firm's profit function and its profit if it makes and sells 150 units of X and 100 units of Y.

(b) Find a function that describes the firm's average profit per unit, and find its average profit per unit if it makes and sells 150 units of X and 100 units of Y.

(c) What can you say about the firm's profit when production and sales of X are held fixed at 150 units? When production and sales of Y are held fixed at 100 units?

(d) What can you say about the firm's average profit per unit when production and sales of X are held fixed at 150 units? When production and sales of Y are held fixed at 100 units?

Life Science

19. Let x denote the number of dogs, y the number of mice, and z the number of cats in a neighborhood. If

$$z = 10 - 2x + 0.2y,$$

(a) Find the number of cats in the neighborhood if there are 6 dogs and 40 mice there.

(b) What can you say about the cat population if the dog population of the neighborhood remains constant? If the mouse population remains constant?

20. The approximate time it will take for a patient to recover from pneumonia is

$$f(t, d) = \frac{1.2t}{d}$$

days, where t is the patient's age in years, $t \geq 21$, and d is the patient's daily dosage of penicillin in millions of units.

(a) Find the approximate recovery time for a patient who is 40 years old and takes 4 million units of penicillin per day.

(b) What can you say about the recovery time for patients who are all the same age?

(c) What can you say about the recovery time for patients all of whom take the same daily dosage of penicillin?

7.2 GRAPHING FUNCTIONS OF TWO VARIABLES

In this section we show how to graph functions of two variables using the **three-dimensional cartesian coordinate system** pictured in Figure 7–1. In the figure the x-axis, y-axis, and z-axis are perpendicular and intersect at the **origin.** Note that only the nonnegative portions of the three axes are shown in Figure 7–1; in Figure 7–2 the axes are extended, with their negative portions shown as dashed lines. Because of the difficulties with perspective that arise when the negative portions of the axes are shown, the three-dimensional cartesian coordinate system is usually drawn as in Figure 7–1.

Just as any ordered pair (a, b) of numbers can be plotted as a point on the two-dimensional cartesian coordinate system, so any ordered triple (a, b, c) of numbers can be plotted as a point in three-dimensional space using the three-dimensional cartesian coordinate system (Figure 7–3).

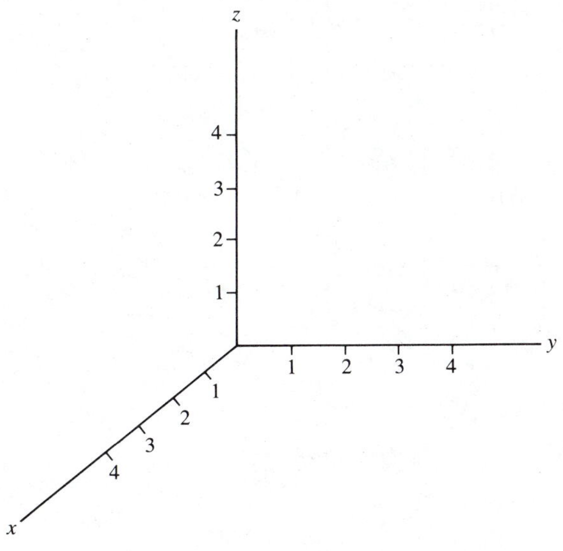

FIGURE 7–1 The Cartesian Coordinate System

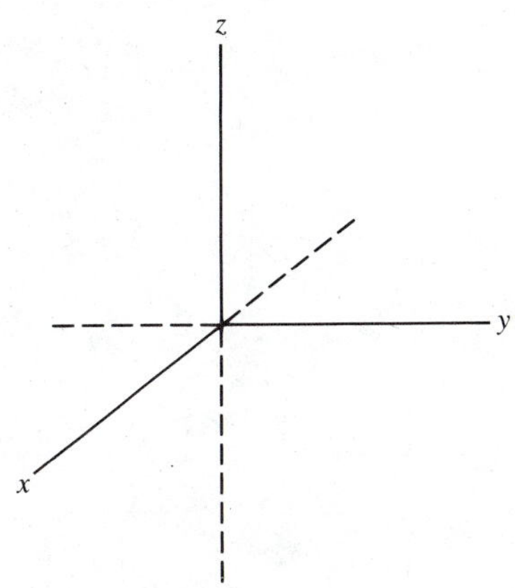

FIGURE 7–2

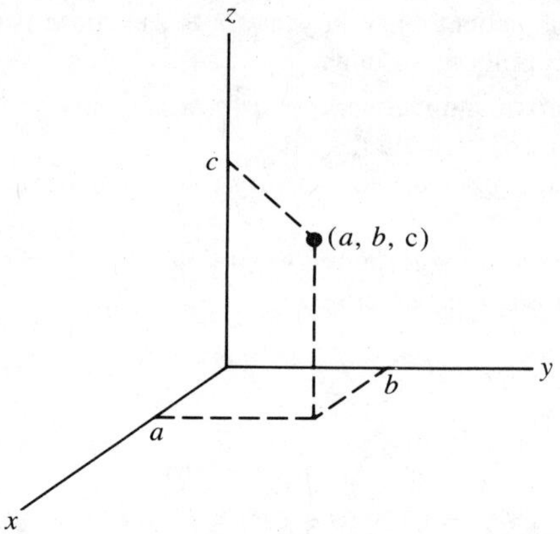

FIGURE 7–3

EXAMPLE 1 In Figure 7–4 we have plotted the origin (0, 0, 0) and the points

$$(3, 0, 0), \quad (0, 2, 0), \quad (0, 0, 4), \quad (3, 2, 0), \quad (0, 2, 3),$$
$$(1, 0, 1) \quad (3, 0, 4), \quad (3, 2, 4), \quad (0, 2, 4), \quad (1, 2, 2).$$

The region consisting of all points (x, y, z) for which x, y, and z are nonnegative is called the **first octant.** All the points of Example 1 lie in the first octant.

The set of all points of the form $(x, y, 0)$ is called the **xy-plane,** the set of all points of the form $(x, 0, z)$ the **xz-plane,** and the set of all points of the form

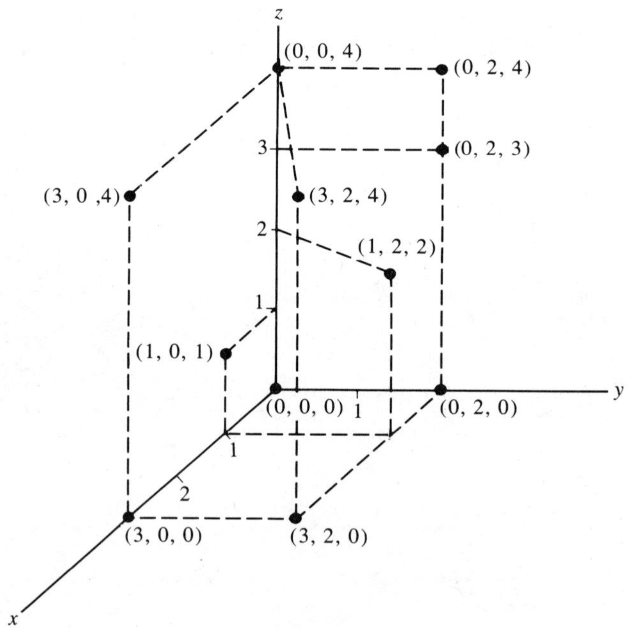

FIGURE 7–4

$(0, y, z)$ the **yz-plane.** In Example 1, $(3, 2, 0)$ and $(0, 2, 0)$ are points in the xy-plane, $(3, 0, 4)$ and $(3, 0, 0)$ are points in the xz-plane, and $(0, 2, 4)$ and $(0, 2, 0)$ are points in the yz-plane. The xy-, xz-, and yz-planes are known collectively as the **coordinate planes.**

We can use the three-dimensional cartesian coordinate system to graph any function f of two variables. The technique is to set $z = f(x, y)$, give values to x and y, calculate $z = f(x, y)$, and plot the resulting point (x, y, z). The graph will be a surface.

EXAMPLE 2 Let us graph the function whose defining equation is

$$f(x, y) = -2x - y + 4.$$

We set

$$z = -2x - y + 4.$$

Then we have

x	y	$z = -2x - y + 4$	(x, y, z)
0	0	4	(0, 0, 4)
1	0	2	(1, 0, 2)
2	0	0	(2, 0, 0)
0	1	3	(0, 1, 3)
1	1	1	(1, 1, 1)
0	2	2	(0, 2, 2)
1	2	0	(1, 2, 0)
0	3	1	(0, 3, 1)
0	4	0	(0, 4, 0)
3	−3	1	(3, −3, 1)
−2	6	2	(−2, 6, 2)

The graph of f is the plane depicted in Figure 7–5.

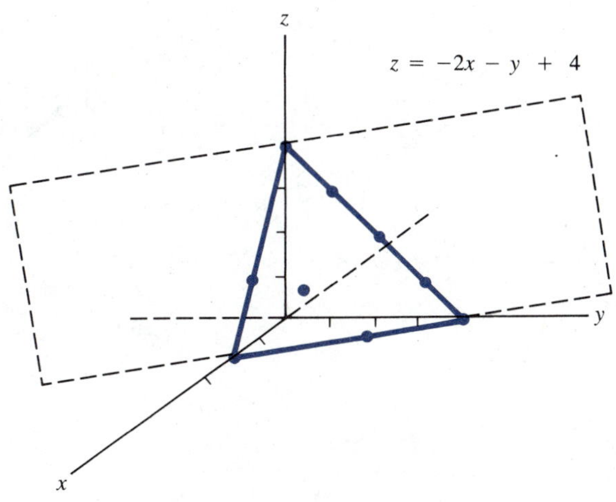

FIGURE 7–5

The intersections of the graph of a function of two variables with the coordinate planes are called the **traces** of the function. The traces of a function are often helpful in drawing its graph. Since in each coordinate plane one variable is always equal to zero, the equations of the traces of a function f may be found by setting each variable, one at a time, equal to zero in the equation $z = f(x, y)$.

Finding the Traces of a Function of Two Variables

If $z = f(x, y)$, then

$f(x, y) = 0$ gives the trace of f in the xy-plane;
$f(x, 0) = z$ gives the trace of f in the xz-plane;
$f(0, y) = z$ gives the trace of f in the yz-plane.

EXAMPLE 3 Consider again the function of Example 2, defined by

$$z = -2x - y + 4.$$

The equations of its traces are as follows:

trace in xy-plane ($z = 0$): $0 = -2x - y + 4$, or $y = -2x + 4$;

trace in xz-plane ($y = 0$): $z = -2x + 4$;

trace in yz-plane ($x = 0$): $z = -y + 4$.

Each trace is a line in its coordinate plane. Figure 7–6 shows how drawing the traces yields a sketch of the portion of the graph that lies in the first octant.

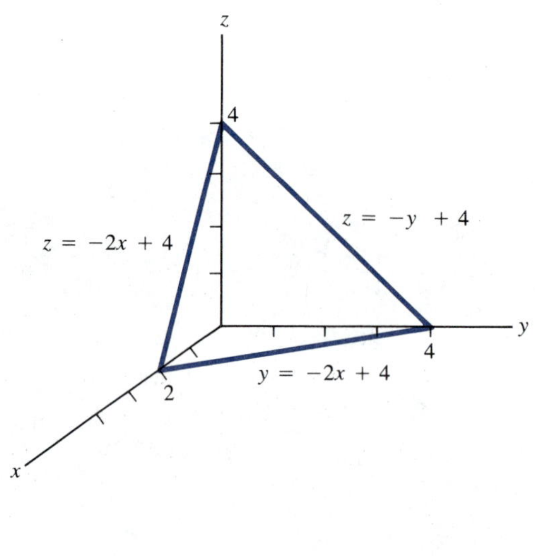

EXAMPLE 4 Let us use traces to sketch the graph of

$$z = x^2 + y^2.$$

We will only sketch the portion of the graph that lies in the first octant.

The equation of the trace in the xy-plane is $0 = x^2 + y^2$, which implies that $x = 0$ and $y = 0$. Hence the xy-trace is the single point $(0, 0, 0)$.

The equation of the trace in the xz-plane is $z = x^2$, and the graph of this equation is a parabola in the xz-plane with vertex at the origin.

The equation of the trace in the yz-plane is $z = y^2$, and the graph of this equation is a parabola in the yz-plane with vertex at the origin.

We have drawn the traces in Figure 7–7. To indicate the three-dimensional aspect of the graph of $z = x^2 + y^2$, we also found its trace on the horizontal plane $z = 4$. This trace is the graph of the equation

$$4 = x^2 + y^2,$$

which we obtained by substituting 4 for z in the equation of the surface. But this is the equation of a circle of radius 2. Thus, the trace of the surface $z = x^2 + y^2$ on the horizontal plane $z = 4$ is a circle of radius 2 with center on the z-axis. The portion of this circle that lies in the first octant is shown in Figure 7–7.

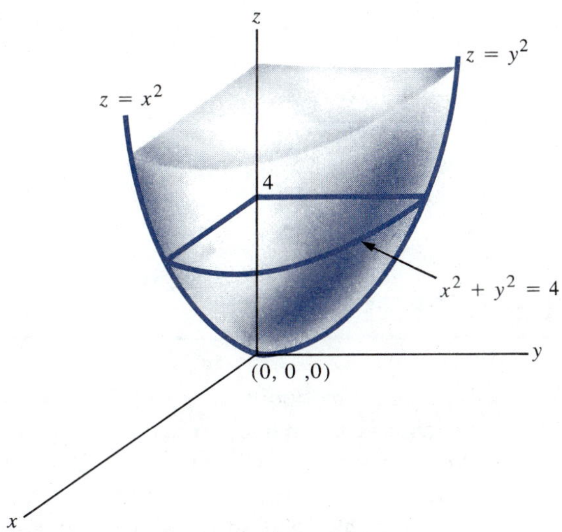

FIGURE 7–7

If the defining equation of a function of two variables is at all complicated, it is not easy to draw its graph. Furthermore, it is simply not possible to graph functions of three or more variables, since we cannot draw a cartesian coordinate system for such functions. Fortunately, it is usually possible to analyze functions of several variables without resorting to their graphs.

7.2 EXERCISES

1. Plot the points (0, 0, 0), (3, 0, 0), (0, 2, 0), (0, 0, 5), (1, 0, 3), (2, 1, 0), (0, 2, 4), (1, 3, 1), (2, 2, 4), (3, 1, 2), (−3, 0, 0), (0, −4, 0), (0, 0, −2), (1, 1, −2), (2, −1, 3), (−2, 2, 2), and (−1, −2, 4) in the three-dimensional cartesian coordinate system.

In Exercises 2 through 17 sketch the graph of the function in the three-dimensional cartesian coordinate system.

2. $f(x, y) = 1 - x - y$
3. $f(x, y) = 8 - 2x - 4y$
4. $z = 6 - x - 3y$
5. $z = 10 + 2x - 5y$
6. $z = 6 - x + 3y$
7. $f(x, y) = 12 + 4x - 3y$
8. $z = -4 + x + y$
9. $f(x, y) = x^2 + y^2 + 1$
10. $f(x, y) = x^2 + y^2 - 4$
11. $z = \sqrt{4 - x^2 - y^2}$
12. $z = \sqrt{9 - x^2 - y^2}$
13. $f(x, y) = y^2$
14. $f(x, y) = x^2$
15. $z = \sqrt{x}$
16. $z = e^{-y}$
17. $f(x, y) = e^x - 1$

7.3 PARTIAL DERIVATIVES

Just as a function of one variable may have a first derivative with respect to its independent variable, so a function of more than one variable may have several first derivatives, one with respect to each of its independent variables. These derivatives, in turn, may themselves have derivatives with respect to each independent variable, and so on. Such derivatives of functions of several variables are known as **partial derivatives,** and the process by which they are found is **partial differentiation.** In this section we show how to find partial derivatives.

First Partial Derivatives

A function of two variables can have two first partial derivatives, one with respect to each of its independent variables.

First Partial Derivatives

Let f be a function of the two variables x and y.

1. We obtain the **first partial derivative of f with respect to x,** denoted by

$$f_x \quad \text{or} \quad \frac{\partial f}{\partial x},$$

by differentiating f with respect to x while treating y as a constant.

2. We obtain the **first partial derivative of f with respect to y,** denoted by

$$f_y \quad \text{or} \quad \frac{\partial f}{\partial y},$$

by differentiating f with respect to y while treating x as a constant.

Partial differentiation for functions of more than two variables is carried out in a similar manner.

EXAMPLE 1 Let us find the first partial derivatives f_x and f_y for the function f defined by

$$f(x, y) = x^2 + y^2 + 2x + 3y.$$

To find f_x, we must differentiate f with respect to x while treating y as a constant. To remind ourselves that y is to be treated as a constant, we replace y by c in $f(x, y)$:

$$f(x, c) = x^2 + c^2 + 2x + 3c,$$

or

$$f(x, c) = x^2 + 2x + (c^2 + 3c);$$

differentiating with respect to x, we obtain

$$f_x = 2x + 2 + 0 = 2x + 2.$$

Similarly, to find f_y we replace x by c, obtaining

$$f(c, y) = y^2 + 3y + (c^2 + 2c),$$

and differentiate with respect to y to find that

$$f_y = 2y + 3.$$

It is not really necessary for us to replace x and y with c when finding partial derivatives. Instead, we need merely keep in mind that when we are differentiating f to find f_x, y is to be treated as a constant, and when we are differentiating f to find f_y, x is to be treated as a constant.

EXAMPLE 2 Let us find f_x and f_y if f is defined by

$$f(x, y) = 2x - 5y + x^2y^3.$$

Differentiating with respect to x while treating y as a constant, we have

$$f_x = 2(1) - 5(0) + 2xy^3 = 2 + 2xy^3.$$

Similarly, differentiating with respect to y while treating x as a constant, we obtain

$$f_y = 2(0) - 5(1) + x^2(3y^2) = -5 + 3x^2y^2.$$

EXAMPLE 3 Let us find the partial derivatives $\partial g/\partial x$, $\partial g/\partial y$, and $\partial g/\partial z$ for the function g of three variables defined by

$$g(x, y, z) = 5x + 3xy - 2x^2z + y^2.$$

To find $\partial g/\partial x$, we differentiate with respect to x while treating y and z as constants:

$$\frac{\partial g}{\partial x} = 5(1) + 3(1)y - 2(2x)z + 0 = 5 + 3y - 4xz.$$

Similarly, to find $\partial g/\partial y$, we differentiate with respect to y while treating x and z as constants:

$$\frac{\partial g}{\partial y} = 5(0) + 3x(1) - 2(0) + 2y = 3x + 2y.$$

To find $\partial g/\partial z$, we differentiate with respect to z while treating x and y as constants:

$$\frac{\partial g}{\partial z} = 5(0) + 3(0) - 2x^2(1) + 0 = -2x^2.$$

EXAMPLE 4 Let us find $f_x(1, 1)$ and $f_y(2, 1)$ if

$$f(x, y) = x^2 - 7x^2y^3 + y + 1.$$

Differentiating with respect to x and treating y as a constant, we have

$$f_x = 2x - 14xy^3.$$

Therefore

$$f_x(1, 1) = 2(1) - 14(1)(1)^3 = -12.$$

Similarly,

$$f_y = -21x^2y^2 + 1,$$

so

$$f_y(2, 1) = -21(2)^2(1)^2 + 1 = -83.$$

EXAMPLE 5 Let

$$h(x, y, z, w) = (2x + 3y^2 + 4z^3 + 5w^4)^{10}.$$

Then

$$h_x = 10(2x + 3y^2 + 4z^3 + 5w^4)^9(2)$$
$$= 20(2x + 3y^2 + 4z^3 + 5w^4)^9,$$
$$h_y = 10(2x + 3y^2 + 4z^3 + 5w^4)^9(6y)$$
$$= 60y(2x + 3y^2 + 4z^3 + 5w^4)^9,$$
$$h_z = 10(2x + 3y^2 + 4z^3 + 5w^4)^9(12z^2)$$
$$= 120z^2(2x + 3y^2 + 4z^3 + 5w^4)^9,$$

and

$$h_w = 10(2x + 3y^2 + 4z^3 + 5w^4)^9(20w^3)$$
$$= 200w^3(2x + 3y^2 + 4z^3 + 5w^4)^9.$$

(Check this.)

Notice that we have shown how to *find* first partial derivatives, but we have not *defined* them. First partial derivatives are defined by difference quotients in a manner analogous to the way we used the difference quotient to define the derivative of a function of one variable in Section 2.2. For our purposes, knowing how to find partial derivatives will be sufficient; we will not need to use their definitions.

Higher Partial Derivatives

Just as a function of one variable may have a first derivative, a second derivative, and so on, a function of several variables may have first partial derivatives, second partial derivatives, and so on. The notation for second partial derivatives of a function of two variables is as follows:

Second Partial Derivatives

Let f be a function of two variables x and y with first partial derivatives f_x and f_y. Then the second partial derivatives of f are f_{xx}, f_{xy}, f_{yx}, and f_{yy}, where

1. f_{xx} is the partial derivative of f_x with respect to x;

2. f_{xy} is the partial derivative of f_x with respect to y;

3. f_{yx} is the partial derivative of f_y with respect to x;

4. f_{yy} is the partial derivative of f_y with respect to y.

Note that if we read the subscripts that label a second partial derivative from left to right, they tell us how to find the second partial; for instance, the notation f_{xy} tells us that we are to find the first partial of f with respect to x and then differentiate it with respect to y.

EXAMPLE 6 Let f be defined by

$$f(x, y) = 5x + y^2 + 2x^3y^4.$$

Then

$$f_x = 5 + 6x^2y^4,$$

so

$$f_{xx} = 12xy^4 \qquad \text{and} \qquad f_{xy} = 24x^2y^3.$$

Similarly,

$$f_y = 2y + 8x^3y^3$$

and thus

$$f_{yx} = 24x^2y^3 \qquad \text{and} \qquad f_{yy} = 2 + 24x^3y^2.$$

Notice that the **mixed partials** f_{xy} and f_{yx} in Example 6 were the same. This is not the case for all functions, but it will be true for all the functions we use in this book.

The extension of the concept of second partial derivatives to functions of more than two variables is straightforward.

EXAMPLE 7 Let

$$g(x, y, z) = x^2y^3 + xyz + z^2.$$

Then

$$g_x = 2xy^3 + yz, \qquad g_y = 3x^2y^2 + xz,$$

and

$$g_z = xy + 2z.$$

Therefore there are nine second partial derivatives:

$$\begin{array}{lll} g_{xx} = 2y^3, & g_{xy} = 6xy^2 + z, & g_{xz} = y, \\ g_{yx} = 6xy^2 + z, & g_{yy} = 6x^2y, & g_{yz} = x, \\ g_{zx} = y, & g_{zy} = x, & g_{zz} = 2. \end{array}$$

Notice that mixed partials that have the same subscripts are equal:

$$g_{xy} = g_{yx}, \quad g_{xz} = g_{zx}, \quad \text{and} \quad g_{yz} = g_{zy}.$$

An alternative notation for second partial derivatives is as follows:

$$f_{xx} = \frac{\partial}{\partial x}\left(\frac{\partial f}{\partial x}\right) = \frac{\partial^2 f}{\partial x^2}, \qquad f_{xy} = \frac{\partial}{\partial y}\left(\frac{\partial f}{\partial x}\right) = \frac{\partial^2 f}{\partial y\, \partial x},$$

$$f_{yx} = \frac{\partial}{\partial x}\left(\frac{\partial f}{\partial y}\right) = \frac{\partial^2 f}{\partial x\, \partial y}, \qquad f_{yy} = \frac{\partial}{\partial y}\left(\frac{\partial f}{\partial y}\right) = \frac{\partial^2 f}{\partial y^2}.$$

Notice that when we use this notation, it is the order of the variables in the denominator from right to left that tells us how to find the second partial:

$$\frac{\partial^2 f}{\partial x\, \partial y} \text{ is the partial derivative of } \frac{\partial f}{\partial y} \text{ with respect to } x.$$

EXAMPLE 8 Let

$$f(x, y) = e^y + xy^2.$$

Then

$$f_y = e^y + 2xy$$

and

$$\frac{\partial^2 f}{\partial x\, \partial y} = \frac{\partial}{\partial x}(f_y) = \frac{\partial}{\partial x}(e^y + 2xy) = 2y.$$

EXAMPLE 9 Let

$$z = x^2 + \ln xy.$$

Then

$$\frac{\partial z}{\partial x} = 2x + \frac{1}{xy}y = 2x + \frac{1}{x},$$

so

$$\frac{\partial^2 z}{\partial x^2} = 2 - \frac{1}{x^2} \quad \text{and} \quad \frac{\partial^2 z}{\partial y\, \partial x} = 0.$$

Third and higher-order partial derivatives are found in the obvious manner.

EXAMPLE 10 Find the third partial derivative f_{xyx} for

$$f(x, y) = 3y - 2xy^2 + y \ln x.$$

The notation f_{xyx} tells us that we must differentiate first with respect to x, then with respect to y, and finally with respect to x again. Thus,

$$f_x = -2y^2 + \frac{y}{x},$$

$$f_{xy} = -4y + \frac{1}{x},$$

and

$$f_{xyx} = -\frac{1}{x^2}.$$

7.3 EXERCISES

First Partial Derivatives

1. Find f_x and f_y if $f(x, y) = 5x - 7y + 12$.

2. Find f_x and f_y if $f(x, y) = x^2y^3 - 2xy^2 - 1$.

3. Find $\frac{\partial f}{\partial x}$ and $\frac{\partial f}{\partial y}$ if $f(x, y) = 3x + \frac{2x^2}{y^2} + 6$.

4. Find $\frac{\partial f}{\partial x}$ and $\frac{\partial f}{\partial y}$ if $f(x, y) = 5x^3 + 7y^4 - 3x^2y^2 + 5xy - 2x + y$.

5. Find g_x and g_y if $g(x, y) = \sqrt{2xy}$.

6. Find $\frac{\partial z}{\partial x}$ and $\frac{\partial z}{\partial y}$ if $z = \sqrt{x^2 - y^2}$.

7. Find $\frac{\partial h}{\partial u}$ and $\frac{\partial h}{\partial v}$ if $h(u, v) = (u^2 + uv + v^3)^4$.

8. Find f_r and f_s if $f(r, s) = e^{2r} - \ln s$.

9. Find $\frac{\partial z}{\partial x}$ and $\frac{\partial z}{\partial y}$ if $z = x \ln y^2$.

10. Find $\frac{\partial s}{\partial p}$ and $\frac{\partial s}{\partial q}$ if $s = 2p^2q^4 - 5pq + 2$.

11. Find $\frac{\partial w}{\partial x}$ and $\frac{\partial w}{\partial y}$ if $w = \frac{2x - y}{x + y}$.

12. Find f_x, f_y, and f_z if $f(x, y, z) = xz + x^2z^2 + yz^3 - 5y$.

13. Find h_x, h_y, and h_z if $h(x, y, z) = e^{2xz} - \ln yz + 1$.

14. Find $f_x(1, 3)$ and $f_y(-2, 1)$ if $f(x, y) = 2xy$.

15. Find $f_x(0, 0)$ and $f_y(1, 1)$ if $f(x, y) = 2x + y + 4xy - x^2 + y^2$.

16. Find $f_x(1, 0)$ and $f_y(0, 1)$ if $f(x, y) = 2e^x - 3e^{-y}$.

17. Evaluate $\frac{\partial f}{\partial x}$ and $\frac{\partial f}{\partial y}$ at (8, 3) if $f(x, y) = \sqrt{2x + 3y}$.

18. Evaluate $\frac{\partial z}{\partial x}$ and $\frac{\partial z}{\partial y}$ at $(0, e)$ if $z = e^{-x} \ln y$.

19. Find $\left.\frac{\partial w}{\partial x}\right|_{(1,-1,1)}$ and $\left.\frac{\partial w}{\partial y}\right|_{(1,-1,1)}$ if $w = \sqrt{x^2 + y^2} + 4xz$.

20. Find $\left.\frac{\partial p}{\partial u}\right|_{(0,0,2)}$, $\left.\frac{\partial p}{\partial v}\right|_{(0,0,2)}$, and $\left.\frac{\partial p}{\partial w}\right|_{(0,0,2)}$ if $p = (3u^2w^2 + 4v^2w - 2u + 3v + w + 1)^4$.

Higher Partial Derivatives

21. If $f(x, y) = 2x + 3y^2 + 1$, find f_{xx}, f_{xy}, f_{yx}, and f_{yy}.

22. If $f(x, y) = 4x^2 + 3xy - y^2$, find f_{xx}, f_{xy}, f_{yx}, and f_{yy}.

23. If $f(x, y) = \frac{2x}{3y^2} + 1$, find f_{xx}, f_{xy}, f_{yx}, and f_{yy}.

24. If $z = y^2\sqrt{x} - 2x^3 + 3$, find $\frac{\partial^2 x}{\partial x^2}$, $\frac{\partial^2 z}{\partial y\, \partial x}$, $\frac{\partial^2 z}{\partial x\, \partial y}$, and $\frac{\partial^2 z}{\partial y^2}$.

25. If $z = \frac{4x^4}{y} + \frac{2y^2}{x^2}$, find $\frac{\partial^2 z}{\partial x^2}$, $\frac{\partial^2 z}{\partial y\, \partial x}$, $\frac{\partial^2 z}{\partial x\, \partial y}$, and $\frac{\partial^2 z}{\partial y^2}$.

26. If $f(x, y, z) = x^2 + 3xyz + 2yz + z^2$, find all second partial derivatives of f.

27. If $f(x, y, z) = ye^x + 2 \ln x + x \ln z$, find all second partial derivatives of f.

28. If $f(x, y) = 10x^2y^3 + x + 2y^3$, find f_{xx}, f_{xyx}, and f_{yyx}.

29. If $f(x, y) = e^{xy}$, find f_{xyxy} and f_{xyyy}.

30. If $z = \ln 2x + 3 \ln y$, find $\frac{\partial^2 z}{\partial x^2}$, $\frac{\partial^3 z}{\partial x\, \partial y\, \partial x}$, and $\frac{\partial^3 z}{\partial x\, \partial y^2}$.

31. If $f(x, y) = 5x^2 - 3xy + y$, find $f_{xx}(1, 1)$, $f_{xy}(2, 2)$, $f_{yx}(3, -5)$, and $f_{yy}(-2, 7)$.

32. If $f(x, y) = 6x + 2ye^x$, find $f_{xx}(2, 0)$, $f_{xy}(1, -1)$, $f_{yx}(-2, 1)$, and $f_{yy}(0, 2)$.

33. If $f(x, y) = x^2y^3 + x^4 + y^4$, find $f_{xyx}(1, 1)$ and $f_{yyx}(-1, -1)$.

34. If $f(x, y) = xy + \ln xy$, find $f_{xxyy}(2, 2)$ and $f_{xyxx}(2, 2)$.

7.4 APPLICATIONS OF PARTIAL DERIVATIVES

Just as the derivative $f'(a)$ of a function of one variable is the slope of the line tangent to the graph of f at the point $(a, f(a))$, so the partial derivatives $f_x(a, b)$ and $f_y(a, b)$ of a function of two variables are slopes of tangent lines to the graph of f at the point $(a, b, f(a, b))$. To be specific:

Slopes of Tangent Lines to Surfaces

Let f be a function of the two variables x and y, and let (a, b) be a point in the domain of f.

1. If $f_x(a, b)$ exists, it is the slope of the line parallel to the xz-plane and tangent to the surface $z = f(x, y)$ at the point $(a, b, f(a, b))$.
2. If $f_y(a, b)$ exists, it is the slope of the line parallel to the yz-plane and tangent to the surface $z = f(x, y)$ at the point $(a, b, f(a, b))$.

See Figure 7–8.

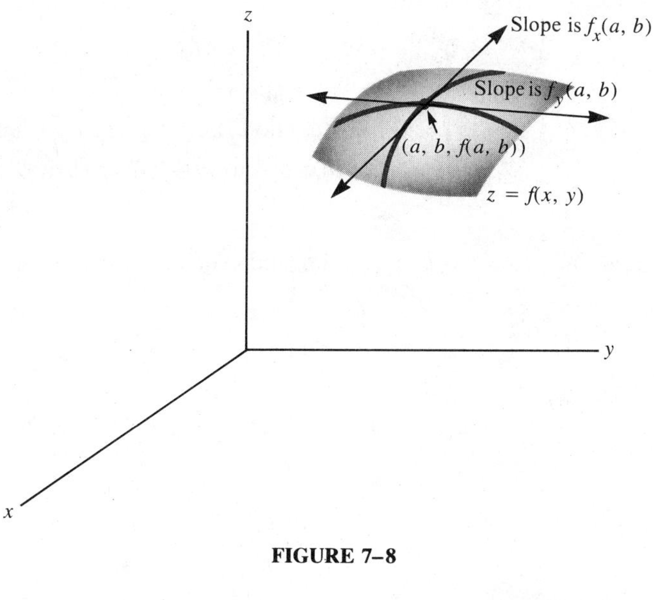

FIGURE 7–8

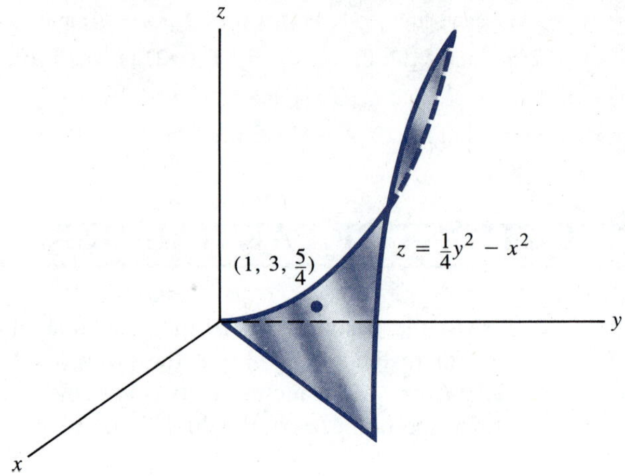

FIGURE 7–9

EXAMPLE 1 Let f be defined by
$$f(x, y) = \frac{1}{4}y^2 - x^2.$$

A portion of the surface
$$z = \frac{1}{4}y^2 - x^2$$

is shown in Figure 7–9. When $x = 1$ and $y = 3$, we have $z = 5/4$, so the point $(1, 3, 5/4)$ is on the surface. Since
$$f_x(x, y) = -2x,$$

the slope of the line tangent to the surface at $(1, 3, 5/4)$ and parallel to the xz-plane is
$$f_x(1, 3) = -2(1) = -2.$$

Thus since $f_x(1, 3)$ is negative, at the point $(1, 3, 5/4)$ the surface bends downward in the positive x-direction. See Figure 7–10. Similarly, since
$$f_y(x, y) = \frac{1}{2}y,$$

the slope of the line tangent to the surface at $(1, 3, 5/4)$ and parallel to the yz-plane is
$$f_y(1, 3) = \frac{1}{2}(3) = \frac{3}{2},$$

which is positive; hence at $(1, 3, 5/4)$ the surface bends upward in the positive y-direction.

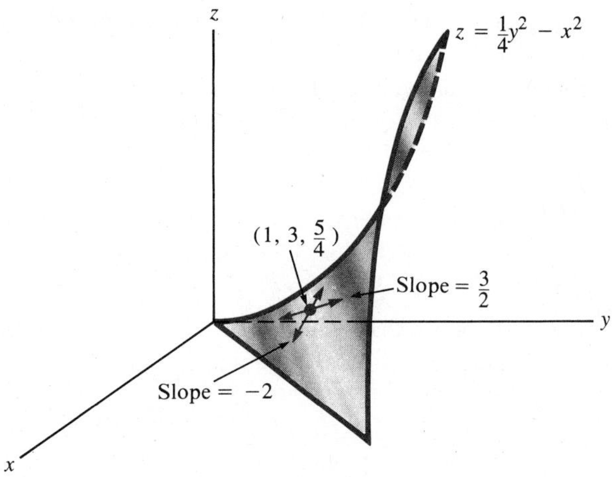

FIGURE 7–10

Consider Figure 7–8 again: as the figure indicates, the slope $f_x(a, b)$ may be interpreted as the instantaneous rate of change of $z = f(x, y)$ at the point $(a, b, f(a, b))$ in the positive x-direction; that is, $f_x(a, b)$ is the instantaneous rate of change of f with respect to x at $(a, b, f(a, b))$ when y is held fixed at the value $y = b$. Similarly, $f_y(a, b)$ is the instantaneous rate of change of f with respect to y at $(a, b, f(a, b))$ when x is held fixed at the value $x = a$. Thus for functions of two variables, partial derivatives can be interpreted as rates of change, and this remains true for functions of more than two variables.

Partial Derivatives as Rates of Change

Let f be a function of several variables, and let x be any one of the independent variables of f. Then f_x gives the **instantaneous rate of change of f with respect to x** when all the other independent variables of f are fixed. Thus, when all the other independent variables of f are fixed, a small change in x will lead to a change of approximately f_x in f.

We illustrate some applications of this point of view with examples.

EXAMPLE 2 Two bactericides are tested in combination, and it is found that when x units of bactericide X and y units of bactericide Y are used together, they will destroy

$$f(x, y) = 5x + 4y$$

million bacteria. Thus if y is held fixed, the rate of change of $f(x, y)$ with respect to x is

$$f_x(x, y) = 5.$$

The interpretation is that if the number of units y of bactericide Y is held fixed, each additional unit of bactericide X used will kill approximately 5 million bacteria. (Actually, this is so no matter whether y is fixed or not, since in this example y does not appear in the partial derivative $f_x(x, y)$.) Similarly, since

$$f_y(x, y) = 4,$$

if the number of units of bactericide X is held fixed, then each additional unit of bactericide Y used will destroy approximately 4 million bacteria.

EXAMPLE 3 Suppose that if a firm produces and sells x units of product A, y units of product B, and z units of product C, its profit is

$$P(x, y, z) = 2\sqrt{x} + 5y + 4\sqrt{z} + xyz$$

dollars. Then

$$P_x(x, y, z) = \frac{1}{\sqrt{x}} + yz$$

is the firm's **marginal profit with respect to x.** If y and z are held fixed, an additional unit of product A made and sold will increase the firm's profit by approximately

$$\frac{1}{\sqrt{x}} + yz$$

dollars. For instance, if $x = 16$, $y = 10$, and $z = 25$, then the firm's profit is

$$P(16, 10, 25) = 4068$$

dollars. If y and z are held fixed at 10 and 25, respectively, then making and selling an additional unit of product A (the 17th one) will increase profit by approximately

$$P_x(16, 10, 25) = \frac{1}{\sqrt{16}} + (10)(25) = \$250.25.$$

Similarly, since

$$P_y(x, y, z) = 5 + xz \qquad \text{and} \qquad P_z(x, y, z) = \frac{2}{\sqrt{z}} + xy,$$

holding x and z fixed at 16 and 25, respectively, and producing an additional unit of B (the 11th one) will contribute approximately

$$P_y(16, 10, 25) = 5 + (16)(25) = \$405$$

to profit; holding x and y fixed at 16 and 10, respectively, and producing an additional unit of C (the 26th one) will contribute approximately

$$P_z(16, 10, 25) = \frac{2}{\sqrt{25}} + (16)(10) = \$160.40$$

to profit. Therefore if the firm is currently making 16 units of A, 10 of B, and 25 of C and has limited resources for expanding production , it would be wise to use its resources to increase the production of B.

EXAMPLE 4 Two products are said to be **competitive** if a *decrease* in the quantity demanded of one can lead to an *increase* in the quantity demanded of the other. (This occurs if the products can be substituted for each other. For instance, beef products and pork products are competitive; if the price of beef goes up while the price of pork remains fixed, demand for beef will decrease and demand for pork will increase, as consumers substitute pork for beef. Similarly, if the price of pork goes up while the price of beef remains unchanged, demand for pork will decrease while demand for beef will increase.) On the other hand, if a *decrease* in the quantity demanded of one product can lead to a *decrease* in the quantity demanded of the other, then the products are said to be **complementary.** (Automobiles and automobile tires are examples of complementary products.)

We can tell whether products are competitive or complementary by analyzing the partial derivatives of their demand functions. To see how this is done, suppose that the price of butter is x dollars per pound and that the price of margarine is y dollars per pound. Let

$$B(x, y) = 10{,}000 - 200x + 400y \qquad \text{and} \qquad M(x, y) = 8000 + 300x - 500y$$

be the demand functions for butter and margarine, respectively. Then

$$B_x = -200, \quad B_y = 400, \quad M_x = 300, \quad \text{and} \quad M_y = -500.$$

Now, B_x is negative, so an increase in the price x of butter while the price y of margarine is held fixed will lead to a decrease in the quantity of butter demanded. But since M_x is positive, an increase in x when y is held fixed will also lead to an increase in the quantity of margarine demanded. Similarly, since M_y is negative and B_y is positive, an increase in y while x is held fixed leads to a decrease in the quantity of margarine demanded and an increase in the quantity of butter demanded. Thus butter and margarine are competitive products.

EXAMPLE 5 If a firm uses x worker-hours per week and y dollars of capital per week, it can produce

$$f(x, y) = 0.1x^{3/4}y^{1/2}$$

units per week. A function such as this which relates output to input is called a **production function.** In this case, if the firm has 1296 worker-hours and \$2500 in capital available this week, it can produce

$$f(1296{,}2500) = 0.1(1296)^{3/4}(2500)^{1/2} = 1080$$

units. The partial derivative f_x is called the **marginal productivity of labor;** the partial derivative f_y is the **marginal productivity of capital.** Here

$$f_x(x, y) = 0.075x^{-1/4}y^{1/2} \qquad \text{and} \qquad f_y(x, y) = 0.05x^{3/4}y^{-1/2}.$$

Thus if $x = 1296$ and $y = 2500$,

$$f_x(1296, 2500) = 0.075(1296)^{-1/4}(2500)^{1/2} = 0.075\left(\frac{1}{6}\right)(50) = 0.625$$

and

$$f_y(1296, 2500) = 0.05(1296)^{3/4}(2500)^{-1/2} = 0.05(216)\left(\frac{1}{50}\right) = 0.216.$$

Therefore if capital remains fixed at \$2500 per week, the availability of an additional worker-hour each week would result in an increase in production of approximately 0.625 units per week. If labor remains fixed at 1296 worker-hours per week, the availability of an additional dollar of capital each week would result in an increase in production of approximately 0.216 units per week.

7.4 EXERCISES

1. If $f(x, y) = 2x^2 + 4y^2 + 10$, find the slope of the line tangent to the graph of f at $(2, 3, 54)$ and parallel to the

(a) xz-plane **(b)** yz-plane

2. If $f(x, y) = 2xy + x^2y$, find the slope of the line tangent to the graph of f at $(1, -1, -3)$ and parallel to the

(a) xz-plane **(b)** yz-plane

3. If $z = xy^2 + 2x + 3y + 1$, find the instantaneous rate of change of z at the point $(0, 2, 7)$ in the positive

(a) x-direction **(b)** y-direction

4. If $z = x^2y^3 + 2x - 3y$, find the instantaneous rate of change of z at the point $(1, -1, 4)$ in the positive

(a) x-direction **(b)** y-direction

Management

5. If a firm makes and sells x units of product A and y units of product B, its cost function is given by

$$C(x, y) = 20x + 14y + 200$$

and its revenue function by

$$R(x, y) = 25x + 20y.$$

(a) Find the firm's marginal cost with respect to x and with respect to y. Interpret your results.

(b) Find the firm's marginal revenue with respect to x and with respect to y. Interpret your results.

(c) Find the firm's marginal profit with respect to x and with respect to y. Interpret your results.

6. A company that makes and sells x units of product X and y units of product Y has cost function given by

$$C(x, y) = \sqrt{xy + 25}$$

and revenue function given by

$$R(x, y) = 2x + y.$$

(a) Find the company's cost, revenue, and profit if $x = 20$ and $y = 10$.

(b) Find the company's marginal cost, marginal revenue, and marginal profit with respect to x and y when $x = 20$ and $y = 10$. If y is held fixed at 10, what is the effect of making and selling one more unit of X? If x is held fixed at 20, what is the effect of making and selling one more unit of Y?

7. If a corporation's cost and revenue functions are given by

$$C(x, y) = 5x^2 + 3y^2 + xy + 100$$

and

$$R(x, y) = 6x^2 + 5y^2,$$

respectively, find and interpret $C_x(100, 200)$, $C_y(100, 200)$; $R_x(100, 200)$, $R_y(100, 200)$; $R_x(100, 200) - C_x(100, 200)$; and $R_y(100, 200) - C_y(100, 200)$.

8. Suppose the demand function for commodity V is given by

$$V(x, y) = 20{,}000 - 500x + 300y$$

and that for commodity W by

$$W(x, y) = 30{,}000 + 800x - 400y,$$

where x is the unit price of V and y is the unit price of W. Find V_x, V_y, W_x, and W_y, and interpret your results. Are the commodities competitive or complementary?

9. Repeat Exercise 8 if

$$V(x, y) = 20{,}000 - 200x - 100y$$

and

$$W(x, y) = 30{,}000 - 400x - 300y.$$

10. Using the notation of Exercises 8 and 9, let

$$V(x, y) = 200{,}000 - 200x^2 + 50xy - 10y^2$$

and

$$W(x, y) = 300{,}000 - 400y^2 + 100xy - 30x^2.$$

(a) Find $V_x(10, 10)$, $V_y(10, 10)$, $W_x(10, 10)$, and $W_y(10, 10)$, and interpret your results. Are the commodities competitive at this price?

(b) Find $V_x(10, 20)$, $V_y(10, 20)$, $W_x(10, 20)$, and $W_y(10, 20)$, and interpret your results. Are the commodities competitive at these prices?

11. A firm's production function is given by

$$f(x, y) = 200\sqrt{xy},$$

where x is labor input in worker-hours per week, y is capital input in dollars per week, and output is in units per week.

(a) Find the firm's weekly output if it uses 1600 worker-hours and \$10,000 of capital per week.

(b) Find $f_x(1600, 10{,}000)$, and $f_y(1600, 10{,}000)$, and interpret your results.

12. Using the notation of Exercise 11, suppose the production function is defined by

$$f(x, y) = 12x^{1/4}y^{3/4}.$$

Find $f(81, 625)$, $f_x(81, 625)$, and $f_y(81, 625)$, and interpret your results.

13. Suppose a production function is given by

$$f(x, y) = xe^{0.01y},$$

where x is labor input in thousands of worker-hours per week, y is capital input in thousands of dollars per week, and output is in thousands of units per week. Evaluate the first partial derivatives of f with respect to x and y at

(a) $x = y = 50$ **(b)** $x = y = 100$ **(c)** $x = y = 200$

and interpret your results.

Life Science

14. If a farmer irrigates a field x times, weeds it y times, and fertilizes it z times during the growing season, the crop yield will be

$$w = 4x + 5y + 2z + 400$$

bushels. Find the first partial derivatives of w with respect to x, y, and z, and interpret your results.

15. The number of bacteria present in a culture t hours after an experiment has begun depends on t, on the amount of nutrient provided, and on the temperature at which the experiment is run. Suppose that

$$p = 100e^{0.2x} + 2xT + 10T^2 + 10{,}000t,$$

where p is the number of bacteria in the culture, x is nutrient provided in grams, and T is temperature in degrees Celsius. Evaluate the first partial derivatives of p with respect to x, T, and t when $x = 2$, $T = 20$, and $t = 3$, and interpret your results.

16. Suppose that if there are x rabbits for coyotes to prey on and y hunters to kill the coyotes, then the size of the coyote population is

$$f(x, y) = 4\frac{\sqrt{x}}{y}$$

individuals. Find $f_x(400, 2)$ and $f_y(400, 2)$ and interpret your results.

Social Science

17. Ridership on a city's public transportation system averages

$$f(x, y) = \frac{0.1x}{1 + \ln y}$$

thousand people per day, where x is the population of the city in thousands and y is the price of a ticket. The population of the city is currently 100,000 people and the price of a ticket is \$1. Find and interpret $f_x(100, 1)$ and $f_y(100, 1)$.

18. It is estimated that the proportion p of the population who apply for welfare benefits in a community during a year depends on the unemployment rate x and the average years of education y in the community according to the equation

$$p(x, y) = e^{0.08x} - e^{0.01y}.$$

Here $5 \leq x \leq 20$ and $10 \leq y \leq 15$. Find $p_x(8, 12)$ and $p_y(8, 12)$ and interpret your results.

7.5 OPTIMIZATION

In this section we discuss relative maxima, relative minima, and saddle points of functions of two variables and show how to find such points using an analog of the second derivative test for functions of one variable that we introduced in Section 3.4. (Most of the results and techniques presented in this section can be extended to functions of more than two variables, but since the two-variable case illustrates the essential ideas with the fewest possible complications, we will confine ourselves to it.)

We begin by indicating what it means for a function of two variables to have a relative maximum or minimum at a point. If f is such a function, we say that f has a **relative maximum** at the point $(a, b, f(a, b))$ if the surface $z = f(x, y)$ has a peak at $(a, b, f(a, b))$, as shown in Figure 7–11; it has a **relative minimum** at $(a, b, f(a, b))$ if the surface has a pit at $(a, b, f(a, b))$, as shown in Figure 7–12. Also, Figure 7–13 depicts a function that has a **saddle point** at $(a, b, f(a, b))$.

Suppose that a function of two variables f has a relative maximum at $(a, b, f(a, b))$ and that the line parallel to the xz-plane and tangent to the surface $z = f(x, y)$ at $(a, b, f(a, b))$ exists, as in Figure 7–14. Then this line must be horizontal, and its slope, which is $f_x(a, b)$, must be zero. Similarly, the line tangent to the surface at $(a, b, f(a, b))$ and parallel to the yz-plane must also have a slope of zero, so $f_y(a, b)$ must be zero. Therefore if f has a relative maximum at $(a, b, f(a, b))$ and $f_x(a, b)$ and $f_y(a, b)$ exist, then

$$f_x(a, b) = 0 \qquad \text{and} \qquad f_y(a, b) = 0.$$

Similar reasoning holds for relative minima and saddle points.

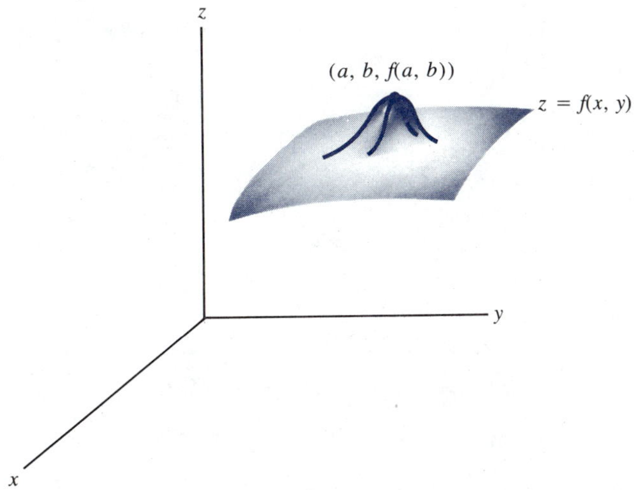

FIGURE 7–11

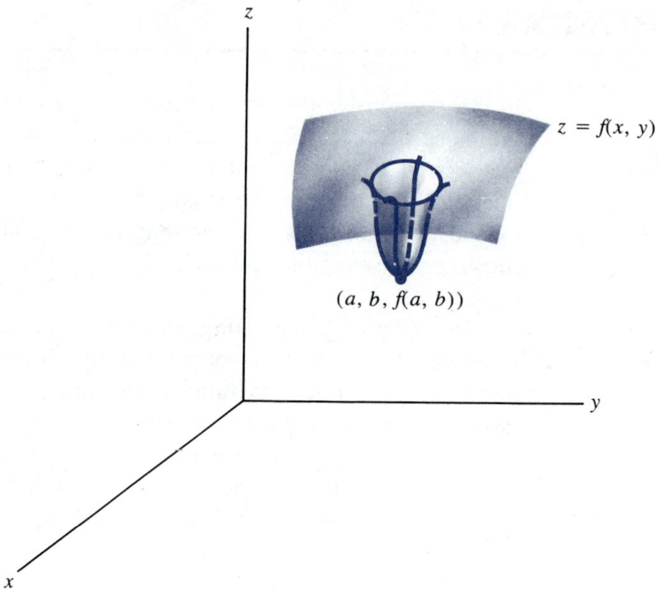

FIGURE 7–12

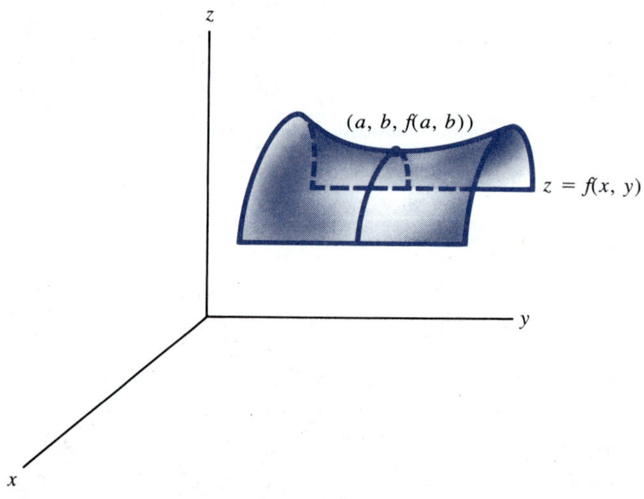

FIGURE 7–13

Relative Maxima, Relative Minima, and Saddle Points

Let f be a function of x and y. If

$$f_x(a, b) = 0 \qquad \text{and} \qquad f_y(a, b) = 0,$$

then f has either a relative maximum, a relative minimum, or a saddle point at $(a, b, f(a, b))$.

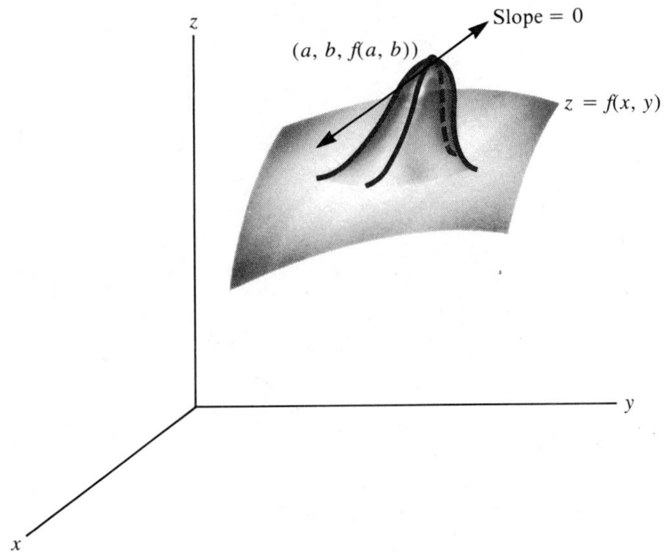

FIGURE 7–14

Thus we can find at least some of the points at which such a function has a relative maximum, a relative minimum, or a saddle point by setting $f_x(x, y) = 0$ and $f_y(x, y) = 0$ and then solving these equations simultaneously for x and y. (We say *some* of the points because there may be relative maxima, relative minima, or saddle points at places where f_x or f_y fail to exist.) Unfortunately, knowing that $x = a$, $y = b$ is a solution of these equations is not enough to tell us whether $(a, b, f(a, b))$ yields a relative maximum, a relative minimum, or a saddle point.

EXAMPLE 1 Consider the function f defined by

$$f(x, y) = x^2 - 2x + y^2 - 4y + 6.$$

Since

$$f_x = 2x - 2 \qquad \text{and} \qquad f_y = 2y - 4,$$

setting $f_x = 0$ and $f_y = 0$ yields

$$2x - 2 = 0 \qquad \text{and} \qquad 2y - 4 = 0,$$

and thus

$$x = 1 \qquad \text{and} \qquad y = 2.$$

Therefore $f(1, 2, f(1, 2)) = (1, 2, 1)$ is either a relative maximum, a relative minimum, or a saddle point, but we cannot tell which of these it is unless we know something else about the function. In this case, the graph of the function is shown in Figure 7–15, so it is clear from the graph that the point (1, 2, 1) is a relative minimum.

Fortunately, there is a method of testing a point (a, b) for which $f_x(a, b) = 0$ and $f_y(a, b) = 0$ to find out whether it yields a relative maximum, a relative minimum, or a saddle point. The method, which we will not prove, is called the **second derivative test.**

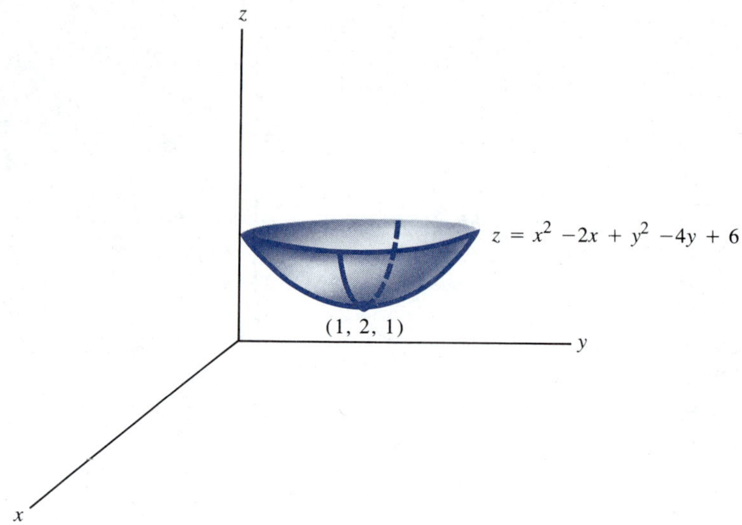

FIGURE 7–15

The Second Derivative Test

To find relative maxima, relative minima, and saddle points for a function f of x and y, proceed as follows:

1. Find the partial derivatives f_x, f_y, f_{xx}, f_{xy}, and f_{yy}.
2. Set $f_x = 0$ and $f_y = 0$ and solve these equations simultaneously for x and y to obtain all pairs of numbers (a, b) such that $f_x(a, b) = 0$ and $f_y(a, b) = 0$.
3. Let $D(x, y) = f_{xy}^2 - f_{xx}f_{yy}$. Then
   a. If $D(a, b) < 0$ and $f_{xx}(a, b) < 0$, the function has a relative maximum at $(a, b, f(a, b))$.
   b. If $D(a, b) < 0$ and $f_{xx}(a, b) > 0$, the function has a relative minimum at $(a, b, f(a, b))$.
   c. If $D(a, b) > 0$, the function has a saddle point at $(a, b, f(a, b))$.
   d. If $D(a, b) = 0$, the test fails to yield information about the behavior of the function at $(a, b, f(a, b))$.

EXAMPLE 2 Let us apply the second derivative test to the function f defined by

$$f(x, y) = -x^2 + 2xy + 2y^2 - 4x - 14y + 1.$$

1. We have

$$f_x = -2x + 2y - 4, \quad f_y = 2x + 4y - 14,$$

$$f_{xx} = -2, \quad f_{xy} = 2, \quad \text{and} \quad f_{yy} = 4.$$

2. We solve the equations

$$-2x + 2y - 4 = 0 \qquad \text{and} \qquad 2x + 4y - 14 = 0$$

simultaneously for x and y: from the first equation,

$$2y = 2x + 4,$$

or

$$y = x + 2.$$

Substituting this expression for y into the second equation, we obtain

$$2x + 4(x + 2) - 14 = 0,$$

or

$$6x = 6.$$

Therefore $x = 1$ and $y = 1 + 2 = 3$, and we must test the single pair $(1, 3)$.

3. To test the pair $(1, 3)$, we must evaluate

$$D(x, y) = f_{xy}^2 - f_{xx}f_{yy} = 2^2 - (-2)(4) = 12$$

at $(1, 3)$. But if $D(x, y) = 12$, then

$$D(1, 3) = 12 > 0,$$

so the function has a saddle point at $(1, 3, f(1, 3)) = (1, 3, -22)$.

EXAMPLE 3 Let us apply the second derivative test to the function whose defining equation is

$$f(x, y) = x^4 - 2x^2 + y^2 - 2y + 1.$$

1. We have

$$f_x = 4x^3 - 4x, \quad f_y = 2y - 2,$$
$$f_{xx} = 12x^2 - 4, \quad f_{xy} = 0, \quad \text{and} \quad f_{yy} = 2.$$

2. We solve

$$4x^3 - 4x = 0$$
$$2y - 2 = 0$$

or

$$4x(x - 1)(x + 1) = 0$$
$$2(y - 1) = 0$$

simultaneously for x and y, obtaining

$$x = 0, \quad x = 1, \quad x = -1, \quad \text{and} \quad y = 1.$$

Therefore we must test the pairs $(0, 1)$, $(1, 1)$, and $(-1, 1)$.

3. We test the pairs (0, 1), (1, 1), and (−1, 1) using

$$D(x, y) = 0^2 - (12x^2 - 4)(2) = -8(3x^2 - 1).$$

Since

$$D(0, 1) = -8(3(0)^2 - 1) = 8 > 0,$$

the function has a saddle point at $(0, 1, f(0, 1)) = (0, 1, 0)$.

Since

$$D(1, 1) = -8(3(1)^2 - 1) = -16 < 0$$

and

$$f_{xx}(1, 1) = 12(1)^2 - 4 = 8 > 0,$$

the function has a relative minimum at $(1, 1, f(1, 1)) = (1, 1, -1)$.

Since

$$D(-1, 1) = -8(3(-1)^2 - 1) = -16 < 0$$

and

$$f_{xx}(-1, 1) = 12(-1)^2 - 4 = 8 > 0,$$

the function has another relative minimum at $(-1, 1, -1)$.

EXAMPLE 4 Let us apply the second derivative test to the function with defining equation

$$f(x, y) = x^4 + (x - y)^4.$$

1. We have

$$f_x = 4x^3 + 4(x - y)^3, \quad f_y = -4(x - y)^3,$$
$$f_{xx} = 12x^2 + 12(x - y)^2, \quad f_{xy} = -12(x - y)^2,$$

and

$$f_{yy} = 12(x - y)^2.$$

2. Solving

$$4x^3 + 4(x - y)^3 = 0$$
$$-4(x - y)^3 = 0$$

simultaneously for x and y, we obtain

$$x = y = 0$$

as the only solution. (Check this.)

3. Since

$$\begin{aligned} D(x, y) &= (-12(x - y)^2)^2 - (12x^2 + 12(x - y)^2)(12(x - y)^2) \\ &= 144(x - y)^4 - (144x^2(x - y)^2 + 144(x - y)^4) \\ &= -144x^2(x - y)^2, \end{aligned}$$

we have $D(0, 0) = 0$. Therefore the second derivative test fails; it cannot tell us whether $(0, 0, f(0, 0)) = (0, 0, 0)$ is a relative maximum, a relative minimum, or a saddle point. In a case such as this we examine the behavior of the function near the point in question. Since

$$f(0, 0) = 0$$

while

$$f(x, y) = x^4 + (x - y)^4 > 0 \qquad \text{if} \qquad (x, y) \neq (0, 0),$$

it follows that the point $(0, 0, 0)$ is a relative minimum.

Now we can use the second derivative test to solve optimization problems involving functions of two variables.

EXAMPLE 5 Suppose we wish to construct a rectangular box with a volume of 1000 cubic inches. Let us minimize the cost of the box if the material for it costs 10 cents per square inch.

The box will look like this:

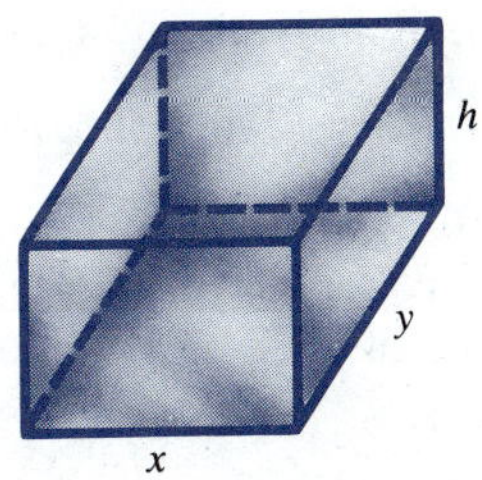

Its volume is $xyh = 1000$ cubic inches, and its cost is

$$C = 2(0.10xy) + 2(0.10xh) + 2(0.10yh)$$

dollars. Since

$$h = \frac{1000}{xy},$$

the cost function may be written as a function of two variables:

$$C(x, y) = 0.2xy + 0.2x\frac{1000}{xy} + 0.2y\frac{1000}{xy},$$

or

$$C(x, y) = 0.2xy + \frac{200}{y} + \frac{200}{x}.$$

The second derivative test shows that $C(x, y)$ has a minimum at $(10, 10, 60)$. (Check this.) Therefore the dimensions of the box of minimum cost are

$$x = 10 \text{ inches}, \quad y = 10 \text{ inches}, \quad h = 10 \text{ inches},$$

and the minimum cost is \$60.

EXAMPLE 6 A producer who sells the same product to different classes of customers at different prices is said to practice **price discrimination.** To see how this can occur, suppose that the demand functions for a certain product are different in two regions. For instance, suppose that in region A the demand function for a product is given by

$$p_A = 100 - \frac{x}{5},$$

but in region B it is given by

$$p_B = 200 - \frac{y}{15}.$$

Here x is quantity demanded in region A, y is quantity demanded in region B, and p_A and p_B are the unit prices in the two regions.

Let us find the producer's total revenue from sales in the two regions. Since x units are sold in region A at price p_A, and y are sold in region B at price p_B, the revenue function is given by

$$R(x, y) = x\left(100 - \frac{x}{5}\right) + y\left(200 - \frac{y}{15}\right).$$

The second derivative test shows that the revenue function has a maximum at

$$(250, 1500, R(250, 1500)) = (250, 1500, 162{,}500).$$

(Check this.) Therefore to obtain maximum revenue of \$162,500, the producer will sell 250 units in region A at a price of $p_A = \$50$ each and 1500 units in region B at a price of $p_B = \$100$ each, and thus will practice price discrimination.

7.5 EXERCISES

In Exercises 1 through 15 use the second derivative test to find relative maxima, relative minima, and saddle points for the function.

1. $f(x, y) = 2xy$
2. $f(x, y) = x^2 + y^2 - 4$
3. $f(x, y) = 9 - 2x^2 - 3y^2$
4. $f(x, y) = x - (x + y)^2$
5. $f(x, y) = 3x^2 + 2y^2 - 18x + 8y - 2$
6. $f(x, y) = 4x^2 + y^2 - 16x - 6y + 3$
7. $f(x, y) = 4x^2 + 5y^2 + 24x - 100y + 50$
8. $f(x, y) = x^3 + y^3 - 3xy$
9. $f(x, y) = x^3 - y^3 - 3xy$
10. $f(x, y) = 4x^3 + 6x^2 - 72x + 3y^4 - 44y^3 + 180y^2$
11. $f(x, y) = x^3 + 2y^2 - 8xy + 5x + 2$
12. $f(x, y) = 2y^3 + 100y + 10xy - x^2$
13. $f(x, y) = x^4 - 8x^2 + 6y^2 - 12y + 3$
14. $f(x, y) = x^4 - (x + y)^4$
15. $f(x, y) = xy(x + y - 1) - x^2$
16. Find the dimensions of the rectangular box, without a top, of minimum cost if the material for the box costs \$2 per square foot and the box must have a volume of 144 cubic feet.

17. Find the dimensions of the rectangular box, without a top, of maximum volume if the surface area of the box must be 12 square meters.

Management

18. If a firm makes and sells x units of one product and y units of another, its profit function is given by

$$P(x, y) = 20x + 30y - 2x^2 - 3y^2.$$

Find the firm's maximum profit.

19. If a company makes x units of one product and y of another, its average cost per unit is given by

$$A(x, y) = 50 + \frac{80}{x} + \frac{216}{y} + 5x + 6y.$$

Find the company's minimum average cost per unit.

20. The demand functions for a commodity in regions A and B are given by

$$p_A = 2000 - \frac{x}{20} \qquad \text{and} \qquad p_B = 1500 - \frac{y}{15},$$

respectively, with x and y the quantities demanded in the regions, and p_A, p_B the unit prices in the regions. Find the prices that will maximize revenue. Will the supplier practice price discrimination?

21. Repeat Exercise 20 if

$$p_A = 500 - \frac{x}{4} \qquad \text{and} \qquad p_B = 500 - \frac{y}{3}.$$

22. A firm can produce

$$f(x, y) = -x^2 - 1.5y^2 + 12x + 9y$$

thousand units of output using x units of labor and y units of capital. Find the firm's maximum output.

Life Science

23. If a farmer applies x tons of fertilizer and y tons of pesticide to a crop, the yield will be

$$f(x, y) = 20x + 50y - 10x^2 - 30y^2$$

bushels. How much fertilizer and how much pesticide should be applied if the crop yield is to be maximized?

24. A patient undergoing chemotherapy who receives x units of drug A and y units of drug B per day suffers z units of discomfort, where

$$z = 4x^2 + 5y^2 - 8x - 6y - 5xy + 15$$

and discomfort is measured on a scale ranging from 0 to 15. What drug dosage will minimize discomfort?

7.6 LAGRANGE MULTIPLIERS

In this section we consider the problem of finding maxima and minima for functions of several variables subject to side conditions known as **constraints.** For instance, suppose we wish to find the minimum value of

$$f(x, y) = x^2 + y^2$$

subject to the condition that

$$x + y - 2 = 0.$$

In this case the **constraint equation** $x + y - 2 = 0$ is easily solved for y in terms of x; if we do so, we find that $y = -x + 2$, and this expression may be substituted for y in $f(x, y)$ to obtain

$$f(x, y) = x^2 + (-x + 2)^2 = 2x^2 - 4x + 4.$$

But now f is a function of the single variable x, so it may be written as

$$f(x) = 2x^2 - 4x + 4,$$

and we can then use either the first or second derivative test of Chapter 3 to conclude that the function has a relative minimum of 2 when $x = 1$.

Unfortunately, when we attempt to maximize or minimize a function of several variables subject to a constraint, it is not always possible to solve the constraint equation for one of the variables in terms of the others, as we did above. Therefore we need a method for finding the maximum or minimum values of $z = f(x, y)$ subject to a constraint equation $g(x, y) = 0$ without having to solve $g(x, y) = 0$ for one of the variables. In this section we present such a method: it is known as the **Lagrange multiplier method.**

Suppose we wish to find the maximum or minimum value of $f(x ,y)$ subject to the constraint $g(x, y) = 0$. We begin by setting

$$h(x, y) = f(x, y) + \lambda g(x, y).$$

The symbol λ is the Greek letter lambda, which in this context is called the **Lagrange multiplier.** If we consider λ to be a specific number, then we know from our work in Section 7.5 that we can find relative maxima and minima for the function h by solving

$$h_x = 0 \qquad \text{and} \qquad h_y = 0$$

simultaneously for x and y. If we do this and at the same time ensure that the solutions satisfy $g(x, y) = 0$, then since

$$h(x, y) = f(x, y) \qquad \text{when} \qquad g(x, y) = 0,$$

it will follow that the solutions that maximize or minimize h and satisfy the constraint equation will also maximize or minimize f and satisfy the constraint equation.

But now let us consider λ to be another independent variable, so that h is a function of x, y, and λ. Then

$$h(x, y, \lambda) = f(x, y) + \lambda g(x, y),$$

so h_x and h_y are the same as before. Taking the first partial derivative of h with respect to λ shows that

$$h_\lambda = g(x, y).$$

Therefore we can ensure that the solutions of $h_x = 0$ and $h_y = 0$ satisfy the constraint equation if we take λ to be an independent variable and require that $h_\lambda = 0$.

The Lagrange Multiplier Method

Let f be a function of x and y. To maximize or minimize f subject to the constraint $g(x, y) = 0$:

1. Let

$$h(x, y, \lambda) = f(x, y) + \lambda g(x, y)$$

and find h_x, h_y, and $h_\lambda = g(x, y)$.

2. Set $h_x = 0$, $h_y = 0$, and $h_\lambda = 0$, and solve these equations simultaneously for x and y.

3. The result of step 2 will be pairs of numbers $x = a$, $y = b$ that maximize or minimize f and satisfy the constraint $g(x, y) = 0$. To decide whether a particular pair (a, b) yields a maximum or a minimum for f, compare the value $f(a, b)$ with values of $f(x, y)$ for points (x, y) that are near (a, b) and satisfy the constraint equation $g(x, y) = 0$.

(The simplest way to solve the equations of step 2 simultaneously is usually to first solve $h_x = 0$ and $h_y = 0$ for λ in terms of x and y, equate the two expressions for λ, solve for either x or y in terms of the other, and substitute the result into the equation $h_\lambda = 0$.)

EXAMPLE 1 Let us find maxima and minima for

$$f(x, y) = x^2 + y^2$$

subject to the constraint

$$x + y - 2 = 0.$$

1. We form $\quad h(x, y, \lambda) = x^2 + y^2 + \lambda(x + y - 2).$

Then $\quad h_x = 2x + \lambda = 0, \qquad h_y = 2y + \lambda = 0,$

and $\quad h_\lambda = x + y - 2 = 0.$

2. We must solve the equations

$$2x + \lambda = 0, \quad 2y + \lambda = 0, \quad \text{and} \quad x + y - 2 = 0$$

simultaneously for x and y. Solving the first two of these equations for λ, we obtain

$$\lambda = -2x \qquad \text{and} \qquad \lambda = -2y.$$

Thus

$$-2x = -2y,$$

or

$$x = y.$$

Substituting x for y in the equation

$$x + y - 2 = 0$$

now yields

$$2x - 2 = 0,$$

so

$$x = 1 \qquad \text{and} \qquad y = 1.$$

3. Therefore the function has either a constrained maximum or a constrained minimum when $x = 1$, $y = 1$. To find out whether $f(1, 1) = 2$ is a constrained maximum or a constrained minimum, we compare it with $f(x, y)$ for points (x, y) near $(1, 1)$ that also satisfy $x + y - 2 = 0$. For instance, we have

$$\begin{aligned} f(1, 1) &= 2, \\ f(1.05, 0.95) &= 2.005, \\ f(1.1, 0.9) &= 2.02, \\ f(0.8, 1.2) &= 2.08, \end{aligned}$$

and so on. Hence it is clear that $f(1, 1) = 2$ is a constrained minimum for f.

EXAMPLE 2 Let us optimize $\qquad f(x, y) = 12 - x^2 - y^2$

subject to the constraint $\qquad xy = 4.$

1. We have $\qquad h(x, y, \lambda) = 12 - x^2 - y^2 + \lambda(xy - 4).$

Note that the constraint equation was not originally in the form $g(x, y) = 0$, and thus we had to rewrite it in this form before we constructed the function h. Now,

$$h_x = -2x + \lambda y, \quad h_y = -2y + \lambda x, \quad \text{and} \quad h_\lambda = xy - 4.$$

2. We must solve the equations

$$-2x + \lambda y = 0, \quad -2y + \lambda x = 0, \quad \text{and} \quad xy - 4 = 0$$

simultaneously for x and y. From the first two equations,

$$\lambda = \frac{2x}{y} = \frac{2y}{x},$$

or

$$x^2 = y^2.$$

Therefore

$$y = \pm x.$$

If we substitute x for y in

$$xy - 4 = 0,$$

we find that

$$x^2 = 4,$$

so $x = \pm 2$, and we obtain the points $(2, 2)$, $(-2, -2)$. If we substitute $-x$ for y in

$$xy - 4 = 0,$$

we find that

$$x^2 = -4,$$

which has no solutions.

3. Thus $(2, 2)$ and $(-2, -2)$ are the only points that yield constrained maxima or minima for f. Since

$$f(\pm 2, \pm 2) = 4, \quad f\left(\pm 2.1, \frac{4}{\pm 2.1}\right) \approx 3.96, \quad f\left(\pm 2.2, \frac{4}{\pm 2.2}\right) \approx 3.85,$$

and so on, it is clear that $(2, 2)$ and $(-2, -2)$ both yield constrained maxima for f.

Many practical situations require constrained optimization of functions of several variables. We illustrate with an example.

EXAMPLE 3 Consider the production function defined by

$$f(x, y) = 300x^{1/3}y^{2/3}.$$

Here x denotes the number of units of labor and y the number of units of capital required to produce $f(x, y)$ units of product. Suppose that each unit of labor costs \$50 and each unit of capital costs \$100, and assume that the production budget is \$15,000. Then we would like to maximize production $f(x, y)$ subject to the budgetary constraint

$$50x + 100y = 15{,}000.$$

Using the Lagrange multiplier method, we have

$$h(x, y, \lambda) = 300x^{1/3}y^{2/3} + \lambda(50x + 100y - 15{,}000).$$

Thus

$$h_x = 100x^{-2/3}y^{2/3} + 50 = 0,$$

$$h_y = 200x^{1/3}y^{-1/3} + 100 = 0,$$

and

$$h_\lambda = 50x + 100y - 15{,}000 = 0,$$

from which we find that

$$\lambda = -2x^{-2/3}y^{2/3} = -2x^{1/3}y^{-1/3}.$$

Therefore

$$x^{-2/3}y^{2/3} = x^{1/3}y^{-1/3}.$$

Multiplying this last equation by $x^{2/3}y^{1/3}$ yields

$$y = x.$$

Substituting x for y in the equation

$$50x + 100y = 15{,}000,$$

we find that

$$x = y = 100.$$

Therefore the maximum production subject to the budgetary constraint is

$$f(100, 100) = 30{,}000$$

units.

Note that $x = y = 100$ implies that

$$\lambda = -2(100)^{1/3}(100)^{-1/3} = -2.$$

The absolute value of λ is called the **marginal productivity of money;** it tells us approximately how many additional units could be produced with one additional dollar. In this example, $|\lambda| = |-2| = 2$, so approximately 2 additional units could be produced if one additional dollar were available for production.

The Lagrange multiplier method can be used with functions of more than two variables: we simply set all first partial derivatives of $h = f + \lambda g$ equal to zero and solve simultaneously.

EXAMPLE 4 Let us find three positive numbers such that the first plus twice the second plus three times the third is 36 and such that their product is a maximum. Thus we wish to maximize

$$f(x, y, z) = xyz$$

subject to the constraint

$$x + 2y + 3z = 36.$$

We have

$$h(x, y, z, \lambda) = xyz + \lambda(x + 2y + 3z - 36),$$

so

$$h_x = yz + \lambda, \quad h_y = xz + 2\lambda, \quad h_z = xy + 3\lambda,$$

and

$$h_\lambda = x + 2y + 3z - 36.$$

Solving

$$h_x = 0, \quad h_y = 0, \quad h_z = 0, \quad \text{and} \quad h_\lambda = 0,$$

we find that

$$y = \frac{x}{2} \quad \text{and} \quad z = \frac{x}{3}.$$

Substituting these expressions for y and z in the constraint equation, we have $3x = 36$, from which it follows that $x = 12$, $y = 6$, and $z = 4$. Therefore the maximum product is $f(12, 6, 4) = 288$.

7.6 EXERCISES

In Exercises 1 through 8 use the Lagrange multiplier method to find either a maximum or a minimum for the function subject to the given constraint. In each case state whether the value you find is a maximum or a minimum.

1. $f(x, y) = x^2 + y^2 + 2$, subject to $x + y - 3 = 0$.
2. $f(x, y) = 2x^2 + 3y^2$, subject to $x + y - 5 = 0$.
3. $f(x, y) = 9 - x^2 - y^2$, subject to $x + y = 6$.
4. $f(x, y) = 100 - 2x^2 - 4y^2$, subject to $6x + 3y = 12$.
5. $f(x, y) = xy$, subject to $x + 2y = 6$.
6. $f(x, y) = xy$, subject to $x^2 + y^2 = 8$.
7. $f(x, y, z) = 2x - 4y + 2z - x^2 - y^2$, subject to $x + y + z = 0$.
8. $f(x, y, z) = x^2 + y^2 + z^2 - x - 2y - 3z$, subject to $10 - x - y - z = 0$.

9. Maximize $f(x, y) = x^3y^2$ subject to $12x + 6y = 24$.

10. Minimize $f(x, y) = x^4y^4$ subject to $2x - y = 4$.

11. Minimize $f(x, y) = x^2 + y^2 - 3$ subject to $xy = 16$.

12. Minimize $f(x, y) = x^3 + y^3$ subject to $xy = 1$.

13. Minimize $f(x, y) = x^3 + y^3$ subject to $x + y = 1$.

14. Find two positive numbers whose product is 25 and whose sum is a minimum.

15. Find two numbers whose sum is 20 and whose product is a maximum.

16. Find three positive numbers whose sum is 20 and whose product is a maximum.

17. Find two numbers whose product is 64 such that the square of one plus twice the other is a minimum.

18. Find two numbers such that 3 times one plus ⅓ of the other equals 24 and the product of their squares is a maximum.

19. Find the minimum distance from the line $x + y = 4$ to the origin (0, 0). (*Hint:* The distance from (x, y) to (0,0) is $\sqrt{x^2 + y^2}$.)

20. Find the rectangle of largest area that can be inscribed in a circle of radius r.

21. Find the rectangular box of largest volume that can be inscribed in a sphere of radius r. (The equation of such a sphere is $x^2 + y^2 + z^2 = r^2$.)

Management

22. A firm can produce

$$f(x, y) = 4000x^{2/5}y^{3/5}$$

units by employing x units of labor and y units of capital. If each unit of labor costs \$200, each unit of capital costs \$300, and the production budget is \$16,000, find the firm's maximum production. Also find and interpret the marginal productivity of money.

23. Repeat Exercise 22 if

$$f(x, y) = x^{1/4}y^{3/4}$$

and each unit of labor costs \$20, each unit of capital costs \$40, and the production budget is \$24,000.

24. Suppose the cost of producing x units of product X and y units of product Y is

$$C(x, y) = x^2 + 2y^2 + xy + 200$$

dollars. If it takes 5 worker-hours to make each unit of X, 7 worker-hours to make each unit of Y, and there are 3456 worker-hours available, how many units of each product must be made in order to minimize cost?

25. If a store purchases x radio commercials and y newspaper ads each week, its advertising reaches

$$z = 20x\sqrt{10y + 20}$$

thousand people. If it costs \$2000 to run each radio commercial and \$1000 to run each newspaper ad, and if the firm budgets \$7000 per week for advertising expenses, find the maximum number of people reached by its ads.

Life Science

26. If a farmer applies x tons of fertilizer and y tons of weedkiller to a crop, the yield is

$$z = 10x + 30y - 2x^2 - 5y^2$$

bushels per acre. If each ton of fertilizer costs \$500 and each ton of weedkiller costs \$800, and if the farmer has exactly \$3650 to spend on these products, find the maximum crop yield.

7.7 DOUBLE INTEGRALS

We have seen how functions of several variables may be differentiated by the process of partial differentiation; often they may also be integrated by a process known as **multiple integration.** In this section we illustrate multiple integration for functions of two variables. The integral of a function of two variables is called a **double integral.** Double integrals have many uses, most of which are beyond the scope of this book. However, we will demonstrate how they may be used to find volumes and areas.

Let f be a function of the two variables x and y whose domain includes a region R in the xy-plane, and such that $f(x, y) \geq 0$ for all (x, y) in R, as in Figure 7–16. Suppose we place a grid of rectangles on R, with each rectangle having width Δx and length Δy. See Figure 7–17. In each rectangle that overlaps R, we choose a point (x_i, y_i) that is also in R, and construct a rectangular box whose base is the rectangle and whose height is $f(x_i, y_i)$. Figure 7–18 depicts the situation for one rectangle. The rectangular box will touch the surface $z = f(x, y)$ at the point $(x_i, y_i, f(x_i, y_i))$, and its volume will be $f(x_i, y_i)\,\Delta x\,\Delta y$. If there are n

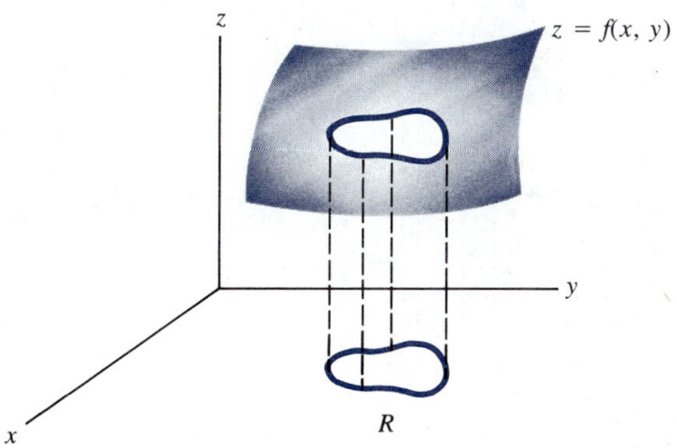

FIGURE 7–16

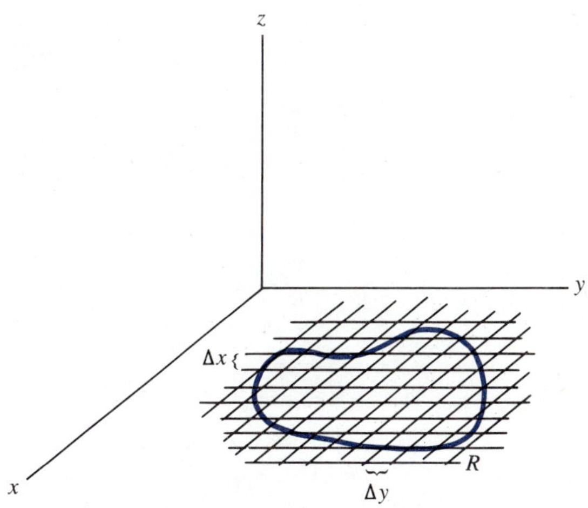

FIGURE 7–17

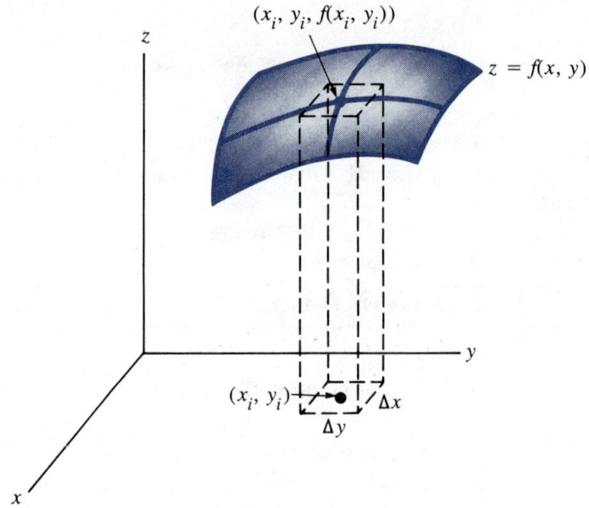

FIGURE 7–18

rectangles that overlap R, then there will be n such rectangular boxes, and the sum of their volumes may be written as the **Riemann sum**

$$[f(x_1, y_1) + f(x_2, y_2) + \cdots + f(x_n, y_n)]\ \Delta x\ \Delta y.$$

This Riemann sum is therefore an approximation to the volume of the solid whose base is R and whose top is the portion of the surface $z = f(x,y)$ that lies over R (Figure 7–19). If we make the grid on R finer and finer by letting Δx and Δy approach 0, the Riemann sum will approach the volume of this solid; in other words,

$$\lim_{\substack{\Delta x \to 0 \\ \Delta y \to 0}} [f(x_1, y_1) + f(x_2, y_2) + \cdots + f(x_n, y_n)]\ \Delta x\ \Delta y$$

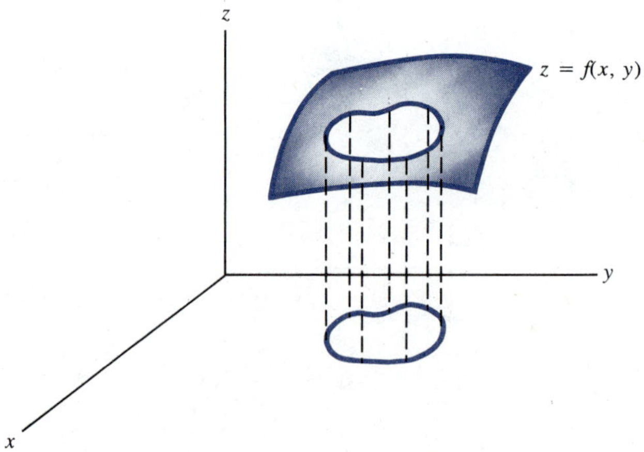

FIGURE 7–19

will exist and will be equal to the volume of the solid. We call this limit the **double integral of f over R.** [We have assumed that $f(x, y) \geq 0$ over the region R, but a similar argument shows that if $f(x, y) \leq 0$ over R then the limit of the Riemann sums as Δx and Δy approach 0 is the negative of the volume of the solid bounded by R and the surface $z = f(x, y)$.]

The Double Integral

If f is a function of x and y whose domain includes the region R in the xy-plane, the **double integral of f over R,** denoted by

$$\iint_R f(x, y)\, dA,$$

is defined to be the limit

$$\lim_{\substack{\Delta x \to 0 \\ \Delta y \to 0}} [f(x_1, y_1) + \cdots + f(x_n, y_n)]\, \Delta x\, \Delta y,$$

if it exists. If $f(x, y) \geq 0$ for all (x, y) in R, then

$$\iint_R f(x, y)\, dA$$

is the volume of the cylindrical solid bounded by R and the surface $z = f(x, y)$.

Fortunately, we will not need to evaluate double integrals by actually taking limits of Riemann sums; instead, we will evaluate them as **iterated integrals** of the form

$$\int_c^d \int_a^b f(x, y)\, dx\, dy \qquad \text{or} \qquad \int_a^b \int_c^d f(x, y)\, dy\, dx.$$

Such iterated integrals are integrated from the inside out. Thus, to evaluate

$$\int_c^d \int_a^b f(x, y)\, dx\, dy,$$

we first find the inside integral

$$\int_a^b f(x, y)\, dx$$

by integrating $f(x, y)$ with respect to x from $x = a$ to $x = b$ while treating y as a constant. Since each occurrence of x in the antiderivative will be replaced by a and b, the result of this integration will be a function of y, which we can then integrate from $y = c$ to $y = d$, as the outside integral requires. We illustrate with an example.

EXAMPLE 1 Let us evaluate the integrated integral

$$\int_0^2 \int_0^1 (2x + y)\, dx\, dy.$$

We first evaluate the inside integral by integrating $f(x, y) = 2x + y$ with respect to x while treating y as a constant. Thus

$$\int_0^1 (2x + y)\, dx = x^2 + yx \Big|_{x=0}^{x=1} = [1^2 + y(1)] - [0^2 + y(0)] = 1 + y.$$

Now

$$\begin{aligned}\int_0^2 \int_0^1 (2x + y)\, dx\, dy &= \int_0^2 \left(\int_0^1 (2x + y)\, dx \right) dy \\ &= \int_0^2 (1 + y)\, dy \\ &= y + \frac{y^2}{2} \Big|_0^2 \\ &= 4.\end{aligned}$$

We can interchange the order of integration in an iterated integral without affecting the value of the integral.

EXAMPLE 2 Let us evaluate the double integral

$$\int_0^1 \int_0^2 (2x + y)\, dy\, dx.$$

This is the integral of Example 1 with the order of integration reversed. We have

$$\begin{aligned}\int_0^1 \int_0^2 (2x + y)\, dy\, dx &= \int_0^1 \left(\int_0^2 (2x + y)\, dy \right) dx \\ &= \int_0^1 \left(2xy + \frac{y^2}{2} \Big|_{y=0}^{y=2} \right) dx \\ &= \int_0^1 \left[2x \cdot 2 + \frac{2^2}{2} \right] - \left[2x \cdot 0 + \frac{0^2}{2} \right] dx \\ &= \int_0^1 (4x + 2)\, dx \\ &= 2x^2 + 2x \Big|_0^1 \\ &= 4.\end{aligned}$$

Now we return to our consideration of double integrals. It turns out that if the region R over which we are integrating is bounded by lines and curves, then we can evaluate the double integral as an iterated integral.

Evaluating Double Integrals

1. If R is the region bounded by the lines $x = a$ and $x = b$ and the graphs of the equations $y = g_1(x)$ and $y = g_2(x)$, as shown in Figure 7–20, then

$$\iint_R f(x, y)\, dA = \int_a^b \int_{g_1(x)}^{g_2(x)} f(x, y)\, dy\, dx.$$

2. If R is the region bounded by the lines $y = c$ and $y = d$ and the graphs of the equations $x = h_1(y)$ and $x = h_2(y)$, as shown in Figure 7–21, then

$$\iint_R f(x, y)\, dA = \int_c^d \int_{h_1(y)}^{h_2(y)} f(x, y)\, dx\, dy.$$

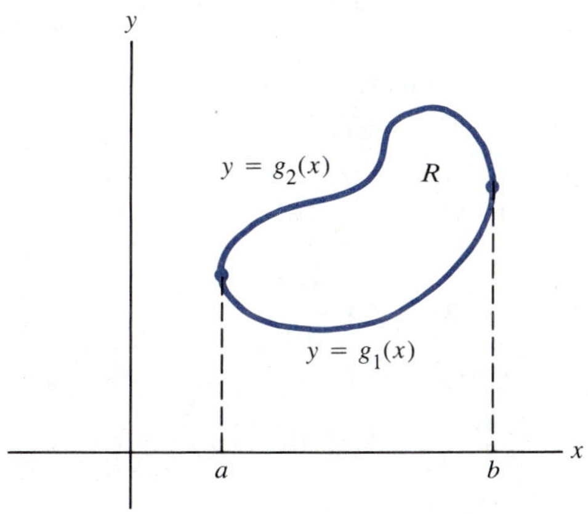

FIGURE 7–20

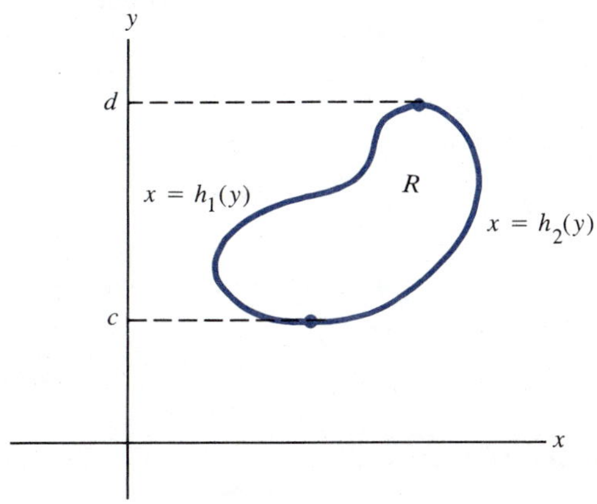

FIGURE 7–21

Note that in the integral

$$\int_a^b \int_{g_1(x)}^{g_2(x)} f(x, y)\, dy\, dx$$

we integrate first with respect to y, and y ranges from the lower boundary $g_1(x)$ to the upper boundary $g_2(x)$ of R. The result of the inner integration will thus be a function of the single variable x, which we then integrate from $x = a$ to $x = b$, as required by the outer integral. Similar remarks apply to the integral

$$\int_c^d \int_{h_1(y)}^{h_2(y)} f(x, y)\, dx\, dy.$$

EXAMPLE 3 Let $f(x, y) = 12 - x - y$, and let R be the rectangle shown in Figure 7–22.

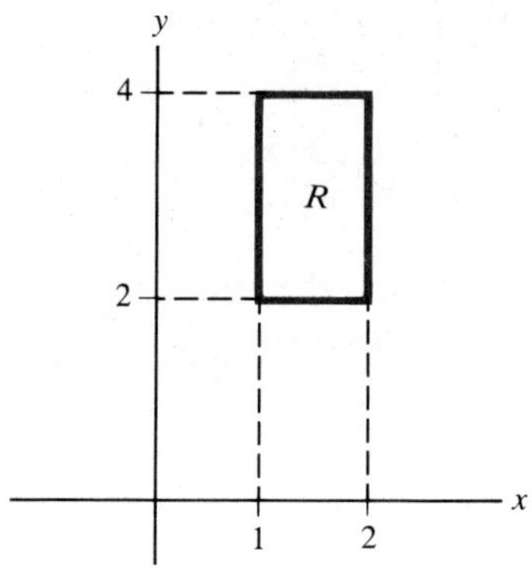

FIGURE 7–22

Since R is bounded by the lines $x = 1$, $x = 2$, $y = 2$, and $y = 4$, we have

$$\iint_R (12 - x - y)\, dA = \int_1^2 \int_2^4 (12 - x - y)\, dy\, dx$$

$$= \int_1^2 \left(12y - xy - \frac{y^2}{2} \Bigg|_{y=2}^{y=4} \right) dx$$

$$= \int_1^2 [(48 - 4x - 8) - (24 - 2x - 2)]\, dx$$

$$= \int_1^2 (18 - 2x)\, dx$$

$$= 18x - x^2 \Bigg|_1^2$$

$$= 15.$$

Thus the volume of the solid pictured in Figure 7–23 is 15. Note that we can also evaluate the double integral of $f(x, y) = 12 - x - y$ over the rectangle R as

$$\iint_R (12 - x - y)\, dA = \int_2^4 \int_1^2 (12 - x - y)\, dx\, dy.$$

This gives the same result. (Check that this is so.)

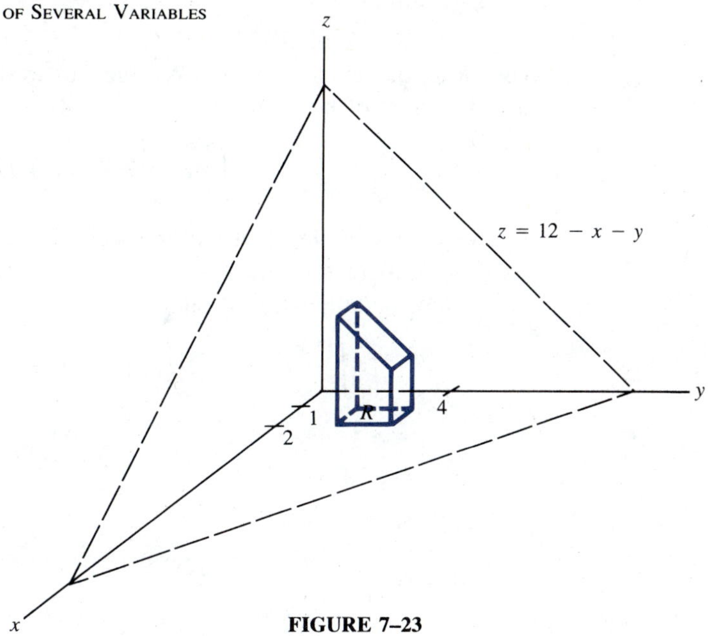

FIGURE 7–23

EXAMPLE 4 Let us find the volume of the solid whose base is the triangle bounded by the x-axis, the y-axis, and the line $x + y = 1$, and whose top is the surface $z = x^2 + y^2$. See Figure 7–24.

The volume of the solid is the double integral

$$\iint_R (x^2 + y^2)\, dA,$$

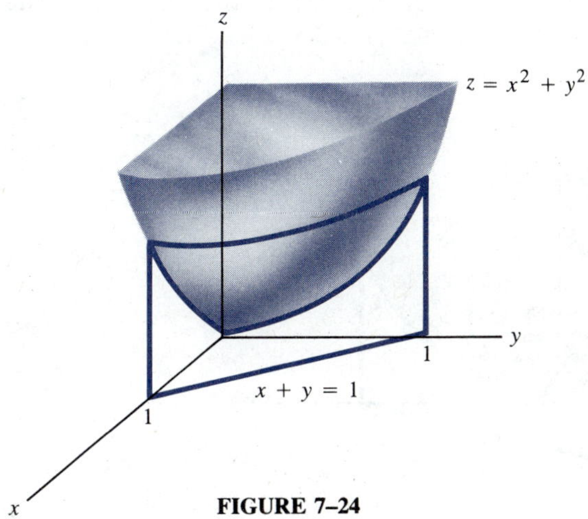

FIGURE 7–24

where R is the given triangle. We seek to evaluate this double integral as an iterated integral of the form

$$\int_a^b \int_{g_1(x)}^{g_2(x)} (x^2 + y^2)\, dy\, dx.$$

Here we are integrating first with respect to y, and hence y must range from the x-axis ($y = 0$) to the line $y = 1 - x$, and x must range from 0 to 1 (Figure 7–25). Thus the volume of the solid is

$$\begin{aligned}\int_0^1 \int_0^{1-x} (x^2 + y^2)\, dy\, dx &= \int_0^1 \left(x^2 y + \frac{1}{3} y^3 \Big|_{y=0}^{y=1-x} \right) dx \\ &= \int_0^1 \left[x^2(1 - x) + \frac{1}{3}(1 - x)^3 \right] dx \\ &= \int_0^1 \frac{1}{3}(-4x^3 + 6x^2 - 3x + 1)\, dx \\ &= \frac{1}{6}.\end{aligned}$$

(Check the integration.)

If we had wished, we could have set up the iterated integral in the previous example so that we integrated first with respect to x and then with respect to y; in this case the integral would have been

$$\int_0^1 \int_0^{1-y} (x^2 + y^2)\, dx\, dy,$$

since now x must range from the y-axis to the line $x = 1 - y$, and y must range from 0 to 1. See Figure 7–26. Sometimes one way of setting up the iterated integral leads to an easier integration than the other.

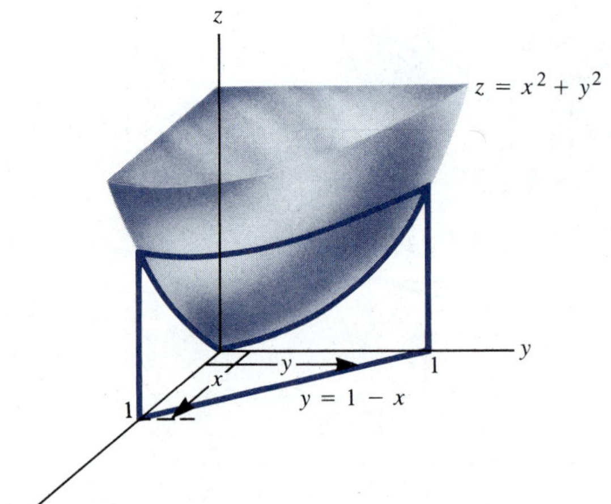

FIGURE 7–25

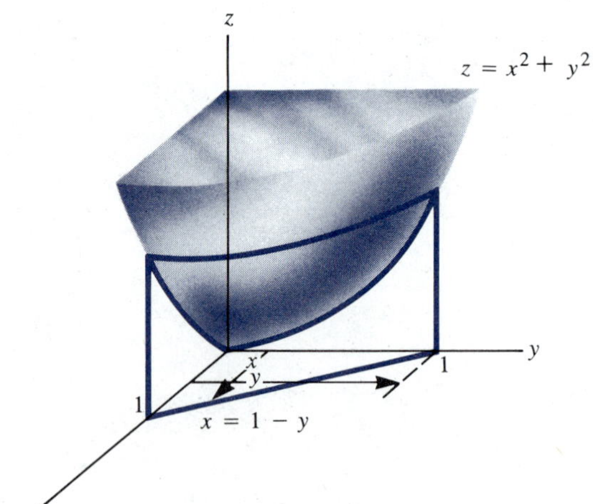

FIGURE 7–26

EXAMPLE 5 Consider the iterated integral

$$\int_0^1 \int_0^x e^{y^2}\, dy\, dx.$$

Here we are integrating first with respect to y, so y ranges from $y = 0$ to $y = x$, and then x ranges from $x = 0$ to $x = 1$. Therefore the region R over which we are integrating is the triangle shown in Figure 7–27. If we attempt to evaluate this integral in the order given, we must first find

$$\int_0^x e^{y^2}\, dy;$$

but since we do not know an antiderivative for e^{y^2}, we cannot carry out the integration. However, if we reverse the order of integration, we have

$$\int_0^1 \int_0^y e^{y^2}\, dx\, dy = \int_0^1 \left(xe^{y^2} \Big|_{x=0}^{x=y} \right) dy$$

$$= \int_0^1 ye^{y^2}\, dy.$$

If we now let $u = y^2$, so that $du = 2y\, dy$, then

$$\int_0^1 ye^{y^2}\, dy = \frac{1}{2} \int_0^1 e^u\, du = \frac{1}{2}(e - 1).$$

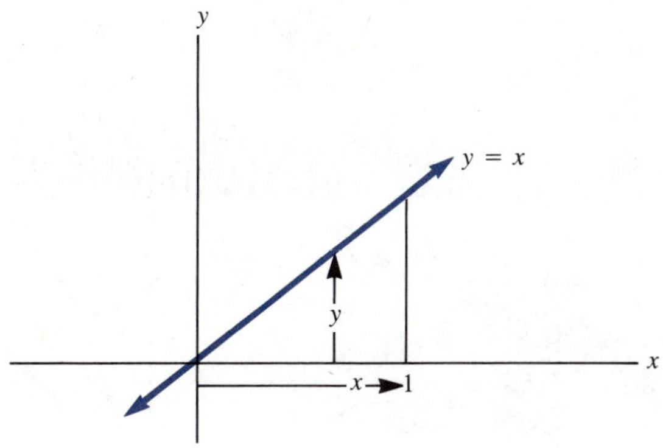

FIGURE 7–27

We conclude this section by remarking that the area of a region R in the xy-plane may be calculated by finding the double integral over R of the constant function defined by $f(x, y) = 1$.

The Double Integral and Area

If R is a region in the xy-plane, the area of R is given by

$$\iint_R dA.$$

EXAMPLE 6 Find the area of the region R pictured in Figure 7–28. The area in question is the value of the double integral of 1 over R, and thus is equal to

$$\begin{aligned}\iint_R dA &= \int_0^3 \int_x^{-x^2+4x} dy\, dx \\ &= \int_0^3 \left(y \Big|_{y=x}^{y=-x^2+4x} \right) dx \\ &= \int_0^3 (-x^2 + 3x)\, dx \\ &= -\frac{x^3}{3} + \frac{3}{2}x^2 \Big|_0^3 \\ &= \frac{9}{2}.\end{aligned}$$

Note that we could have found this area by the method of Section 5.4, in which case we would have obtained the same result by evaluating the single integral

$$\int_0^3 (-x^2 + 4x - x)\, dx.$$

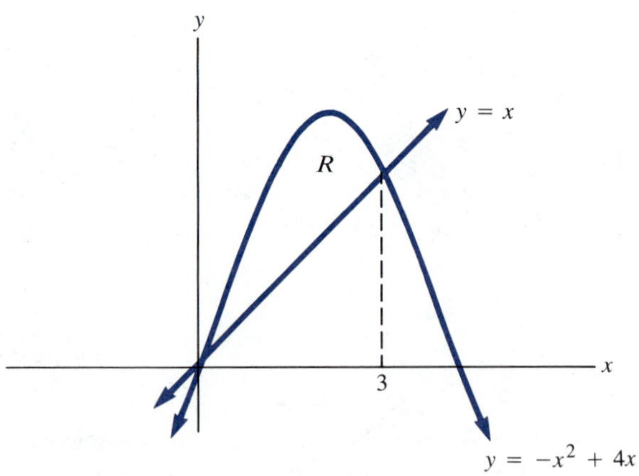

FIGURE 7–28

7.7 EXERCISES

In Exercises 1 through 6 evaluate the iterated integral.

1. $\int_0^1 \int_0^1 x\,dy\,dx$

2. $\int_0^1 \int_0^1 2y\,dx\,dy$

3. $\int_0^2 \int_{-1}^1 (2x + 3y^2)\,dx\,dy$

4. $\int_{-2}^3 \int_{-2}^1 xy\,dy\,dx$

5. $\int_0^1 \int_0^2 xe^{-y}\,dy\,dx$

6. $\int_0^2 \int_0^1 ye^{-x}\,dx\,dy$

In Exercises 7 through 12 evaluate the double integral over the indicated region R in the xy-plane.

7. $\iint_R (x + y)\,dA$, R the rectangle bounded by the x-axis, the y-axis, and the lines $x = 1$, $y = 2$.

8. $\iint_R (2x - 4y)\,dA$, R the square bounded by the lines $x = 1$, $x = 3$, $y = 1$, $y = 3$.

9. $\iint_R (x - 2y)\,dA$, R the triangle bounded by the y-axis and the lines $x - y = 0$, $y = 2$.

10. $\iint_R (x - 2y)\,dA$, R the triangle bounded by the x-axis and the lines $x - y = 0$, $x = 2$.

11. $\iint_R x^2y\,dA$, R the region bounded by $y = x^2$ and $y = 1 - x^2$.

12. $\iint_R \frac{1}{xy}\,dA$, R the region bounded by $y = e^{-x}$ and the lines $y = 1$, $x = 1$, $x = 2$.

In Exercises 13 through 18 find the volume of the solid bounded by the given surface and the region R.

13. The plane $z = 2$; the rectangle R bounded by the x-axis, the y-axis, and the lines $x = 2$ and $y = 4$.

14. The plane $z = 8 - x - y$; the rectangle R bounded by the lines $x = 1$, $y = 1$, $x = 2$, and $y = 3$.

15. The plane $z = y$; the square R bounded by $x = 0$, $y = 0$, $x = 2$, and $y = 2$.

16. The plane $z = 6 - 2x - 3y$; the triangle R bounded by $x = 0$, $y = 0$, and $-2x - 3y = 6$.

17. The plane $z = 4x$; the region R bounded by the nonnegative x-axis, the nonnegative y-axis, and the circle $x^2 + y^2 = 1$.

18. The surface $z = e^{-y}$; the region R bounded by the x-axis, $x = e$, and $y = \ln x$.

In Exercises 19 through 22 use a double integral to find the area of the region R.

19. R is bounded by the x-axis and $y = 2x - x^2$.

20. R is bounded by the y-axis, $x = \sqrt{y}$, and $y = 1$.

21. R is bounded by $y = x^2$ and $y = x^3$.

22. R is bounded by $y = x^2 - 4x$ and $y = -x^2 + 4x$.

7 REVIEW EXERCISES

Answers to all review exercises are in the back of the book.

1. Find the domain of the function f defined by
$$f(x, y) = x + \frac{2xy}{y - x},$$
and find $f(0, 0)$, $f(1, 0)$, $f(0, -1)$, and $f(3, 4)$, if possible.

2. Find the domain of the function defined by
$$w = e^{-xyz} + \ln 2xy,$$
and, if possible, find w when
(a) $x = y = z = 0$ **(b)** $x = y = z = 2$
(c) $x = -1, y = 3, z = 2$ **(d)** $x = 2, y = 1, z = -3$

3. Reddin Toothenclaw's pet store sells both quails and snails. If quails are priced at p_1 dollars each, Reddin can sell
$$q_1 = 200 - 5p_1$$
of them; if snails are priced at p_2 dollars each, Reddin can sell
$$q_2 = 750 - 20p_2$$
of them. Assuming that the prices of quails and snails do not affect one another, express Reddin's revenue as a function of p_1 and p_2 and find the revenue if quails sell for \$10 each and snails for \$2 each.

4. If a firm makes x units of A and y units of B, its profit is
$$P(x, y) = -8000 + 40x + 30y - 0.05xy$$
dollars. Find the firm's profit if it makes and sells 400 units of A and 300 units of B. If x is fixed at 400, what can you say about the contribution to profit made by each unit of B?

5. Graph $f(x, y) = 24 - 3x - 6y$.

6. Graph $z = x^2 + 2y^2$.

7. Graph $x = e^z - 1$ in the three-dimensional cartesian coordinate system.

8. Find f_x, f_y, f_{xx}, f_{xy}, f_{yx}, and f_{yy} if
$$f(x, y) = 8x^2y - 2xy^3 + 12x + 2y - 5.$$

9. Find $f_x(2, 1)$, $f_y(-1, 0)$, $f_{xx}(2, 2)$, $f_{yy}(-3, 4)$ for the function f of Exercise 8.

10. Let $g(p, q, r) = e^{pq} + e^{-qr} + 2p + r^4$. Find g_{pq}, g_{qqpp}, and g_{qppp}.

11. Find $\frac{\partial z}{\partial x}$, $\frac{\partial z}{\partial y}$, $\frac{\partial z^2}{\partial x \partial y}$, $\frac{\partial z^2}{\partial y \partial x}$, $\frac{\partial z^2}{\partial x^2}$, and $\frac{\partial z^2}{\partial y^2}$, if
$$z = (x^2 + 2xy - y^3)^6.$$

12. Let $f(x, y) = 2x + 3y + 5x^2y^2$. Find the slope of the line tangent to the graph of f at the point (2, 2, 90) and parallel to the
(a) xz-plane **(b)** yz-plane

13. If the number of wolves in a region is given by

$$f(x, y) = x^{1/4}y^{-2},$$

where x is the number of animals the wolves can prey on and y is the number of animals that can prey on the wolves, find $f_x(16, 1)$ and $f_y(16, 1)$ and interpret your results.

14. Suppose a firm's revenue from selling x units of X and y units of Y is

$$R(x, y) = 20x + 25y + 0.1\sqrt{xy}$$

dollars. Find $R_x(25, 100)$ and $R_y(25, 100)$ and interpret your results.

15. Find the relative maxima, relative minima, and saddle points of f if

$$f(x, y) = x^2 + 2y^2 - 1.$$

16. Repeat Exercise 15 if

$$f(x, y) = -2x^4 + 16x^2 + y^2 - 4y + 8.$$

17. Find the dimensions of a rectangular box that is to have a volume of 10.4 cubic meters if the bottom is to be hardwood, which costs \$10 per square meter, the sides are to be softwood, which costs \$5 per square meter, and the top is to be plastic, which costs \$3 per square meter, and the cost of the box is to be minimized.

18. The demand functions for a commodity in regions X and Y are given by

$$p_X = 800 - \frac{x}{10} \quad \text{and} \quad p_Y = 500 - \frac{y}{5},$$

respectively. Find the prices that maximize revenue.

19. Find the maximum or minimum value of $f(x, y) = x^3y^3$ subject to the constraint $6x + 2y = 12$, and state whether your answer is a maximum or a minimum.

20. Repeat Exercise 19 for the function defined by

$$f(x, y, z) = x^2y + yz + z^2$$

subject to the constraint $xyz = 64$.

21. Find two positive numbers such that the product of twice the first and three times the second is 384 and their sum is a minimum.

22. A company can produce $f(x, y) = 1000x^{1/5}y^{4/5}$ units of its product, where x is units of labor and y is units of capital. If each unit of labor costs \$20 and each unit of capital costs \$100, and if the company's production budget is \$1000, find the maximum production. Also find and interpret the marginal productivity of money.

23. Evaluate the integral

$$\iint_R x^2 \, dA,$$

where R is the region bounded by the graphs of $y = x$, $y = 5 - x$, $y = 1$, and the x-axis.

24. Find the volume of the cylindrical solid whose base is the triangle in the xy-plane bounded by the lines $y = 2 - x$, $y = x$, and $x = 0$, and whose top is the plane $z = 4 - 2x$.

7 SUPPLEMENT: LEAST SQUARES

Students sometimes ask how the functions used to model practical situations are "made up." We have seen throughout this book that such functions can often be constructed from an analysis of the situation. For instance, as we saw in Chapter 1, if it costs a firm v dollars to make each unit of its product then its cost function is linear with slope v, and since fixed cost F is the cost of making 0 units, it follows that the cost C of making x units is given by $C = vx + F$.

Often, however, functional models cannot be developed through analysis of a situation, but must be created by fitting curves to experimental data. To illustrate, suppose we want to find a function that models annual family savings as a function of annual family income. To begin with we have no information on the relationship, and thus no reason to prefer any particular type of function (e.g., linear, quadratic, exponential, etc.) as a model. If we collected some data on the relationship between income and savings, perhaps we would be able to decide on the type of function to use as a model. Let us assume we have somehow obtained the following data:

X Annual Family Income (in 1000s of \$)	Y Annual Family Savings (in 1000s of \$)
42	5.0
48	6.0
37	4.5
52	6.1
60	7.1
54	6.1
44	4.9
48	5.6

Let us plot the data as points (x_i, y_i). The result is the diagram of Figure 7–29, which is called a **scatter diagram** for the data. We now guess the type of curve that would best fit the scatter diagram in the sense that it would come as close as possible to all the points of the diagram. The type of curve that best fits the diagram depends on the pattern formed by the points of the diagram: it might be a line, a parabola, an exponential curve, or some other curve. In this case it looks as if a line would fit the scatter diagram fairly well. See Figure 7–30. Thus our data lead us to choose a linear function to model the relationship between income and savings.

Our next task is to determine the equation $Y = mX + b$ of our linear model by finding the values of m and b that make the line come as close as possible to all the points of the scatter diagram simultaneously. For any scatter diagram and any line, the closeness of the line to the points could be measured by the sum of the deviations of the points from the line: see Figure 7–31, which is

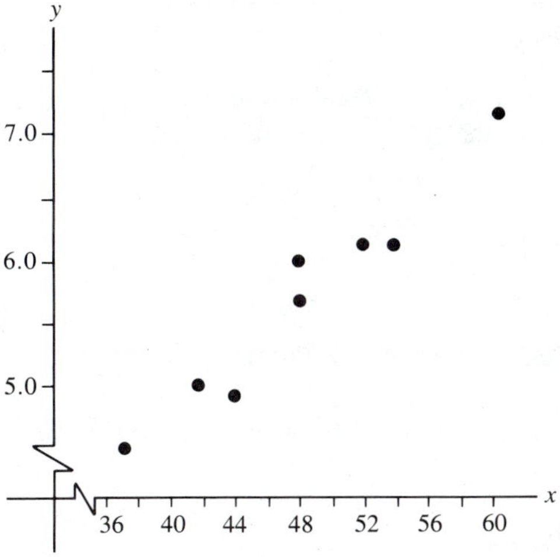

FIGURE 7–29

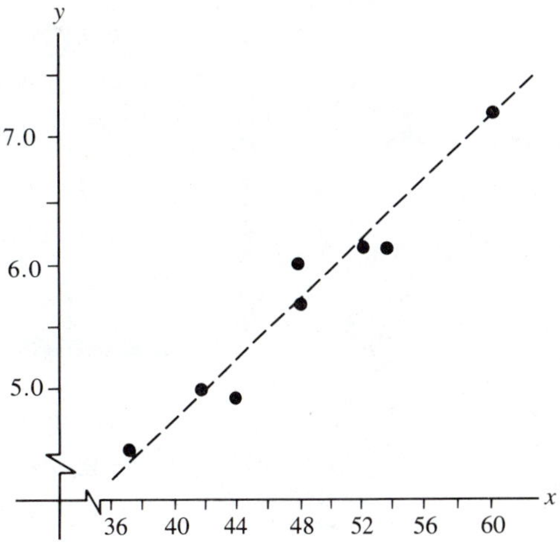

FIGURE 7–30

a picture of a general scatter diagram and line $Y = mX + b$ with the deviations $d_1, d_2, \ldots, d_n$, where

$$d_i = (mx_i + b) - y_i$$

for each i. However, since some of the d_i's (those corresponding to points that lie below the line) will be positive while others (those corresponding to points that lie above the line) will be negative, there will be some "canceling out" in the sum $d_1 + d_2 + \cdots + d_n$, and this could make the line appear to be closer to the points than it really is. Therefore instead of using the d_i's to measure deviation from the line, we use their squares d_i^2. The line for which $d_1^2 + d_2^2 + \cdots + d_n^2$ is a minimum is called the **least squares line** for the data.

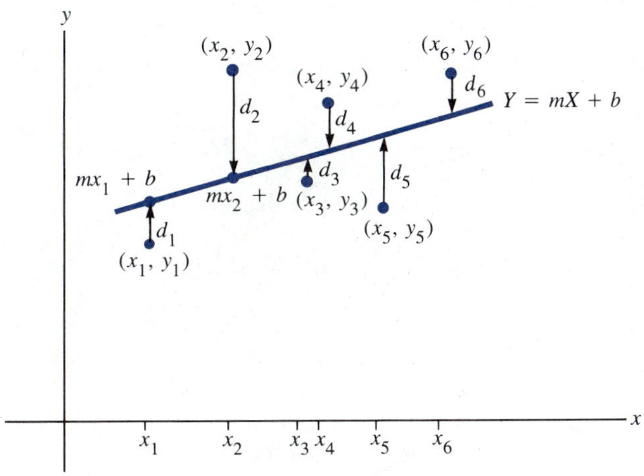

FIGURE 7–31

The Least Squares Line

Given n data points $(x_1, y_1), \ldots, (x_n, y_n)$, the **least squares line** for the data is the line that best fits the data in the sense that in its equation

$$Y = mX + b,$$

m and b are chosen so as to minimize the sum

$$(mx_1 + b - y_1)^2 + \cdots + (mx_n + b - y_n)^2.$$

How can we find the values of m and b that yield the least squares line for a set of data? Let us put

$$f(m, b) = (mx_1 + b - y_1)^2 + \cdots + (mx_n + b - y_n)^2.$$

(Thus f is a function of the variables m and b: $x_1, \ldots, x_n$ and $y_1, \ldots, y_n$ are known numbers.) We must find the values of m and b that minimize f. We will use the second derivative test of Section 7.5 to do so.

We have

$$f_m = 2(mx_1 + b - y_1)x_1 + \cdots + 2(mx_n + b - y_n)x_n$$

and

$$f_b = 2(mx_1 + b - y_1) + \cdots + 2(mx_n + b - y_n),$$

or

$$f_m = 2[m(x_1^2 + \cdots + x_n^2) + b(x_1 + \cdots + x_n) - (x_1y_1 + \cdots + x_ny_n)]$$

and

$$f_b = 2[m(x_1 + \cdots + x_n) + bn - (y_1 + \cdots + y_n)].$$

Setting $f_b = 0$ and solving for b yields

$$b = \frac{1}{n}[(y_1 + \cdots + y_n) - m(x_1 + \cdots + x_n)].$$

Substituting for b in the equation $f_m = 0$ now yields

$$0 = 2[m(x_1^2 + \cdots + x_n^2) + \frac{1}{n}\{(y_1 + \cdots + y_n) - m(x_1 + \cdots + x_n)\}(x_1 + \cdots + x_n) - (x_1y_1 + \cdots + x_ny_n)],$$

or

$$0 = m\left[(x_1^2 + \cdots + x_n^2) - \frac{1}{n}(x_1 + \cdots + x_n)^2\right] - (x_1y_1 + \cdots + x_ny_n) + \frac{1}{n}(y_1 + \cdots + y_n)(x_1 + \cdots + x_n).$$

When solved for m, this last equation gives

$$m = \frac{n(x_1y_1 + \cdots + x_ny_n) - (y_1 + \cdots + y_n)(x_1 + \cdots + x_n)}{n(x_1^2 + \cdots + x_n^2) - (x_1 + \cdots + x_n)^2}.$$

We have solved the equations $f_m = 0$ and $f_b = 0$ simultaneously for m and b. To complete the second-derivative test we must show that these values of m and b yield a minimum for f. We will leave this for the exercises (see Exercise 1). Hence

Least Squares Line Formulas

Given a set of n data points $(x_1, y_1), \ldots, (x_n, y_n)$, let

$$m = \frac{n(x_1y_1 + \cdots + x_ny_n) - (x_1 + \cdots + x_n)(y_1 + \cdots + y_n)}{n(x_1^2 + \cdots + x_n^2) - (x_1 + \cdots + x_n)^2}$$

and

$$b = \frac{1}{n}[(y_1 + \cdots + y_n) - m(x_1 + \cdots + x_n)].$$

Then $Y = mX + b$ is the equation of the least squares line for the data.

Let us use these formulas to find the least squares line for the example at the beginning of this supplement. We need the sum of the x_i's, the sum of the y_i's, the sum of the x_iy_i's, and the sum of the x_i^2's. We have

X	Y	XY	X^2
42	5.0	210.0	1764
48	6.0	288.0	2304
37	4.5	166.5	1369
52	6.1	317.2	2704
60	7.1	426.0	3600
54	6.1	329.4	2916
44	4.9	215.6	1936
48	5.6	268.8	2304
385	45.3	2221.5	18,897

Thus

$$m = \frac{8(2221.5) - (385)(45.3)}{8(18{,}897) - (385)^2} \approx 0.112$$

and

$$b = \frac{1}{8}[(45.3) - 0.112(385)] \approx 0.273.$$

Hence the least squares line that models the relationship between annual family income and annual family savings has equation

$$Y = 0.112X + 0.273,$$

where X is family income in thousands of dollars and Y is family savings in thousands of dollars. See Figure 7–32.

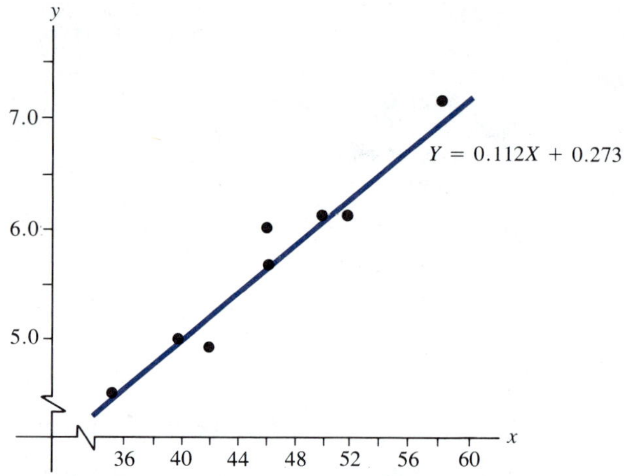

FIGURE 7–32

The technique we have employed to find the linear function that best fits a set of data points can be extended to other types of functions. Thus, if the scatter diagram leads us to believe that a parabola best fits the data, we could find the coefficients a, b, and c that yield the least squares parabola $Y = aX^2 + bX + c$ by minimizing the function f defined by

$$f(a, b, c) = (ax_1 + bx_1 + c - y_1)^2 + \cdots + (ax_n + bx_n + c - y_n)^2,$$

and similarly for other functions. In this way many models can be created from experimental and other data.

Computer Manual: Program LSTSQRS.

SUPPLEMENTARY EXERCISES

1. Show that the values for m and b do indeed minimize f.
2. A firm has the following data on its profit from making and selling x units of product:

X Units Made and Sold	Y Profit $
5	−155
10	−114
15	−77
20	−40
25	−2
30	44
35	83
40	120
45	162
50	194

Make a scatter diagram and find the equation of the least squares line for the data. Can you estimate the firm's fixed cost and the contribution to profit made by each unit?

3. A new drug that shrinks tumors is being tested in various concentrations. The results of the tests are as follows:

X Drug Dosage (Units)	Y Time until Tumor Disappears (Days)
1	19.0
3	17.5
4	16.6
6	15.4
7	14.4
8	13.7
10	12.1
12	10.8
13	9.6
14	9.0

Make a scatter diagram and find the least squares line for the data. Can you estimate the effect that increasing the dosage by 1 unit has on the tumor?

APPENDIX

Trigonometric Functions

In this appendix we introduce a new class of functions, the **trigonometric functions.** Trigonometric functions can be used to model the motion of a satellite in orbit, the movement of a weight at the end of a spring as the spring expands and contracts, the flow of current in an electrical circuit, and many other phenomena. In fact, virtually every situation of a repetitive or cyclical nature requires the use of trigonometric functions in its mathematical description; hence such functions are extremely important in modeling the natural world.

Since trigonometric functions are defined in terms of angles, before we can begin our study of them we must briefly discuss the question of how angles are measured. We then define the trigonometric functions and describe some of their properties. The last two sections of the appendix consider the differentiation and integration of trigonometric functions.

A.1 MEASURING ANGLES

In this section we discuss the measurement of angles. Angles may be measured either by **degrees** or by **radians.** We briefly review degree measure and then introduce radian measure, which we use in the rest of the appendix.

Degrees

Every angle has two **sides** and a **vertex.** Figure A–1 shows an angle *AOB:* its sides are *OA* and *OB*, and its vertex is the point *O*. An angle is measured from its **initial side** to its **terminal side,** the direction of measurement being indicated by an arrow. Thus the fact that the arrow in Figure A–1 goes from *OA* to *OB* tells us that *OA* is the initial side of the angle and *OB* is its terminal side. We attach a degree measure to an angle by placing the vertex of the angle at the center of a circle (Figure A–2); then as we move from the initial side to the terminal side of the angle we sweep out some fraction of the circle. Since by ancient tradition every circle contains 360 degrees (360°), this movement will sweep out some fraction of 360°. This fraction of 360° is the degree measure of the angle. The degree measure of the angle is positive if movement from the initial side to the terminal side is in the counterclockwise direction; it is negative if it is in the clockwise direction.

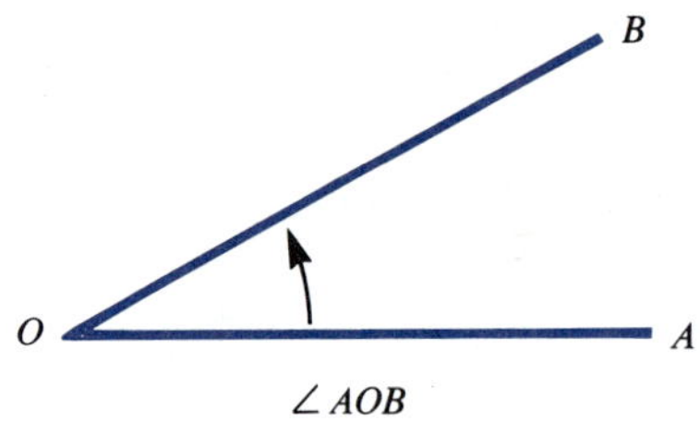

FIGURE A–1

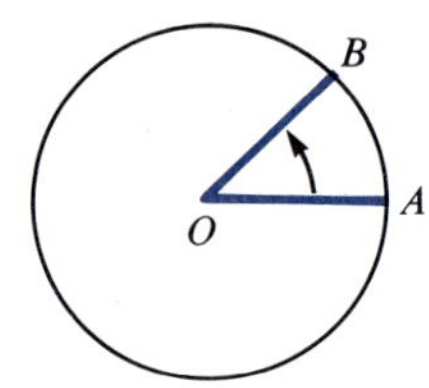

FIGURE A–2

EXAMPLE 1 Angle *AOB* of Figure A–3(a) sweeps out one-fourth of the circle in a counterclockwise direction; therefore

$$\text{degree measure of angle } AOB = \frac{1}{4}(360°) = 90°.$$

Similarly, angle *EOF* of Figure A–3(b) sweeps out one-eighth of the circle in a counterclockwise direction, so

$$\text{angle } EOF = \frac{1}{8}(360°) = 45°.$$

On the other hand, angle *POQ* of Figure A–3(c) sweeps out one-half of the circle in a clockwise direction, so its degree measure is negative, and thus

$$\text{angle } POQ = -\frac{1}{2}(360°) = -180°.$$

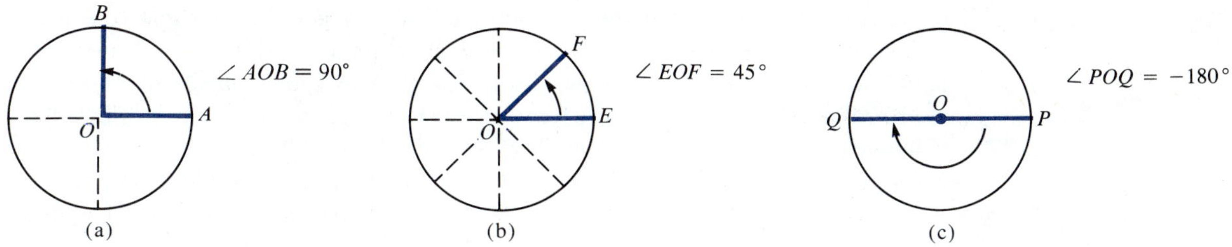

FIGURE A–3

EXAMPLE 2 Suppose we are asked to construct an angle of 120°. Since 120° is positive and 120°/360° = ⅓, an angle of 120° must sweep out one-third of a circle in a counterclockwise direction. Figure A–4(a) shows an angle of 120°. Similarly, an angle of 60° must sweep out 60°/360° = ⅙ of a circle in a counterclockwise direction [Figure A–4(b)]. But since −270° is negative, an angle of −270° must sweep out 270°/360° = ¾ of a circle in a counterclockwise direction [Figure A–4(c)].

Figure A–5 shows some common angles and their degree measures.

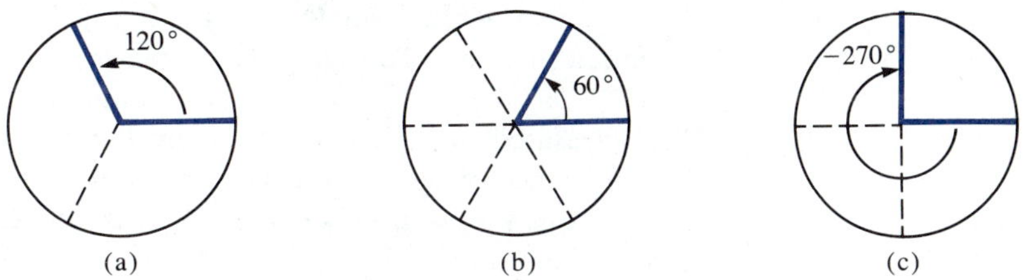

FIGURE A–4

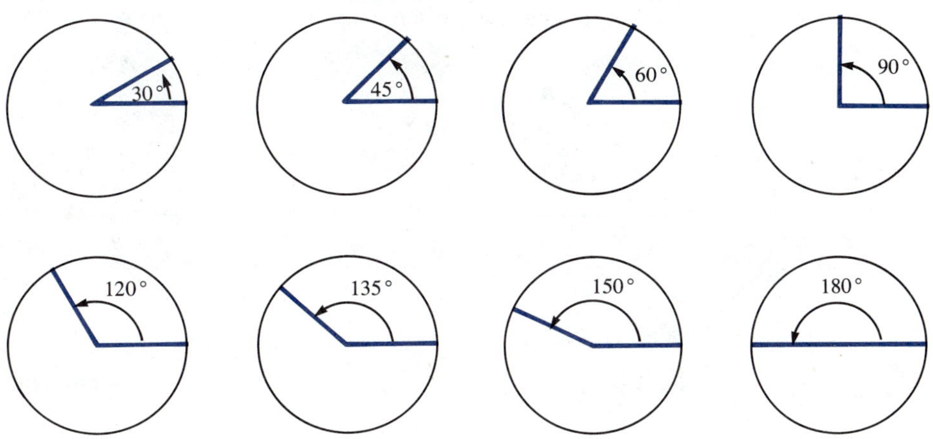

FIGURE A–5

Angles can measure more than 360°, in either a positive or negative sense. When this occurs, the angles are equivalent to angles having degree measure between $-360°$ and $360°$. For instance, Figure A–6 shows angles of 480° and $-810°$. Since $480° = 360° + 120°$, to construct an angle of 480°, we must proceed counterclockwise through a full circle (360°) plus another 120°, as the figure indicates. Thus an angle of 480° is equivalent to an angle of 120°. Similarly, since $-810° = 2(-360°) - 90°$, an angle of $-810°$ is equivalent to an angle of $-90°$.

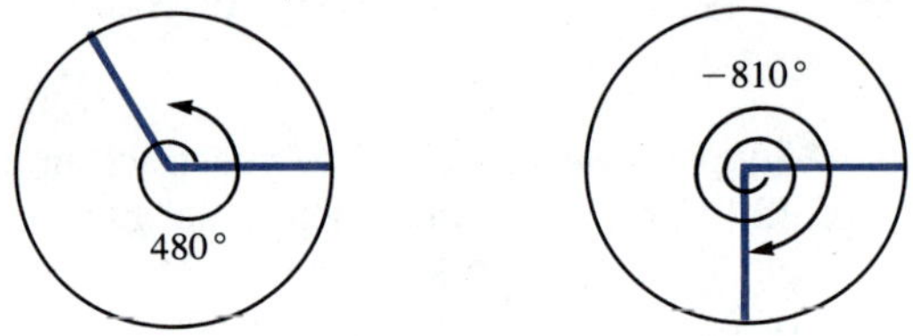

FIGURE A–6

Radians

We can also measure angles in **radians,** and in fact radian measure will be our preferred measure. To obtain the radian measure of an angle AOB, we place its vertex O at the center of a circle of radius 1, as in Figure A–7. The angle is then measured by the length of the arc PQ which is traced as we move along the circumference of the circle from the initial side to the terminal side of the angle. If the arc is traced in the counterclockwise direction as we move from the initial side to the terminal side, we take the radian measure of the angle to be positive; if it is traced in the clockwise direction, we take the radian measure to be negative. Thus in Figure A–8 the radian measure of angle EOF will be r radians while that of angle GOH will be $-r$ radians. Since the circumference of a circle of radius 1 is 2π, the length of the arc PQ will be some fraction of 2π, and hence radian measures are often expressed in terms of the number π.

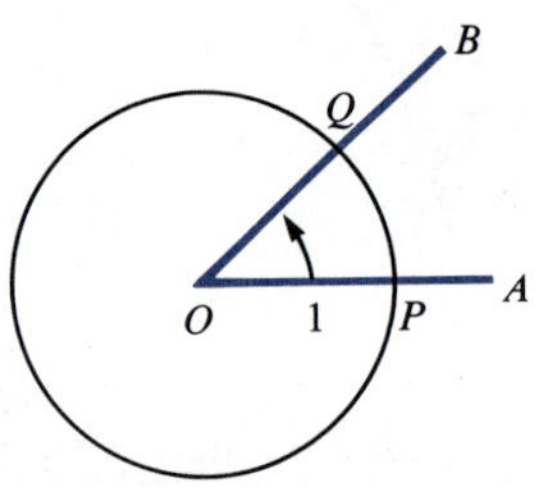

$\angle AOB$ measured by length of arc PQ

FIGURE A–7

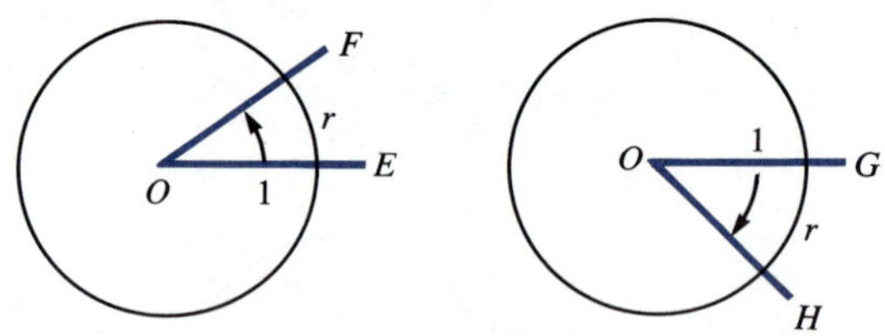

$\angle EOF = r$ radians $\qquad \angle GOH = -r$ radians

FIGURE A–8

EXAMPLE 3 Consider the angle AOB of Figure A–9(a): as we move from the initial side to the terminal side of the angle we trace one fourth of the circumference of the circle in a counterclockwise direction. Therefore

$$\text{radian measure of angle } AOB = \frac{1}{4}(2\pi) = \frac{\pi}{2} \text{ radians.}$$

Similarly, for the angle EOF of Figure A–9(b), as we move from the initial side to the terminal side of the angle we trace one-half of the circumference of the circle in a counterclockwise direction, so

$$\text{angle } EOF = \frac{1}{2}(2\pi) = \pi \text{ radians.}$$

On the other hand, for the angle POQ of Figure A–9(c), as we move from the initial to the terminal side of the angle we trace out one third of the circumference of the circle in a clockwise direction, and thus

$$\text{angle } POQ = -\frac{1}{3}(2\pi) = -\frac{2\pi}{3} \text{ radians.}$$

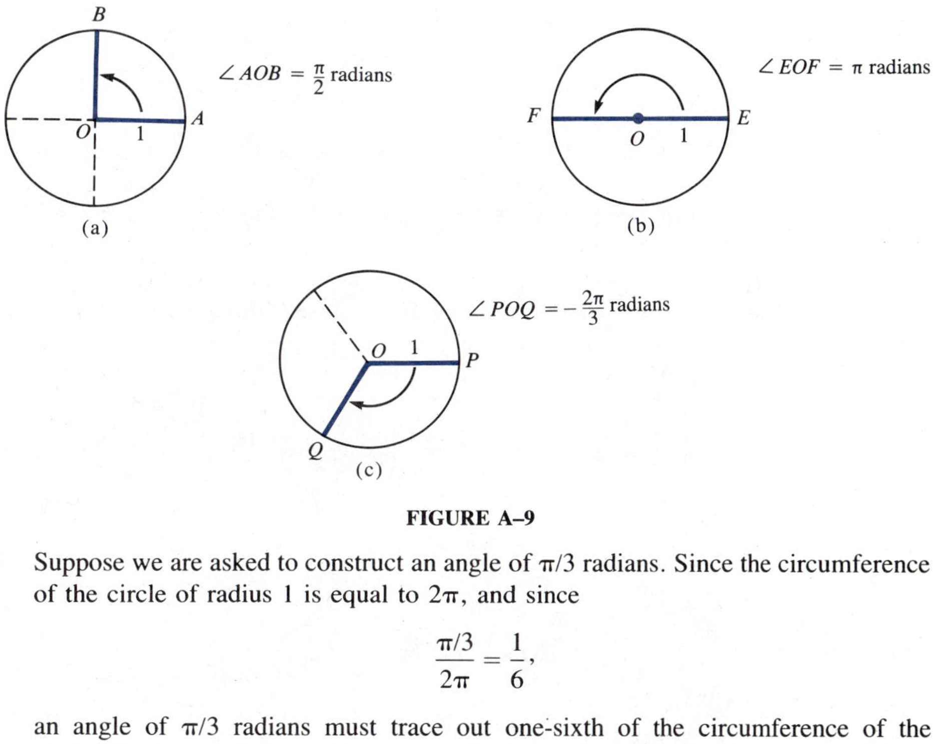

FIGURE A–9

EXAMPLE 4 Suppose we are asked to construct an angle of $\pi/3$ radians. Since the circumference of the circle of radius 1 is equal to 2π, and since

$$\frac{\pi/3}{2\pi} = \frac{1}{6},$$

an angle of $\pi/3$ radians must trace out one-sixth of the circumference of the circle. Because $\pi/3$ is positive, the tracing must move in a counterclockwise direction. Figure A–10(a) shows an angle of $\pi/3$ radians. Similarly, an angle of $-3\pi/2$ radians must trace out

$$\frac{3\pi/2}{2\pi} = \frac{3}{4}$$

of the circumference of the circle. Because $-3\pi/2$ is negative, the tracing must move in a clockwise direction. Figure A–10(b) depicts an angle of $-3\pi/2$ radians.

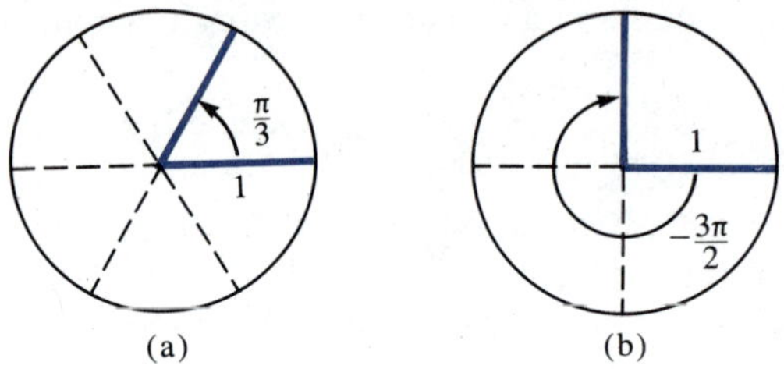

FIGURE A–10

Any angle that measures more than 2π radians is equivalent to one that measures between 0 and 2π radians. For instance, since

$$\frac{7\pi}{2} = 2\pi + \frac{3\pi}{2},$$

an angle of $7\pi/2$ radians is equivalent to one of $3\pi/2$ radians. See Figure A–11(a). Similarly, any angle that measures less than -2π radians is equivalent to one that measures between -2π and 0 radians. Figure A–11(b) shows an angle of -5π radians.

Since we have two ways of measuring angles, we must be able to convert from one to the other. But if angle AOB has degree measure $d°$ and radian measure r radians, then d is the same proportion of 360 as r is of 2π. See Figure A–12. Thus

$$\frac{d}{360} = \frac{r}{2\pi}.$$

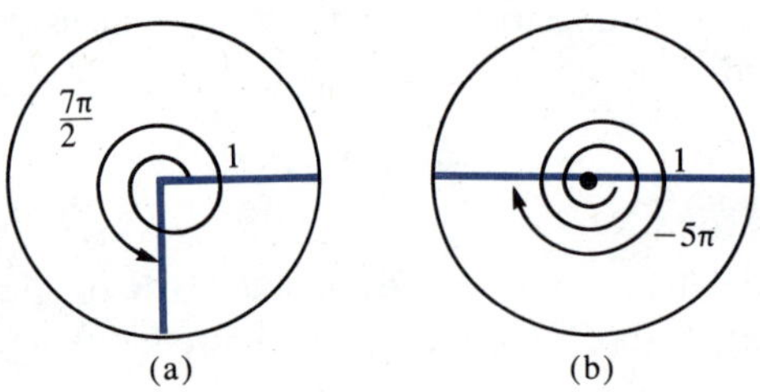

FIGURE A–11

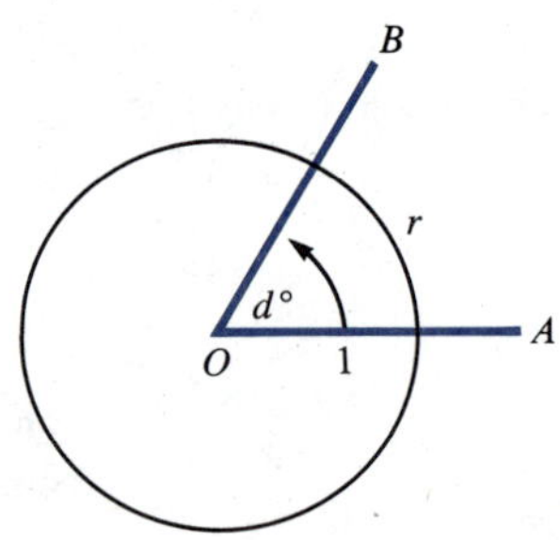

FIGURE A–12

The Conversion Formula for Angle Measurement

If an angle measures $d°$ and r radians, then d and r are related by the equation

$$r = \frac{\pi d}{180}.$$

Note that

$$r = \frac{\pi d}{180} \quad \text{implies that} \quad d = \frac{180 r}{\pi}.$$

EXAMPLE 5 Let us convert 1°, 135°, −30°, and 68° to radians.

$$1°: \quad r = \frac{\pi(1)}{180} = \frac{\pi}{180} \text{ radians;}$$

$$135°: \quad r = \frac{\pi(135)}{180} = \frac{3\pi}{4} \text{ radians;}$$

$$-30°: \quad r = \frac{\pi(-30)}{180} = -\frac{\pi}{6} \text{ radians;}$$

$$68°: \quad r = \frac{\pi(68)}{180} = \frac{17\pi}{45} \text{ radians.}$$

EXAMPLE 6 Let us convert 1, $\pi/3$, $-5\pi/4$, and 3.8 radians to degrees.

$$1 \text{ radian:} \quad d = \frac{180(1)}{\pi} = \frac{180}{\pi} \approx 57.3°;$$

$$\pi/3 \text{ radians:} \quad d = \frac{180(\pi/3)}{\pi} = 60°;$$

$$-5\pi/4 \text{ radians:} \quad d = \frac{180(-5\pi/4)}{\pi} = -225°;$$

$$3.8 \text{ radians:} \quad d = \frac{180(3.8)}{\pi} = \frac{684}{\pi} \approx 217.7°.$$

Figure A–13 shows some common angles with their radian measures. Note that the angles of Figure A–13 are exactly the same as the angles of Figure A–5.

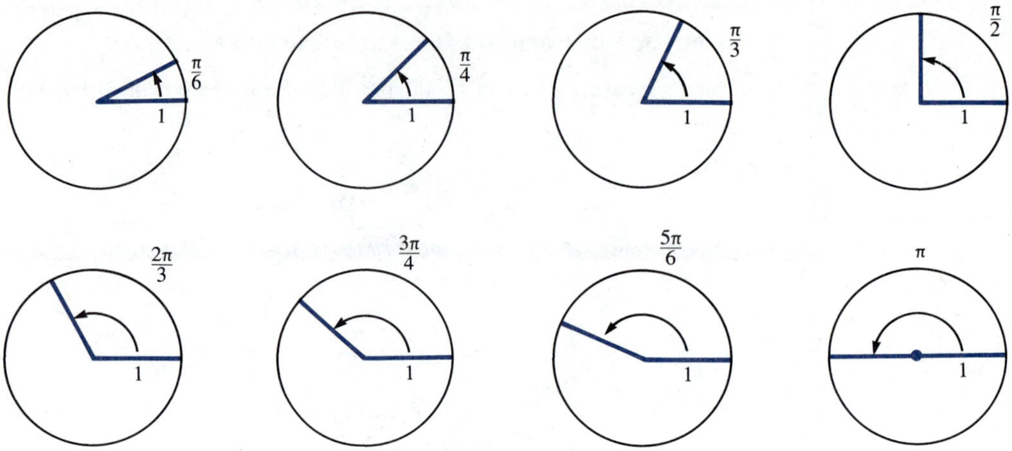

FIGURE A–13

A.1 EXERCISES

Degrees

In Exercises 1 through 6 state the degree measurement of the angles pictured.

1.
2.
3.

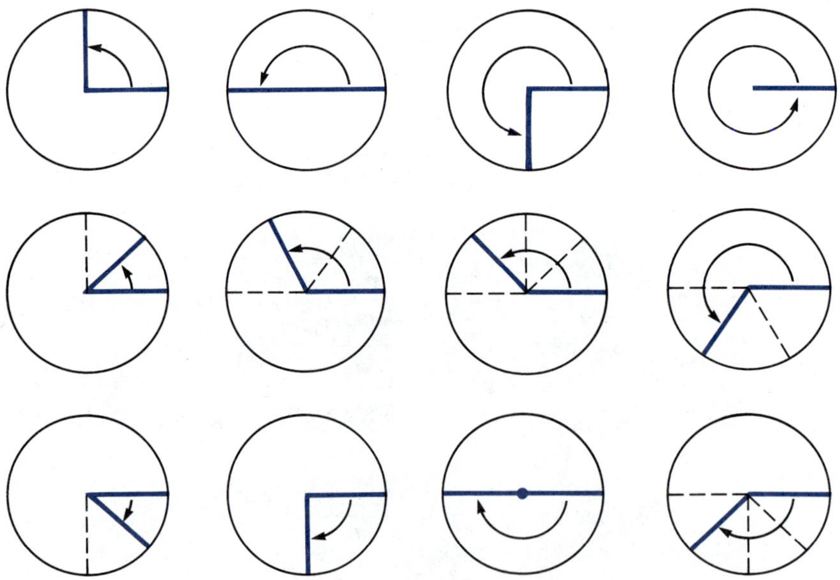

4.

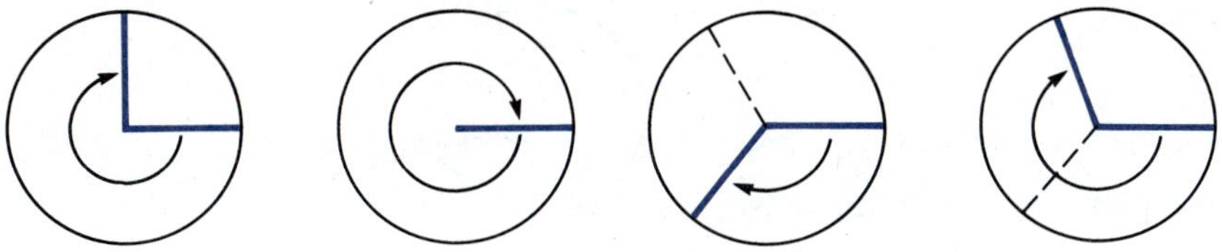

5.

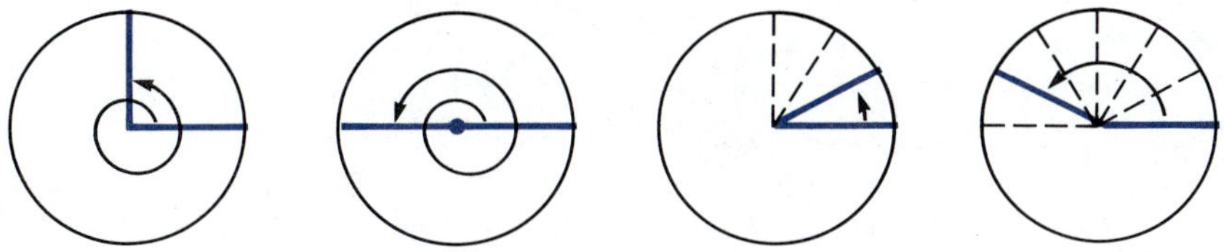

6.

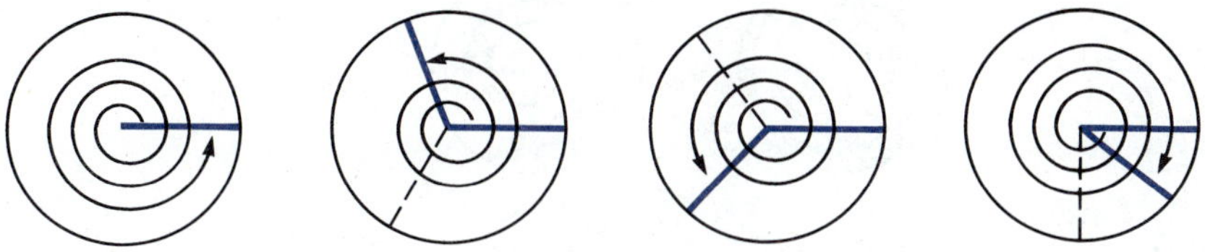

In Exercises 7 through 12 construct the angles whose degree measures are given.

7. 90°, 30°, 180°, 720°

8. 180°, 540°, 315°, −60°

9. 45°, 135°, 60°, 150°

10. 270°, 225°, −120°, 0°, 360°

11. −900°, −30°, 630°, −150°, 15°

12. 495°, 840°, −765°, 22.5°, 40°, 7.5°

Radians

In Exercises 13 through 18 state the radian measure of the angles pictured. Do not use the conversion formula.

13.

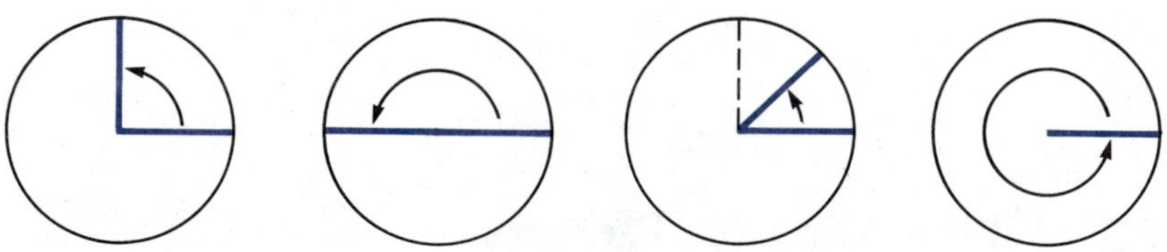

14.

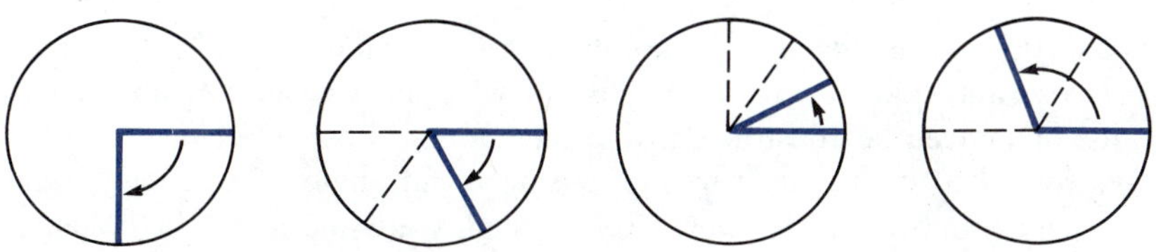

15.

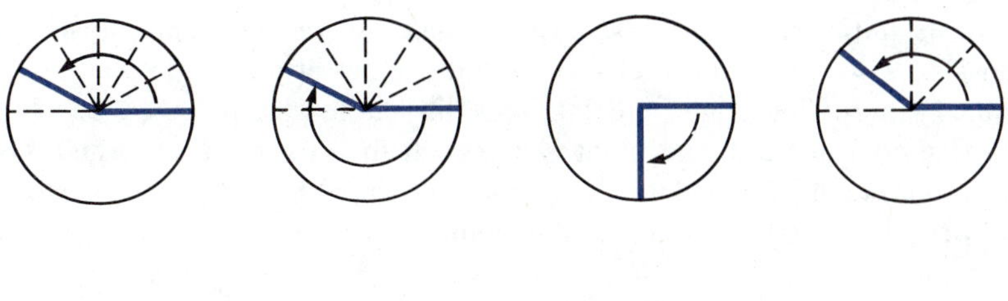

16.

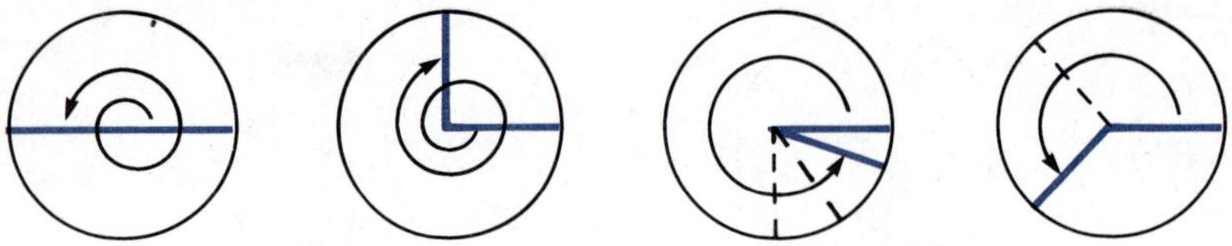

17.

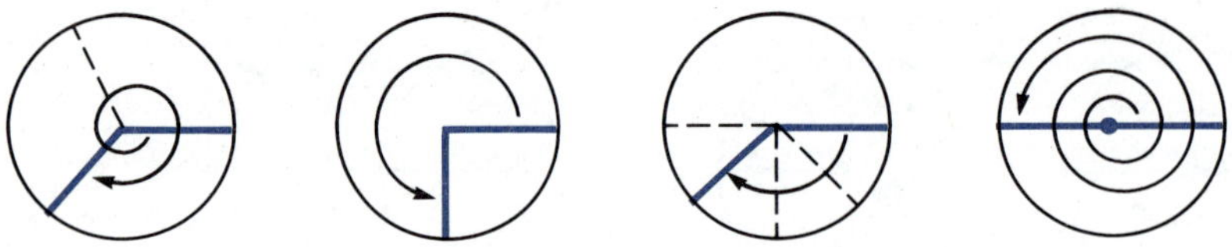

18.

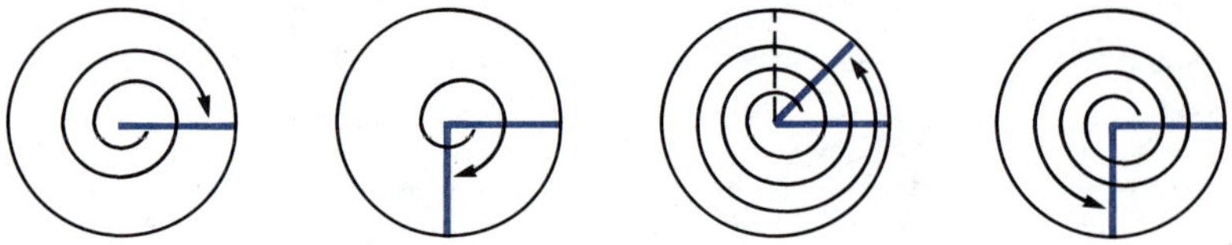

In Exercises 19 through 24 construct the angles whose radian measures are given.

19. $\frac{3\pi}{2}, \frac{5\pi}{2}, -\frac{\pi}{2}, \pi$

20. $2\pi, \frac{\pi}{3}, -\frac{\pi}{6}, \frac{2\pi}{3}$

21. $-\frac{5\pi}{4}, \frac{3\pi}{4}, \frac{11\pi}{4}, -\frac{3\pi}{2}$

22. $\frac{7\pi}{4}, -\frac{5\pi}{6}, \frac{11\pi}{6}, 4\pi, \frac{\pi}{8}$

23. $\frac{8\pi}{3}, \frac{\pi}{6}, \frac{9\pi}{2}, -\frac{\pi}{8}, \frac{2\pi}{9}$

24. $-3\pi, -\frac{2\pi}{3}, \frac{11\pi}{8}, \frac{\pi}{12}, \frac{26\pi}{3}$

25. Convert to radians: 90°, 30°, 125°, −60°

26. Convert to radians: 45°, 12°, 120°, −900°

27. Convert to radians: 135°, −150°, 350°, 624°

28. Convert to degrees: $\frac{5\pi}{6}, \frac{3\pi}{2}, -\frac{8\pi}{3}, \frac{9\pi}{2}$ radians.

29. Convert to degrees: $-6\pi, \frac{4\pi}{9}, 5\pi, \frac{3\pi}{8}$ radians.

30. Convert to degrees: $\frac{7\pi}{6}, -\frac{5\pi}{12}, \frac{22}{7}$, 2.5 radians.

A.2 THE TRIGONOMETRIC FUNCTIONS

In this section we define the six trigonometric functions: sine, cosine, tangent, secant, cosecant, and cotangent. We begin with the sine and cosine functions, since the others can be constructed from them.

We intend to define the sine and cosine of an angle of x radians, where x may be any number. (From now on, all angles are measured in radians.) To accomplish this, we take an angle AOB of x radians and place its vertex at the origin and its initial side along the positive u-axis in the uv-plane, as shown in Figure A–14. If we now construct a circle with center at the origin and radius 1, the terminal side of the angle will intersect this circle at a point Q. See Figure A–15. The u-coordinate of the point Q is called the **cosine of x,** written $\cos x$, and its v-coordinate is the **sine of x,** written $\sin x$: thus $Q = (\cos x, \sin x)$. Figure A–16 shows the location of the point Q for some typical values of x.

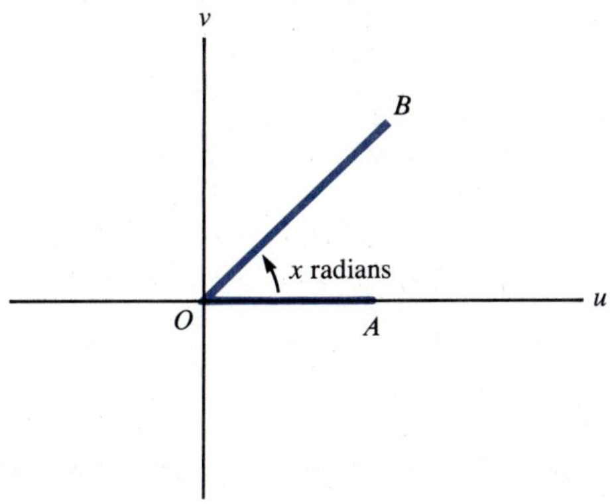

FIGURE A–14

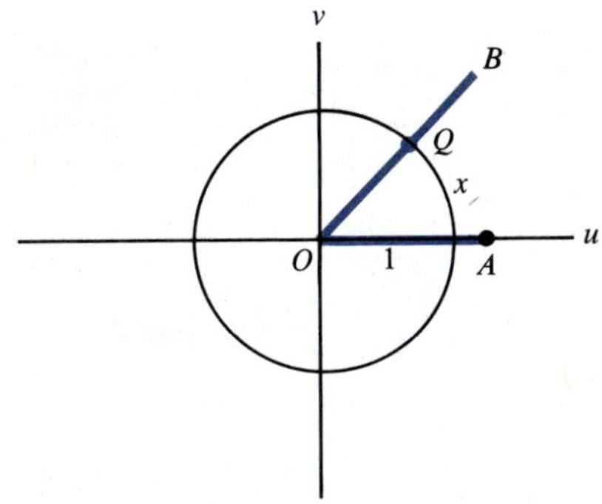

FIGURE A–15

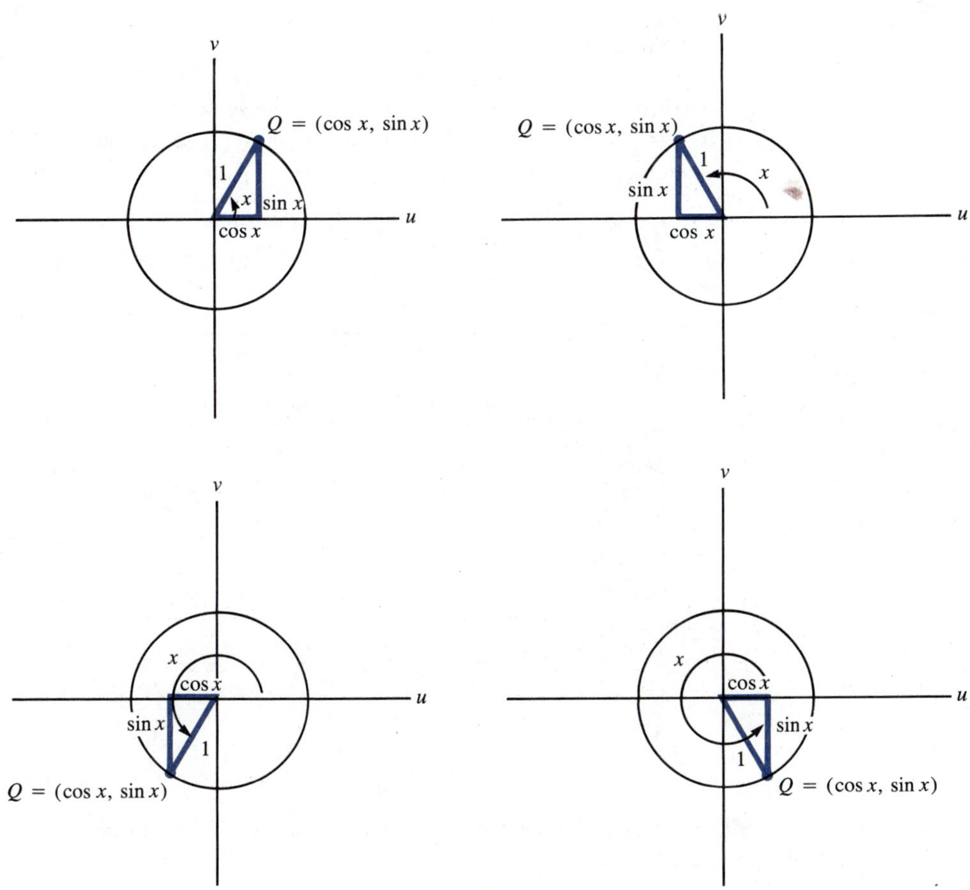

FIGURE A–16

Note that

$\cos x$ is positive when $0 < x < \pi/2$ and $3\pi/2 < x < 2\pi$;

$\cos x$ is negative when $\pi/2 < x < 3\pi/2$;

$\sin x$ is positive when $0 < x < \pi$;

$\sin x$ is negative when $\pi < x < 2\pi$.

Note also that

$$-1 \leq \cos x \leq 1 \qquad \text{and} \qquad -1 \leq \sin x \leq 1,$$

since the absolute values of $\cos x$ and $\sin x$ can never be larger than the radius of the circle, which is 1.

Let us find $\cos x$ and $\sin x$ for some selected values of x.

a. If $x = 0$, then $Q = (\cos 0, \sin 0) = (1,0)$. See Figure A–17(a). Therefore $\cos 0 = 1$ and $\sin 0 = 0$.
b. If $x = \pi/2$, then $Q = (\cos \pi/2, \sin \pi/2) = (0,1)$. See Figure A–17(b). Therefore $\cos \pi/2 = 0$ and $\sin \pi/2 = 1$.
c. If $x = \pi$, then $Q = (\cos \pi, \sin \pi) = (-1,0)$. See Figure A–17(c). Therefore $\cos \pi = -1$ and $\sin \pi = 0$.
d. If $x = 3\pi/2$, then $Q = (\cos 3\pi/2, \sin 3\pi/2) = (0,-1)$. See Figure A–17(d). Therefore $\cos 3\pi/2 = 0$ and $\sin 3\pi/2 = -1$.
e. If $x = 2\pi$, then $Q = (\cos 2\pi, \sin 2\pi) = (1,0)$. See Figure A–17(e). Therefore $\cos 2\pi = 1$ and $\sin 2\pi = 0$.

We summarize in Tables A–1 and A–2 the information we have obtained thus far.

TABLE A–1

	Sign of	
Radians x	$\sin x$	$\cos x$
$0 < x < \pi/2$	+	+
$\pi/2 < x < \pi$	+	−
$\pi < x < 3\pi/2$	−	−
$3\pi/2 < x < 2\pi$	−	+

TABLE A–2

	Radians				
	0	$\pi/2$	π	$3\pi/2$	2π
sin	0	1	0	−1	0
cos	1	0	−1	0	1

Since the equation of the circle with center at the origin and radius 1 in the uv-plane is

$$u^2 + v^2 = 1,$$

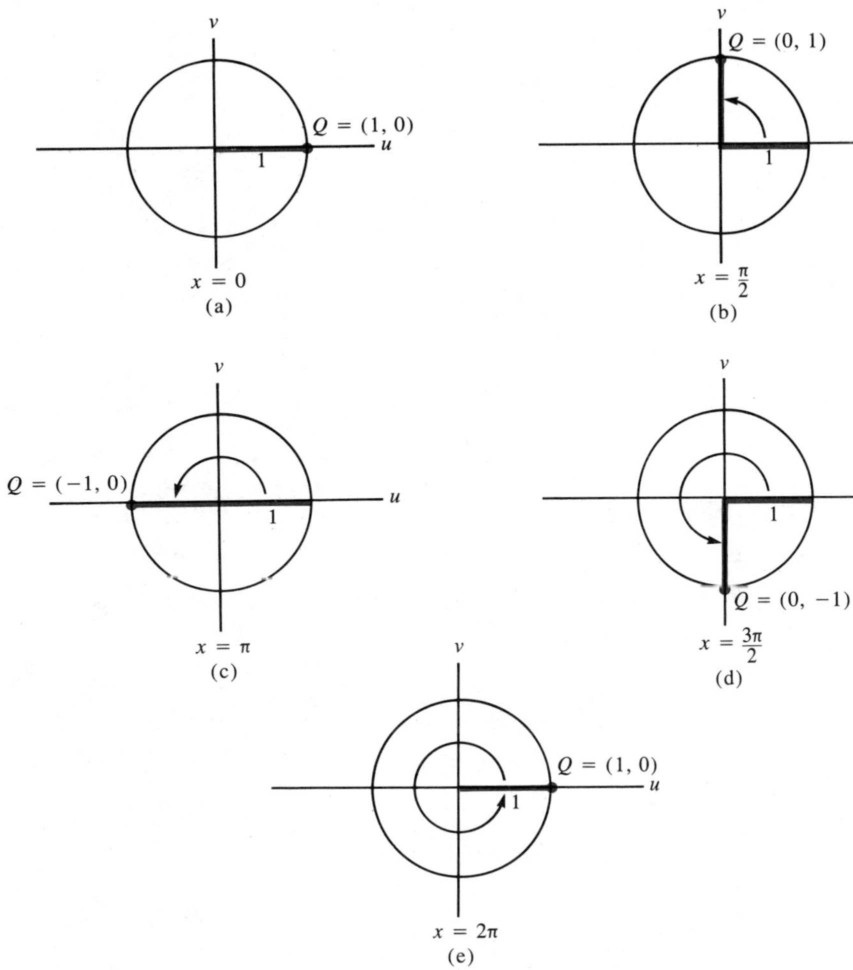

FIGURE A–17

and since $Q = (\cos x, \sin x)$ is a point on this circle, we must have

$$(\cos x)^2 + (\sin x)^2 = 1$$

for all numbers x. An equation such as this which is true for all x is called a **trigonometric identity.** It is common practice to write $\cos^2 x$ for $(\cos x)^2$ and $\sin^2 x$ for $(\sin x)^2$. Hence this trigonometric identity is usually expressed in the form

$$\cos^2 x + \sin^2 x = 1.$$

There are other trigonometric identities that follow from the definitions of the sine and cosine. For instance,

$$\cos(x + 2\pi) = \cos x$$

for all x and

$$\sin(x + 2\pi) = \sin x$$

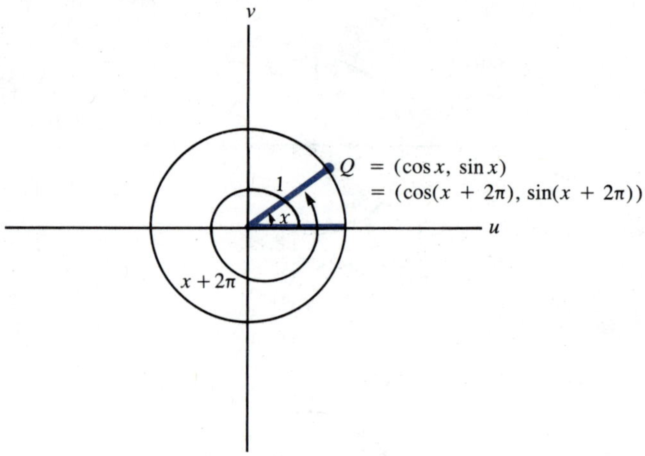

FIGURE A–18

for all x because adding 2π to x yields the same point Q on the circle, as Figure A–18 shows. Similarly,

$$\cos(x - 2\pi) = \cos x \qquad \text{and} \qquad \sin(x - 2\pi) = \sin x$$

for all x. Furthermore, Figure A–19 shows that

$$\cos(-x) = \cos x \qquad \text{and} \qquad \sin(-x) = -\sin x$$

for all x.

The additive identities

$$\sin(x_1 + x_2) = \sin x_1 \cos x_2 + \sin x_2 \cos x_1$$

and

$$\cos(x_1 + x_2) = \cos x_1 \cos x_2 - \sin x_1 \sin x_2$$

will be useful to us. Their proofs are outlined in Exercise 37 at the end of this section. As special cases of the additive identities, we have

$$\begin{aligned}\sin\left(x + \frac{\pi}{2}\right) &= \sin x \cos \frac{\pi}{2} + \sin \frac{\pi}{2} \cos x \\ &= (\sin x)(0) + (1)(\cos x) \\ &= \cos x\end{aligned}$$

and

$$\begin{aligned}\cos\left(x + \frac{\pi}{2}\right) &= \cos x \cos \frac{\pi}{2} - \sin x \sin \frac{\pi}{2} \\ &= (\cos x)(0) - (\sin x)(1) \\ &= -\sin x.\end{aligned}$$

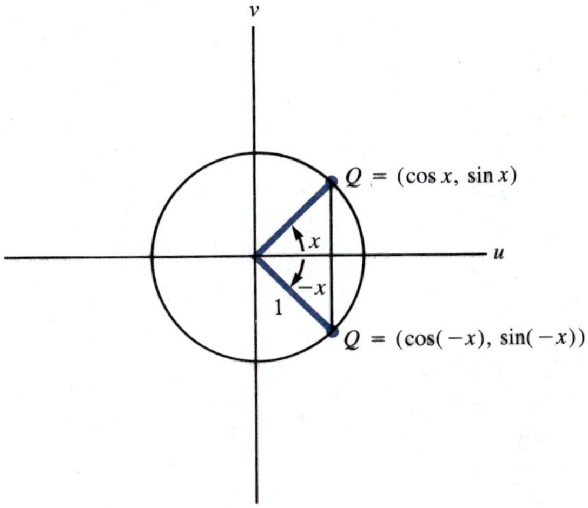

FIGURE A–19

Summary: Trigonometric Identities

The following are true for all x, x_1, and x_2:

1. $\cos^2 x + \sin^2 x = 1$
2. $\cos(x \pm 2\pi) = \cos x, \quad \sin(x \pm 2\pi) = \sin x$
3. $\cos(-x) = \cos x, \quad \sin(-x) = -\sin x$
4. $\sin(x_1 + x_2) = \sin x_1 \cos x_2 + \sin x_2 \cos x_1$
   $\cos(x_1 + x_2) = \cos x_1 \cos x_2 - \sin x_1 \sin x_2$
5. $\sin\left(x + \dfrac{\pi}{2}\right) = \cos x, \quad \cos\left(x + \dfrac{\pi}{2}\right) = -\sin x$

These trigonometric identities in combination with the information collected in Tables A–1 and A–2 will allow us to evaluate the cosine and sine for additional values of x.

EXAMPLE 1 Let us find $\cos 3\pi$ and $\sin(-5\pi/2)$. Using the identity

$$\cos(x + 2\pi) = \cos x,$$

we have

$$\cos 3\pi = \cos(\pi + 2\pi) = \cos \pi = -1;$$

using the identities

$$\sin(-x) = -\sin x \qquad \text{and} \qquad \sin(x + 2\pi) = \sin x,$$

we have $\sin\left(-\frac{5\pi}{2}\right) = -\sin\frac{5\pi}{2} = -\sin\left(\frac{\pi}{2} + 2\pi\right) = -\sin\frac{\pi}{2} = -1.$

EXAMPLE 2 Let us find $\cos \pi/4$ and $\sin \pi/4$. Since the terminal side of an angle of $\pi/4$ radians bisects the first quadrant (Figure A–20), the point $(\cos \pi/4, \sin \pi/4)$ lies on the line $v = u$ in the uv-plane, so

$$\cos\frac{\pi}{4} = \sin\frac{\pi}{4}.$$

But then the identity

$$\cos^2 x + \sin^2 x = 1$$

implies that

$$\sin^2\frac{\pi}{4} + \sin^2\frac{\pi}{4} = 1,$$

so

$$\sin^2\frac{\pi}{4} = \frac{1}{2}.$$

Therefore

$$\sin\frac{\pi}{4} = \pm\frac{1}{\sqrt{2}};$$

however, since $\pi/4$ is between 0 and π, its sine is positive. Hence

$$\sin\frac{\pi}{4} = \cos\frac{\pi}{4} = \frac{1}{\sqrt{2}}.$$

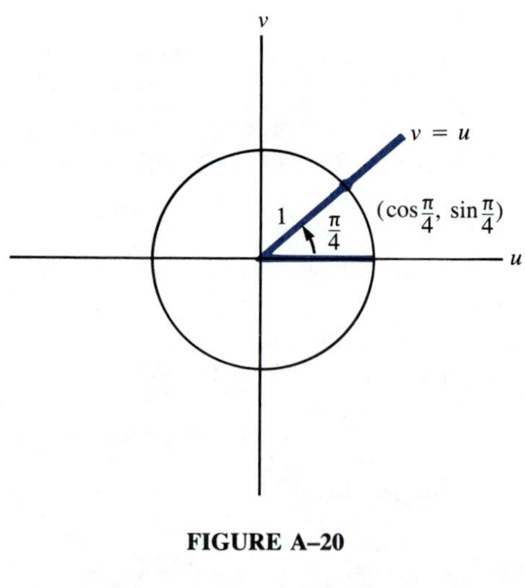

FIGURE A–20

EXAMPLE 3 Let us use the results of Example 2 and the identities

$$\sin\left(x + \frac{\pi}{2}\right) = \cos x \qquad \text{and} \qquad \cos\left(x + \frac{\pi}{2}\right) = -\sin x$$

to find

$$\sin\frac{3\pi}{4}, \quad \cos\frac{3\pi}{4}, \quad \sin\frac{5\pi}{4}, \quad \text{and} \quad \cos\frac{5\pi}{4}.$$

We have

$$\sin\frac{3\pi}{4} = \sin\left(\frac{\pi}{4} + \frac{\pi}{2}\right) = \cos\frac{\pi}{4} = \frac{1}{\sqrt{2}}$$

and

$$\cos\frac{3\pi}{4} = \cos\left(\frac{\pi}{4} + \frac{\pi}{2}\right) = -\sin\frac{\pi}{4} = -\frac{1}{\sqrt{2}}.$$

Then

$$\sin\frac{5\pi}{4} = \sin\left(\frac{3\pi}{4} + \frac{\pi}{2}\right) = \cos\frac{3\pi}{4} = -\frac{1}{\sqrt{2}}$$

and

$$\cos\frac{5\pi}{4} = \cos\left(\frac{3\pi}{4} + \frac{\pi}{2}\right) = -\sin\frac{3\pi}{4} = -\frac{1}{\sqrt{2}}.$$

Fortunately we will not often have to evaluate sines and cosines as we did in Examples 2 and 3. It will be helpful, however, to know the values of $\sin x$ and $\cos x$ when x is between 0 and 2π and is an integral multiple of $\pi/6$, $\pi/4$, or $\pi/3$. Table A–3 gives the values of $\sin x$ and $\cos x$ at these points:

TABLE A–3

	Radians											
	$\frac{\pi}{6}$	$\frac{\pi}{4}$	$\frac{\pi}{3}$	$\frac{2\pi}{3}$	$\frac{3\pi}{4}$	$\frac{5\pi}{6}$	$\frac{7\pi}{6}$	$\frac{5\pi}{4}$	$\frac{4\pi}{3}$	$\frac{5\pi}{3}$	$\frac{7\pi}{4}$	$\frac{11\pi}{6}$
sin	$\frac{1}{2}$	$\frac{1}{\sqrt{2}}$	$\frac{\sqrt{3}}{2}$	$\frac{\sqrt{3}}{2}$	$\frac{1}{\sqrt{2}}$	$\frac{1}{2}$	$-\frac{1}{2}$	$-\frac{1}{\sqrt{2}}$	$-\frac{\sqrt{3}}{2}$	$-\frac{\sqrt{3}}{2}$	$-\frac{1}{\sqrt{2}}$	$-\frac{1}{2}$
cos	$\frac{\sqrt{3}}{2}$	$\frac{1}{\sqrt{2}}$	$\frac{1}{2}$	$-\frac{1}{2}$	$-\frac{1}{\sqrt{2}}$	$-\frac{\sqrt{3}}{2}$	$-\frac{\sqrt{3}}{2}$	$-\frac{1}{\sqrt{2}}$	$-\frac{1}{2}$	$\frac{1}{2}$	$\frac{1}{\sqrt{2}}$	$\frac{\sqrt{3}}{2}$

(See Exercises 9 through 13 at the end of this section.)

For our purposes it will not be necessary to find sines or cosines of any angles other than those we have already considered. In practice, sines and cosines are usually found with the aid of an electronic calculator that has sin and cos keys. Mathematical handbooks also have extensive tables of sines and cosines.

Now we are ready to define the trigonometric functions. For any number x, we define the **sine function** by

$$y = \sin x$$

and the **cosine function** by

$$y = \cos x.$$

The information in Tables A–1, A–2, and A–3 allows us to sketch the graphs of these functions over the interval $[0, 2\pi]$; see Figure A–21. (Check that these graphs reflect the information contained in the tables.) However, the identities

$$\sin(x \pm 2\pi) = \sin x \qquad \text{and} \qquad \cos(x \pm 2\pi) = \cos x$$

imply that the graphs of these functions must look the same over the interval $[2\pi, 4\pi]$ as they do over $[0, 2\pi]$, the same over $[4\pi, 6\pi]$ as over $[2\pi, 4\pi]$, the same over $[-2\pi, 0]$ as over $[0, 2\pi]$, and so on. Hence the graphs of the sine and cosine functions over the entire x-axis are obtained by repeating their graphs over the interval $[0, 2\pi]$, and thus are as depicted in Figure A–22.

A function whose graph repeats itself is said to be **periodic;** its **period** is the length of the interval of repetition. Thus $y = \sin x$ and $y = \cos x$ are periodic functions with period 2π.

We can create functions whose graphs have the same shapes as those of $y = \sin x$ and $y = \cos x$, but with different periods, by multiplying the independent variable x by a constant before we take its sine or cosine. Also, if we multiply $\sin x$ and $\cos x$ by a constant, we will change the maximum and minimum values of y; that is, we will change the height of the "waves" of the graph. We illustrate with an example.

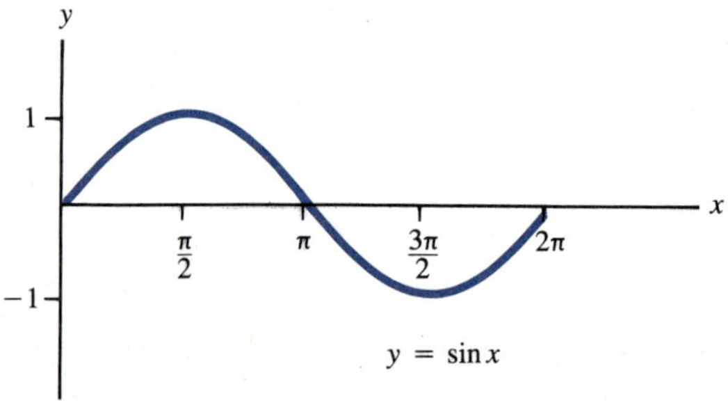

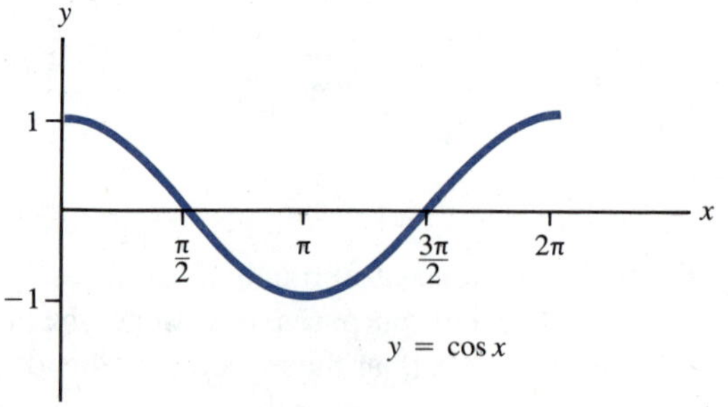

FIGURE A–21

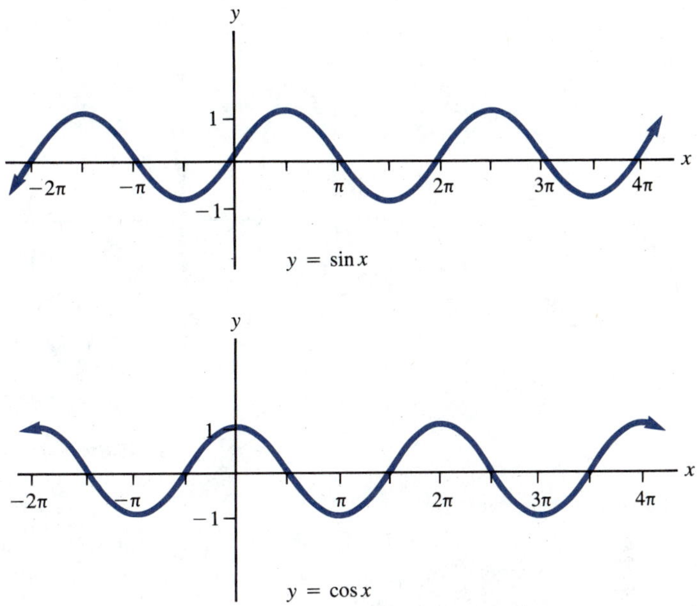

FIGURE A–22

EXAMPLE 4 Let us graph the function $y = 3 \sin 2x$. The graph will have the same shape as that of $y = \sin x$, but its period will be π rather than 2π. The reason for this is that as x ranges from 0 to π, $2x$ ranges from 0 to 2π; hence $\sin 2x$ will begin to repeat as soon as x reaches π. Furthermore, since

$$-1 \leq \sin 2x \leq 1,$$

we must have

$$-3 \leq 3 \sin 2x \leq 3.$$

A table of values for $y = 3 \sin 2x$ over the interval $[0, \pi]$ is as follows:

x:	0	$\frac{\pi}{6}$	$\frac{\pi}{4}$	$\frac{\pi}{3}$	$\frac{\pi}{2}$	$\frac{2\pi}{3}$	$\frac{3\pi}{4}$	$\frac{5\pi}{6}$	π
$2x$:	0	$\frac{\pi}{3}$	$\frac{\pi}{2}$	$\frac{2\pi}{3}$	π	$\frac{4\pi}{3}$	$\frac{3\pi}{2}$	$\frac{5\pi}{3}$	2π
$\sin 2x$:	0	$\frac{\sqrt{3}}{2}$	1	$\frac{\sqrt{3}}{2}$	0	$-\frac{\sqrt{3}}{2}$	-1	$-\frac{\sqrt{3}}{2}$	0
$3 \sin 2x$:	0	$\frac{3\sqrt{3}}{2}$	3	$\frac{3\sqrt{3}}{2}$	0	$-\frac{3\sqrt{3}}{2}$	-3	$-\frac{3\sqrt{3}}{2}$	0

The graph of $y = 3 \sin 2x$ is shown in Figure A–23.

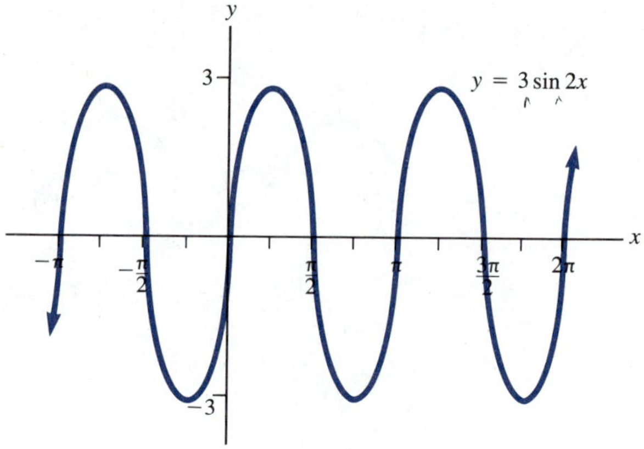

FIGURE A–23

Now we define the remaining four trigonometric functions and display their graphs. For an angle of x radians, x not an odd multiple of $\pi/2$, we define the **tangent** of x, written $\tan x$, by

$$\tan x = \frac{\sin x}{\cos x}.$$

(The tangent of x is not defined if and only if x is an odd multiple of $\pi/2$, because $\cos x = 0$ if and only if x is an odd multiple of $\pi/2$.) Similarly, we define the **cotangent, secant,** and **cosecant** of an angle of x radians, denoted by $\cot x$, $\sec x$, and $\csc x$, respectively, as follows:

$$\cot x = \frac{\cos x}{\sin x}$$

for all x not an integral multiple of π;

$$\sec x = \frac{1}{\cos x}$$

for all x not an odd multiple of $\pi/2$; and

$$\csc x = \frac{1}{\sin x}$$

for all x not an integral multiple of π.

Now we define the **tangent function** by

$$y = \tan x,$$

for all x not an odd multiple of $\pi/2$. The graph of $y = \tan x$ is shown in Figure A–24. Note the vertical asymptotes at the odd multiples of $\pi/2$. The portion of the graph over the interval $(-\pi/2, \pi/2)$ repeats, and hence the tangent function is periodic of period π. Similarly, we define the **cotangent function,** the **secant function** and the **cosecant function** by

$$y = \cot x,$$

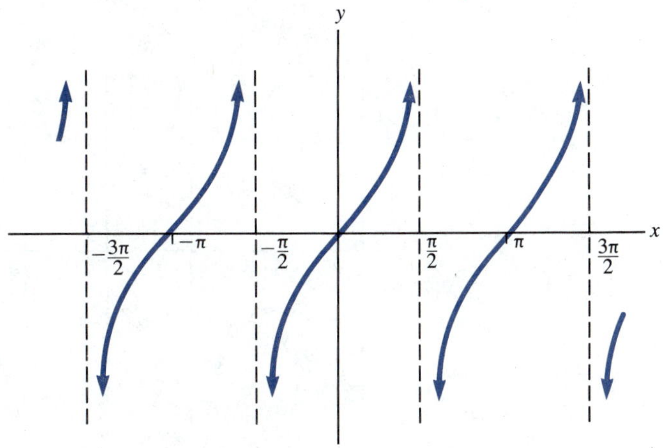

$y = \tan x$

FIGURE A–24

for all x not an integral multiple of π;

$$y = \sec x,$$

for all x not an odd multiple of $\pi/2$; and

$$y = \csc x,$$

for all x not an integral multiple of π, respectively. The graphs of these functions are shown in Figures A–25, A–26, and A–27. Note that the period of the cotangent function is π, and that of the secant and cosecant functions is 2π.

We conclude this section with a brief discussion of the relationship between the trigonometric functions and sides of right triangles. Consider the right triangle

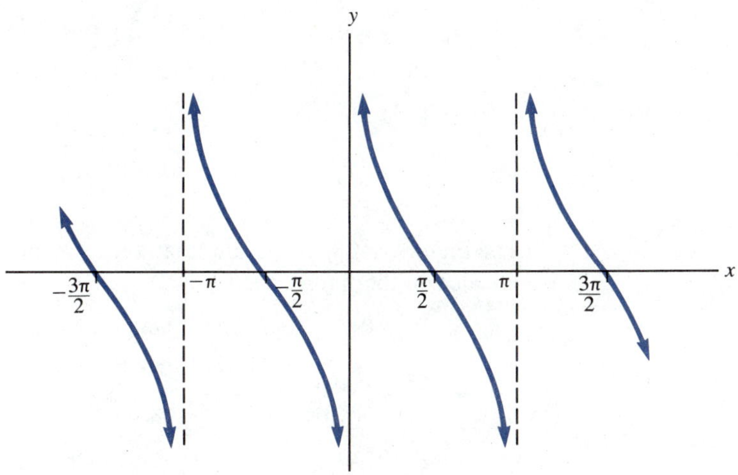

$y = \cot x$

FIGURE A–25

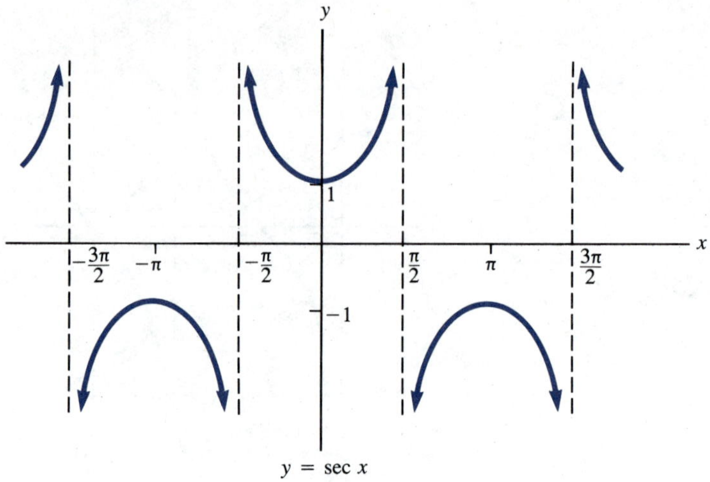

FIGURE A–26

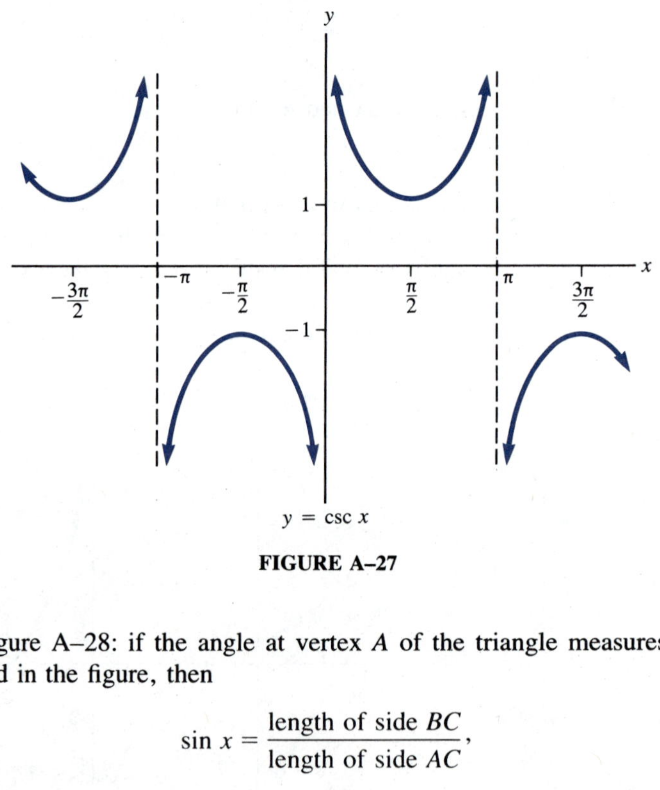

FIGURE A–27

ABC of Figure A–28: if the angle at vertex *A* of the triangle measures x radians as indicated in the figure, then

$$\sin x = \frac{\text{length of side } BC}{\text{length of side } AC},$$

which we write less formally as

$$\sin x = \frac{\text{opposite side}}{\text{hypotenuse}}.$$

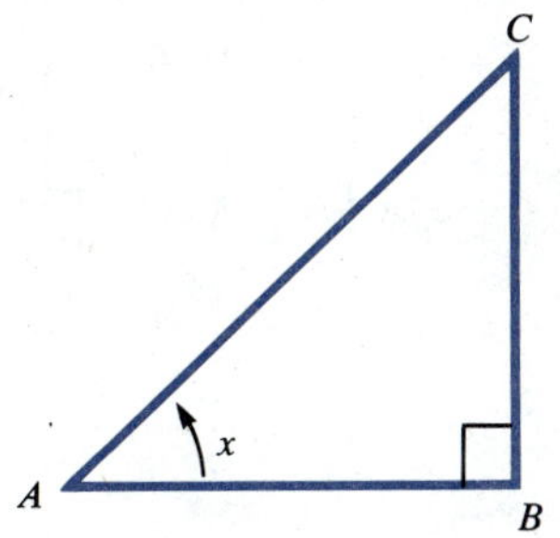

FIGURE A–28

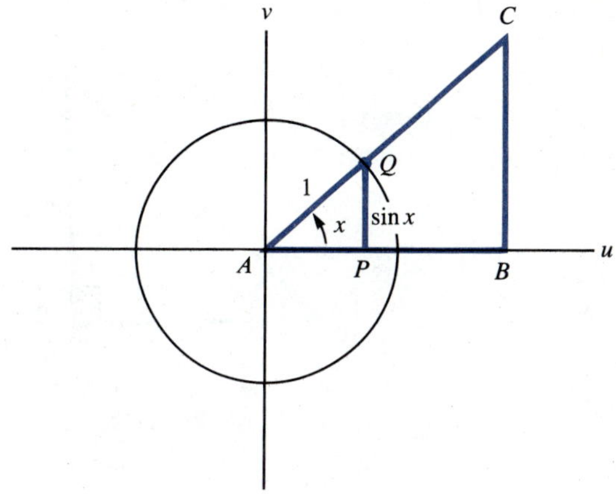

FIGURE A–29

This is so because if we place the vertex A of the triangle at the origin, as in Figure A–29, then triangle APQ is similar to triangle ABC, and hence

$$\frac{\text{length of } PQ}{\text{length of } AQ} = \frac{\text{length of } BC}{\text{length of } AC},$$

or

$$\frac{\sin x}{1} = \frac{\text{length of } BC}{\text{length of } AC}.$$

Similar reasoning shows that

$$\cos x = \frac{\text{length of } AB}{\text{length of } AC} = \frac{\text{adjacent side}}{\text{hypotenuse}},$$

$$\tan x = \frac{\text{opposite side}}{\text{adjacent side}}, \qquad \cot x = \frac{\text{adjacent side}}{\text{opposite side}},$$

$$\sec x = \frac{\text{hypotenuse}}{\text{adjacent side}}, \qquad \csc x = \frac{\text{hypotenuse}}{\text{opposite side}}.$$

EXAMPLE 5 Let us find $\sin x$, $\cos x$, $\tan x$, $\sec x$, $\csc x$, and $\cot x$ for the angle of Figure A–30. The length of the hypotenuse of this triangle is $\sqrt{2^2 + 3^2} = \sqrt{13}$. Thus

$$\sin x = \frac{\text{opposite side}}{\text{hypotenuse}} = \frac{3}{\sqrt{13}},$$

$$\cos x = \frac{\text{adjacent side}}{\text{hypotenuse}} = \frac{2}{\sqrt{13}},$$

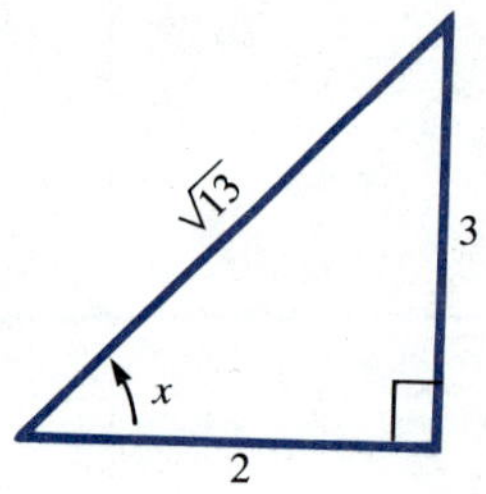

FIGURE A–30

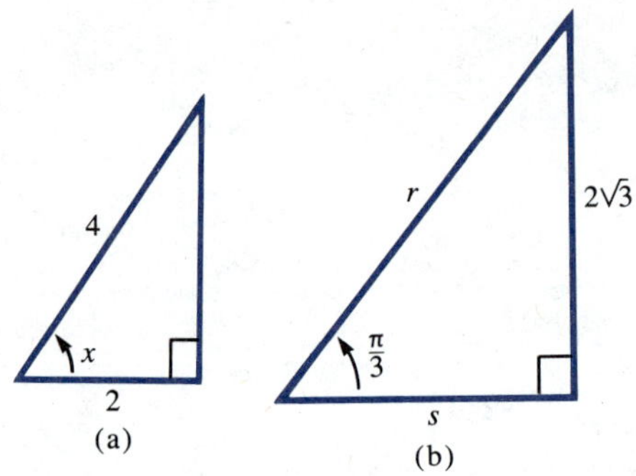

FIGURE A–31

$$\tan x = \frac{\text{opposite side}}{\text{adjacent side}} = \frac{3}{2},$$

$$\sec x = \frac{\text{hypotenuse}}{\text{adjacent side}} = \frac{\sqrt{13}}{2},$$

$$\csc x = \frac{\text{hypotenuse}}{\text{opposite side}} = \frac{\sqrt{13}}{3},$$

and

$$\cot x = \frac{\text{adjacent side}}{\text{opposite side}} = \frac{2}{3}.$$

EXAMPLE 6 Let us find the radian measure x of the angle of Figure A–31(a) and the lengths r and s of Figure A–31(b). From Figure A–31(a),

$$\cos x = \frac{2}{4} = \frac{1}{2}.$$

The angle between 0 and $\pi/2$ whose cosine is ½ is $\pi/3$. Thus $x = \pi/3$.

From Figure A–31(b),

$$\sin\frac{\pi}{3} = \frac{2\sqrt{3}}{r},$$

or

$$\frac{\sqrt{3}}{2} = \frac{2\sqrt{3}}{r},$$

which implies that $r = 4$, and hence that

$$s = \sqrt{r^2 - (2\sqrt{3})^2} = \sqrt{16 - 12} = 2.$$

Computer Manual: Program TRIGFNS.

A.2 EXERCISES

1. Find $\cos 3\pi$, $\sin 3\pi$, $\cos\left(-\frac{3\pi}{2}\right)$, and $\sin\left(-\frac{3\pi}{2}\right)$.

2. Find $\cos(-\pi)$, $\sin(-\pi)$, $\cos\frac{5\pi}{2}$, and $\sin\frac{5\pi}{2}$.

3. Find $\sin\frac{7\pi}{2}$, $\cos\frac{7\pi}{2}$, $\sin(-2\pi)$, and $\cos(-2\pi)$.

4. Find $\sin\left(-\frac{9\pi}{2}\right)$, $\cos\left(-\frac{9\pi}{2}\right)$, $\sin 6\pi$, and $\cos 6\pi$.

5. **(a)** For what values of x is $\sin x = 0$?
(b) For what values of x is $\cos x = 0$?

6. **(a)** For what values of x is $\sin x = 1$?
(b) For what values of x is $\sin x = -1$?

7. **(a)** For what values of x is $\cos x = 1$?
(b) For what values of x is $\cos x = -1$?

8. Use the results of Examples 2 and 3 and trigonometric identities to show that $\cos\frac{7\pi}{4} = \frac{1}{\sqrt{2}}$ and $\sin\frac{7\pi}{4} = -\frac{1}{\sqrt{2}}$.

9. Let ABC be an equilateral triangle each of whose sides has length 2. Since the angles of a triangle add up to 180° and the angles of an equilateral triangle are equal, it follows that each angle of triangle ABC measures 60° or $\pi/3$ radians:

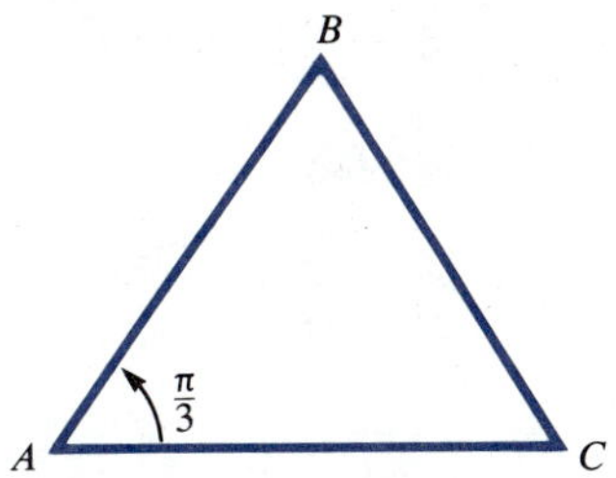

If we construct a perpendicular from vertex B to side AC, it will bisect AC and also bisect the angle at B. Thus we obtain the triangle ABP:

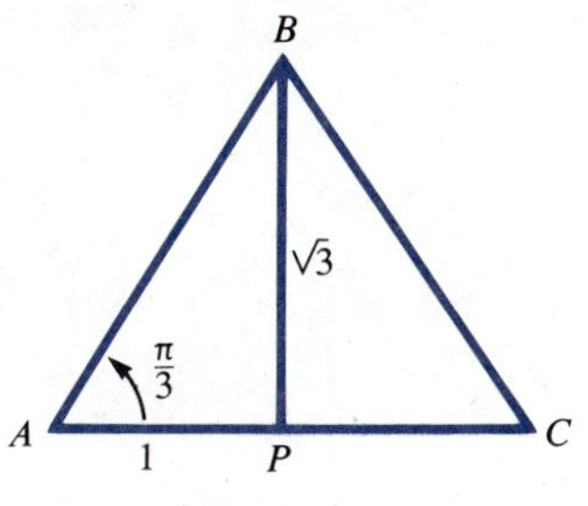

(The length of side BP is equal to $\sqrt{3}$ because ABP is a right triangle; hence $2^2 = 1^2 + (\text{length of } BP)^2$.) Show that

(a) $\sin\frac{\pi}{3} = \frac{\sqrt{3}}{2}$ and $\cos\frac{\pi}{3} = \frac{1}{2}$;

(b) $\sin\frac{\pi}{6} = \frac{1}{2}$ and $\cos\frac{\pi}{6} = \frac{\sqrt{3}}{2}$.

10. Use trigonometric identities and the results of Exercise 9 to show that

(a) $\sin\frac{2\pi}{3} = \frac{\sqrt{3}}{2}$ and $\cos\frac{2\pi}{3} = -\frac{1}{2}$; **(b)** $\sin\frac{5\pi}{6} = \frac{1}{2}$ and $\cos\frac{5\pi}{6} = -\frac{\sqrt{3}}{2}$.

11. Use trigonometric identities and the results of previous exercises to show that

(a) $\sin\frac{4\pi}{3} = -\frac{\sqrt{3}}{2}$ and $\cos\frac{4\pi}{3} = -\frac{1}{2}$; **(b)** $\sin\frac{7\pi}{6} = -\frac{1}{2}$ and $\cos\frac{7\pi}{6} = -\frac{\sqrt{3}}{2}$.

12. Use trigonometric identities and the results of previous exercises to show that

(a) $\sin\frac{5\pi}{3} = -\frac{\sqrt{3}}{2}$ and $\cos\frac{5\pi}{3} = \frac{1}{2}$; **(b)** $\sin\frac{11\pi}{6} = -\frac{1}{2}$ and $\cos\frac{11\pi}{6} = \frac{\sqrt{3}}{2}$.

13. Use trigonometric identities and the results of previous exercises to find $\sin\frac{\pi}{12}$ and $\cos\frac{\pi}{12}$.

14. If possible, find

(a) $\tan 0$, $\sec 0$, $\csc 0$, $\cot 0$; **(b)** $\tan\frac{\pi}{6}$, $\sec\frac{\pi}{6}$, $\csc\frac{\pi}{6}$, $\cot\frac{\pi}{6}$;

(c) $\tan\frac{\pi}{4}$, $\sec\frac{\pi}{4}$, $\csc\frac{\pi}{4}$, $\cot\frac{\pi}{4}$.

15. If possible, find

(a) $\tan\frac{2\pi}{3}$, $\sec\frac{2\pi}{3}$, $\csc\frac{2\pi}{3}$, $\cot\frac{2\pi}{3}$; **(b)** $\tan\frac{\pi}{2}$, $\sec\frac{\pi}{2}$, $\csc\frac{\pi}{2}$, $\cot\frac{\pi}{2}$;

(c) $\tan\pi$, $\sec\pi$, $\csc\pi$, $\cot\pi$.

In Exercises 16 through 26 graph the function and state its period.

16. $y = \sin 3x$ **17.** $y = \cos\frac{1}{2}x$ **18.** $y = 2\cos 3x$

19. $y = 5\sin\frac{x}{3}$ **20.** $y = \frac{1}{2}\sin 4x$ **21.** $y = \frac{4}{3}\cos\frac{2x}{3}$

22. $y = 2\tan x$ **23.** $y = 3\tan 2x$ **24.** $y = 4\sec\frac{1}{2}x$

25. $y = 2\csc 3x$ **26.** $y = 3\cot\frac{x}{4}$

27. If $a > 0$ and $b > 0$, sketch the graphs of

$$y = a\sin bx \quad \text{and} \quad y = a\cos bx.$$

What are the periods of these functions?

In Exercises 28 through 30 find $\sin x$, $\cos x$, $\tan x$, $\sec x$, $\csc x$, and $\cot x$.

28.

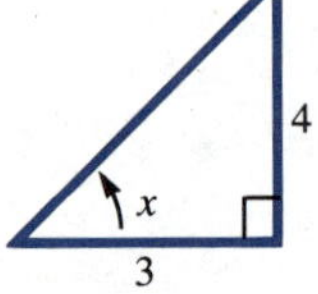

29.

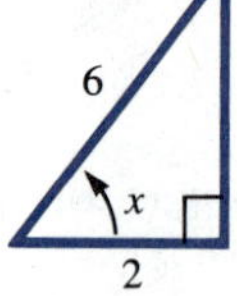

30.

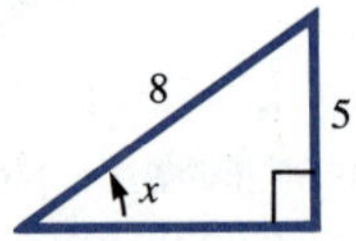

In Exercises 31 through 33 find the lengths r and s.

31.

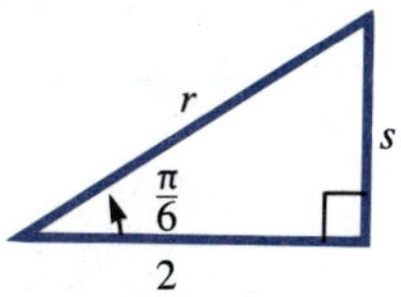

32.

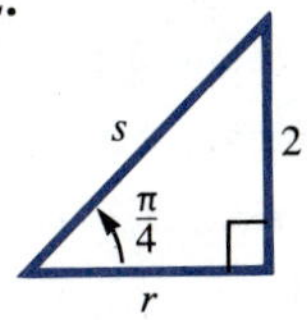

33.

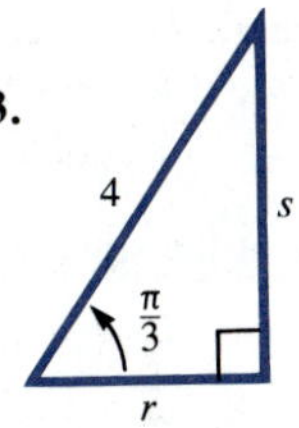

In Exercises 34 through 36 find the radian measure x of the angle.

34.

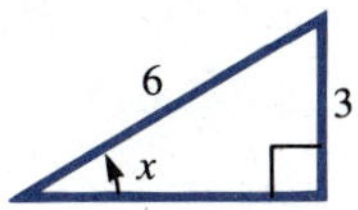

35.

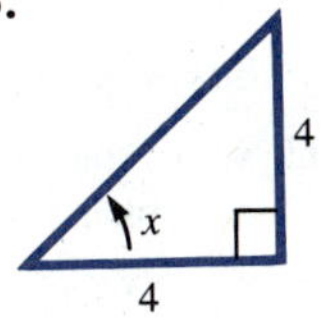

36.

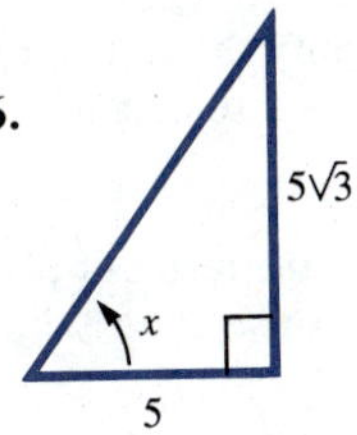

37. In this exercise we outline a proof of the additive trigonometric identities

$$\sin(x + y) = \sin x \cos y + \sin y \cos x$$

and

$$\cos(x + y) = \cos x \cos y - \sin x \sin y.$$

Let angle POQ measure x radians, angle POR measure y radians, and angle POS measure $x + y$ radians:

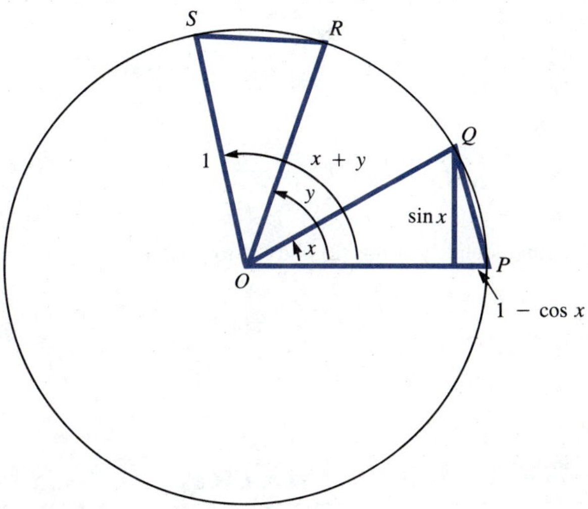

By definition,

$$Q = (\cos x, \sin x), \qquad R = (\cos y, \sin y),$$

and

$$S = (\cos(x + y), \sin(x + y)).$$

Also, since angle ROS measures $x + y - y = x$ radians and angle POQ also measures x radians, it follows that

$$\text{length of segment } SR = \text{length of segment } QP.$$

(a) Show that the square of the length of SR is equal to

$$(\cos y - \cos(x + y))^2 + (\sin y - \sin(x + y))^2$$

and that the square of the length of QP is equal to

$$\sin^2 x + (1 - \cos x)^2.$$

(b) Simplify the expressions of part (a), set them equal to each other, and solve for $\cos x$ to obtain

$$\cos x = \cos y \cos(x + y) + \sin y \sin(x + y).$$

(c) Since the result of part (b) holds for all x and all y, we may substitute $x + y$ for x and $-x$ for y in it. Do this and show that the result is the trigonometric identity

$$\cos(x + y) = \cos x \cos y - \sin x \sin y. \tag{1}$$

(d) Now substitute $-\frac{\pi}{2}$ for x in identity (1) and show that the result is the identity

$$\cos\left(y - \frac{\pi}{2}\right) = \sin y. \tag{2}$$

(e) Now substitute $y - \frac{\pi}{2}$ for y in (2) and expand the result using (1) to obtain the identity

$$\sin\left(y - \frac{\pi}{2}\right) = -\cos y. \tag{3}$$

(f) Now substitute $x + y$ for y in (2) to get

$$\sin(x + y) = \cos\left(x + y - \frac{\pi}{2}\right) = \cos\left(x + \left(y - \frac{\pi}{2}\right)\right);$$

expand $\cos\left(x + \left(y - \frac{\pi}{2}\right)\right)$ using (1) and simplify using (2) and (3) to obtain the identity

$$\sin(x + y) = \sin x \cos y + \sin y \cos x.$$

A.3 DIFFERENTIATION OF TRIGONOMETRIC FUNCTIONS

In this section we develop and use the differentiation formulas for the trigonometric functions. We begin by seeking the derivative of the sine function $y = \sin x$. As we shall see, this will require a knowledge of the limits

$$\lim_{h \to 0} \frac{\sin h}{h} \quad \text{and} \quad \lim_{h \to 0} \frac{\cos h - 1}{h}.$$

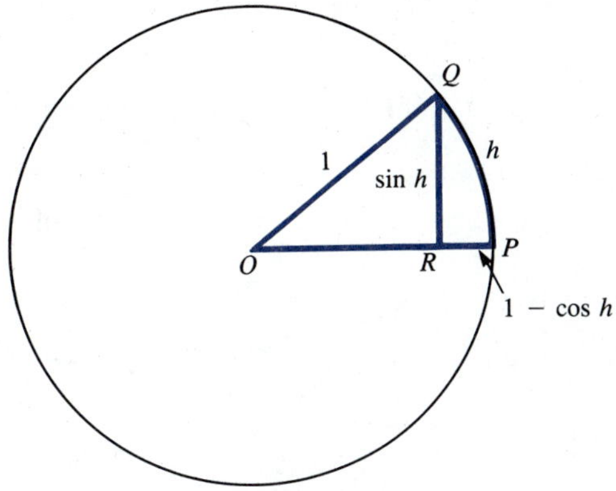

FIGURE A–32

We can determine these limits with the aid of Figure A–32, which depicts an angle POQ of h radians. From the figure we see that

$$\frac{\sin h}{h} = \frac{\text{length of segment } RQ}{\text{length of arc } PQ}.$$

As h approaches 0, that is, as the angle POQ becomes smaller and smaller, the segment RQ and the arc PQ come closer and closer to coinciding, and hence the ratio of their lengths tends toward 1. Therefore

$$\lim_{h \to 0} \frac{\sin h}{h} = 1.$$

Figure A–32 also shows that the length of the segment RP is very small compared to the length of the arc PQ, and the ratio

$$\frac{1 - \cos h}{h} = \frac{\text{length of segment } RP}{\text{length of arc } PQ}$$

becomes smaller and smaller as h approaches 0. Therefore

$$\lim_{h \to 0} \frac{1 - \cos h}{h} = 0,$$

and hence

$$\lim_{h \to 0} \frac{\cos h - 1}{h} = \lim_{h \to 0} \frac{-(1 - \cos h)}{h} = -\lim_{h \to 0} \frac{1 - \cos h}{h} = 0.$$

Now we are ready to find the derivative of $\sin x$. To do so, we must evaluate

$$\lim_{h \to 0} \frac{\sin(x + h) - \sin x}{h}.$$

But

$$\lim_{h\to 0}\frac{\sin(x+h)-\sin x}{h} = \lim_{h\to 0}\frac{\sin x\cos h+\sin h\cos x-\sin x}{h}$$

$$= \lim_{h\to 0}\frac{\sin x(\cos h-1)+\sin h\cos x}{h}$$

$$= \lim_{h\to 0}\sin x\frac{\cos h-1}{h}+\lim_{h\to 0}\cos x\frac{\sin h}{h}$$

$$= (\sin x)\lim_{h\to 0}\frac{\cos h-1}{h}+(\cos x)\lim_{h\to 0}\frac{\sin h}{h}$$

$$= (\sin x)(0)+(\cos x)(1)$$

$$= \cos x.$$

Hence the derivative of the sine is the cosine.

EXAMPLE 1 If $y = 3\sin x$, then

$$\frac{dy}{dx} = \frac{d}{dx}(3\sin x) = 3\frac{d}{dx}(\sin x) = 3\cos x.$$

Now consider a composite function of the form

$$y = \sin u,$$

where u is a function of x. By the chain rule,

$$\frac{dy}{dx} = \frac{d}{dx}(\sin u) = \frac{d}{du}(\sin u)\frac{du}{dx} = (\cos u)\frac{du}{dx}.$$

Derivative of the Sine

$$\frac{d}{dx}(\sin x) = \cos x$$

$$\frac{d}{dx}(\sin u) = (\cos u)\frac{du}{dx}$$

EXAMPLE 2 **a.** If $y = \sin 2x$, then

$$\frac{dy}{dx} = \frac{d}{dx}(\sin 2x) = (\cos 2x)\frac{d}{dx}(2x) = (\cos 2x)(2) = 2\cos 2x.$$

b. If $y = \sin(x^2 - 3x + 5)$, then

$$\begin{aligned}\frac{dy}{dx} &= \frac{d}{dx}(\sin(x^2 - 3x + 5)) \\ &= [\cos(x^2 - 3x + 5)]\frac{d}{dx}(x^2 - 3x + 5) \\ &= [\cos(x^2 - 3x + 5)](2x - 3) \\ &= (2x - 3)\cos(x^2 - 3x + 5).\end{aligned}$$

Now let us find the derivative of the cosine function. Since

$$\cos x = \sin\left(x + \frac{\pi}{2}\right) \qquad \text{and} \qquad \cos\left(x + \frac{\pi}{2}\right) = -\sin x$$

for all x, we have

$$\begin{aligned}\frac{d}{dx}(\cos x) &= \frac{d}{dx}\left(\sin\left(x + \frac{\pi}{2}\right)\right) \\ &= \left[\cos\left(x + \frac{\pi}{2}\right)\right]\frac{d}{dx}\left(x + \frac{\pi}{2}\right) \\ &= \left[\cos\left(x + \frac{\pi}{2}\right)\right](1) \\ &= \cos\left(x + \frac{\pi}{2}\right) \\ &= -\sin x.\end{aligned}$$

The chain rule now shows that if

$$y = \cos u,$$

then

$$\frac{dy}{dx} = \frac{d}{dx}(\cos u) = \frac{d}{du}(\cos u)\frac{du}{dx} = -(\sin u)\frac{du}{dx}.$$

Derivative of the Cosine

$$\frac{d}{dx}(\cos x) = -\sin x$$

$$\frac{d}{dx}(\cos u) = -(\sin u)\frac{du}{dx}$$

EXAMPLE 3 **a.** If $y = 5 \cos x$, then

$$\frac{dy}{dx} = \frac{d}{dx}(5 \cos x) = 5\frac{d}{dx}(\cos x) = 5(-\sin x) = -5 \sin x.$$

b. If $y = \cos(x^2 + 1)$, then

$$\begin{aligned}\frac{dy}{dx} &= \frac{d}{dx}(\cos(x^2 + 1)) \\ &= -[\sin(x^2 + 1)]\frac{d}{dx}(x^2 + 1) \\ &= -[\sin(x^2 + 1)](2x) \\ &= -2x \sin(x^2 + 1).\end{aligned}$$

EXAMPLE 4 **a.** If $y = \sin 2x \cos 3x$, then by the product rule

$$\begin{aligned}\frac{dy}{dx} &= (\sin 2x)\frac{d}{dx}(\cos 3x) + \left[\frac{d}{dx}(\sin 2x)\right](\cos 3x) \\ &= (\sin 2x)(-\sin 3x)\frac{d}{dx}(3x) + (\cos 2x)\left[\frac{d}{dx}(2x)\right](\cos 3x) \\ &= -3 \sin 2x \sin 3x + 2 \cos 2x \cos 3x.\end{aligned}$$

b. If $y = \sin^2 x$, then [since $\sin^2 x$ means $(\sin x)^2$] application of the general power rule shows that

$$\begin{aligned}\frac{dy}{dx} &= \frac{d}{dx}(\sin^2 x) \\ &= \frac{d}{dx}[(\sin x)^2] \\ &= 2(\sin x)\frac{d}{dx}(\sin x) \\ &= 2 \sin x \cos x.\end{aligned}$$

c. If $y = e^{\cos^2 x}$, then

$$\begin{aligned}\frac{dy}{dx} &= \frac{d}{dx}(e^{\cos^2 x}) \\ &= e^{\cos^2 x}\left[\frac{d}{dx}(\cos^2 x)\right] \\ &= e^{\cos^2 x}(2 \cos x)\frac{d}{dx}(\cos x) \\ &= e^{\cos^2 x}(2 \cos x)(-\sin x) \\ &= -2(\sin x \cos x)e^{\cos^2 x}.\end{aligned}$$

EXAMPLE 5 Let us use the curve-sketching procedure to graph the function f defined by

$$f(x) = \sin x + \cos x$$

on the interval $[0, 2\pi]$.

Note that

$$f'(x) = \cos x - \sin x.$$

Setting $f'(x) = 0$ yields

$$\cos x - \sin x = 0$$

or

$$\cos x = \sin x.$$

The only solutions of this equation in the interval $[0, 2\pi]$ are $x = \pi/4$ and $x = 5\pi/4$. Thus we have the diagram

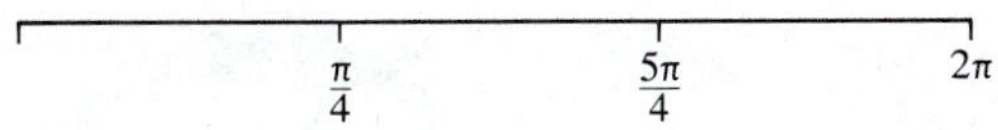

Choosing 0 from the interval $[0, \pi/4)$, π from the interval $(\pi/4, 5\pi/4)$, and 2π from the interval $(5\pi/4, 2\pi]$, we have

$$f'(0) = \cos 0 - \sin 0 = 1 - 0 = 1 > 0,$$

$$f'(\pi) = \cos \pi - \sin \pi = -1 - 0 = -1 < 0,$$

and

$$f'(2\pi) = \cos 2\pi - \sin 2\pi = 1 - 0 = 1 > 0.$$

Therefore the diagram becomes

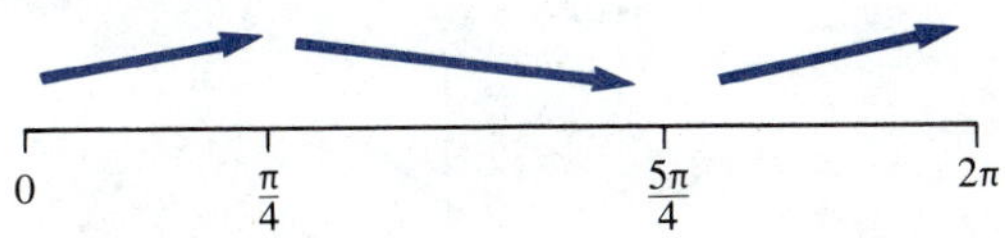

and f has a relative maximum at

$$\left(\frac{\pi}{4}, f\left(\frac{\pi}{4}\right)\right) = \left(\frac{\pi}{4}, \frac{2}{\sqrt{2}}\right) = \left(\frac{\pi}{4}, \sqrt{2}\right)$$

and a relative minimum at

$$\left(\frac{5\pi}{4}, f\left(\frac{5\pi}{4}\right)\right) = \left(\frac{5\pi}{4}, -\frac{2}{\sqrt{2}}\right) = \left(\frac{5\pi}{4}, -\sqrt{2}\right).$$

Since

$$f''(x) = -\sin x - \cos x,$$

setting $f''(x) = 0$ yields

$$\sin x = -\cos x,$$

the solutions of which on $[0, 2\pi]$ are $x = 3\pi/4$ and $x = 7\pi/4$. Thus we have

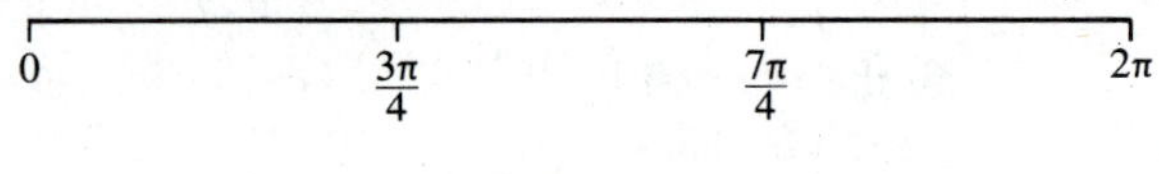

Again selecting 0, π, and 2π as our test numbers, we have

$$f''(0) = -\sin 0 - \cos 0 = 0 - 1 = -1 < 0,$$

$$f''(\pi) = -\sin \pi - \cos \pi = 0 - (-1) = 1 > 0,$$

and

$$f''(2\pi) = -\sin 2\pi - \cos 2\pi = 0 - 1 = -1 < 0.$$

Therefore the diagram for the second derivative becomes

so there are points of inflection at

$$\left(\frac{3\pi}{4}, f\left(\frac{3\pi}{4}\right)\right) = \left(\frac{3\pi}{4}, 0\right) \quad \text{and} \quad \left(\frac{7\pi}{4}, f\left(\frac{7\pi}{4}\right)\right) = \left(\frac{7\pi}{4}, 0\right).$$

The graph of f is shown in Figure A–33.

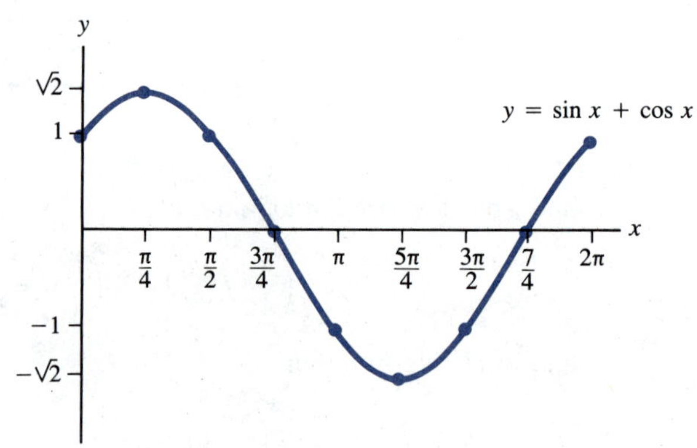

FIGURE A–33

Now that we know the derivatives of the sine and cosine functions, we can easily find the derivatives of the rest of the trigonometric functions. For instance,

$$\frac{d}{dx}(\tan x) = \frac{d}{dx}\left(\frac{\sin x}{\cos x}\right)$$

$$= \frac{(\cos x)\dfrac{d}{dx}(\sin x) - (\sin x)\dfrac{d}{dx}(\cos x)}{(\cos x)^2}$$

$$= \frac{(\cos x)(\cos x) - (\sin x)(-\sin x)}{\cos^2 x}$$

$$= \frac{\cos^2 x + \sin^2 x}{\cos^2 x}$$

$$= \frac{1}{\cos^2 x}$$

$$= \sec^2 x.$$

Similarly,
$$\frac{d}{dx}(\sec x) = \sec x \tan x,$$

$$\frac{d}{dx}(\csc x) = -\csc x \cot x,$$

and
$$\frac{d}{dx}(\cot x) = -\csc^2 x.$$

(See Exercises 23 through 25 at the end of this section.)

Derivatives of the Trigonometric Functions

$$\frac{d}{dx}(\sin u) = (\cos u)\frac{du}{dx} \qquad \frac{d}{dx}(\cos u) = -(\sin u)\frac{du}{dx}$$

$$\frac{d}{dx}(\tan u) = (\sec^2 u)\frac{du}{dx} \qquad \frac{d}{dx}(\cot u) = -(\csc^2 u)\frac{du}{dx}$$

$$\frac{d}{dx}(\sec u) = (\sec u \tan u)\frac{du}{dx} \qquad \frac{d}{dx}(\csc u) = -(\csc u \cot u)\frac{du}{dx}$$

EXAMPLE 6

a.
$$\frac{d}{dx}(\tan^3 x) = (3\tan^2 x)\frac{d}{dx}(\tan x) = 3\tan^2 x \sec^2 x.$$

b.
$$\frac{d}{dx}(\sec e^{2x}) = (\sec e^{2x} \tan e^{2x})\frac{d}{dx}(e^{2x})$$

$$= (\sec e^{2x} \tan e^{2x}) e^{2x} \frac{d}{dx}(2x)$$

$$= 2e^{2x} \sec e^{2x} \tan e^{2x}.$$

Since we can now take derivatives of trigonometric functions, we can analyze the behavior of angles as they change and the behavior of quantities that are described by trigonometric functions. We illustrate with examples.

EXAMPLE 7 A helicopter takes off and rises straight up. If an observer 100 feet away watches the helicopter, and if its altitude t seconds after takeoff is

$$y = 50t$$

feet, find the rate at which the observer's angle of observation is changing 2 seconds after takeoff.

Let x be the radian measure of the angle of observation. See Figure A–34. The rate at which the angle of observation is changing at time t is the derivative of x with respect to t. From Figure A–34 and the relationship between trigonometric functions and the sides of right triangles discussed at the end of Section A.2, we see that

$$\tan x = \frac{\text{opposite side}}{\text{adjacent side}} = \frac{y}{100} = \frac{50t}{100} = \frac{t}{2}.$$

Therefore

$$\frac{d}{dt}(\tan x) = \frac{d}{dt}\left(\frac{t}{2}\right),$$

or

$$(\sec^2 x)\frac{dx}{dt} = \frac{1}{2}.$$

Hence

$$\frac{dx}{dt} = \frac{1}{2\sec^2 x} = \frac{\cos^2 x}{2}.$$

But when $t = 2$, $y = 50(2) = 100$, and Figure A–34 becomes Figure A–35, which shows that

$$\cos x = \frac{100}{100\sqrt{2}} = \frac{1}{\sqrt{2}}.$$

Therefore

$$\left.\frac{dx}{dt}\right|_{t=2} = \frac{(1/\sqrt{2})^2}{2} = \frac{1}{4},$$

so the observer's angle of observation is increasing at a rate of ¼ radian per second when $t = 2$.

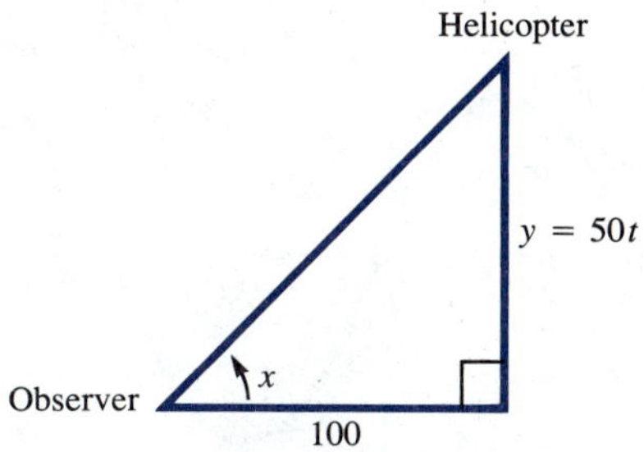

FIGURE A–34

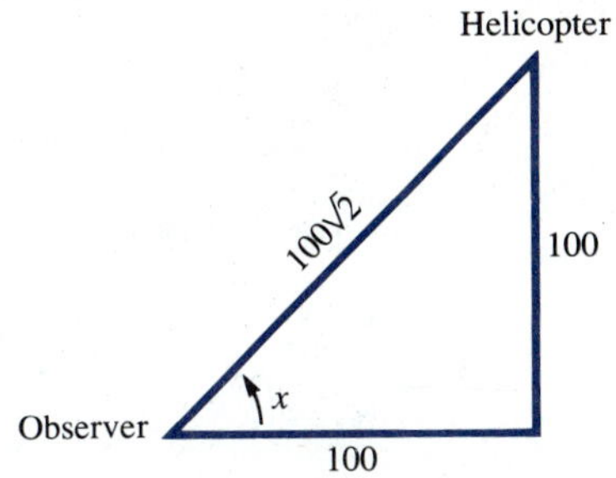

FIGURE A–35

EXAMPLE 8 At time $t = 0$ an electromotive force (generator, battery, etc.) is turned on and begins to supply current to an electrical circuit. At time t the current I in the circuit is given by

$$I = \sin t - \sqrt{3}\cos t + \sqrt{3}e^{-10t},$$

where t is in seconds and I is in amperes. Let us analyze the flow of current in the circuit.

Note that $I(0) = 0$, so initially there is no current in the circuit. Note also that I is the sum of the **steady-state** current

$$S = \sin t - \sqrt{3}\cos t$$

and the **transient** current

$$T = \sqrt{3}e^{-10t}.$$

The transient current quickly becomes negligible (Figure A–36). Thus long-term current flow in the circuit is described by the steady-state current. Applying the curve-sketching procedure to

$$S = \sin t - \sqrt{3}\cos t$$

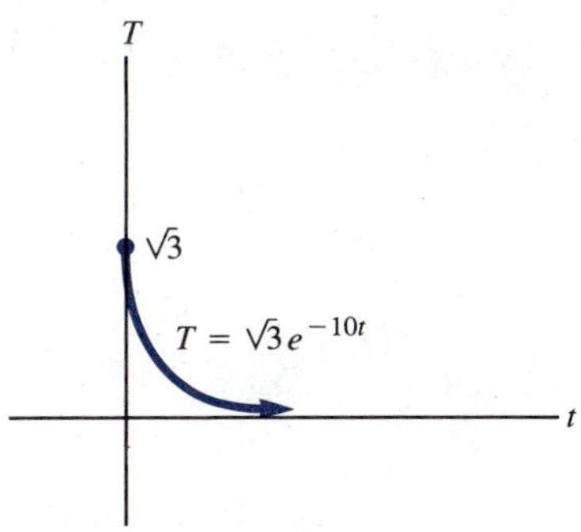

FIGURE A–36

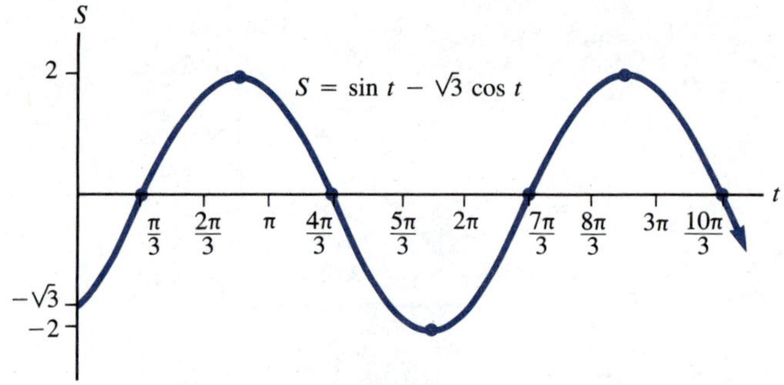

FIGURE A–37

yields the graph of Figure A–37. (Check this.) From the graph we see that the steady-state current alternates, building up to a peak of 2 amperes in one direction, then switching and building to a peak of 2 amperes in the other. The time between successive peaks in the same direction is $t = 2\pi$ seconds.

Computer Manual: Programs LIMIT and DFQUOTNT.

A.3 EXERCISES

In Exercises 1 through 16 find $\frac{dy}{dx}$.

1. $y = 3 \cos x$

2. $y = -5 \sin x$

3. $y = 5 \sin 3x$

4. $y = 6 \cos \pi x$

5. $y = \sin(x^2 + 8x)$

6. $y = -4 \sin \sqrt{2x + 1}$

7. $y = 2 \cos x \sin 2x$

8. $y = -3 \cos 4x \sin(-6x)$

9. $y = \cos^2 7x$

10. $y = \sin^3 x^2$

11. $y = \sin^2 x \cos^3 x$

12. $y = \dfrac{\sin x}{\sqrt{\cos x}}$

13. $y = \left(\dfrac{\cos 4x}{\sin 3x}\right)^4$

14. $y = e^{2 \sin x}$

15. $y = e^{-x} \sin 2x$

16. $y = \ln(\cos x)$

In Exercises 17 through 22 sketch the graph of the function over the interval.

17. $y = \sin x - \cos x, \quad [0, 2\pi]$

18. $y = 2 \sin x + 3 \cos x, \quad [0, 2\pi]$

19. $y = \sqrt{3} \sin 2x + \cos 2x, \quad [0, \pi]$

20. $y = 2 \cos \frac{1}{2}x + 2\sqrt{3} \sin \frac{1}{2}x, \quad [0, 4\pi]$

21. $y = x + \sin x, \quad [0, +\infty)$

22. $y = x - \cos x, \quad [0, +\infty)$

23. Using only the chain rule and the fact that $\frac{d}{dx}(\cos x) = -\sin x$, show that

$$\frac{d}{dx}(\sec x) = \sec x \tan x.$$

24. Using only the chain rule and the fact that $\frac{d}{dx}(\sin x) = \cos x$, show that

$$\frac{d}{dx}(\csc x) = -\csc x \cot x.$$

25. Using only the chain rule and the fact that $\frac{d}{dx}(\tan x) = \sec^2 x$, show that

$\frac{d}{dx}(\cot x) = -\csc^2 x$.

In Exercises 26 through 35 find $\frac{dy}{dx}$.

26. $y = \tan 3x$

27. $y = \sec x^2$

28. $y = 3\csc(x^3 + x^2 + x + 1)$

29. $y = \cot e^x$

30. $y = e^{\cot x}$

31. $y = \sec x \tan x$

32. $y = \csc 3x \sec^2 2x$

33. $y = 1 + \tan^2 x$

34. $y = \ln(\csc x)$

35. $y = \ln(\sec x + \tan x)$

36. A rocket rises straight up from its launch pad and t seconds after liftoff is at an altitude of $y = 100t^2$ meters. An observer watches the rocket from a point 2000 meters from the launch pad. Find the rate at which the observer's angle of observation is changing 3 seconds after liftoff.

37. At time $t = 0$ an airplane passes directly over an observer on the ground. If the plane flies at an altitude of 6000 feet and its ground speed is 600 feet per second, find the rate at which the observer's angle of observation is changing when $t = 10$.

38. Suppose the current I flowing in a circuit at time t is given by

$$I = \sin 2t - \cos 2t + e^{-20t},$$

I in amperes and t in seconds. Analyze current flow in the circuit as we did in Example 8.

39. Repeat Exercise 38 if $I = 2\sqrt{3}\sin 4t - 2\cos 4t + 2e^{-5t}$.

40. If a weight is hung on the end of a spring, allowed to come to rest, then pulled down from its rest position and released it will begin to oscillate about its rest position as the spring expands and contracts. Let

$$y = \sin 3t + \cos 3t$$

be the displacement of the weight from its rest position at time t. Here t is in seconds since the weight was released and y is in inches; a positive value of y denotes displacement below the rest position and a negative value of y displacement above the rest position.

(a) Find the weight's position, relative to its rest position, when $t = 0$, $t = \frac{\pi}{6}$, and $t = \frac{\pi}{4}$.

(b) Find the velocity and acceleration of the weight when $t = 0$, $t = \frac{\pi}{6}$, and $t = \frac{\pi}{4}$.

(c) Sketch the graph of y. What is the maximum displacement of the weight? How long does it take the weight to make one complete oscillation?

41. The model of Exercise 40 is unrealistic because it assumes that the weight will never stop oscillating about its rest position. A more realistic description of the movement of the weight is given by the equation

$$y = e^{-t}(2\sin 3t + \cos 3t).$$

Answer the questions of Exercise 40 using this model.

A.4 INTEGRATION OF TRIGONOMETRIC FUNCTIONS

In this section we discuss the integration of trigonometric functions. Since every differentiation formula yields an integration formula, we immediately have the following:

Trigonometric Integration Formulas

$$\int \sin u\, du = -\cos u + c, \qquad \int \cos u\, du = \sin u + c,$$

$$\int \sec^2 u\, du = \tan u + c, \qquad \int \csc^2 u\, du = -\cot u + c,$$

$$\int \sec u \tan u\, du = \sec u + c, \qquad \int \csc u \cot u\, du = -\csc u + c.$$

EXAMPLE 1 Let us find

$$\int \cos 2x\, dx.$$

Since we know from the formulas preceding the example that $\sin u$ is an antiderivative of $\cos u$, we can guess that an antiderivative of $\cos 2x$ must involve $\sin 2x$. However, since

$$\frac{d}{dx}(\sin 2x) = 2 \cos 2x,$$

it is clear that in fact $\frac{1}{2} \sin 2x$ is an antiderivative of $\cos 2x$. Thus

$$\int \cos 2x\, dx = \frac{1}{2} \sin 2x + c.$$

EXAMPLE 2 Let us find

$$\int_0^{\pi/4} 2 \sec^2 3x\, dx.$$

We have

$$\int 2 \sec^2 3x\, dx = 2 \int \sec^2 3x\, dx = 2\left(\frac{1}{3} \tan 3x\right) + c = \frac{2}{3} \tan 3x + c.$$

Therefore

$$\int_0^{\pi/4} 2 \sec^2 3x\, dx = \frac{2}{3} \tan 3x \Big|_0^{\pi/4}$$

$$= \frac{2}{3}\left(\tan \frac{3\pi}{4} - \tan 0\right)$$

$$= \frac{2}{3}(-1 - 0)$$

$$= -\frac{2}{3}.$$

EXAMPLE 3 We will use integration by substitution to find

$$\int x \sin(x^2 + 1)\, dx.$$

We recognize that, except for a constant factor of 2, x is the derivative of $x^2 + 1$. Therefore the substitution $u = x^2 + 1$ will simplify the integral. If we set

$$u = x^2 + 1,$$

we have

$$du = 2x\, dx,$$

or

$$\frac{1}{2}\, du = x\, dx.$$

Thus

$$\int x \sin(x^2 + 1)\, dx = \frac{1}{2}\int \sin u\, du$$

$$= -\frac{1}{2}\cos u + c$$

$$= -\frac{1}{2}\cos(x^2 + 1) + c.$$

Integration by substitution is often useful when it is necessary to integrate products and quotients of trigonometric functions. We illustrate with examples.

EXAMPLE 4 Find

$$\int \sin^2 x \cos x\, dx.$$

Since $\cos x$ is the derivative of $\sin x$, the substitution $u = \sin x$ will simplify the integral. We have

$$u = \sin x, \qquad du = \cos x\, dx,$$

and thus

$$\int \sin^2 x \cos x\, dx = \int u^2\, du = \frac{u^3}{3} + c = \frac{\sin^3 x}{3} + c.$$

EXAMPLE 5 Find

$$\int \frac{\sin 2x}{1 + \cos 2x}\,dx.$$

Note that $\sin 2x$ is the derivative of $1 + \cos 2x$ (except for a constant factor), and thus the substitution $u = 1 + \cos 2x$ will simplify the integral. We have

$$u = 1 + \cos 2x, \qquad du = (-2 \sin 2x)\,dx,$$

and therefore

$$\begin{aligned}\int \frac{\sin 2x}{1 + \cos 2x}\,dx &= -\frac{1}{2}\int \frac{du}{u}\\ &= -\frac{1}{2}\ln|u| + c\\ &= -\frac{1}{2}\ln|1 + \cos 2x| + c.\end{aligned}$$

We can, of course, use definite integrals of trigonometric functions as we previously used definite integrals of other functions: to find areas, to find average values, and so on.

EXAMPLE 6 Consider the shaded region in Figure A–38. The area of this region is

$$\begin{aligned}\int_0^{\pi/4} (\cos x - \sin x)\,dx &= \sin x + \cos x \Big|_0^{\pi/4}\\ &= \left(\sin\frac{\pi}{4} + \cos\frac{\pi}{4}\right) - (\sin 0 + \cos 0)\\ &= \left(\frac{1}{\sqrt{2}} + \frac{1}{\sqrt{2}}\right) - (0 + 1)\\ &= \frac{2}{\sqrt{2}} - 1.\end{aligned}$$

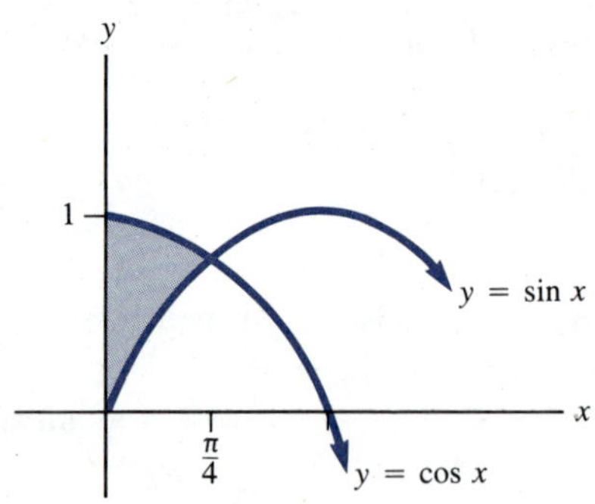

FIGURE A–38

EXAMPLE 7 Let us find the average value of the function $y = \sin x$ over the interval $[0, \pi/2]$. We have

$$\begin{aligned}\text{average value} &= \frac{1}{\pi/2 - 0}\int_0^{\pi/2} \sin x\,dx \\ &= \frac{2}{\pi}\left(-\cos x\Big|_0^{\pi/2}\right) \\ &= \frac{2}{\pi}\left(-\cos\frac{\pi}{2} + \cos 0\right) \\ &= \frac{2}{\pi}(0 + 1) \\ &= \frac{2}{\pi}.\end{aligned}$$

See Figure A–39.

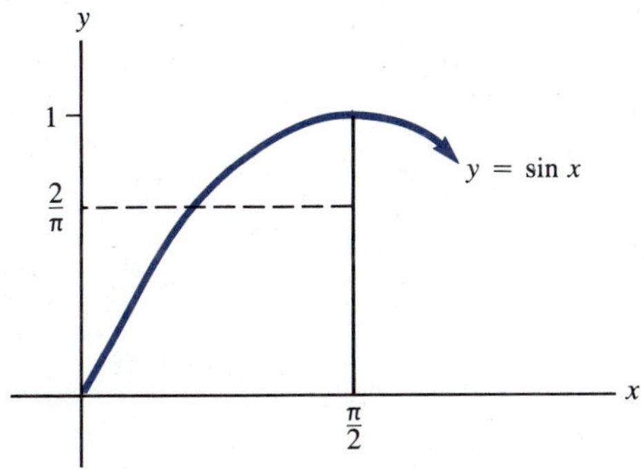

FIGURE A–39

We know that

$$\int \sin u\,du = \cos u + c$$

and

$$\int \cos u\,du = -\sin u + c,$$

but we do not yet possess integration formulas for $\tan u$, $\cot u$, $\sec u$, and $\csc u$. We will now derive the integration formulas for these functions.

To find

$$\int \tan x\,dx,$$

we write

$$\tan x = \frac{\sin x}{\cos x}$$

and make the substitution $u = \cos x$, since then $du = -\sin x\,dx$ and we obtain

$$\int \tan x\, dx = \int \frac{\sin x}{\cos x}\, dx$$
$$= -\int \frac{du}{u}$$
$$= -\ln|u| + c$$
$$= -\ln|\cos x| + c.$$

Similarly,

$$\int \cot x\, dx = \ln|\sin x| + c.$$

Finding

$$\int \sec x\, dx$$

is not so easy. The derivation of the integration formula for sec x rests upon the observation that

$$\frac{d}{dx}(\sec x + \tan x) = \sec x \tan x + \sec^2 x$$
$$= \sec x\, (\sec x + \tan x).$$

Thus if we write

$$\int \sec x\, dx = \int \frac{\sec x\, (\sec x + \tan x)}{\sec x + \tan x}\, dx,$$

the numerator of the integrand on the right side of the equation is the derivative of its denominator. Hence if

$$u = \sec x + \tan x,$$

then

$$du = \sec x\, (\sec x + \tan x)\, dx,$$

and

$$\int \sec x\, dx = \int \frac{\sec x\, (\sec x + \tan x)}{\sec x + \tan x}\, dx,$$
$$= \int \frac{du}{u}$$
$$= \ln|u| + c$$
$$= \ln|\sec x + \tan x| + c.$$

Similarly,

$$\int \csc x\, dx = -\ln|\csc x + \cot x| + c.$$

Integrals of the Trigonometric Functions

$$\int \sin x\,dx = -\cos x + c,$$

$$\int \cos x\,dx = \sin x + c,$$

$$\int \tan x\,dx = -\ln|\cos x| + c,$$

$$\int \cot x\,dx = \ln|\sin x| + c,$$

$$\int \sec x\,dx = \ln|\sec x + \tan x| + c,$$

$$\int \csc x\,dx = -\ln|\csc x + \cot x| + c.$$

EXAMPLE 8 Let us find the area of the region bounded by the x-axis and the graph of $y = \tan x$ from $x = 0$ to $x = \pi/3$. See Figure A–40. We have

$$\begin{aligned}\int_0^{\pi/3} \tan x\,dx &= -\ln|\cos x|\Big|_0^{\pi/3}\\ &= -\ln\left|\cos\frac{\pi}{3}\right| + \ln|\cos 0|\\ &= -\ln\left|\frac{1}{2}\right| + \ln|1|\\ &= -\ln\frac{1}{2}\\ &= \ln 2.\end{aligned}$$

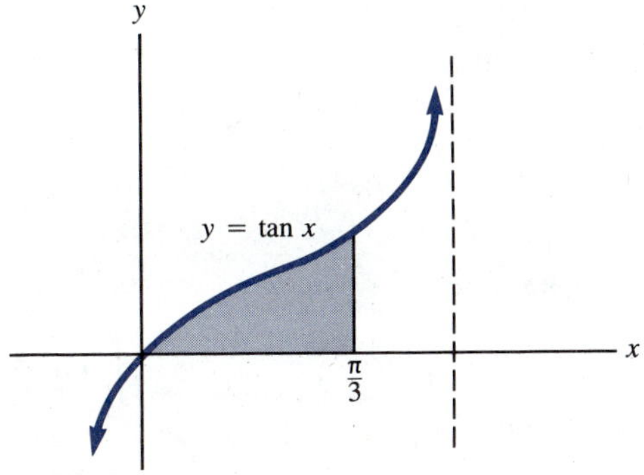

FIGURE A–40

Finally we consider integrals which "mix" trigonometric functions with other types of functions, such as

$$\int x \sin x \, dx \qquad \text{and} \qquad \int e^x \cos x \, dx.$$

Integrals such as these can often be handled with the aid of the integration by parts formula

$$\int u \, dv = uv - \int v \, du$$

which we introduced in Section 6.5. We illustrate with examples.

EXAMPLE 9 Find

$$\int x \sin x \, dx.$$

If we let

$$u = x \qquad \text{and} \qquad dv = \sin x \, dx,$$

then

$$du = dx \qquad \text{and} \qquad v = \int \sin x \, dx = -\cos x.$$

Therefore

$$\begin{aligned} \int x \sin x \, dx &= \int u \, dv \\ &= uv - \int v \, du \\ &= x(-\cos x) - \int -\cos x \, dx \\ &= -x \cos x + \int \cos x \, dx \\ &= -x \cos x + \sin x + c. \end{aligned}$$

EXAMPLE 10 Let us find

$$\int e^x \cos x \, dx.$$

We will use integration by parts with

$$u = e^x \qquad \text{and} \qquad dv = \cos x \, dx,$$

so that

$$du = e^x \, dx \qquad \text{and} \qquad v = \int \cos x \, dx = \sin x.$$

Then

$$\int u \, dv = uv - \int v \, du$$

becomes

$$\int e^x \cos x \, dx = e^x \sin x - \int e^x \sin x \, dx.$$

Now we try integration by parts on $\int e^x \sin x\, dx$. If we let

$$u = e^x \qquad \text{and} \qquad dv = \sin x\, dx,$$

then

$$du = e^x\, dx \qquad \text{and} \qquad v = \int \sin x\, dx = -\cos x,$$

so

$$\int e^x \sin x\, dx = e^x(-\cos x) - \int -e^x \cos x\, dx.$$

Therefore

$$\int e^x \cos x\, dx = e^x \sin x - [-e^x \cos x + \int e^x \cos x\, dx],$$

or

$$\int e^x \cos x\, dx = e^x \sin x + e^x \cos x - \int e^x \cos x\, dx.$$

Notice that the integral

$$\int e^x \cos x\, dx$$

appears on the right side of the last equation preceded by a minus sign. Therefore if we add it to both sides of the equation we obtain

$$2\int e^x \cos x\, dx = e^x(\sin x + \cos x).$$

Hence

$$\int e^x \cos x\, dx = \frac{1}{2} e^x(\sin x + \cos x) + c.$$

Computer Manual: Program DEFINT.

A.4 EXERCISES

In Exercises 1 through 16 find the integral.

1. $\int \sin 3x\, dx$

2. $\int 2\cos(-4x)\, dx$

3. $\int \csc^2(-\pi x)\, dx$

4. $\int_{\pi/6}^{\pi/3} \sec x \tan x\, dx$

5. $\int_0^{\pi} x \cos x^2\, dx$

6. $\int x^2 \sec^2 x^3\, dx$

7. $\int \frac{\sin\sqrt{x}}{\sqrt{x}}\, dx$

8. $\int \frac{\cos x}{2 + \sin x}\, dx$

9. $\int_{-\pi/4}^{\pi/4} \sin 2x \cos 2x\, dx$

10. $\int_{\pi}^{2\pi} \cos^2 x \sin x\, dx$

11. $\int \sin^3 2x \cos 2x\, dx$

12. $\int \tan x \sec^2 x\, dx$

13. $\int_0^{\pi/2} e^{-\cos x} \sin x\, dx$

14. $\int \frac{1}{x^2} \sec \frac{1}{x} \tan \frac{1}{x}\, dx$

15. $\int \sqrt{1 - \cos^2 x}\, dx$

16. $\int_0^{\pi/6} \frac{\csc x \cot x}{2 + \cot x}\, dx$

17. Sketch and find the area of

(a) the region bounded by the x-axis and the graph of $y = \sin x$ from $x = 0$ to $x = \pi$;

(b) the region bounded by the x-axis and the graph of $y = \cos x$ from $x = 0$ to $x = \frac{\pi}{2}$.

18. Sketch and find the area of

(a) the region bounded by $y = \sin x$ and $y = \cos x$ from $x = \frac{\pi}{4}$ to $x = \frac{5\pi}{4}$;

(b) the region bounded by $y = \sin x$ and $y = \cos x$ from $x = \frac{5\pi}{4}$ to $x = 2\pi$.

19. Over the interval $[0, \pi]$, find the average value of

(a) $y = \sin x$ **(b)** $y = \cos x$

20. Let $I = \sin 2t - \cos 2t$ be the steady-state current in an electrical circuit at time $t \geq 0$, where t is in seconds and I is in amperes. Find the average current in the circuit over the interval

(a) $\left[0, \frac{\pi}{4}\right]$ **(b)** $\left[0, \frac{\pi}{2}\right]$ **(c)** $[0, \pi]$

In Exercises 21 through 26 find the integral.

21. $\int \tan 5x\, dx$ **22.** $\int_{\pi/4}^{\pi/3} \cot 2x\, dx$ **23.** $\int x \csc x^2\, dx$ **24.** $\int e^{2x} \sec e^{2x}\, dx$

25. $\int_0^{\pi/4} (1 + \tan^2 x)\, dx$

(*Hint:* Dividing through the identity $\sin^2 x + \cos^2 x = 1$ by $\cos^2 x$ shows that $1 + \tan^2 x = \sec^2 x$.)

26. $\int \tan^3 x\, dx$ (See the hint for Exercise 25.)

27. Sketch and find the area of the region bounded by $y = \sec x$ and $y = 1$ from $x = -\frac{\pi}{3}$ to $x = \frac{\pi}{3}$.

28. Sketch and find the area of the region bounded by $y = \tan x$, the y-axis, and $y = \sqrt{3}$.

29. Find the average value of $y = \sec x$ over the interval $\left[-\frac{\pi}{4}, \frac{\pi}{4}\right]$.

30. Find the average value of $y = \tan x$ over the interval $\left[\frac{\pi}{6}, \frac{\pi}{4}\right]$.

In Exercises 31 through 37 use integration by parts to find the integral.

31. $\int x \cos x\, dx$ **32.** $\int x \sec^2 x\, dx$ **33.** $\int_0^{\pi/2} x^2 \sin x\, dx$ **34.** $\int_0^{\pi/4} x^3 \cos x\, dx$

35. $\int_0^{\pi} e^{-x} \sin x\, dx$ **36.** $\int e^{2x} \cos 2x\, dx$ **37.** $\int \sin^2 x\, dx$

APPENDIX REVIEW EXERCISES

Answers to all review exercises are in the back of the book.

In Exercises 1 through 4 state the degree measure and the radian measure of the given angles. Do not use the conversion formula.

1. 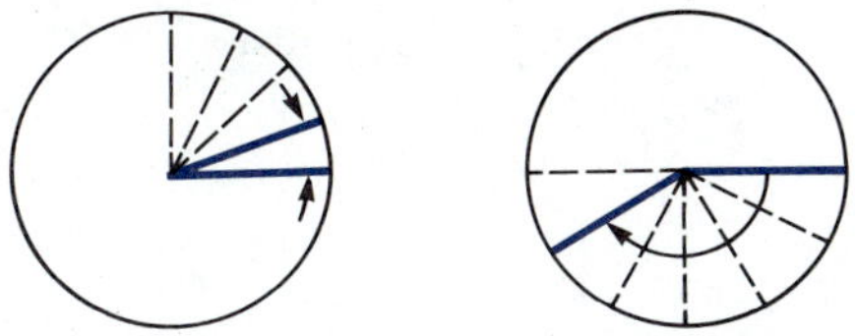

2. 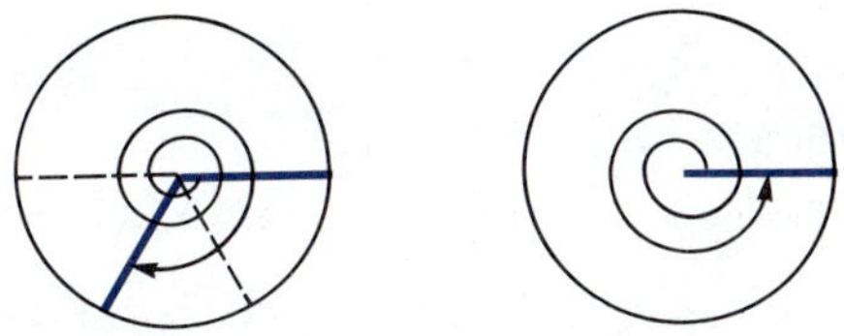

3. 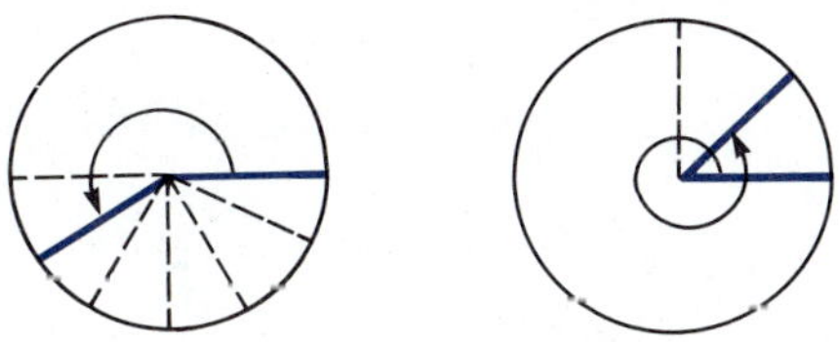

4. 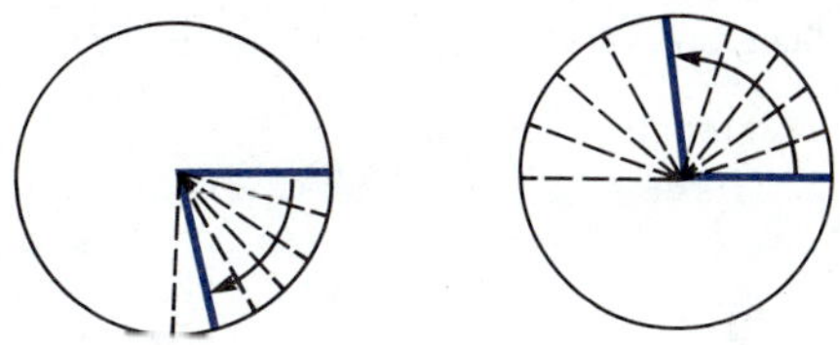

In Exercises 5 through 8 construct the angles whose measures are given.

5. $-60°, 225°, 240°, -300°$

6. $20°, 50°, 25°, 15°$

7. $-5\pi, \frac{5\pi}{3}, \frac{11\pi}{4}, -\frac{7\pi}{6}$ radians

8. $\frac{3\pi}{8}, \frac{3\pi}{16}, \frac{5\pi}{12}, \frac{\pi}{10}$ radians

9. Convert to radians: $135°, -240°, 22.5°, -4°$

10. Convert to degrees: $\frac{4\pi}{3}, -\frac{5\pi}{4}, \frac{\pi}{15}$, 3.25 radians

In Exercises 11 through 16 find sin x, cos x, tan x, csc x, sec x, and cot x for the given value of x, if possible.

11. $x = \frac{3\pi}{2}$ **12.** $x = \frac{5\pi}{4}$ **13.** $x = -4\pi$ **14.** $x = \frac{7\pi}{3}$ **15.** $x = \frac{17\pi}{6}$ **16.** $x = -\frac{11\pi}{6}$

17. Graph the function $y = \frac{1}{2}\cos(-2x)$ and state its period.

18. Graph the function $y = 2\sin\frac{3x}{4}$ and state its period.

19. Graph the functions

(a) $y = \tan\frac{x}{2}$ **(b)** $y = 2\csc(x + \pi)$

20. Find

(a) sin x, cos x, tan x, csc x, sec x, and tan x if 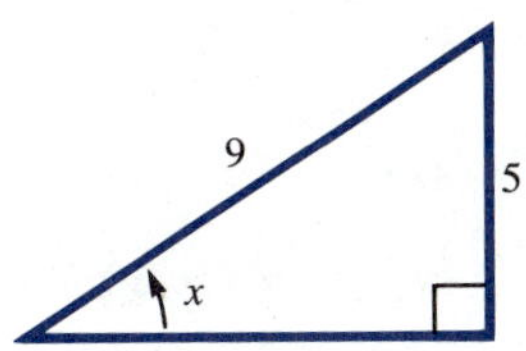

(b) a and b if 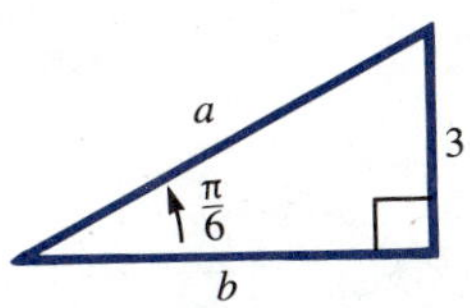

(c) the radian measure of x if 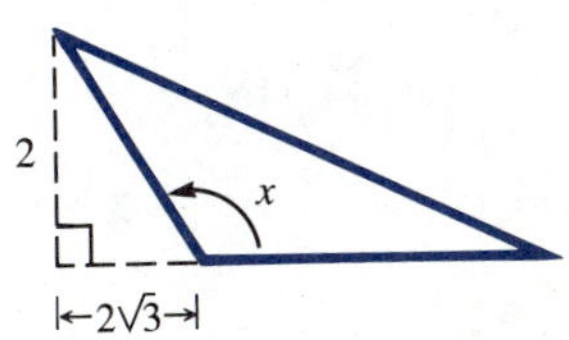

In Exercises 21 through 30 find $\frac{dy}{dx}$.

21. $y = \sin 3x$

22. $y = \cos 2\sqrt{x}$

23. $y = \frac{\sin x}{x}$

24. $y = \frac{1}{6} \sin 3x \cos 2x$

25. $y = e^{\tan x}$

26. $y = 2 \cot 3x^2$

27. $y = \frac{\tan 4x}{\csc 2x}$

28. $y = \sin^3(-4x)$

29. $y = \sec^2 (\tan x)$

30. $y = \ln(\sin x + \csc x)$

31. Sketch the graph of $y = \sin 2x + \cos 2x$ over the interval $[0, \pi]$.

32. Sketch the graph of $y = \sin 3x - \sqrt{3} \cos 3x$ over the interval $\left[0, \frac{2\pi}{3}\right]$.

33. Sketch the graph of $y = x - \tan x$ over the interval $(-\pi/2, \pi/2)$.

In Exercises 34 through 46 find the integral.

34. $\int_{-\pi/6}^{\pi/2} \cos 5x \, dx$

35. $\int 2x \sin x^2 \, dx$

36. $\int 2 \csc 5x \cot 5x \, dx$

37. $\int \cos^4 x \sin x \, dx$

38. $\int_{\pi/8}^{\pi/6} 3 \tan 2x \, dx$

39. $\int \sqrt{1 - \sin^2 2x} \, dx$

40. $\int \cos x \ln(\sin x) \, dx$

41. $\int \sec 2x \tan 2x \, e^{\sec 2x} \, dx$

42. $\int \frac{\cos \sqrt{x}}{\sqrt{x}} \, dx$

43. $\int_0^{\pi/4} 3x \sec(x^2 + 2) \, dx$

44. $\int x \sin 2x \, dx$

45. $\int_{2\pi/3}^{\pi} \frac{1 - \sin x}{\cos x + x} \, dx$

46. $\int e^{3x} \cos 2x \, dx$

47. Find the average value of

(a) $y = \sin x \cos x$ over $[0, \pi]$;

(b) $y = \sec^2 x$ over $\left[-\frac{\pi}{3}, \frac{\pi}{3}\right]$.

48. Sketch and find the area of the region bounded by $y = \cos x$, $y = x$, the y-axis, and $x = \frac{\pi}{6}$.

Tables

TABLE 1 Powers of e

x	e^x	e^{-x}	x	e^x	e^{-x}	x	e^x	e^{-x}	x	e^x	e^{-x}	x	e^x	e^{-x}	x	e^x	e^{-x}
0.00	1.00000	1.00000	0.42	1.52196	0.65705	0.84	2.31637	0.43171	1.26	3.52542	0.28365	2.40	11.0232	0.09072	5.80	330.300	0.00303
0.01	1.01005	0.99005	0.43	1.53726	0.65051	0.85	2.33965	0.42741	1.27	3.56085	0.28083	2.45	11.5883	0.08629	5.90	365.038	0.00274
0.02	1.02020	0.98020	0.44	1.55271	0.64404	0.86	2.36316	0.42316	1.28	3.59664	0.27804	2.50	12.1825	0.08209	6.00	403.429	0.00248
0.03	1.03045	0.97045	0.45	1.56831	0.63763	0.87	2.38691	0.41895	1.29	3.63279	0.27527	2.55	12.8071	0.07808	6.10	445.858	0.00224
0.04	1.04081	0.96079	0.46	1.58407	0.63128	0.88	2.41090	0.41478	1.30	3.66930	0.27253	2.60	13.4637	0.07427	6.20	492.749	0.00203
0.05	1.05127	0.95123	0.47	1.59999	0.62500	0.89	2.43513	0.41066	1.31	3.70617	0.26982	2.65	14.1540	0.07065	6.30	544.572	0.00184
0.06	1.06184	0.94176	0.48	1.61607	0.61878	0.90	2.45960	0.40657	1.32	3.74342	0.26714	2.70	14.8797	0.06721	6.40	601.845	0.00166
0.07	1.07251	0.93239	0.49	1.63232	0.61263	0.91	2.48432	0.40252	1.33	3.78104	0.26448	2.75	15.6426	0.06393	6.50	665.142	0.00150
0.08	1.08329	0.92312	0.50	1.64872	0.60653	0.92	2.50929	0.39852	1.34	3.81904	0.26185	2.80	16.4447	0.06081	6.60	735.09	0.00136
0.09	1.09417	0.91393	0.51	1.66529	0.60050	0.93	2.53451	0.39455	1.35	3.85743	0.25924	2.85	17.2878	0.05784	6.70	812.41	0.00123
0.10	1.10517	0.90484	0.52	1.68203	0.59452	0.94	2.55998	0.39063	1.36	3.89619	0.25666	2.90	18.1741	0.05502	6.80	897.85	0.00111
0.11	1.11628	0.89583	0.53	1.69893	0.58861	0.95	2.58571	0.38674	1.37	3.93535	0.25411	2.95	19.1060	0.05234	6.90	992.27	0.00101
0.12	1.12750	0.88692	0.54	1.71601	0.58275	0.96	2.61170	0.38289	1.38	3.97490	0.25158	3.00	20.0855	0.04979	7.00	1096.63	0.00091
0.13	1.13883	0.87810	0.55	1.73325	0.57695	0.97	2.63794	0.37908	1.39	4.01485	0.24908	3.05	21.1153	0.04736	7.10	1211.97	0.00083
0.14	1.15027	0.86936	0.56	1.75067	0.57121	0.98	2.66446	0.37531	1.40	4.05520	0.24660	3.10	22.1980	0.04505	7.20	1339.43	0.00075
0.15	1.16183	0.86071	0.57	1.76827	0.56553	0.99	2.69123	0.37158	1.41	4.09596	0.24414	3.15	23.3361	0.04285	7.30	1480.30	0.00068
0.16	1.17351	0.85214	0.58	1.78604	0.55990	1.00	2.71828	0.36788	1.42	4.13712	0.24171	3.20	24.533	0.04076	7.40	1635.98	0.00061
0.17	1.18531	0.84366	0.59	1.80399	0.55433	1.01	2.74560	0.36422	1.43	4.17870	0.23931	3.30	27.113	0.03688	7.50	1808.04	0.00055
0.18	1.19722	0.83527	0.60	1.82212	0.54881	1.02	2.77320	0.36059	1.44	4.22070	0.23693	3.40	29.964	0.03337	7.60	1998.20	0.00050
0.19	1.20925	0.82696	0.61	1.84043	0.54335	1.03	2.80107	0.35701	1.45	4.26312	0.23457	3.50	33.115	0.03020	7.70	2208.35	0.00045
0.20	1.22140	0.81873	0.62	1.85893	0.53794	1.04	2.82922	0.35345	1.46	4.30596	0.23224	3.60	36.598	0.02732	7.80	2440.60	0.00041
0.21	1.23368	0.81058	0.63	1.87761	0.53259	1.05	2.85765	0.34994	1.47	4.34924	0.22993	3.70	40.447	0.02472	7.90	2697.28	0.00037
0.22	1.24608	0.80252	0.64	1.89648	0.52729	1.06	2.88637	0.34646	1.48	4.39295	0.22764	3.80	44.701	0.02237	8.00	2980.96	0.00034
0.23	1.25860	0.79453	0.65	1.91554	0.52205	1.07	2.91538	0.34301	1.49	4.43710	0.22537	3.90	49.402	0.02024	8.10	3294.47	0.00030
0.24	1.27125	0.78663	0.66	1.93479	0.51685	1.08	2.94468	0.33960	1.50	4.4817	0.22313	4.00	54.598	0.01832	8.20	3640.95	0.00027
0.25	1.28403	0.77880	0.67	1.95424	0.51171	1.09	2.97427	0.33622	1.55	4.7115	0.21225	4.10	60.340	0.01657	8.30	4023.87	0.00025
0.26	1.29693	0.77105	0.68	1.97388	0.50662	1.10	3.00417	0.33287	1.60	4.9530	0.20190	4.20	66.686	0.01500	8.40	4447.07	0.00022
0.27	1.30996	0.76338	0.69	1.99372	0.50158	1.11	3.03436	0.32956	1.65	5.2070	0.19205	4.30	73.700	0.01357	8.50	4914.77	0.00020
0.28	1.32313	0.75578	0.70	2.01375	0.49659	1.12	3.06485	0.32628	1.70	5.4739	0.18268	4.40	81.451	0.01228	8.60	5431.66	0.00018
0.29	1.33643	0.74826	0.71	2.03399	0.49164	1.13	3.09566	0.32303	1.75	5.7546	0.17377	4.50	90.017	0.01111	8.70	6002.91	0.00017
0.30	1.34986	0.74082	0.72	2.05443	0.48675	1.14	3.12677	0.31982	1.80	6.0496	0.16530	4.60	99.484	0.01005	8.80	6634.25	0.00015
0.31	1.36343	0.73345	0.73	2.07508	0.48191	1.15	3.15819	0.31664	1.85	6.3598	0.15724	4.70	109.947	0.00910	8.90	7331.97	0.00014
0.32	1.37713	0.72615	0.74	2.09594	0.47711	1.16	3.18993	0.31349	1.90	6.6859	0.14957	4.80	121.510	0.00823	9.00	8103.08	0.00012
0.33	1.39097	0.71892	0.75	2.11700	0.47237	1.17	3.22199	0.31037	1.95	7.0287	0.14227	4.90	134.290	0.00745	9.10	8955.29	0.00011
0.34	1.40495	0.71177	0.76	2.13828	0.46767	1.18	3.25437	0.30728	2.00	7.3891	0.13534	5.00	148.413	0.00674	9.20	9897.13	0.00010
0.35	1.41907	0.70469	0.77	2.15977	0.46301	1.19	3.28708	0.30422	2.05	7.7679	0.12873	5.10	164.022	0.00610	9.30	10938.0	0.00009
0.36	1.43333	0.69768	0.78	2.18147	0.45841	1.20	3.32012	0.30119	2.10	8.1662	0.12246	5.20	181.272	0.00552	9.40	12088.4	0.00008
0.37	1.44773	0.69073	0.79	2.20340	0.45384	1.21	3.35348	0.29820	2.15	8.5849	0.11648	5.30	200.337	0.00499	9.50	13359.7	0.00007
0.38	1.46228	0.68386	0.80	2.22554	0.44933	1.22	3.38719	0.29523	2.20	9.0250	0.11080	5.40	221.406	0.00452	9.60	14764.8	0.00007
0.39	1.47698	0.67706	0.81	2.24791	0.44486	1.23	3.42123	0.29229	2.25	9.4877	0.10540	5.50	244.692	0.00409	9.70	16317.6	0.00006
0.40	1.49182	0.67032	0.82	2.27050	0.44043	1.24	3.45561	0.28938	2.30	9.9742	0.10026	5.60	270.427	0.00370	9.80	18033.7	0.00006
0.41	1.50682	0.66365	0.83	2.29332	0.43605	1.25	3.49034	0.28650	2.35	10.4856	0.09537	5.70	298.867	0.00335	9.90	19930.4	0.00005

TABLE 2 Natural Logarithms

x	$\ln x$	x	$\ln x$	x	$\ln x$	x	$\ln x$	x	$\ln x$	x	$\ln x$	x	$\ln x$	x	$\ln x$	x	$\ln x$
0.01	−4.6052	0.51	−0.6733	1.01	0.0100	1.51	0.4121	2.01	0.6981	2.55	0.9361	5.05	1.6194	7.55	2.0215	10.05	2.3076
0.02	−3.9120	0.52	−0.6539	1.02	0.0198	1.52	0.4187	2.02	0.7031	2.60	0.9555	5.10	1.6292	7.60	2.0281	10.10	2.3125
0.03	−3.5066	0.53	−0.6349	1.03	0.0296	1.53	0.4253	2.03	0.7080	2.65	0.9746	5.15	1.6390	7.65	2.0347	10.15	2.3175
0.04	−3.2189	0.54	−0.6162	1.04	0.0392	1.54	0.4318	2.04	0.7130	2.70	0.9933	5.20	1.6487	7.70	2.0412	10.20	2.3224
0.05	−2.9957	0.55	−0.5978	1.05	0.0488	1.55	0.4383	2.05	0.7178	2.75	1.0116	5.25	1.6582	7.75	2.0477	10.25	2.3273
0.06	−2.8134	0.56	−0.5798	1.06	0.0583	1.56	0.4447	2.06	0.7227	2.80	1.0296	5.30	1.6677	7.80	2.0541	10.30	2.3321
0.07	−2.6593	0.57	−0.5621	1.07	0.0677	1.57	0.4511	2.07	0.7275	2.85	1.0473	5.35	1.6771	7.85	2.0605	10.35	2.3370
0.08	−2.5257	0.58	−0.5447	1.08	0.0770	1.58	0.4574	2.08	0.7324	2.90	1.0647	5.40	1.6864	7.90	2.0669	10.40	2.3418
0.09	−2.4079	0.59	−0.5276	1.09	0.0862	1.59	0.4637	2.09	0.7372	2.95	1.0318	5.45	1.6956	7.95	2.0732	10.45	2.3466
0.10	−2.3026	0.60	−0.5108	1.10	0.0953	1.60	0.4700	2.10	0.7419	3.00	1.0986	5.50	1.7047	8.00	2.0794	10.50	2.3514
0.11	−2.2073	0.61	−0.4943	1.11	0.1044	1.61	0.4762	2.11	0.7467	3.05	1.1151	5.55	1.7138	8.05	2.0857	10.55	2.3561
0.12	−2.1203	0.62	−0.4780	1.12	0.1133	1.62	0.4824	2.12	0.7514	3.10	1.1314	5.60	1.7228	8.10	2.0919	10.60	2.3609
0.13	−2.0402	0.63	−0.4620	1.13	0.1222	1.63	0.4886	2.13	0.7561	3.15	1.1474	5.65	1.7317	8.15	2.0980	10.65	2.3656
0.14	−1.9661	0.64	−0.4463	1.14	0.1310	1.64	0.4947	2.14	0.7608	3.20	1.1632	5.70	1.7405	8.20	2.1041	10.70	2.3702
0.15	−1.8971	0.65	−0.4308	1.15	0.1398	1.65	0.5008	2.15	0.7655	3.25	1.1787	5.75	1.7492	8.25	2.1102	10.75	2.3749
0.16	−1.8326	0.66	−0.4155	1.16	0.1484	1.66	0.5068	2.16	0.7701	3.30	1.1939	5.80	1.7579	8.30	2.1163	10.80	2.3795
0.17	−1.7720	0.67	−0.4005	1.17	0.1570	1.67	0.5128	2.17	0.7747	3.35	1.2090	5.85	1.7664	8.35	2.1223	10.85	2.3842
0.18	−1.7148	0.68	−0.3857	1.18	0.1655	1.68	0.5188	2.18	0.7793	3.40	1.2238	5.90	1.7750	8.40	2.1282	10.90	2.3888
0.19	−1.6607	0.69	−0.3711	1.19	0.1740	1.69	0.5247	2.19	0.7839	3.45	1.2384	5.95	1.7834	8.45	2.1342	10.95	2.3933
0.20	−1.6094	0.70	−0.3567	1.20	0.1823	1.70	0.5306	2.20	0.7885	3.50	1.2528	6.00	1.7918	8.50	2.1401	11.00	2.3979
0.21	−1.5606	0.71	−0.3425	1.21	0.1906	1.71	0.5365	2.21	0.7930	3.55	1.2669	6.05	1.8001	8.55	2.1459	11.05	2.4024
0.22	−1.5141	0.72	−0.3285	1.22	0.1989	1.72	0.5423	2.22	0.7975	3.60	1.2809	6.10	1.8083	8.60	2.1518	11.10	2.4069
0.23	−1.4697	0.73	−0.3147	1.23	0.2070	1.73	0.5481	2.23	0.8020	3.65	1.2947	6.15	1.8165	8.65	2.1576	11.15	2.4114
0.24	−1.4271	0.74	−0.3011	1.24	0.2151	1.74	0.5539	2.24	0.8065	3.70	1.3083	6.20	1.8245	8.70	2.1633	11.20	2.4159
0.25	−1.3863	0.75	−0.2877	1.25	0.2231	1.75	0.5596	2.25	0.8109	3.75	1.3218	6.25	1.8326	8.75	2.1691	11.25	2.4204
0.26	−1.3471	0.76	−0.2744	1.26	0.2311	1.76	0.5653	2.26	0.8154	3.80	1.3350	6.30	1.8405	8.80	2.1748	11.30	2.4248
0.27	−1.3039	0.77	−0.2614	1.27	0.2390	1.77	0.5710	2.27	0.8198	3.85	1.3481	6.35	1.8485	8.85	2.1804	11.35	2.4292
0.28	−1.2730	0.78	−0.2485	1.28	0.2469	1.78	0.5766	2.28	0.8242	3.90	1.3610	6.40	1.8563	8.90	2.1861	11.40	2.4336
0.29	−1.2379	0.79	−0.2357	1.29	0.2546	1.79	0.5822	2.29	0.8286	3.95	1.3737	6.45	1.8641	8.95	2.1917	11.45	2.4380
0.30	−1.2040	0.80	−0.2231	1.30	0.2624	1.80	0.5878	2.30	0.8329	4.00	1.3863	6.50	1.8718	9.00	2.1972	11.50	2.4423
0.31	−1.1712	0.81	−0.2107	1.31	0.2700	1.81	0.5933	2.31	0.8372	4.05	1.3987	6.55	1.8795	9.05	2.2028	13.0	2.5649
0.32	−1.1394	0.82	−0.1985	1.32	0.2776	1.82	0.5988	2.32	0.8416	4.10	1.4110	6.60	1.8871	9.10	2.2083	17.0	2.8332
0.33	−1.1087	0.83	−0.1863	1.33	0.2852	1.83	0.6043	2.33	0.8459	4.15	1.4231	6.65	1.8946	9.15	2.2138	19.0	2.9444
0.34	−1.0788	0.84	−0.1744	1.34	0.2927	1.84	0.6098	2.34	0.8502	4.20	1.4351	6.70	1.9021	9.20	2.2192	23.0	3.1355
0.35	−1.0498	0.85	−0.1625	1.35	0.3001	1.85	0.6152	2.35	0.8544	4.25	1.4469	6.75	1.9095	9.25	2.2246	29.0	3.3673
0.36	−1.0217	0.86	−0.1508	1.36	0.3075	1.86	0.6206	2.36	0.8587	4.30	1.4586	6.80	1.9169	9.30	2.2300	31.0	3.4340
0.37	−0.9943	0.87	−0.1393	1.37	0.3148	1.87	0.6259	2.37	0.8629	4.35	1.4702	6.85	1.9242	9.35	2.2354	37.0	3.6109
0.38	−0.9676	0.88	−0.1278	1.38	0.3221	1.88	0.6313	2.38	0.8671	4.40	1.4816	6.90	1.9315	9.40	2.2407	41.0	3.7136
0.39	−0.9416	0.89	−0.1165	1.39	0.3293	1.89	0.6366	2.39	0.8713	4.45	1.4929	6.95	1.9387	9.45	2.2460	43.0	3.7612
0.40	−0.9163	0.90	−0.1054	1.40	0.3365	1.90	0.6419	2.40	0.8755	4.50	1.5041	7.00	1.9459	9.50	2.2513	47.0	3.8501
0.41	−0.8916	0.91	−0.0943	1.41	0.3436	1.91	0.6471	2.41	0.8796	4.55	1.5151	7.05	1.9530	9.55	2.2565	53.0	3.9703
0.42	−0.8675	0.92	−0.0834	1.42	0.3507	1.92	0.6523	2.42	0.8838	4.60	1.5261	7.10	1.9601	9.60	2.2618	59.0	4.0775
0.43	−0.8440	0.93	−0.0726	1.43	0.3577	1.93	0.6575	2.43	0.8879	4.65	1.5369	7.15	1.9671	9.65	2.2670	61.0	4.1109
0.44	−0.8210	0.94	−0.0619	1.44	0.3646	1.94	0.6627	2.44	0.8920	4.70	1.5476	7.20	1.9741	9.70	2.2721	67.0	4.2047
0.45	−0.7985	0.95	−0.0513	1.45	0.3716	1.95	0.6678	2.45	0.8961	4.75	1.5581	7.25	1.9810	9.75	2.2773	71.0	4.2627
0.46	−0.7765	0.96	−0.0408	1.46	0.3784	1.96	0.6729	2.46	0.9002	4.80	1.5686	7.30	1.9879	9.80	2.2824	73.0	4.2905
0.47	−0.7550	0.97	−0.0305	1.47	0.3853	1.97	0.6780	2.47	0.9042	4.85	1.5790	7.35	1.9947	9.85	2.2875	79.0	4.3694
0.48	−0.7340	0.98	−0.0202	1.48	0.3920	1.98	0.6831	2.48	0.9083	4.90	1.5892	7.40	2.0015	9.90	2.2925	83.0	4.4188
0.49	−0.7134	0.99	−0.0101	1.49	0.3988	1.99	0.6881	2.49	0.9123	4.95	1.5994	7.45	2.0082	9.95	2.2976	89.0	4.4886
0.50	−0.6931	1.00	0.0000	1.50	0.4055	2.00	0.6931	2.50	0.9163	5.00	1.6094	7.50	2.0149	10.00	2.3026	97.0	4.5747

Solutions to Exercises

1.1 EXERCISES, PAGE 11

1.

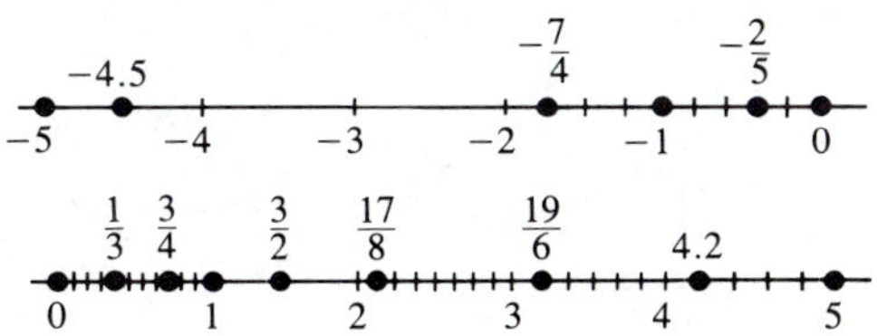

3. true **5.** true

7. true **9.** false

11. true **13.** true

15. true

17.

19.

21.

23.

25. 4 **27.** -8

29. $\frac{4}{5}$ **31.** 5

33. $2 - \sqrt{2}$ **35.** $\frac{1}{81}$

37. $\frac{8}{125}$ **39.** 6

41. 27 **43.** $\frac{1}{10}$

45. -2 **47.** $\dfrac{27x^6y^9}{z^3}$

49. $\dfrac{x^{3/2}y}{z^{9/4}}$ **51.** $\dfrac{1}{9xy^5}$

53. $2^{5/4}|x|y^{9/4}$

1.2 EXERCISES, PAGE 22

1. **(a)** -2 **(b)** 2
(c) -14 **(d)** -1
(e) 4 **(f)** $4b - 2$
(g) $4b + 6$ **(h)** all x

3. **(a)** 0 **(b)** 2
(c) 4 **(d)** $2\sqrt{3}$
(e) $2\sqrt{b}$ **(f)** $\{x | x \geq -1\}$

5. **(a)** $-\$200{,}000$ **(b)** $-\$75{,}000$
(c) \$0 **(d)** \$25,000
(e) \$0 **(f)** $-\$75{,}000$

7. **(a)** $-15°$ C **(b)** 0° C
(c) 25° C **(d)** 100° C

9. 200,000; 207,143; 208,571

11. **(a)** 1 **(b)** 2
(c) 3 **(d)** 4
(e) 5

13.

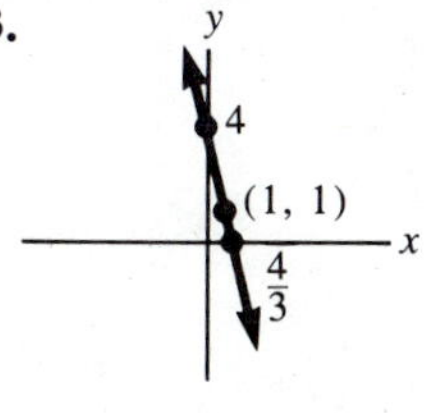

15.

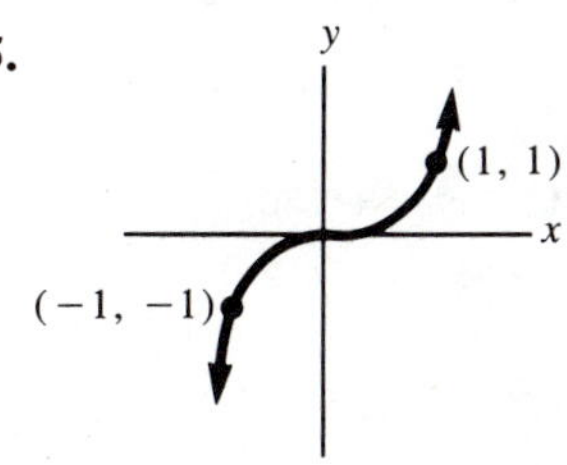

17.

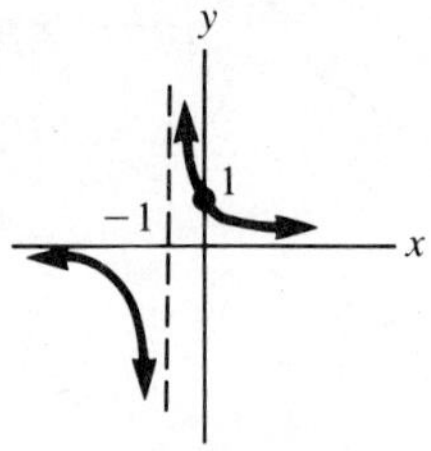

19. not a function

21. function; domain is all real numbers

23. not a function

25. **(a)**

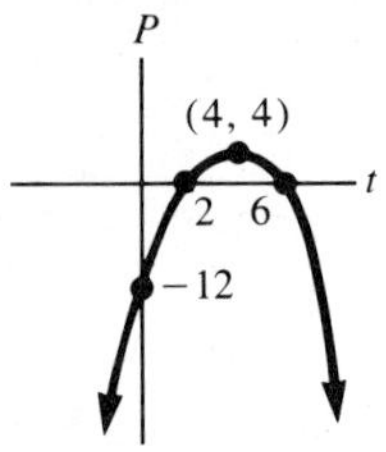

(b) 1983 − 1985 ($2 < t < 6$); before 1982 ($t < 2$) and after 1986 ($t > 6$); \$40,000 in 1984 ($t = 4$)

27. **(a)**

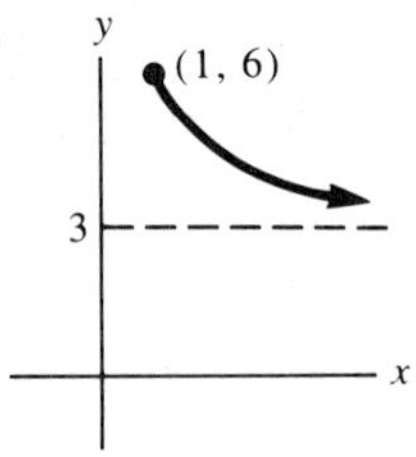

(b) 6 min; no, but tends towards 3 min

29.

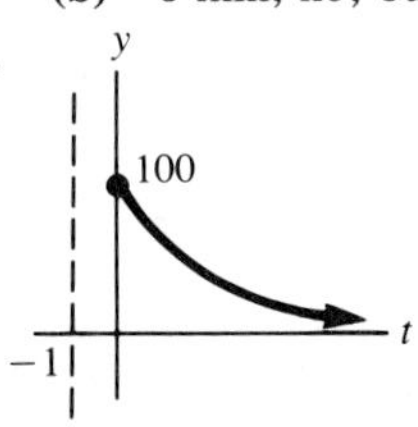

no

1.3 EXERCISES, PAGE 34

1. $5x + 4$, $-x - 14$, $6x^2 + 3x - 45$, $\frac{2x-5}{3x+9}$; 14, -14, 18, $-\frac{7}{6}$; all x, $\{x|x \neq -3\}$

3. $-3x^4 - 16x^3 + 29x^2 - 98x + 130$, $\frac{x^2 + 6x - 10}{-3x^2 + 2x - 13}$; 42, $\frac{10}{13}$

5. $\frac{3x^2 + 9x + 10}{x^2 + 4x + 3}$, $\frac{-x^2 - 11x - 14}{x^2 + 4x + 3}$, $\frac{2x^2 - 8}{x^2 + 4x + 3}$, $\frac{x^2 - x - 2}{2x^2 + 10x + 12}$; $\frac{11}{4}$, $-\frac{1}{6}$; $\{x|x \neq -3, -1\}$, $\{x|x \neq -3, -1\}$

7. $\frac{x^3 - 6x^2 + 15x - 26}{-2x^2 + 2x + 4}$, $\frac{3x^3 - 8x^2 - 13x + 30}{-2x^2 + 2x + 4}$, $\frac{-6x^2 + 22x - 20}{x^3 - 7x - 6}$; $\{x|x \neq -1, 2\}$, $\{x|x \neq -1, 2\}$, $\{x|x \neq -2, -1, 3\}$

9. **(a)** $P = 30x - 10{,}000$
(b) $-\$7000$; \$30

11. Population size

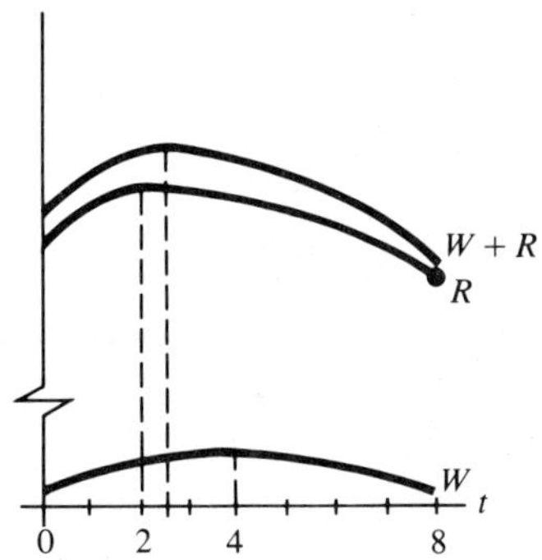

$t = 4$, $t = 2$, $t = 2.5$

13. $3x^2 + 18$, $9x^2 + 36x + 40$

15. $x^4 + 4hx^3 + (6h^2 + 5)x^2 + (4h^3 + 10)x + h^4 + 5h^2 + 1$, $x^4 + 5x^2 + h + 1$

17. $\frac{1}{x+h}$, $\frac{1}{x} + h$

19. $\frac{-3x^2 + 10x + 5}{7x^2 + 10x - 5}$, $\frac{24x^2 + 68x - 56}{-40(x + 1)}$

21. $y = 4t$

23. $t = y/4$

25. $c(t) = 0.1t^2 + 2.1t + 12$, 35.2 ppm

27. -5, -3

29. -3, 0, 3

31. 0, 3

33. 2, 7, 0

35. 0, 1

37. $-\frac{1}{2}$, 1

39. 2002 ($t = 12$)

1.4 EXERCISES, PAGE 47

1.

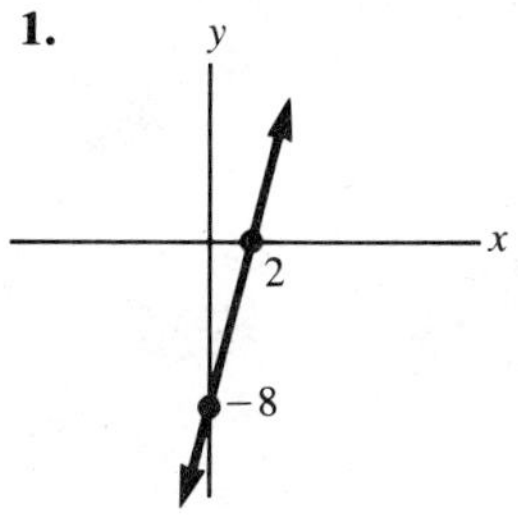

3.

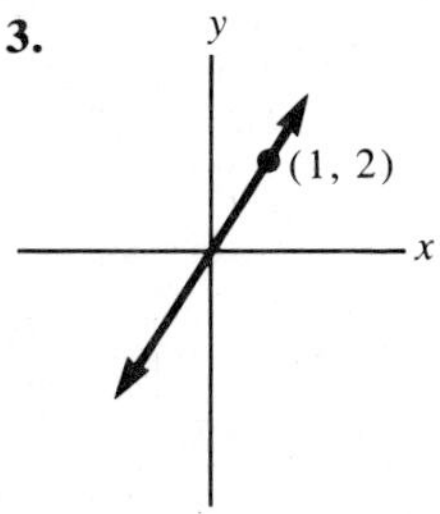

5.

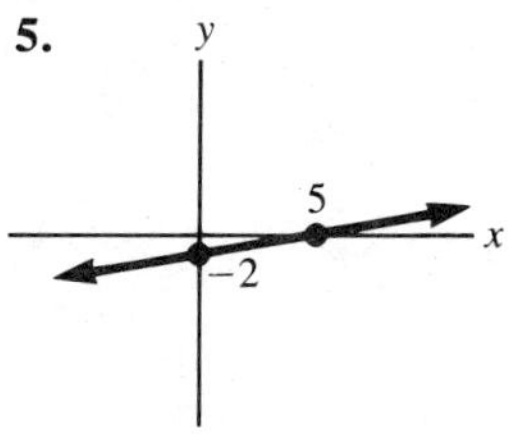

7.

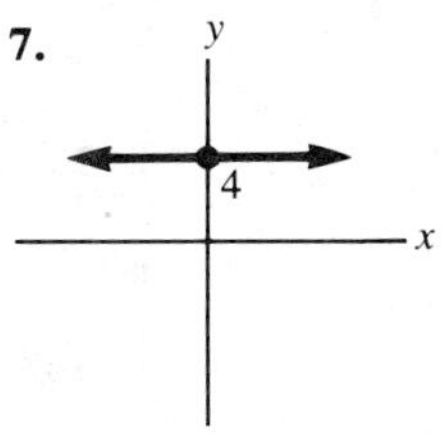

9.

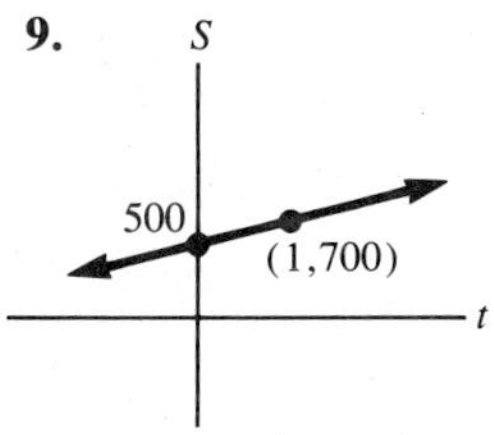

; \$500,000; increasing; by \$200,000

11.

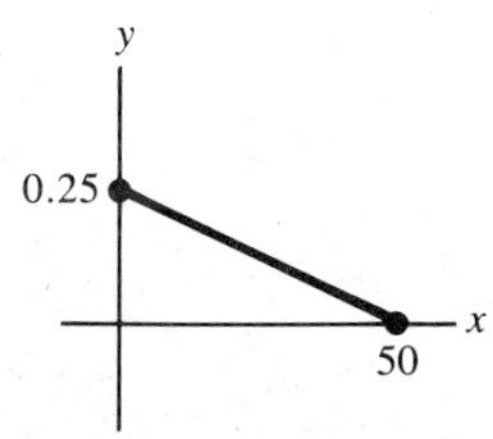

; 0.005; 50 mg/L

13. $y = -3x + 2$ **15.** $y = 6x - 16$

17. $y = 2x + 2$ **19.** $y = 2$

21. $C = \frac{5}{9}F - \frac{160}{9}$

23. $V = 1000t + 10{,}000$
(a) \$16,000 **(b)** 7 yrs

25. $y = 4.75x + 2{,}025{,}000$; each additional \$1 spent on advertising increases audience size by 4.75 people; \$626,316; 6,775,000 people

27. **(a)** $y = -45t + 645$ **(b)** 645 ppm
(c) $14\frac{1}{3}$ mos

29. $(-4, -13)$ **31.** $(36, 20)$

33. none, lines are parallel.

35. 30 small, 22 large **37.** 20.2 yrs

39. **(a)** end of 1987 ($t \approx 7.9$)
(b) mid-1988 ($t \approx 8.6$)
(c) end of 1994 ($t \approx 14.9$)

41.

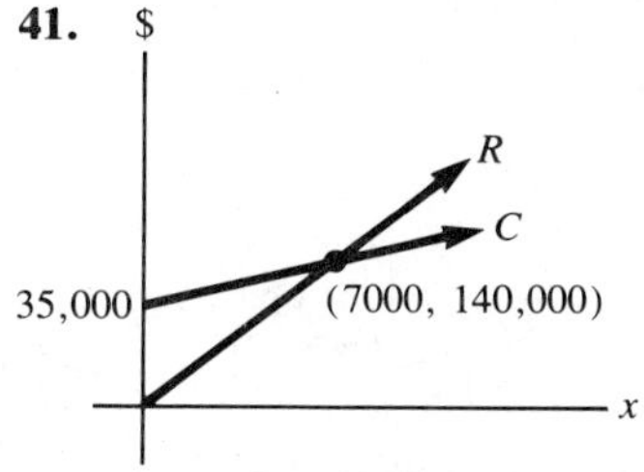

cost of making each unit is \$15, cost of making 0 units is \$35,000; each unit sells for \$20, revenue from selling 0 units is \$0; (7000, \$140,000); $P(x) = 5x - 35{,}000$; each unit made and sold contributes \$5 toward profit, making and selling 0 units results in a loss of \$35,000, making and selling 7000 units results in a profit of \$0

43.

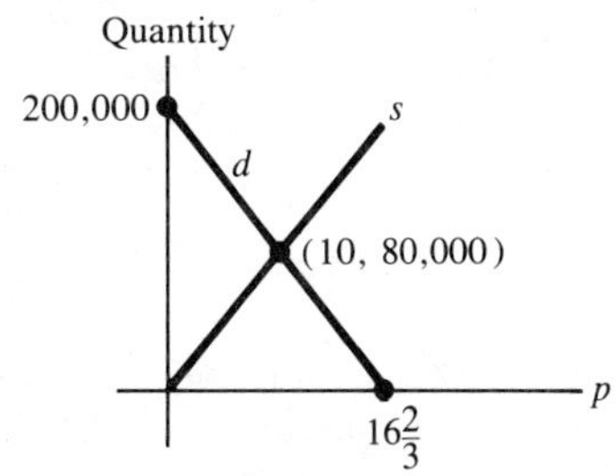

(\$10, 80,000); each \$1 rise in price brings 8000 more units onto the market; no units are supplied to the market when the price is \$0; each \$1 rise in price causes quantity demanded to decrease by 12,000 units; when price is \$0, 200,000 units are demanded; no units are demanded when price is ≈ \$16.67.

45. $S = 2500p - 10{,}000$; $d = -3000p + 120{,}000$; ≈ (\$23.64, 49,091)

1.5 EXERCISES, PAGE 57

1.

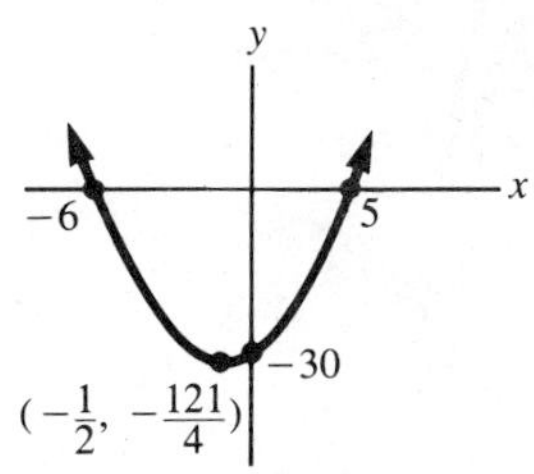

3.

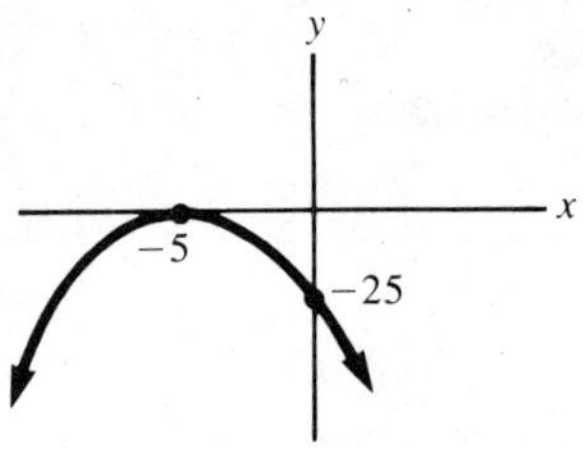

5.

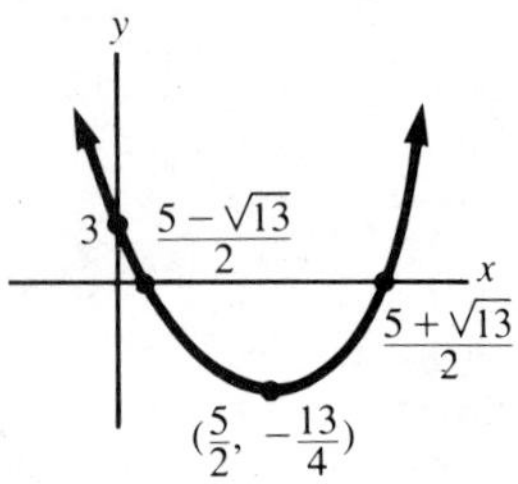

7.

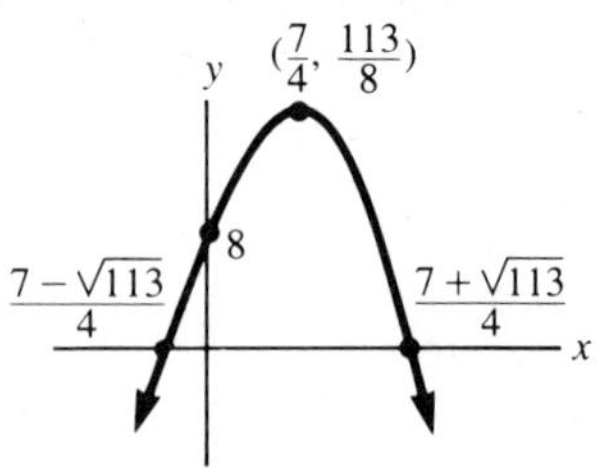

9.

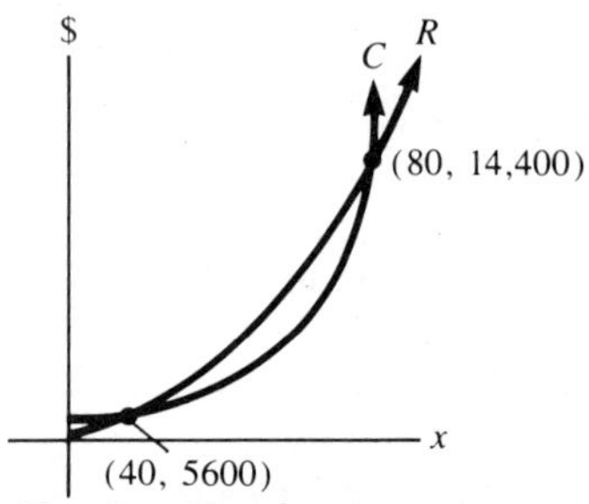

40 units, 80 units

11. **(a)** $P = -x^2 + 100x$

(b)

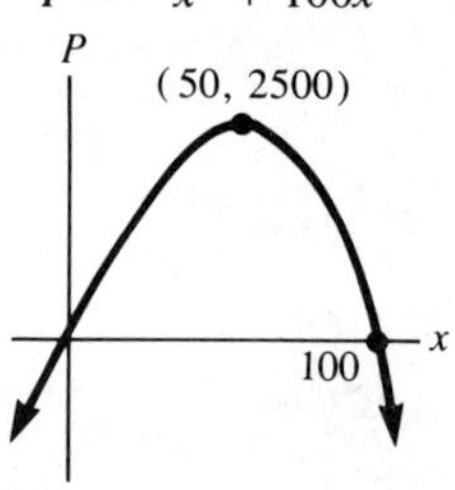

(c) $2500

13.

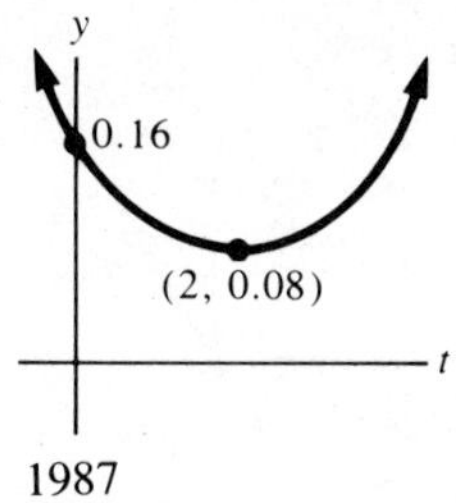

17.

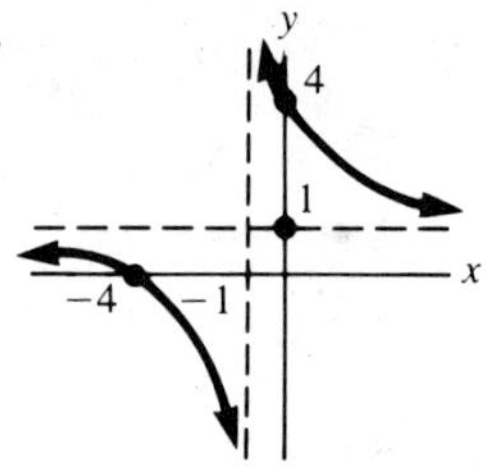

19.

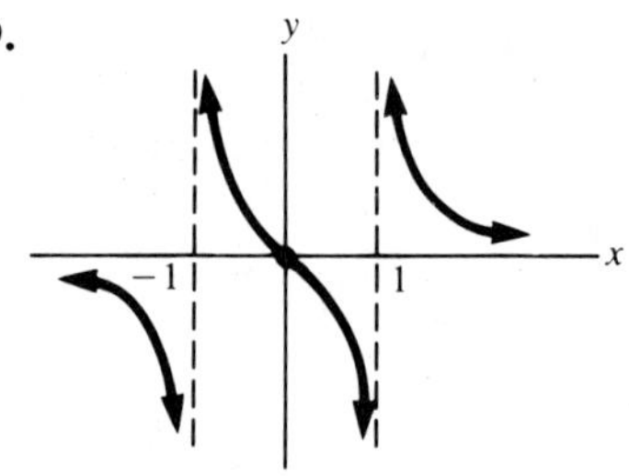

21.

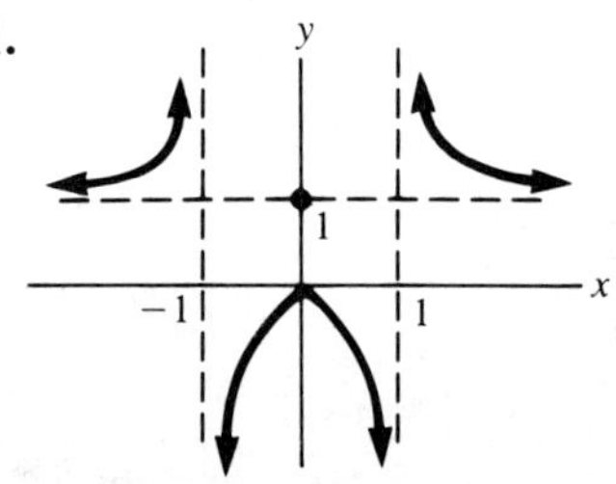

23.

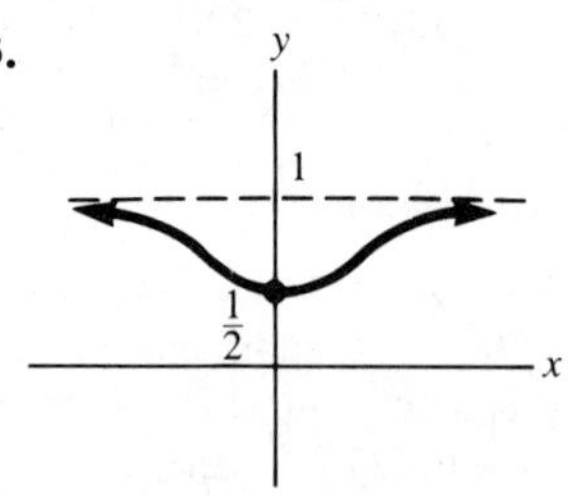

25. $A(x) = 2x + 8 + 20/x;$

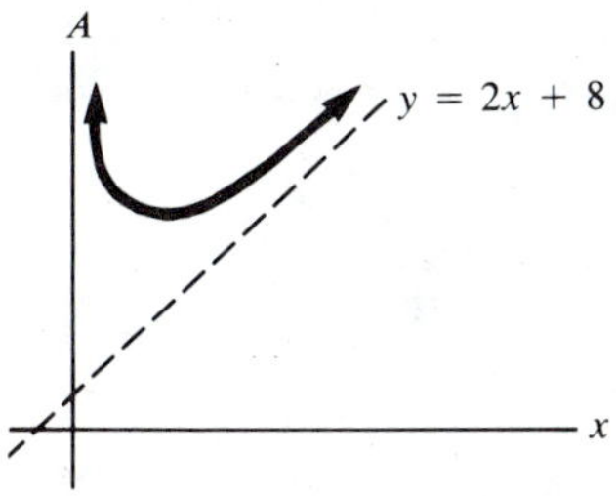

avg. cost per unit decreases to a minimum then increases without bound

27.

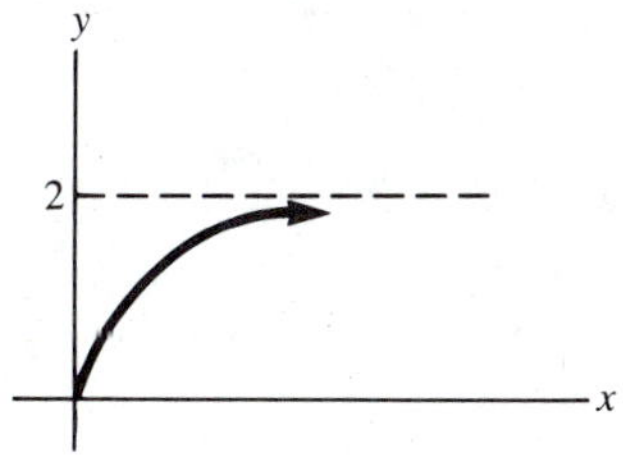

increases, approaches \$2 million

26. 4

27. $-\frac{1}{2}$

28. 0

29. undefined

30.

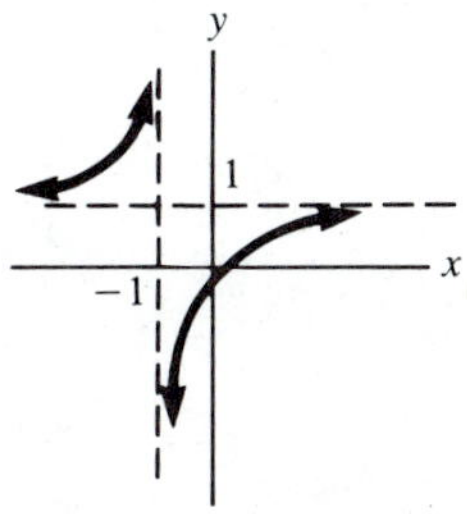

31. $2x^3 + 5x^2 - 3x - 2$, $-6x^3 + 5x^2 + 9x - 10$, $-8x^6 + 20x^5 + 24x^4 - 62x^3 + 2x^2 + 48x - 24$, $\frac{-2x^3 + 5x^2 + 3x - 6}{4x^3 - 6x + 4}$

32. $\frac{-x^3 + 5x^2 + 6x - 36}{-2x^2 + 12x - 18}$, $\frac{-x^3 + x^2 - 2x + 24}{-2x^2 + 12x - 18}$, $\frac{x^3 + 5x^2 - 2x - 10}{-2x^2 + 12x - 18}$, $\frac{-x^3 + 3x^2 + 2x - 6}{2x^2 + 4x - 30}$

33. $10x + 7$, $10x + 17$, $\frac{3x + 1}{x - 1}$, $\frac{2}{5x + 1}$

34. $C(w) = 4.5w + 6\sqrt{w + 1} + 106.5$

35. −8, 8

36. −8, 4

37. 4

38. −1, 0, 1

39. −3, 0, 4

40. $-\frac{3}{2}, \frac{1}{2}$

41.

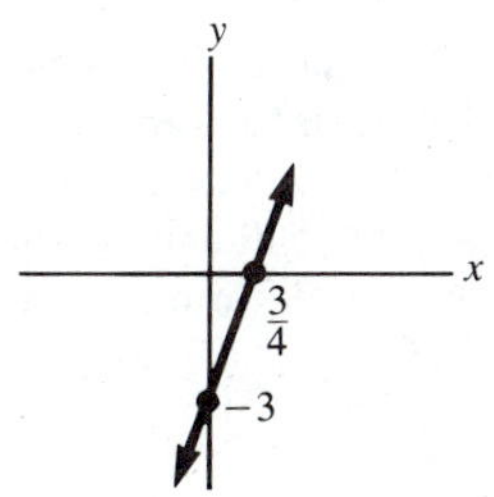

4

42.

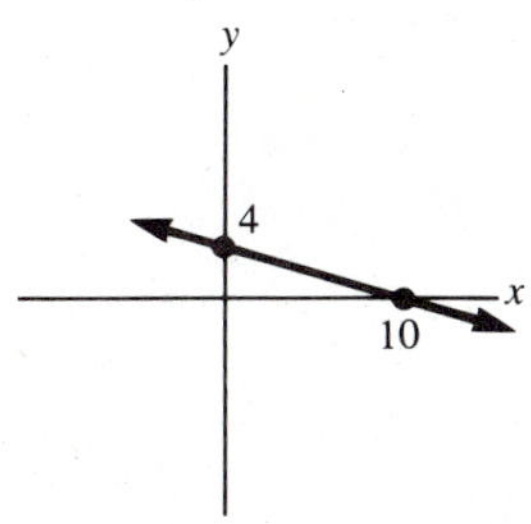

$-\frac{2}{5}$

1 REVIEW EXERCISES, PAGE 59

1. true

2. false

3. false

4. false

5. true

6. false

7. false

8. true

9. true

10. true

11. true

12. false

13.

14.

15.

16.

17. **(a)** $\frac{x^2 z}{y^2}$ **(b)** $\frac{x}{y^{6/5} z^2}$

18. 5

19. −1

20. 8

21. $5 + 3a$

22. −14

23. $\frac{7}{4}$

24. 0

25. $2 - b$

43.

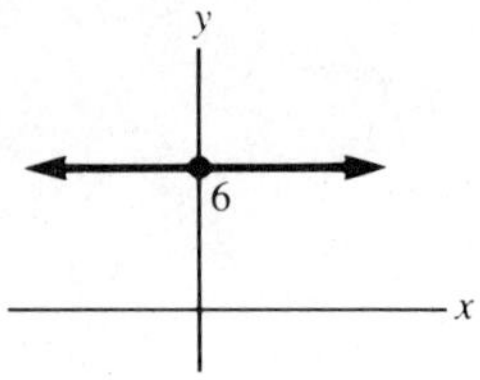

0

44.

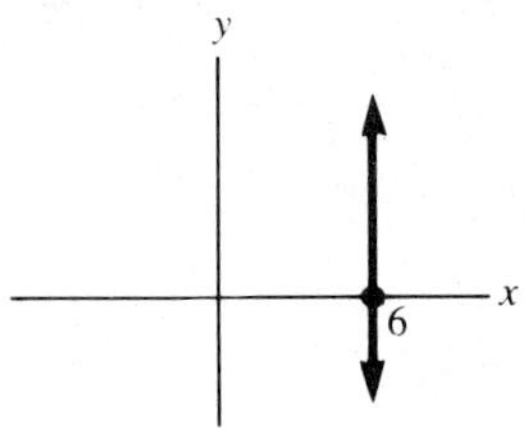

no slope

45. $y = \frac{1}{3}x + \frac{16}{3}$

46. $y = -3x - 1$

47. $y = 12$

48. $x = 5$

49. $y = -6000t + 120{,}000$; 20 hrs

50. $S = 124{,}000t + 368{,}000$; \$1,360,000; 1992

51. **(a)** (9, 33) **(b)** $\left(\frac{63}{26}, \frac{3}{13}\right)$

52. 2.5 lb and 1.4 lb/week

53. $C = 46x + 92{,}000$; $R = 62x$;
$P = 16x - 92{,}000$; (5750, \$356,500)

54. each \$1 rise in price brings 6250 units onto the market; when $p = \$2.50$ (or less), no units are supplied to the market; each \$1 rise in price causes quantity demanded to decrease by 8500 units; at a price of \$0, 183,500 units would be demanded; when $p \approx \$21.59$ (or more), no units will be demanded; \$13.50

55.

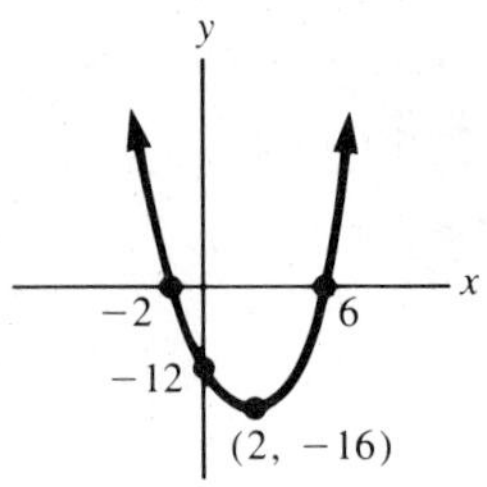

56.

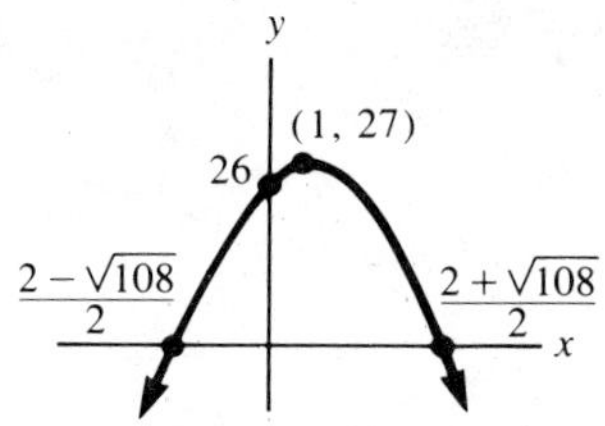

57.

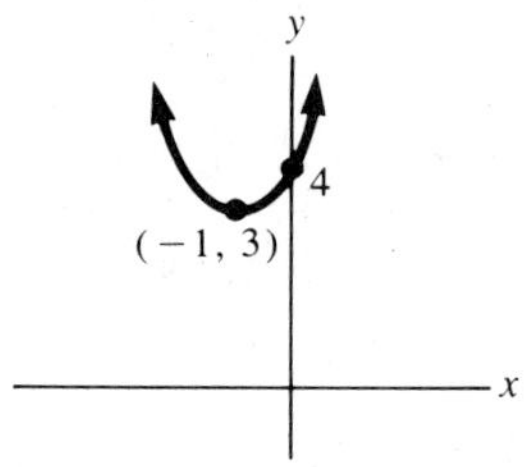

58.

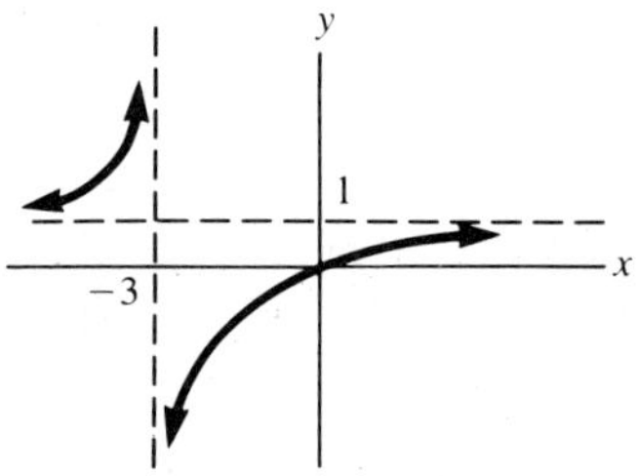

59.

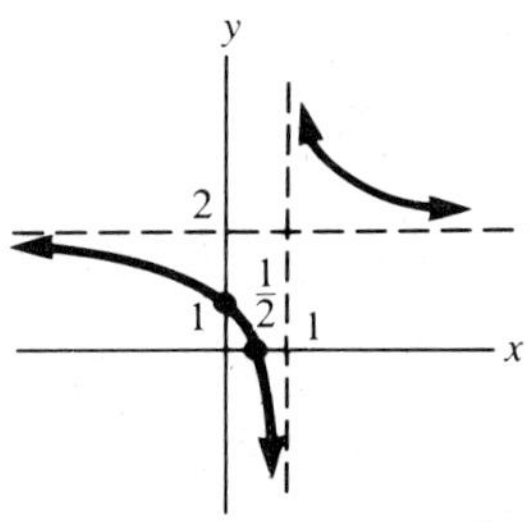

60.

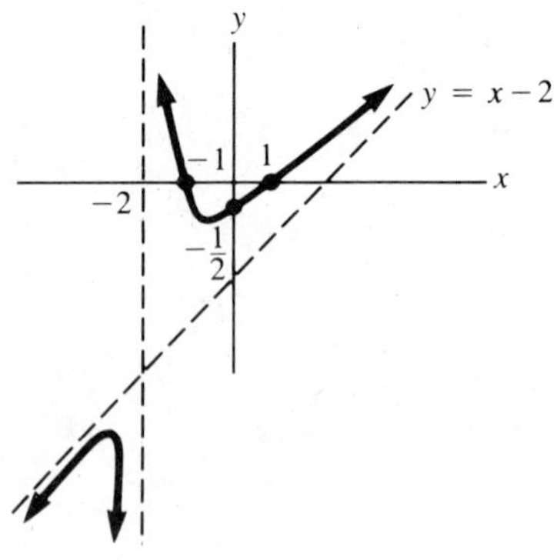

61. **(a)** $d = -1000p + 50{,}000$
(b) $P = -1000p^2 + 60{,}000p - 500{,}000$;
(c) \$400,000; \$30

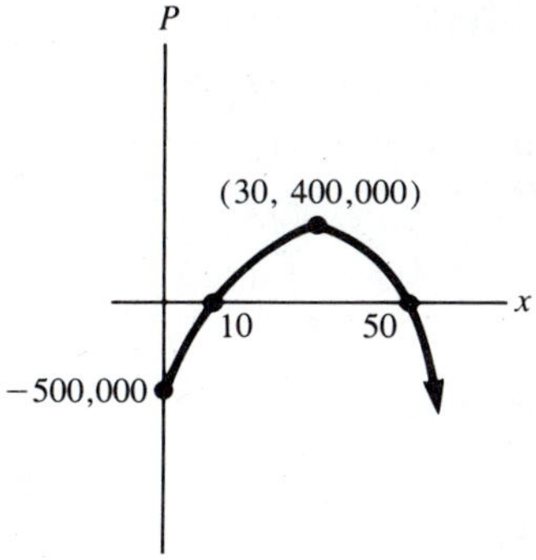

62.

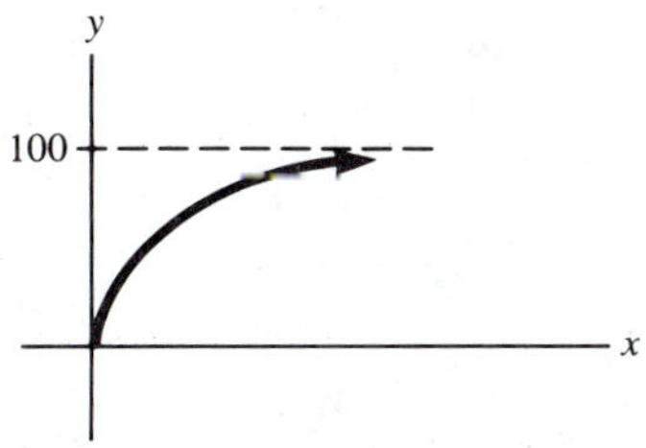

\$40 million; \$90 million; \$9.99 billion

1 SUPPLEMENTARY EXERCISES, PAGE 67

1. \$80; 16 regular, 8 deluxe
3. 2000 medium, 5600 large
5. transfer 200 from East to West, 100 from West to East

2.1 EXERCISES, PAGE 80

1.

x	2.8	2.9	2.99	2.999	2.9999
$f(x)$	−1.16	−0.59	−0.0599	−0.005999	−0.0005999

x	3.2	3.1	3.01	3.001	3.0001
$f(x)$	1.24	0.61	0.0601	0.006001	0.0006000

$\lim_{x \to 3} (x^2 - 9) = 0$

3.

x	−0.1	−0.01	−0.001	−0.0001	−0.00001
$f(x)$	−10	−100	−1000	−10,000	−100,000

x	0.1	0.01	0.001	0.0001	0.00001
$f(x)$	10	100	1000	10,000	100,000

$\lim_{x \to 0} \frac{1}{x}$ does not exist.

5. does not exist
7. 2
9. 2
11. does not exist
13. 2
15. −1
17. 5
19. 0
21. 243
23. 18
25. 25
27. $\frac{5}{4}$
29. −10
31. $\frac{11}{3}$
33. $\frac{1}{7}$
35. does not exist
37. 2
39. −1, 2
41. 0
43. does not exist
45. $\frac{2}{3}$
47. 0
49. $x = 3,\ y = 2$;

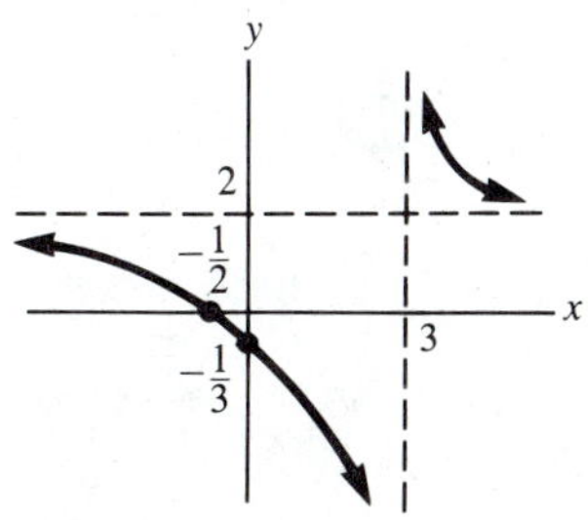

51. $x = 0,\ y = 0$;

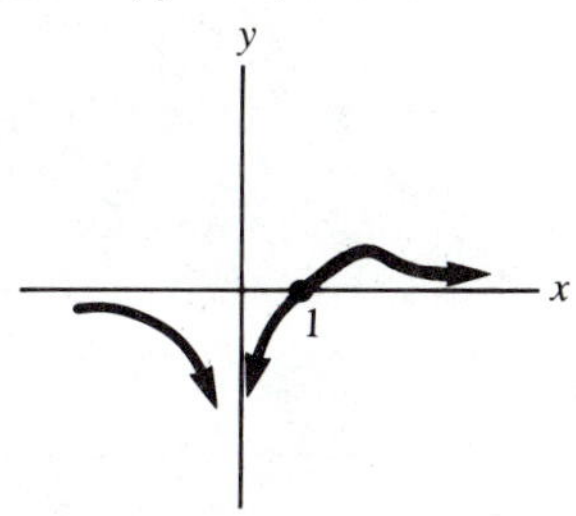

53. $x = 0, y = 2x$;

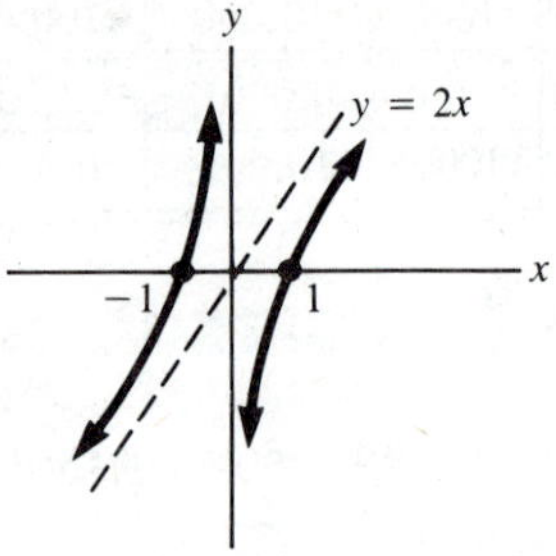

55. $\text{Lim}_{x \to +\infty} A(x) = 5$; as number of units produced increases, average cost per unit approaches \$5

57. $\text{Lim}_{t \to +\infty} y = \frac{1}{4}$; as time passes, concentration of poison decreases toward 0.25 ppm

2.2 EXERCISES, PAGE 93

1. $f'(x) = 3$ **3.** $f'(x) = 3x^2$

5. $f'(x) = -\frac{2}{x^3}$ **7.** $y = 3x - 2$

9. $y = 3x + 2$ **11.** $y = -\frac{2}{27}x + \frac{1}{3}$

13. $-1, 2$ **15.** $-2, 2$

17. no discontinuities **19.** $-2, 1$

21. **(a)**

$$C(x) = \begin{cases} 1.8x, & 0 \le x \le 2000, \\ 2.2x - 800, & 2000 < x \le 3500, \\ 3.5x - 5350, & 3500 < x \le 4300 \end{cases}$$

(b)

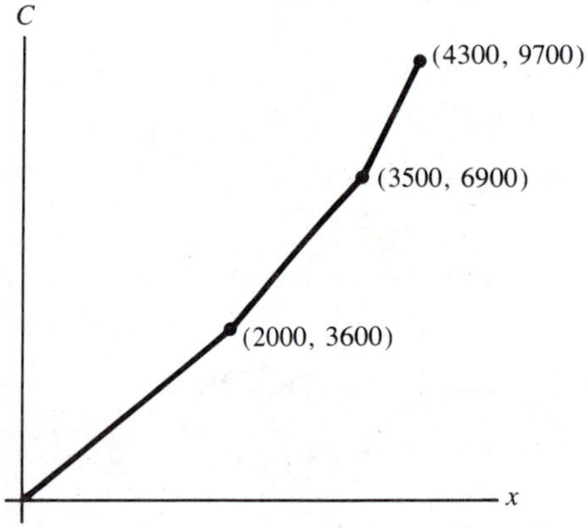

nondifferentiable at production levels where another machine is put into production

2.3 EXERCISES, PAGE 100

1. $f'(x) = 0$ **3.** $f'(x) = 6x$

5. $f'(x) = 16x - 12$

7. $f'(x) = 100x^{99} - 99x^{98}$

9. $f'(x) = -60x^{11} + x^6 - 26x^3 + 6600x^2 + 0.04x + 200{,}000$

11. $f'(x) = \frac{4}{3x^{2/3}}$ **13.** $f'(x) = \frac{64}{x^3}$

15. $f'(x) = -\frac{1}{x^2} + \frac{2}{x^3} - \frac{3}{x^4} + \frac{4}{x^5}$

17. $y = 5x - 3$ **19.** $y = 12x + 15$

21. $y = -\frac{13}{3}x + 6$ **23.** $y = \frac{1}{2}x - 1$

25. **(a)**

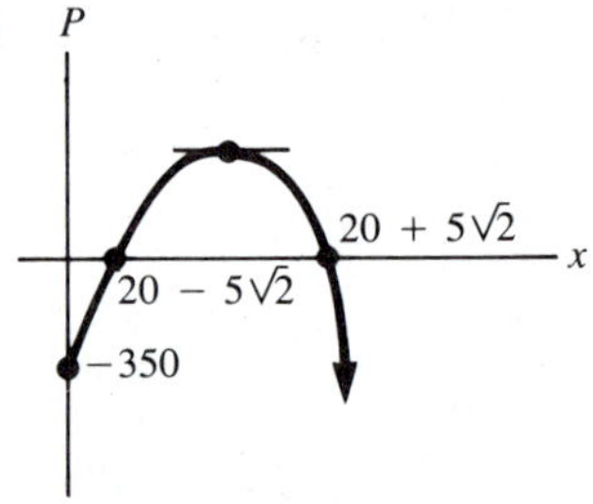

(b) \$50 when $x = 20$

27.

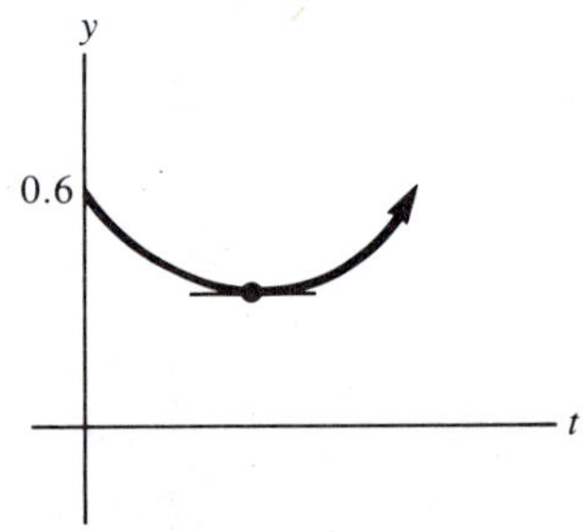

0.475 when $t = 25$

2.4 EXERCISES, PAGE 106

1. $v(t) = 40$ mph, $a(t) = 0$ mph^2

3. **(a)** 134 fps, 206 fps **(b)** 24 fps^2, 24 fps^2

5. $a(t) = 1.6$ mps^2; $d(t) = 0.8t^2$ m

7. −\$75, −\$15, \$0, and \$2.78 per thousand units; decreasing, decreasing, neither increasing nor decreasing, increasing

9. −50,000 and −130,000 units per dollar

11. 36, 135, and 420 bacteria/hour
13. 1.875, 2.5, and 3.875 percent/day
15. −0.001, −0.004, −0.10 of the voters/day

2.5 EXERCISES, PAGE 115

1. (a) \$105,000, \$60,000, −\$45,000
   (b) \$100, \$600, \$500
   (c) \$80,000 when $x = 600$
3. (a) \$100, \$600, \$500; \$200, \$500, \$300
   (b) \$22,500, 350 units
   (c) \$525
5. \$3,750,000, 1,250,000, \$9875
7. (a) −0.6; 75,000, \$1,125,000; 79,500, \$1,073,250; 70,500, \$1,163,250
   (b) −1.0; 60,000, \$1,200,000; 63,000, \$1,197,000; 57,000, \$1,197,000
   (c) −3.0; 30,000, \$900,000; 39,000, \$1,053,000; 21,000, \$693,000
9. (a) accounting (b) neither
   (c) marketing

2.6 EXERCISES, PAGE 121

1. 1.414214 3. 0.292893, 1.707107
5. 1.189207
7. with $r_0 = 1$, r_1 is undefined; with $r_0 = 0$, root is ≈ 0.206299
9. 12,166 and 25,654
11. ≈ 2.283 sec

2.7 EXERCISES, PAGE 124

1. $\dfrac{dy}{dx} = 6x^2 - 3$ 3. $\dfrac{df}{dt} = -8t^{-3} + 6t + 2$
5. $\dfrac{dg}{dx} = -\dfrac{2}{x^2} + \dfrac{12}{x^4}$ 7. $\left.\dfrac{dy}{dx}\right|_{x=4} = 1265$
9. $\left.\dfrac{dq}{dp}\right|_{p=1} = -400$
11. $\dfrac{dy}{dt} = -\dfrac{1}{4t^2}, \dfrac{dt}{dy} = -\dfrac{1}{4y^2}$
13. $f''(x) = 6x + 2, f'''(x) = 6$
15. $\dfrac{d^2f}{dx^2} = 12x^2 - 24x, \dfrac{d^3f}{dx^3} = 24x - 24$
17. $g''(x) = 12x^2 - \dfrac{1}{2x^{3/2}}, g'''(4) = \dfrac{12{,}291}{128}$

2 REVIEW EXERCISES, PAGE 125

1. (a) does not exist (b) 2
   (c) does not exist (d) 0
   (e) −1 (f) 0
   (g) does not exist (h) 1
2. 0 3. 4
4. 2 5. $\dfrac{106}{9}$
6. −1 7. 1
8. does not exist 9. $\dfrac{5}{4}$
10. does not exist 11. $\dfrac{1}{5}$
12. 0
13. $\dfrac{7}{2}$
14. does not exist
15. $\dfrac{1}{4}$
16. 0
17. $\dfrac{1}{4}$
18. (a) $x = 2, y = -\dfrac{3}{2}$;

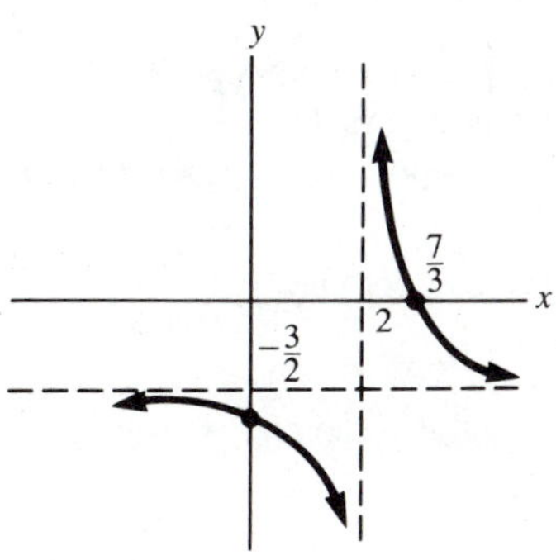

(b) $x = 0, y = x - 3;$

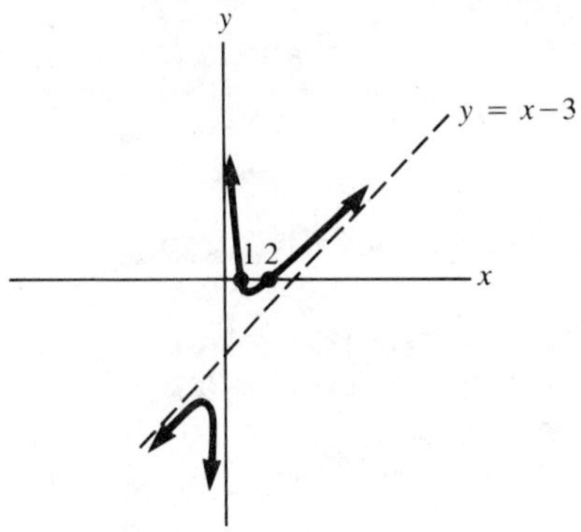

19. $f'(x) = 3x^2 + 2x;\ y = 16x - 20$

20. $g'(x) = \frac{1}{3}x^{-2/3}$; does not exist at $x = 0$

21. $-1, 0, 2, 3, 5$ **22.** $-1, 2, 3$

23. $f'(x) = 0$ **24.** $f'(x) = \sqrt{3}$

25. $f'(x) = 6x + 6$

26. $f'(x) = -6x^5 + 20x^4 - 36x^2 + 32x - 8$

27. $f'(x) = -\frac{6}{x^4} + \frac{1}{x^2}$

28. $f'(x) = -8x^{-3} - 6x^{-4}$

29. $f'(x) = \frac{2}{\sqrt{x}}$

30. $f'(x) = 6\sqrt{x} - \frac{3}{2\sqrt{x}}$

31. $f'(x) = 4x^{-1/5} + x^{-3/2} - \frac{2}{x^{5/3}}$

32. $y = 18x - 34$ **33.** $y = -\frac{1}{2}x + \frac{7}{2}$

34. $v(3) = 120$ mps, $v(6) = 90$ mps, $a(3) = a(6) = -10$ mps^2, 1125 m, 30 sec

35. $-\$96{,}000$, $-\$48{,}000$, and $-\$48/\sqrt{2} \approx -\$33{,}941$ per year

36. \$72, 14 units, \$42 per unit

37. **(a)** $-\frac{1}{3}, -1, -3$

(b) \$1,800,000, \$2,400,000, \$1,800,000

(c) \$1,914,000, \$2,376,000, \$1,386,000

38. 2.154435 **39.** \$1054.32

40. $\frac{df}{dx} = 8x^3 - 10x,\ \frac{d^2f}{dx^2} = 24x^2 - 10,\ \left.\frac{d^3f}{dx^3}\right|_{x=2} = 96$

41. $\frac{dy}{dx} = \frac{1}{\sqrt{x}} - \frac{1}{x^{3/2}},\ \frac{d^2y}{dx^2} = -\frac{1}{2x^{3/2}} + \frac{3}{2x^{5/2}}$

42. $g'(x) = 9x^2 - \frac{16}{5}x^{-1/5},\ g''(x) = 18x + \frac{16}{25}x^{-6/5},\ g'''(1) = \frac{2154}{125}$

43. $\frac{dw}{du} = 4u + 3b + \frac{1}{\sqrt{u}},\ \frac{d^2w}{du^2} = 4 - \frac{1}{2u^{3/2}}$

44. $\frac{dy}{dx} = \frac{8}{11},\ \frac{dx}{dy} = \frac{11}{8}$

3.1 EXERCISES, PAGE 138

1. $k'(x) = 12x + 7$

3. $k'(x) = 56x^6 + 12x^5$

5. $k'(x) = -\frac{1}{x^2}$

7. $k'(x) = -\frac{22}{(5x - 1)^2}$

9. $k'(x) = \frac{2x(x + 1)}{(2x + 1)^2}$

11. $k'(x) = 10(5x + 3)$

13. $k'(x) = 1 - 6x^{-4}$

15. $k'(x) = -\frac{3x^3 - 8}{x^5}$

17. $k'(x) = -5x^{3/2} + 7x^{4/3} + 9\sqrt{x} - 12x^{1/3} - \frac{1}{\sqrt{x}} + x^{-2/3}$

19. $k'(x) = \frac{5}{6}x^{-1/6} + \frac{5}{3}x^{-8/3}$

21. $k'(x) = 6x^5 + 5x^4 + 4x^3 + 6x^2 + 1$

23. $k'(x) = \frac{3x^6 - 8x^5 + x^4 - 4x^3 - 4x^2 - 4x + 4}{x^2(x - 2)^2}$

25. $C'(1) = \$0.75$ = approximate cost of making 2nd unit,

$C'(10) = \$\frac{120}{121} \approx \$0.99$ = approximate cost of making 11th unit,

$C'(20) = \$\frac{440}{441} \approx \$1.00$ = approximate cost of making 21st unit

27. $\frac{2000}{121}, \frac{80}{9}$, and 5 bacteria per hr; $\lim_{t \to +\infty} y = 200$, so number of bacteria in culture will approach 200

29. $70, \frac{70}{9}$, and $\frac{35}{18}$ percent per repetition

3.2 EXERCISES, PAGE 145

1. $k(x) = f(g(x))$, where $f(x) = x^{10}$ and $g(x) = 3x + 5$
3. $k(x) = f(g(x))$ where $f(x) = x^5$ and $g(x) = x^2 - 2x$
5. $k(x) = f(g(x))$, where $f(x) = \sqrt{x}$ and $g(x) = 4x - 1$
7. $k'(x) = 30(3x + 5)^9$
9. $k'(x) = 10(x - 1)(x^2 - 2x)^4$
11. $k'(x) = \dfrac{2}{\sqrt{4x - 1}}$
13. $k'(x) = 100(10x - 3)(5x^2 - 3x + 1)^{99}$
15. $k'(x) = \dfrac{x}{3}(5x^3 - 6)(x^5 - 3x^2)^{-2/3}$
17. $k'(x) = (2x^2 + 1)^{10}(x^3 - 1)^8(98x^4 + 27x^2 - 44x)$
19. $k'(x) = -6(4\sqrt{x} - 3x^{-2})^{-4}(3x^{4/3} + 10)^{-4}$ $(11x^{5/6} + 10x^{-1/2} + 3x^{-5/3} + 30x^{-3})$
21. $k'(x) = \dfrac{1}{2}(x^2 + 3)^{1/2}(2x - 3)^{-2/3}\left(\dfrac{58}{3}x^3 - 27x^2 + 4\right)$
23. $k'(x) = -\dfrac{16(x + 2)^3}{(x - 2)^5}$
25. $k'(x) = \dfrac{x^2(14x + 9)}{\sqrt{4x + 3}}$
27. $k'(x) = \dfrac{9 - x}{(x^2 + 3)^{3/2}}$
29. $\dfrac{dy}{dt} = 8t - 16$ 31. $\dfrac{du}{dv} = \dfrac{v}{|v|}$
33. (a) $\dfrac{dy}{dt} = -4(4t - 5)^{-2}$
    (b) $\dfrac{dy}{dt} = 750(15{,}625t^6 + 75t^2 - 1)^4$ $(625t^4 + 1)$
35. $R'(1) = \$0.9405$ = approximate revenue from selling 2nd unit,
$R'(10) = \$0$ = approximate revenue from selling 11th unit,
$R'(20) = \$57$ = approximate revenue from selling 21st unit;
$P'(1) = \$0.1905$ = approximate profit from making and selling 2nd unit,
$P'(10) \approx -\$0.99$ = approximate profit from making and selling 11th unit,
$P'(20) \approx \$56$ = approximate profit from making and selling 21st unit
37. $\dfrac{34}{13^{3/2}18^{1/2}} \approx 171$ and $\dfrac{34}{15^{3/2}26^{1/2}} \approx 115$ individuals per yr; $\text{Lim}_{t\to+\infty}\, y = 2$, so population size will approach 2000

3.3 EXERCISES, PAGE 156

1. rel max: $(-1, 2)$, $(2, 1)$; rel min: $(1, -1)$, $(3, -2)$
3. rel max: $(-2, 1)$, $(1, 3)$; rel min: $(0, 0)$, $(2, 2)$
5. min: $(0, -1)$ 7. min: $(2, -1)$
9. max: $\left(3, -\dfrac{1}{2}\right)$
11. max: $(0, 0)$, min: $\left(\dfrac{4}{3}, -\dfrac{32}{27}\right)$
13. min: $(-2, -16)$, max: $(2, 16)$
15. saddle pt: $(2, 10)$
17. max: $(-1, 11)$ saddle pt: $(1, -5)$
19. min: $(0, 0)$
21. max: $\left(-3, \dfrac{1}{6}\right)$, min: $\left(1, -\dfrac{1}{2}\right)$
23. \$400 per unit 25. \$166.20, $t = 18$
27. ≈ 1174, mid-May ($t = 5.5$)
29. max of 6,628,100 in 1993, min of 1,962,500 in 2005

3.4 EXERCISES, PAGE 162

1. min: $(4, -14)$ 3. min: $(5, -74)$
5. min: $(0, 0)$, max: $(5, 125)$
7. max: $(-4, 521)$, min: $(9, -1676)$
9. max: $(1, 9)$, min: $(2, 8)$, max: $(3, 9)$
11. max: $(-3, 162)$, min: $(3, -162)$
13. max: $(0, 0)$, min: $(1, -21)$
15. min: $(0, 0)$
17. \$3,125,000
21. 75 per thousand in 1985

3.5 EXERCISES, PAGE 170

1. (a) abs max is 3 at $x = 0$, abs min is -2 at $x = 3$
   (b) abs max is 4 at $x = 6$, abs min is -2 at $x = 3$
   (c) abs max is 3 at $x = 0$, no abs min
   (d) no abs max, abs min is -2 at $x = 3$

3. **(a)** abs max is -5 at $x = 0$, abs min is -32 at $x = 3$
(b) abs max is 43 at $x = 8$, abs min is -5 at $x = 6$
(c) no abs max, abs min is -32 at $x = 3$
(d) no abs max, abs min is -32 at $x = 3$

5. \$314,000; \$179,000

7. \$50,000,000, \$500; \$48,000,000, \$600

9. 200 boxes, 1 order every other week, \$52

13. 34,225 eggs, 185 chickens

15. maximum of $\approx$ 1333 in 1985, minimum of $\approx$ 771 in 1970

17. velocity is greatest on the central axis

19. $\dfrac{20\pi}{4 + \pi}$ cm from one end; the longer piece is bent into a circle

21. minimum sum is 12 when both numbers equal 6

23. 200 ft by 200 ft

25. 100 m by 100 m

27. side parallel to wall 160 ft, other two sides 80 ft

29. base 2 ft by 4 ft, height $4\frac{4}{9}$ ft

31. 128 in^3

33. either 40 radio and 0 newspaper ads or 0 radio and 20 newspaper ads

3.6 EXERCISES, PAGE 182

1.

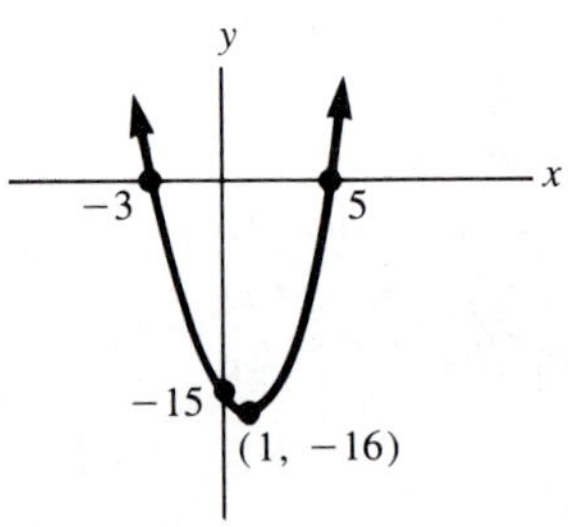

3.

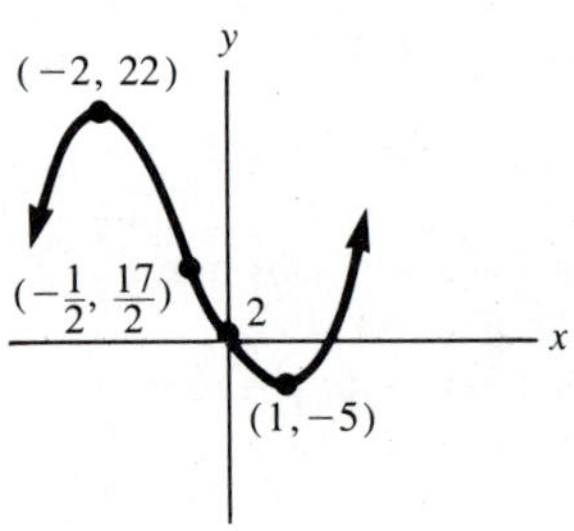

5.

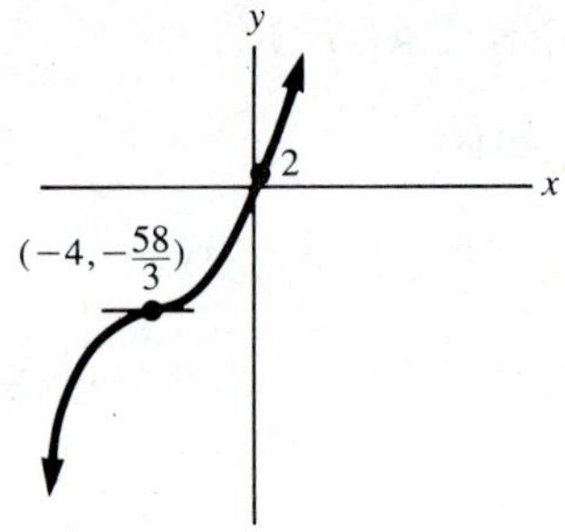

7.

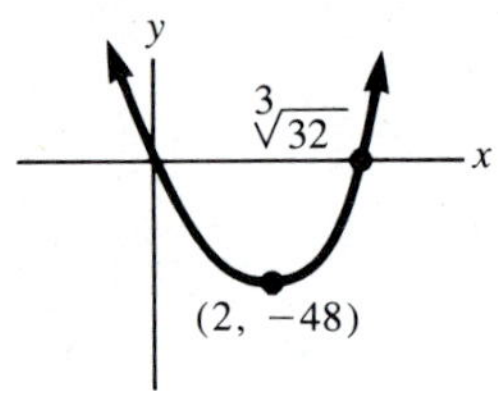

9.

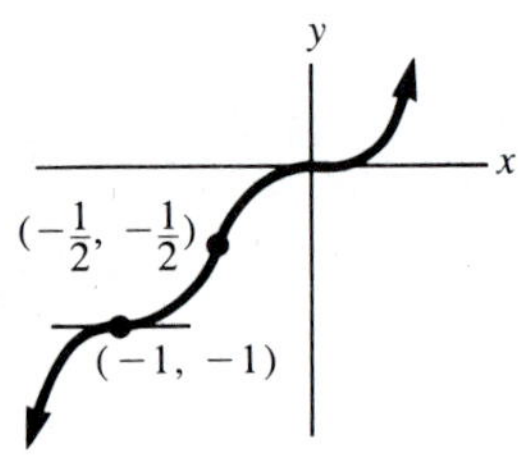

11.

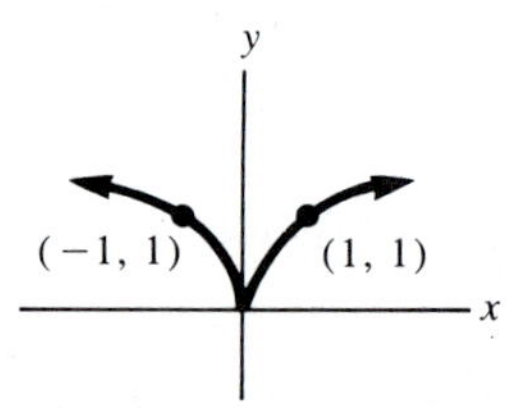

13.

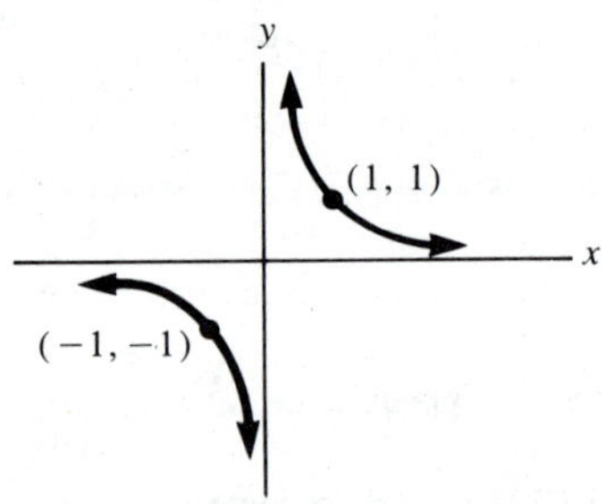

15.

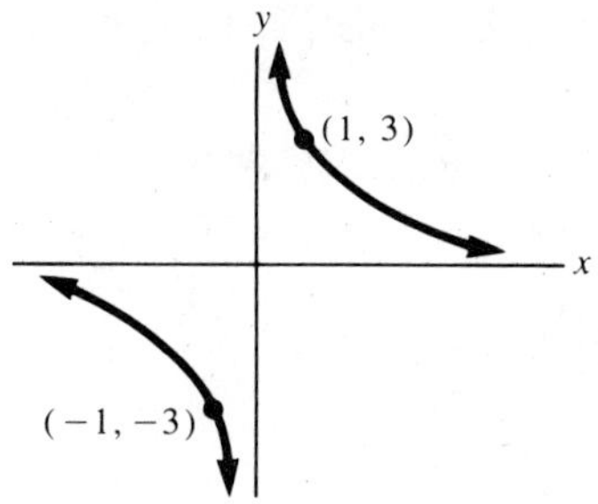

17.

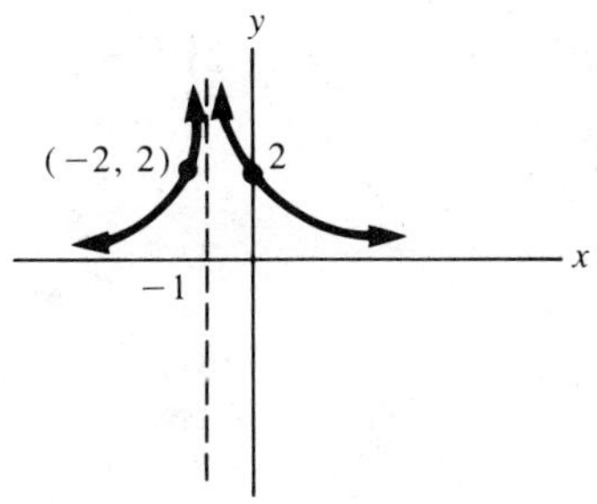

19.

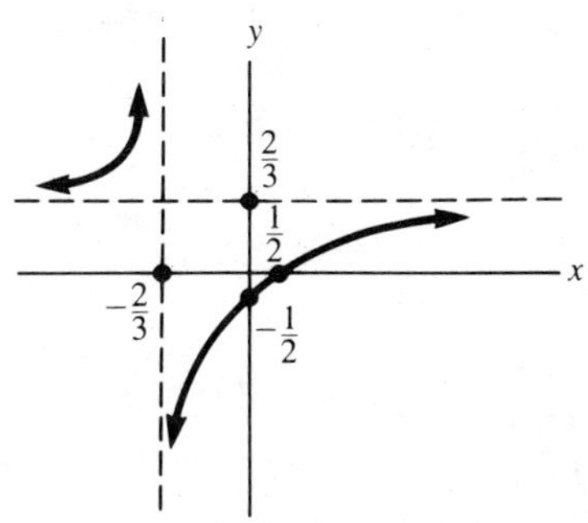

21.

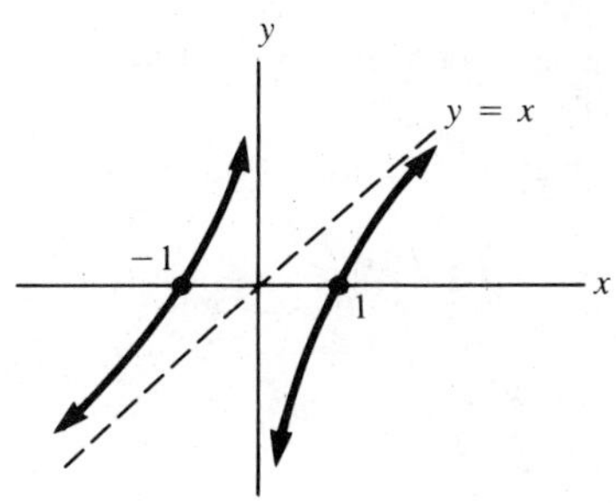

23.

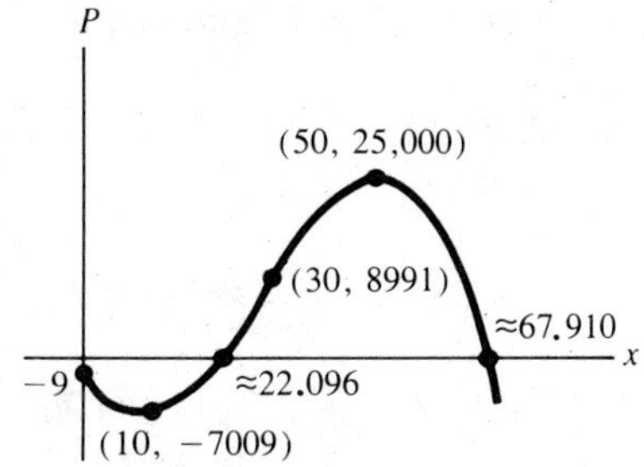

25.

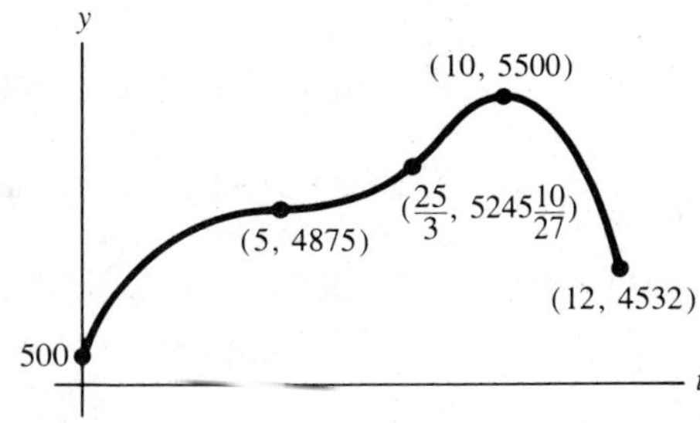

27.

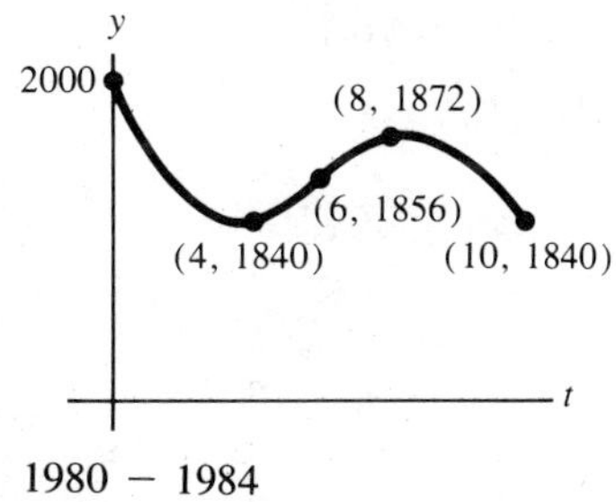

29.

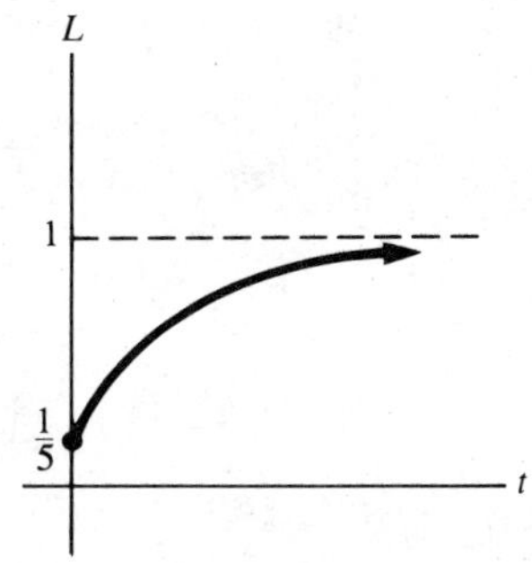

3.7 EXERCISES, PAGE 189

1. $dy = (6x - 2)\,dx$
3. $dy = (3x^2 - 8x + 3x^{-7/4})\,dx$
5. $dy = 2(x^2 + 4)^3(15x^3 - 9x^2 + 16x - 4)\,dx$
7. $\frac{31}{8}$
9. $\frac{82}{27}$
11. 2.005
13. **(a)** \$99.80 **(b)** \$83.56
15. \$13,500
17. 5650
19. 618
21. perimeter: ± 0.005, ± 0.5 percent; area: ± 0.01, ± 1 percent
23. volume: ± 0.012, ± 1.2 percent; surface area: ± 0.008, ± 0.8 percent
25. 0.03, 3 percent
27. -0.03, -3 percent
29. 0.08, 8 percent; -0.04, -4 percent

3.8 EXERCISES, PAGE 194

1. $\frac{dy}{dx} = -2$
3. $\frac{dy}{dx} = -\frac{1}{x^2}$
5. $\frac{dy}{dx} = \frac{x}{2y}$
7. $\frac{dy}{dx} = -\frac{2y}{x}$
9. $\frac{dy}{dx} = 5$
11. $\frac{dy}{dx} = -\frac{y}{x + 2\sqrt{xy}}$
13. $y = -\frac{1}{3}x + \frac{10}{3}$
15. $y = \frac{1}{4}x - 2$
17. $y = 2x - 4$
19. $y = -1$
21. $\frac{dy}{dt} = -\frac{x}{y}\frac{dx}{dt}$
23. $\frac{dy}{dt} = -\frac{y(2x + y)}{x(2y + x)}\frac{dx}{dt}$
25. $\frac{dy}{dt} = -\frac{y^2}{2xy - 3y - 6}\frac{dx}{dt}$

3.9 EXERCISES, PAGE 199

1. 128π cm^2/sec
3. -1 ft/sec
5. 500 mph
7. $5/\pi \approx 1.59$ m/min
9. -28.5 units/month
11. $\$\frac{25}{7} \approx \$3.57$ thousand/ad
13. 0.05625 ppm/day
15. ≈ 33.5 individuals/yr

3 REVIEW EXERCISES, PAGE 201

1. $f'(x) = 40x^4 + 76x^3 - 15x^2 + 2x + 2$
2. $f'(x) = -\frac{44}{(5x - 3)^2}$
3. $f'(x) = \frac{4x^2 + 12x - 3}{2(2x + 3)^2}$
4. $f'(x) = 9x^2 + 7x^{5/2} - x^{-1/2}$
5. $f'(x) = \frac{5(2x^4 - 3x^2 - 4x + 3)}{(2x^2 + 1)^2}$
6. $f'(x) = 12(6x + 5)(3x^2 + 5x + 1)^{11}$
7. $f'(x) = \frac{9x^2 - 1}{\sqrt{6x^3 - 2x + 3}}$
8. $f'(x) = 4(2x + 3)^5(x^2 - 5x)^7(11x^2 - 23x - 30)$
9. $f'(x) = 20(2x - 2)^{19}(x^2 + 1)^4(3x^2 - x + 2)$
10. $f'(x) = \frac{7(3x - 1)^6(-3x^2 + 2x + 6)}{(x^2 + 2)^8}$
11. $f'(x) = -\frac{5(3x + 4)^4(9x^2 + 32x + 3)}{(5x^2 - 1)^5}$
12. $f'(x) = -\frac{2x}{(x^2 + 1)^{1/2}(x^2 - 1)^{3/2}}$
13. $\frac{dy}{dt} = 128t - 88$
14. $\frac{du}{dw} = 11\left(7 - \frac{3}{2\sqrt{w}}\right)(4v^3 - 2)(v^4 - 2v + 1)^{10}$
15. $\frac{118125(120)^{-1/4}}{2} \approx 17{,}845$ units/dollar
16. rel max: (1, 5), (5, 3), (7, 5); rel min: (0, 4), (4, 2); saddle pt: $(-1, 5)$
17. critical values: -10, 6; max: $(-10, 200)$, min: $(6, -1848)$
18. critical values: -3, 0; min: $(-3, -27)$, saddle pt: (0, 0)
19. critical value: $-\frac{1}{2}$; no max or min
20. \$78.50 per unit
21. $\frac{7}{12}$ of the trees
22. max: $(-5, 100)$, min: $(1, -8)$
23. max: (0, 1), min: $\left(2, -\frac{43}{5}\right)$
24. \$80,000,000,000
25. **(a)** abs max is 4 at $x = 5$, abs min is 1 at $x = 1$

(b) abs max is 3 at $x = 3$, abs min is 1 at $x = 1$

(c) abs max is 4 at $x = 5$, abs min is -2 at $x = -2$

(d) abs max is 3 at $x = 3$, abs min is -2 at $x = -2$

(e) abs max is 4 at $x = 5$, no abs min

(f) abs max is 4 at $x = 5$, no abs min

(g) abs max is 0 at $x = -3$, no abs min

(h) abs max is 2 at $x = 0$, no abs min

26. **(a)** no abs max, abs min is -256 at $x = \pm 4$

(b) abs max is 0 at $x = 0$, abs min is -31 at $x = \pm 1$

(c) abs max is 144 at $x = 6$, abs min is -256 at $x = \pm 4$

(d) abs max is 144 at $x = 6$, abs min is -256 at $x = 4$

27. \$10,000; \$64,000

28. 20 cars, \$4,840,000

29. $r = \dfrac{1}{\sqrt{\pi}}, h = \dfrac{2}{\sqrt{\pi}}$

30.

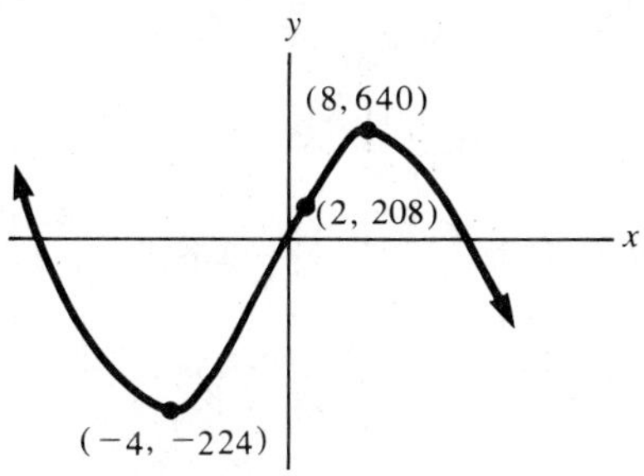

31.

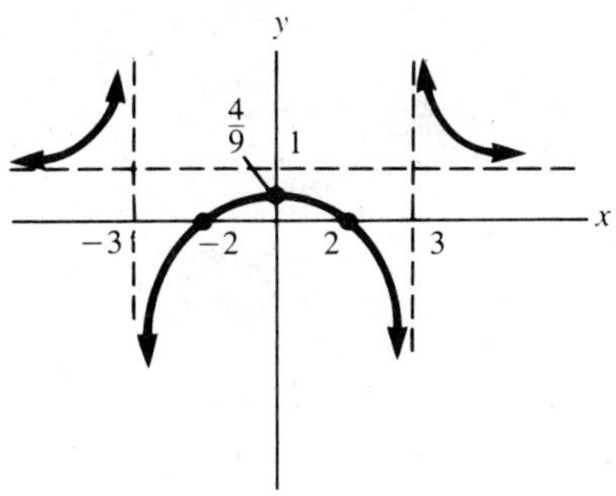

32.

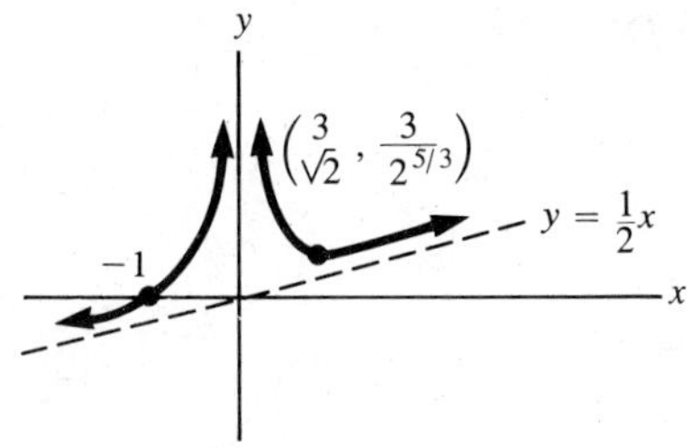

33.

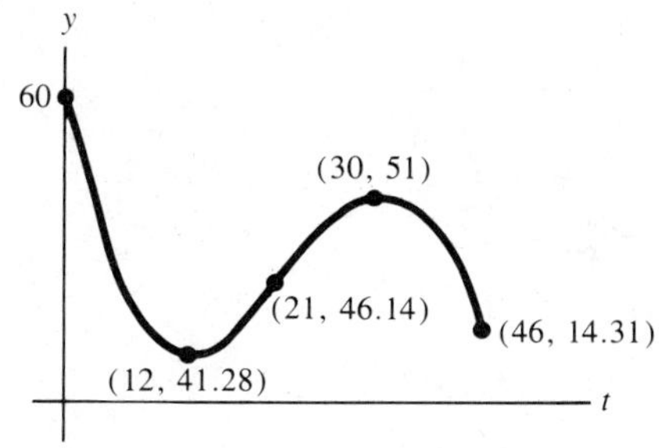

34. $4\dfrac{1}{48}$

35. -0.05, -5 percent

36. \$135

37. $\dfrac{dy}{dx} = -\dfrac{y(8xy^4 - 1)}{2x(4xy^4 + 1)}$

38. $\dfrac{dy}{du} = \dfrac{y(2 - 3x^2y)}{2x(3x^2y + 1)}\dfrac{dx}{du}$

39. -50 units/month

40. $\dfrac{2815}{\sqrt{1489}}$ mph

3 SUPPLEMENTARY EXERCISES, PAGE 206

1. $\dfrac{dC}{dY} = 0.6$; at every income level, 60 cents of each additional dollar of income is spent

3. \$2.50; at every income level, each additional \$1 of savings generates an additional \$2.50 in income

5. **(a)** 0.40; at every income level, 40 cents of each additional dollar of income is saved

(b) 0.75; at an income level of \$410 billion, 75 cents of each additional dollar of income is saved; 0.90; at an income level of \$2510 billion, 90 cents of each additional dollar of income is saved

4.1 EXERCISES, PAGE 217

1.

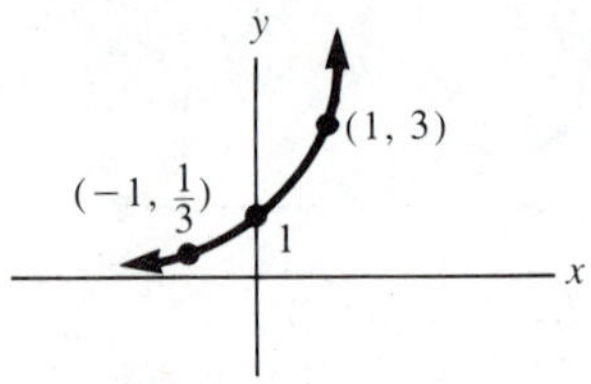

3.

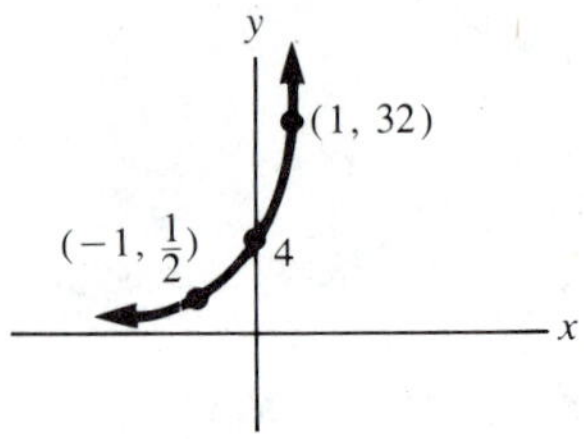

5.

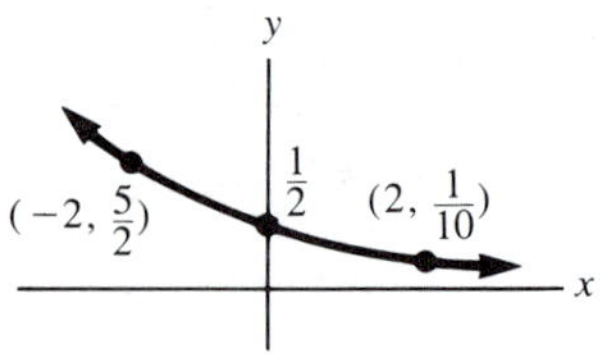

7.

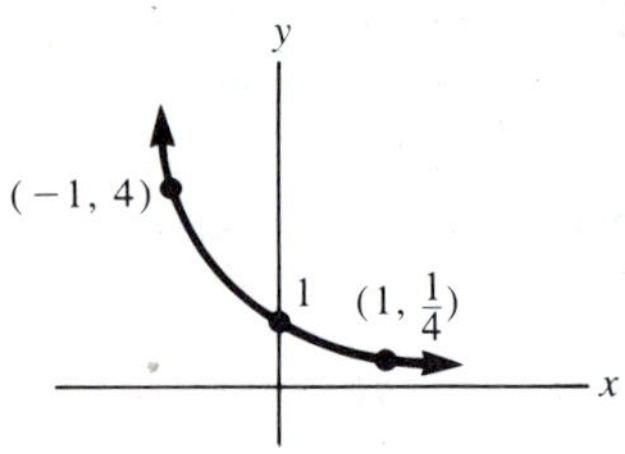

9. **(a)** \$106,000 **(b)** \$106,090

11. **(a)** \$10,250 **(b)** \$10,506.25
(c) \$12,184.03 **(d)** \$15,986.50

13. \$6736.25 **17.** \$106,183.65

19. **(a)** \$1123.60 **(b)** \$1125.51
(c) \$1126.83 **(d)** \$1127.50

21.

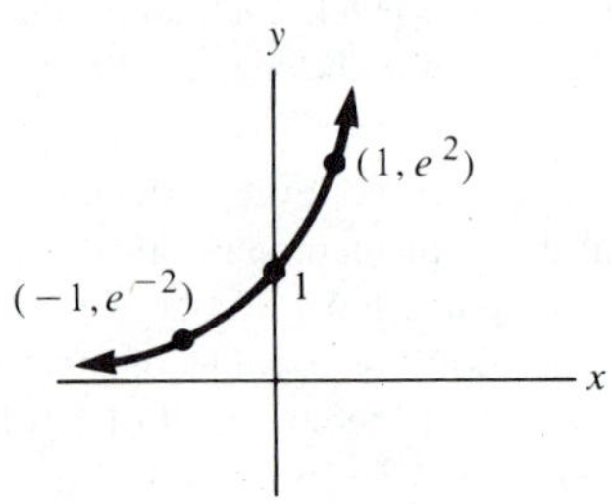

23.

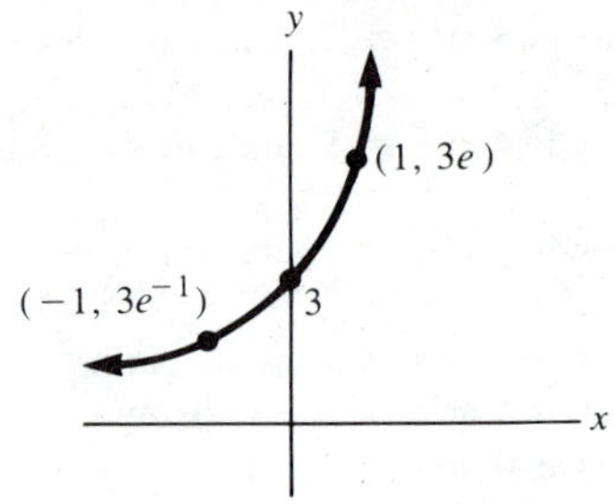

25.

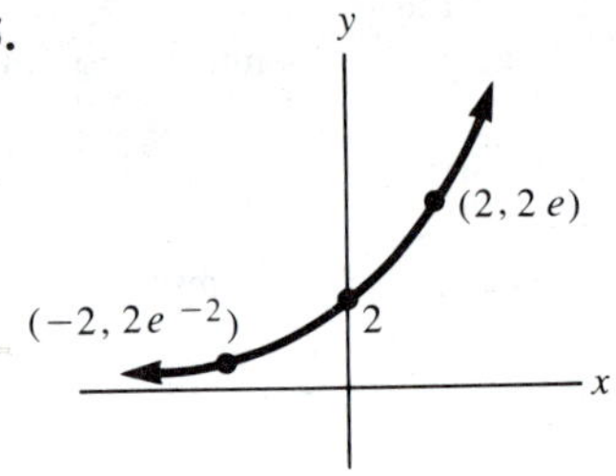

4.2 EXERCISES, PAGE 225

1. 4 **3.** −4

5. 3 **7.** 1

9. 5 **11.** 2.85

13. 3

15. **(a)** 0.9542 **(b)** 2.097
(c) 1.8751 **(d)** 0.5104

19.

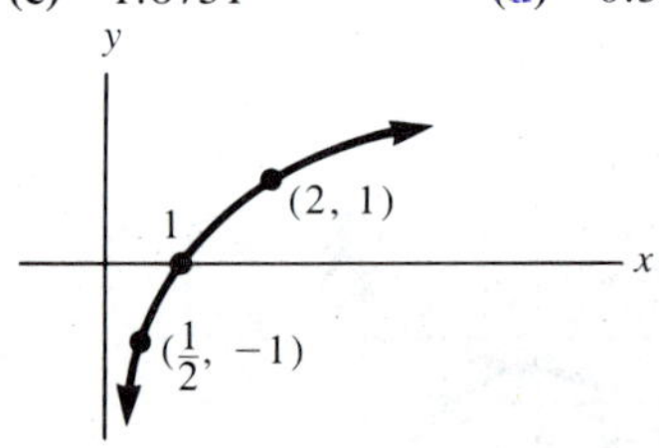

21.

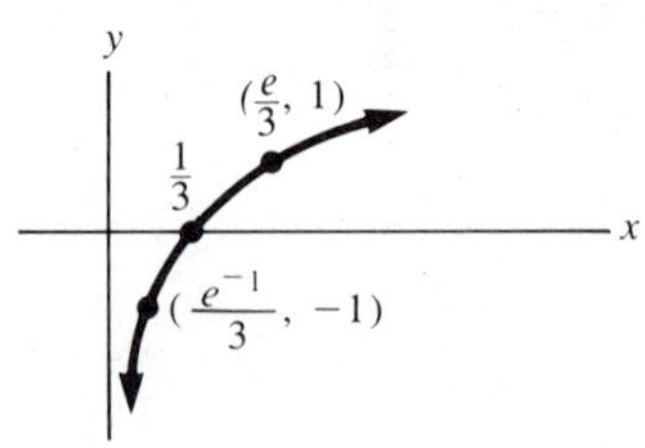

23.

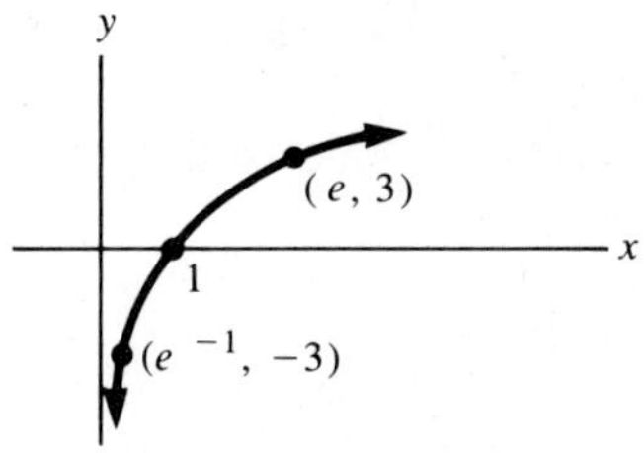

25.	$x = 0.6931$	**27.**	$x = 0.6$
29.	**(a)** 16.09 yr	**(b)**	16.29 yr
31.	5.76 yr	**33.**	74 percent; 2006 ($t \approx 36.9$)

4.3 EXERCISES, PAGE 234

1. $f'(x) = \dfrac{3}{x}$ **3.** $f'(x) = \dfrac{7}{7x + 2}$

5. $f'(x) = x(2 \ln 2x + 1)$ **7.** $f'(x) = \dfrac{3(\ln x)^2}{x}$

9. $f'(x) = [x(\ln x)(\ln(\ln x))]^{-1}$

11.

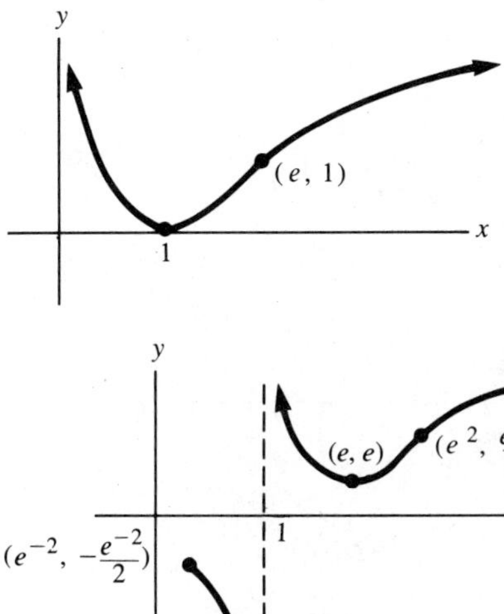

13.

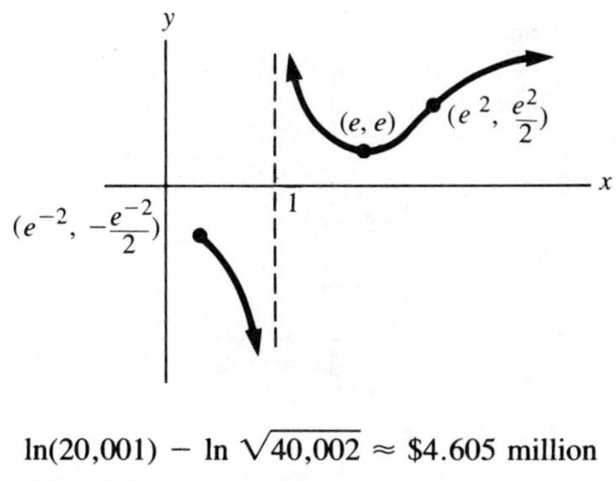

17. $\ln(20{,}001) - \ln \sqrt{40{,}002} \approx \$4.605$ million

19. 0.2 ppb/yr, 0.15 ppb/yr

21. $f'(x) = -2e^{-x}$ **23.** $f'(x) = 24e^{3x}$

25. $f'(x) = 5(2x + 3)e^{x^2+3x+4}$

27. $f'(x) = 6xe^{2x}(1 + x)$

29. $f'(x) = \dfrac{e^x(x \ln x - 1)}{x(\ln x)^2}$

31.

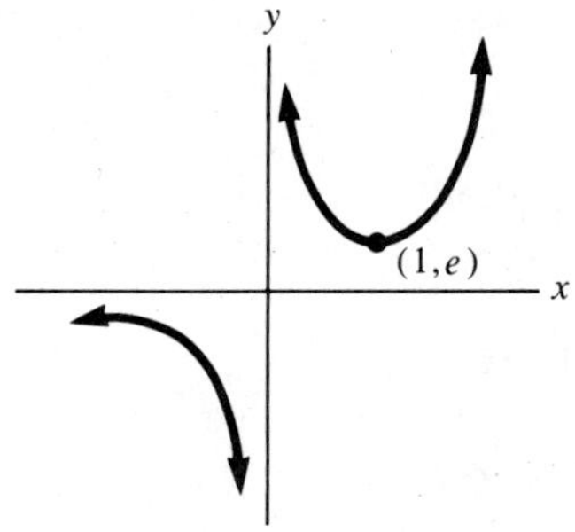

33. **(a)**

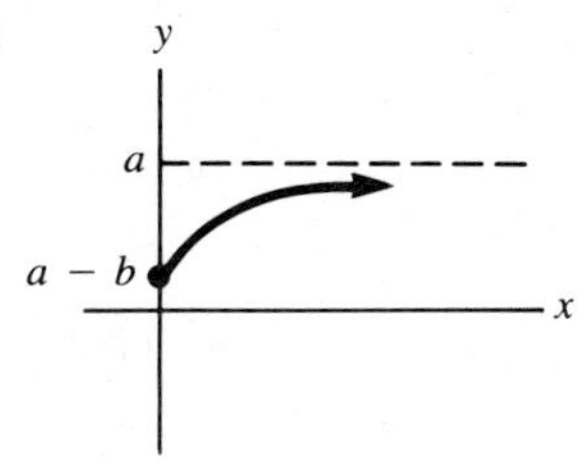

(b)

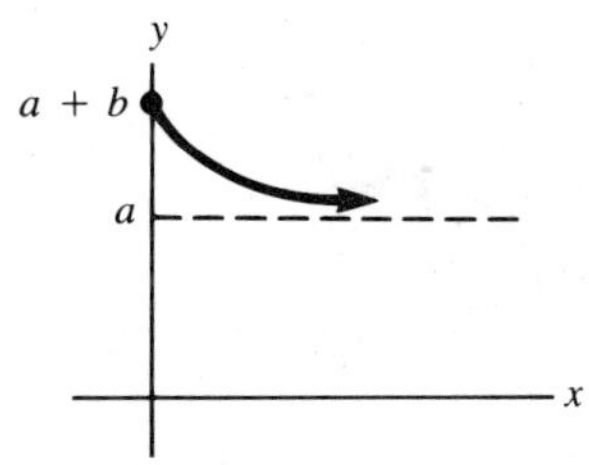

35. **(a)** $f'(x) = (\ln 2)2^x$
(b) $f'(x) = -6(\ln 5)5^{-2x}$
(c) $f'(x) = (\ln 10)e^x 10^{e^x}$

37. $\approx -\$0.11$, $\$0.027$, $-\$0.083$ per unit

39. $f'(x) = x(x^3 + 1)^3(x^5 + 1)^5 \left[\dfrac{1}{x} + \dfrac{9x^2}{x^3 + 1} + \dfrac{25x^4}{x^5 + 1}\right]$

41. $f'(x) = 2(\ln x + 1)x^{2x}$

43. $f'(x) = x^{x-1}x^{x^x}[x(\ln x)(\ln x + 1) + 1]$

4.4 EXERCISES, PAGE 243

1. **(a)** $y = 100{,}000e^{0.4t}$
(b) $1.644 million
(c) mid-1992 ($t \approx 12.5$)

3. **(a)** $y = 2.5e^{-0.193t}$
(b) $0.363 million
(c) end of 1992 ($t \approx 11.9$)

5. 2008 ($t \approx 28$)

7. $q = 100{,}000e^{-0.05p}$, 77,880 units

9. **(a)** $y = ce^{-0.043t}$
(b) ≈ 65.1 percent
(c) ≈ 53.55 days

11. $\approx 211{,}458$

13. $y = 4.85e^{0.015t}$, 13.05 billion, 2185 ($t \approx 201.7$)

15. $y = \ln(697.7t + 1096.6)$
(a) $\approx$ 10.28 million tons
(b) 2034 ($t \approx 84.2$)

17. $y = \ln(317.3t + 121.5)$, 1970 ($t \approx 25.2$)

4.5 EXERCISES, PAGE 252

1.

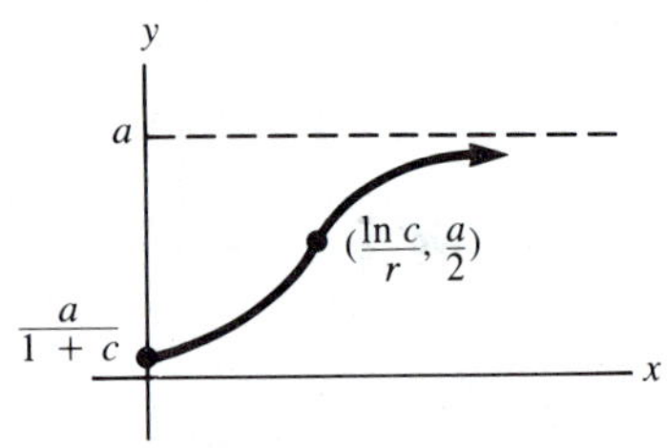

3. $y = 200 - 140e^{-0.113t}$;

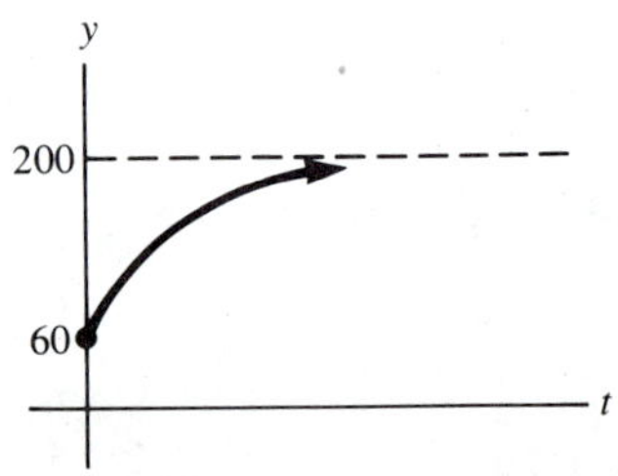

≈ 7.5 weeks

5. $y = 2000 + 900e^{-0.405x}$;

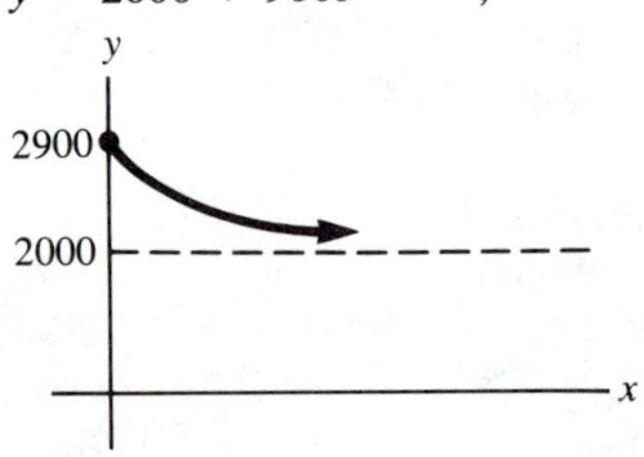

9445 hr

7. $y = 120 - 62.5e^{-0.223t}$, y in thousands; 120,000 units/yr

9. **(a)** 2 million units; 52.7 million units
(b) 98.4 million units
(c)

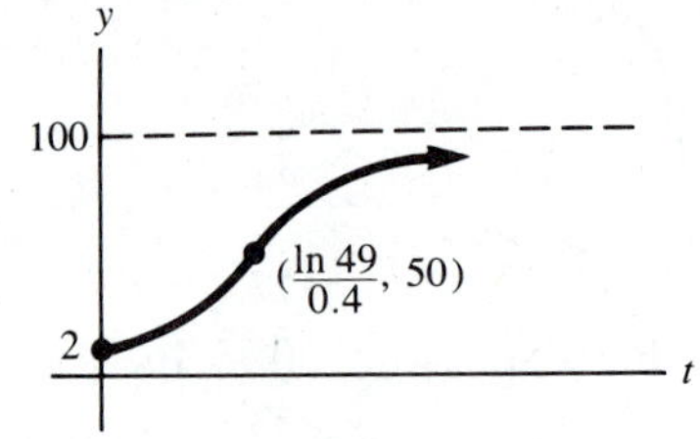

11. $y = 40 - 38e^{-0.589t}$, y in thousands

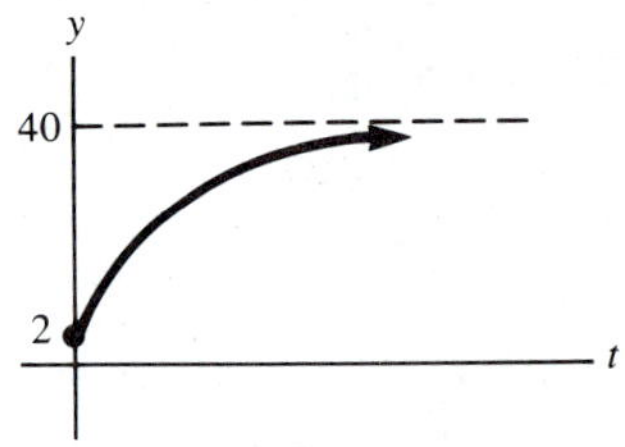

≈ 5.5 min ($t \approx 0.092$ hr)

13. $y = 40 + 60e^{-0.693t}$; ≈ 43.75 percent

15.

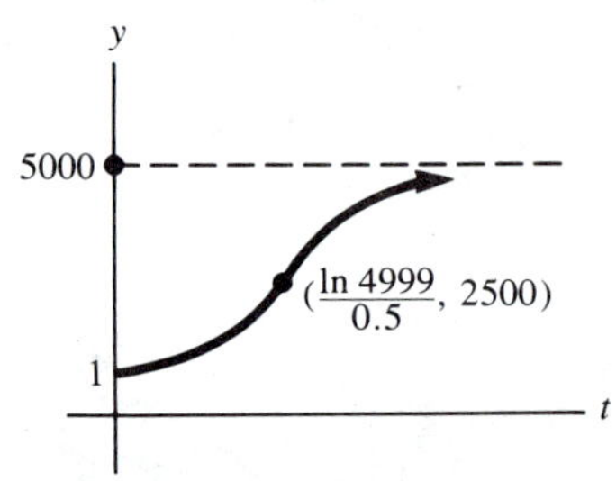

$\frac{5}{12}$

17.

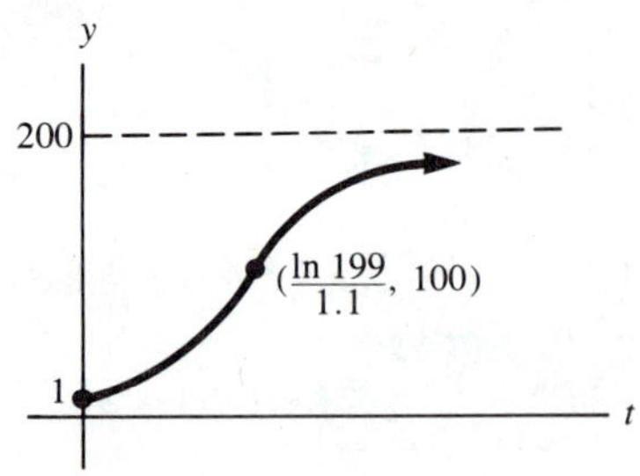

≈ 9.62 hr

REVIEW EXERCISES, PAGE 254

1. **(a)**

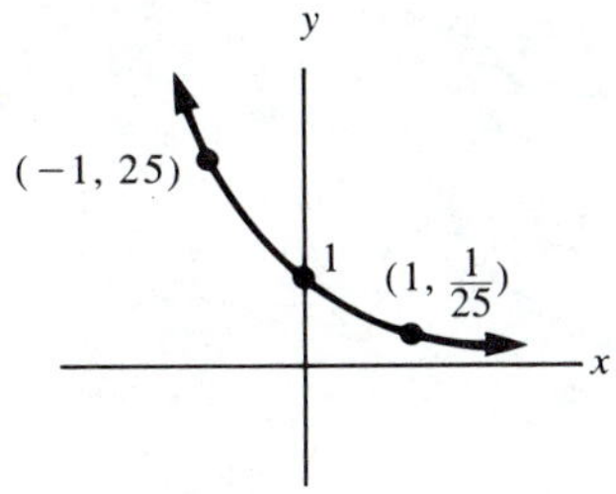

(b)

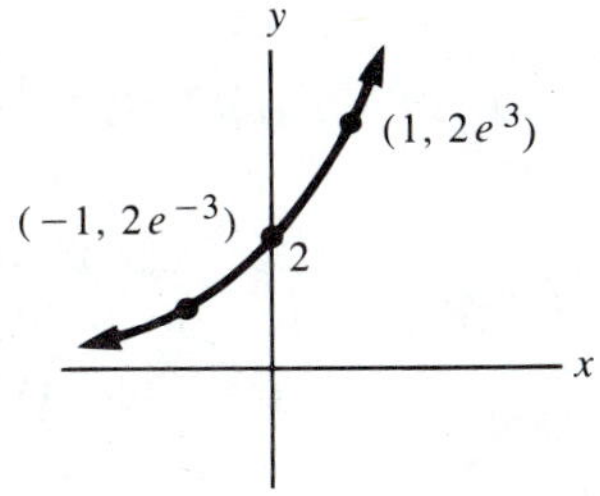

(c)

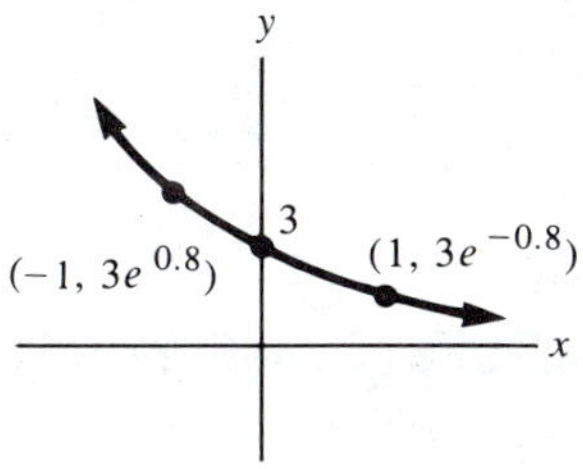

(d)

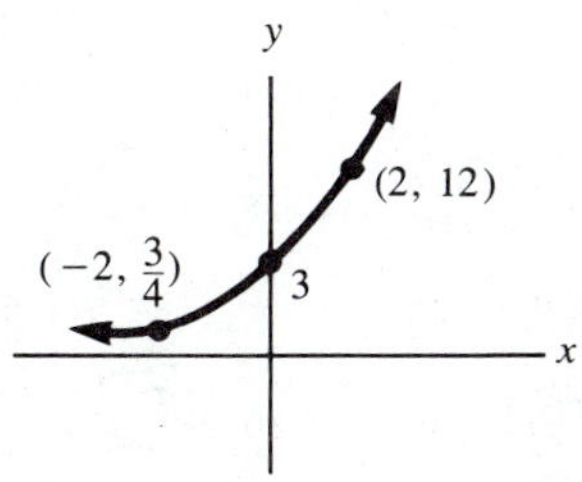

2. **(a)** \$7024.64 **(b)** \$7128.80
(c) \$7153.84 **(d)** \$7166.65

3. **(a)** \$17,799.93 **(b)** \$17,754.22
(c) \$17,743.71 **(d)** \$17,738.41

4. **(a)** 2 **(b)** −3
(c) 1 **(d)** 0
(e) 1.2 **(f)** 4
(g) 9.57498 **(h)** −0.26706

5. **(a)** 2 **(b)** 5.6148
(c) −4.7550 **(d)** 5.1293
(e) 0.2514

6. **(a)**

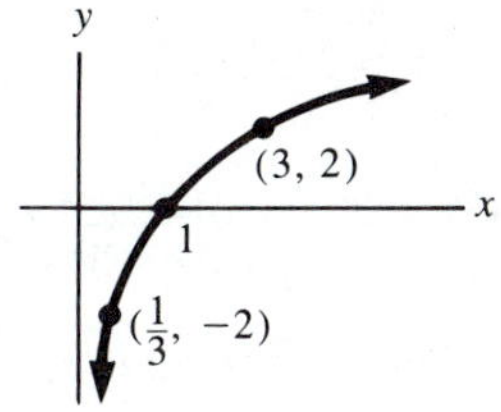

(b)

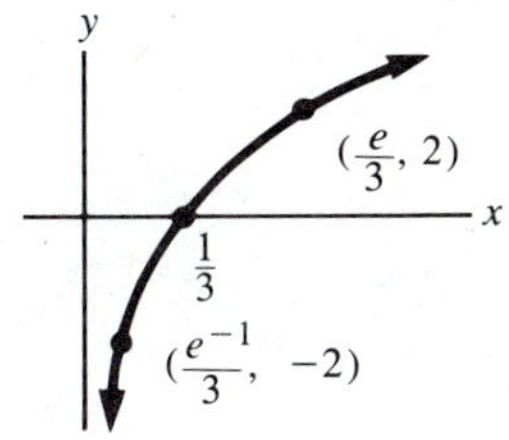

(c)

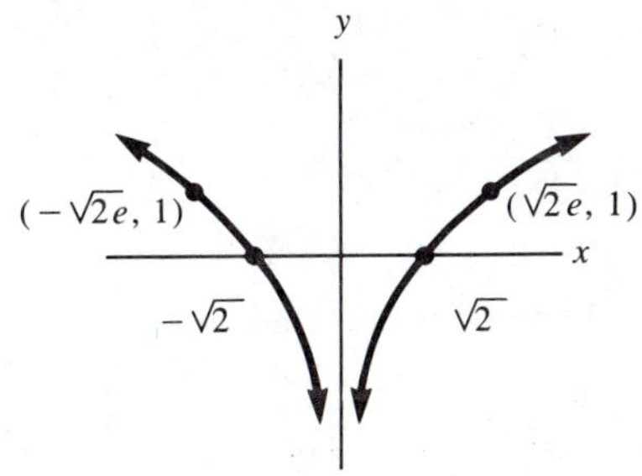

7. **(a)** $x = -0.536$ **(b)** $x = 6.288$

8. ≈ 13.86 min **9.** ≈ 9.4 yr

10. $f'(x) = \dfrac{3(x^2 + 1)}{x^3 + 3x - 2}$

11. $f'(x) = \dfrac{4x}{x^2 + 1}$

12. $f'(x) = 2\left(\dfrac{1}{x} - \dfrac{1}{2x - 1}\right)$

13. $f'(x) = \dfrac{2[(2x - 1)\ln(2x - 1) - 2x \ln x]}{x(2x - 1)(\ln(2x - 1))^2}$

14. $f'(x) = 10e^{5x}$

15. $f'(x) = 4xe^{-x}(2 - x)$

16. $f'(x) = 2e^{x^2}\left(2x \ln x + \dfrac{1}{x}\right)$

17. $f'(x) = \dfrac{2e^x}{(e^x + 1)^2}$

18. $f'(x) = -2e^{-2x}e^{e^{-2x}}$

19. $f'(x) = -x^{1-x^2}(2 \ln x + 1)$

20.

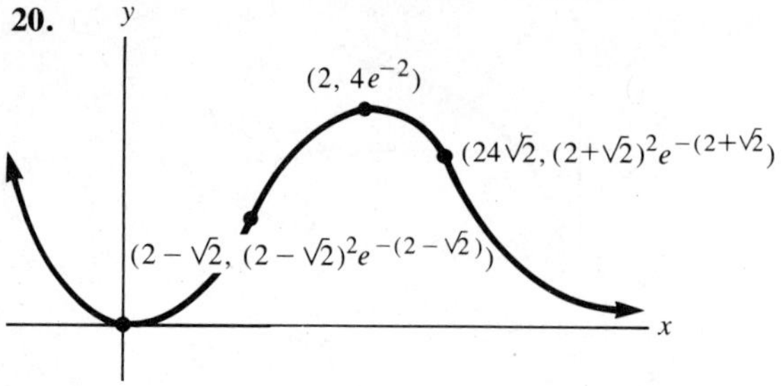

21.

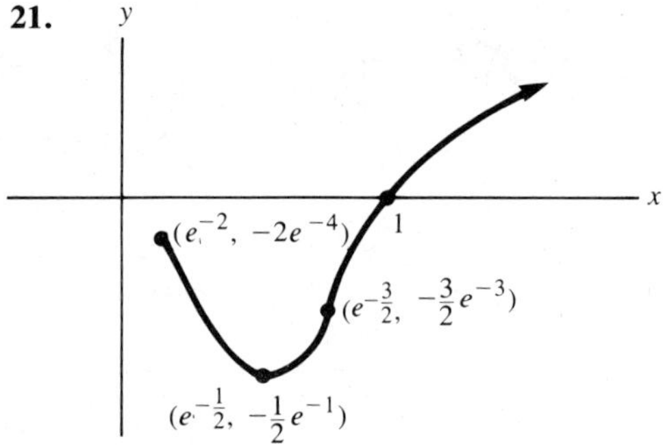

22. $\frac{3}{16}(\ln 8)^{-1/2} \approx \$0.13$ million/yr

23. $-40e^{-1} \approx -14.7$ ppm/yr

24. $y = 2000e^{-0.1t}$; ≈ 736; ≈ 23 days

25. \$5.29; mid-1994 ($t \approx 14.5$)

26. \$167,900; 2028

27. $y = 2000 - 400e^{-0.208t}$;

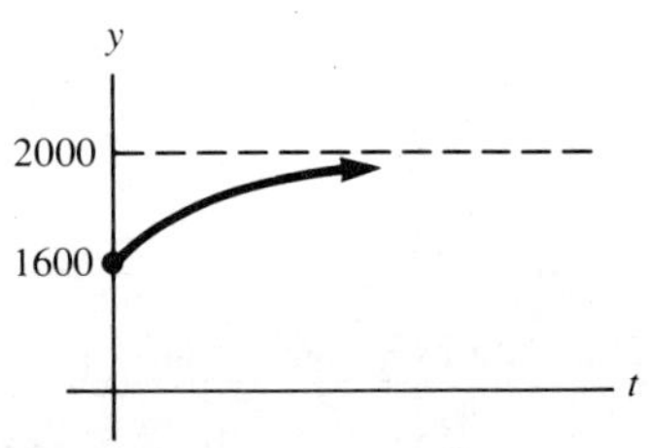

1885 units/month

28. $y = 250 + 173.1e^{-0.143x}$;

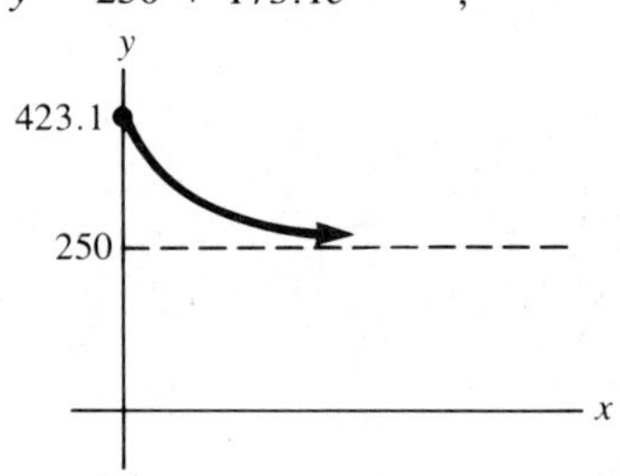

1490.4 hr

29. $y = 305 + 67.5e^{-0.405x}$; 305 sec

30.

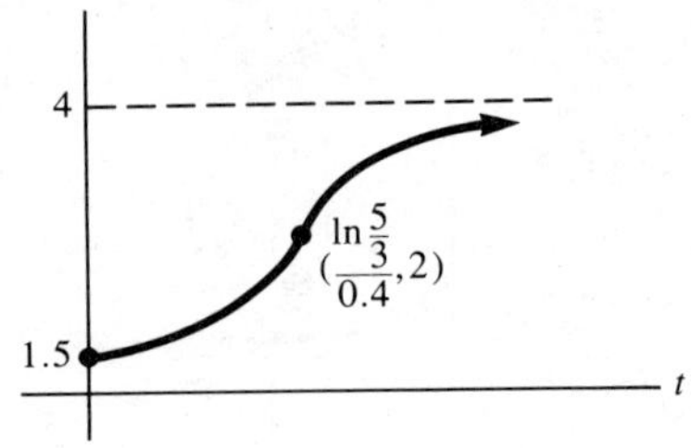

4 SUPPLEMENTARY EXERCISES, PAGE 261

1. \$13,781.56

3. **(a)** Y **(b)** X **(c)** neither

5. \$572,749.99

7. \$7065.63

9. X and Z; Z

11. $S = \dfrac{P(1 + r)[(1 + r)^{mt} - 1]}{r}$

13. $V = \dfrac{P(1 + r)[1 - (1 + r)^{-mt}]}{r}$

5.1 EXERCISES, PAGE 270

1. $3x + c$

3. $\dfrac{u^5}{5} + c$

5. $-\dfrac{t^{-3}}{3} + c$

7. $\dfrac{x^4}{2} + c$

9. $-2y^{3/2} + c$

11. $-4\ln|x| + c$

13. $-3e^y + c$

15. $\frac{1}{2}x^2 + 2x + c$

17. $\frac{1}{4}x^4 + x^3 - x^2 + c$

19. $\frac{\pi}{4}x^4 - \sqrt[3]{3}\,x + c$

21. $-\dfrac{2}{t} - \dfrac{1}{2t^2} + 3t + c$

23. $x^{100} + x^{99} + x^{98} + \cdots + x^2 + x + c$

25. $\frac{2}{3}x^3 + \frac{7}{2}x^2 - 15x + c$

27. 600 km

29. $C(x) = x^2 + x + 5000$, $R(x) = 15x$, $P(x) = -x^2 + 14x - 5000$

31. $P(x) = \frac{1}{3}x^3 - 50x^2 + 1600x - 18$

33. $f(t) = -0.4t^{3/2} + 68$

5.2 EXERCISES, PAGE 275

1. $\frac{2}{3}(x+1)^{3/2}+c$ **3.** $\frac{2}{25}(5t+2)^{5/2}+c$

5. $\frac{1}{91}(7x-3)^{13}+c$ **7.** $-\frac{6}{5}\ln|5x+2|+c$

9. $-\frac{3}{7}e^{-7x}+c$ **11.** $15(2x-3)^{1/3}+c$

13. $\frac{1}{10}(x^2+1)^5+c$ **15.** $\frac{1}{20}(2x^5+10)^2+c$

17. $\frac{1}{24}(2x^2-8x)^6+c$ **19.** $-\frac{7}{3}\ln|t^3+2|+c$

21. $\frac{1}{2}e^{2x^2+2x}+c$ **23.** $-\ln|x^3-2x|+c$

25. $\frac{1}{3}(\ln x)^3+c$

27. $C(x)=(x^2+1)^{3/2}+999$, $R(x)=(x^2+1)^{7/4}-1$

29. $f(t)=-100e^{-2t}+200$

31. $40\ln\frac{5t+4}{4}$

5.3 EXERCISES, PAGE 282

1. 8 **3.** 6 **5.** -2

7. $\frac{14}{3}$ **9.** $\frac{581{,}639}{24}$ **11.** $\frac{3}{32}$

13. $3(e^3-1)$

15. cost: \$226,500; revenue: \$543,000

17. \$5833.33 **19.** 4590 g

21. $84+4(\sqrt{24}-\sqrt{10})\approx 90.9$ in

23. $\frac{19}{3}$ **25.** $1-e^{-4}$ **27.** 6

29. $\frac{250{,}048}{9}$ **31.** $\frac{1}{2}\ln\frac{20}{3}$

33. $\approx 512{,}711$; ≈ 18.2 yr

35. ≈ 579.7 cc **37.** $\approx 14{,}456$

5.4 EXERCISES, PAGE 295

1. $\frac{92}{3}$ **3.** $\frac{124}{3}$ **5.** $\frac{81}{4}$

7. $1+\ln 2$ **9.** $\frac{1}{2}(1-e^{-9})$

11. **(a)**

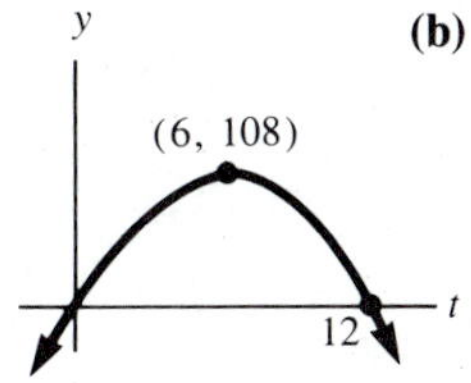

(b) 640

13. **(a)**

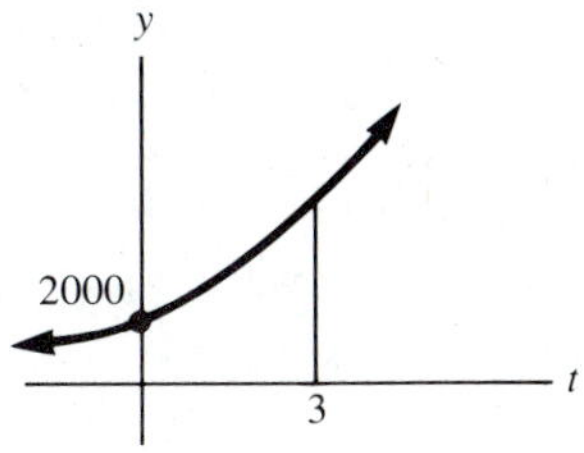

(b) ≈ 6184 **(c)** ≈ 119.9 hr

15. $\frac{1}{12}$ **17.** $\frac{1}{3}$ **19.** $\frac{1}{6}$

21. 12 **23.** 6

25. $2(1-e^{-1})$

27. **(a)** $\approx 191{,}606$ **(b)** $\approx -536{,}711$

5 REVIEW EXERCISES, PAGE 297

1. **(a)** $-\frac{2}{3}x^6+x^3-4x^2+14x+c$

(b) $2x^{-1}+\frac{15}{9}x^{9/5}+4x+c$

(c) $t^{-2}-\frac{2}{t}+\frac{8}{3}t^{3/2}+t+c$

(d) $2e^x+c$

(e) $-3\ln|x|+c$

(f) $2\ln|x|-3e^x+c$

2. $C(x)=3x^2+2x+5000$, $R(x)=10x^2$

3. $f(t)=200t^{3/2}+400$

4. $\frac{1}{6}(4x+1)^{3/2}+c$ **5.** $\frac{3}{8}(x^2+4)^4+c$

6. $e^{x^3}+c$ **7.** $\frac{1}{4}\ln(4x^2+1)+c$

8. $-3(3x^2+2x+1)^{-1/3}+c$

9. $-\frac{5}{2}e^{-x^2}+c$ **10.** $-(\ln x)^{-1}+c$

11. $-\frac{4}{3}$ **12.** $-\frac{49}{3}$ **13.** $\frac{7}{3}$

14. $\frac{461}{36}$ **15.** 2 **16.** $1 - e^{-1/2}$

17. $\frac{1}{3}\ln 9$ **18.** ln 2 **19.** 112

20. ≈ \$177.16 **21.** $\frac{15}{8}$ **22.** $\frac{1}{2}\ln 5$

23. $\frac{125}{6}$ **24.** $\frac{863}{6}$ **25.** $\frac{125}{6}$

26. **(a)** $50 \ln \frac{b+2}{2}$; total increase in junk bonds over period from 1984 to 1984 + b
(b) last half of 1996 ($t \approx 12.8$)

6.1 EXERCISES, PAGE 311

1. $\frac{1}{4}$ **3.** $\frac{1}{2}(1 - e^{-2})$

5. 64 fps **9.** 407.8 units

11. 135 units **13.** $16\frac{2}{3}$ °C

15. ≈ \$7.45 billion

6.2 EXERCISES, PAGE 320

1. PS = \$1452, CS = \$22,748
3. PS = \$41.67, CS = \$625
5. PS = \$1.45, CS = \$3.87
7. ≈ \$5.83 million
9. ≈ \$73.25 million
13. \$1,344,000
17. 1033
19. $\pi c/2$, $8\pi c$ cc/sec
21. 20 mos: no (population ≈ 3490); 21 mos: yes (population ≈ 3514)
23. ≈ 166,028

6.3 EXERCISES, PAGE 330

1. midpoint rule: 2.331019; true value: 2.333333
3. midpoint rule: 0.691908; true value: ln 2 = 0.693147
5. trapezoidal rule: 0.343750; true value: 0.333333
7. trapezoidal rule: 30; true value: 30
9. 3.131177/4
11. ≈ \$2070
13. ≈ 185 concepts
15. Simpson's rule: 0.25; true value: 0.25
17. Simpson's rule: 0.766830; true value: 0.75
19. 0.746855
21. 326.67 ppm/day

6.4 EXERCISES, PAGE 337

1. convergent; 1 **3.** convergent; $\frac{1}{8}$

5. divergent **7.** divergent

9. convergent; 1 **11.** convergent; $\frac{1}{2}$

15. \$400,000

6.5 EXERCISES, PAGE 343

1. $e^x(x - 1) + c$
3. $6 - 16e^{-1}$
5. $\frac{1}{2}x^2\left(\ln x - \frac{1}{2}\right) + c$
7. $\frac{2}{3}x^{3/2}\left(\ln x - \frac{2}{3}\right) + c$
9. $x[(\ln x)^3 - 3(\ln x)^2 + 6 \ln x - 6] + c$
11. $\frac{19{,}022}{21}$
13. $C(x) = \frac{2}{15}(x + 1)^{3/2}(3x - 2) + \frac{4}{15}$
15. $-\frac{1}{9}\ln\left|\frac{9 + \sqrt{81 - x^2}}{x}\right| + c$
17. $\frac{x}{3} - \frac{2}{9}\ln|3x + 2| + c$
19. $-\frac{1}{8}\ln\left|\frac{x - 2}{x + 2}\right| + c$
21. $\frac{2}{3}\ln\left|3x + \sqrt{9x^2 - \frac{1}{4}}\right| + c$
23. $2[x\sqrt{x^2 + 5} + 5\ln|x + \sqrt{x^2 + 5}|] + c$
25. $\frac{2^{3x}}{3 \ln 2} + c$

6 REVIEW EXERCISES, PAGE 344

1. $\frac{1}{4}(1 - e^{-4})$
2. (a) $68,908 (b) $47,730
3. 582.3 units
4. 51,600
5. PS = $250, CS = $1666.67
6. $1,160,563,000
8. (a) $\frac{\pi}{3}$ (b) $\frac{31\pi}{5}$ (c) $\frac{\pi}{2}(1 - e^{-4})$
9. true value: 3.8; midpoint rule: 3.804607; trapezoidal rule: 3.790767; Simpson's rule: 3.799897
10. 3554
11. diverges
12. converges; $\frac{1}{24}$
13. converges; $\frac{5}{6}$
14. $-e^{-4x}\left(x + \frac{1}{4}\right) + c$
15. $\frac{3}{56}(7x - 5)(x + 5)^7 + c$
16. $\frac{1}{2}\ln\left|\frac{x}{x + 2}\right| + c$
17. $-\frac{1}{16}\ln\left|\frac{x - 2}{x + 2}\right| + c$

6 SUPPLEMENTARY EXERCISES, PAGE 348

1. (a) ≈ $9.024 million (b) $20 million
3. (a) ≈ $15.523 million (b) ≈ $38.167 million
   (c) ≈ $45.833 million
5. (a) 3 yrs: ≈ $12.776 million; forever: ≈ $45.455 million
   (b) 3 yrs: $15 million; forever: unlimited amount
7. 12.77 yr
9. (a) $57.69 (b) $171.79 (c) $175

7.1 EXERCISES, PAGE 355

1. all ordered pairs (x, y), 0, 16
3. $\{(p, q) \mid p \geq 0, q \geq 0\}$, 4, 11
5. $\{(u, v) \mid uv \geq 0\}$, 1, undefined
7. all ordered triples (x, y, z), 6, 29
9. $\{(x, y, z) \mid x > 0, y > 0, z > 0\}$, 0, 14
11. 2; 11
13. (a) $C = 0.4x + 0.5y + 100{,}000$
    (b) $118,000
15. (a) $R = 400x + 100y + 6xy - 6x^2 - 10y^2$
    (b) $2270
17. (a) $2000 (b) $10,000
    (c) each additional unit of Y produced costs $20; each additional unit of X produced costs $40
    (d) $A(x, y) = \frac{2000 + 40x + 20y}{x + y}$; $40
    (e) avg cost of making 150 units of X and y units of Y is $\frac{8000 + 20y}{150 + y}$ dollars; avg cost of making 100 units of Y and x units of X is $\frac{4000 + 40x}{100 + x}$ dollars
19. (a) 6
    (b) increases by 2 for every 10 additional mice; decreases by 2 for every additional dog

7.2 EXERCISES, PAGE 362

1.

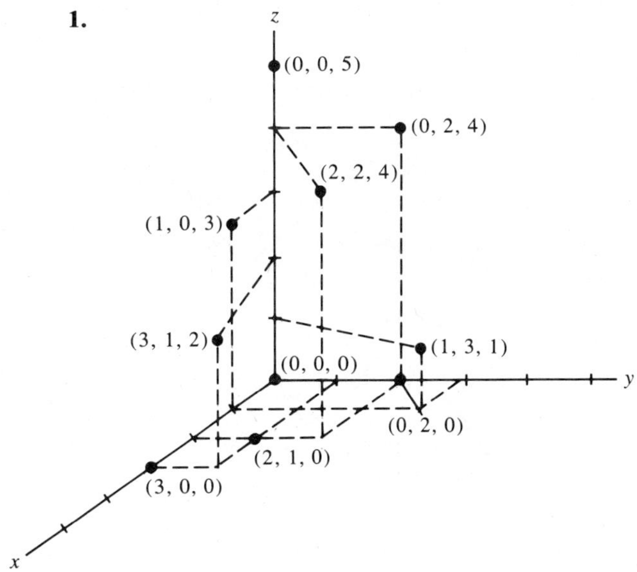

(1. *continued*)

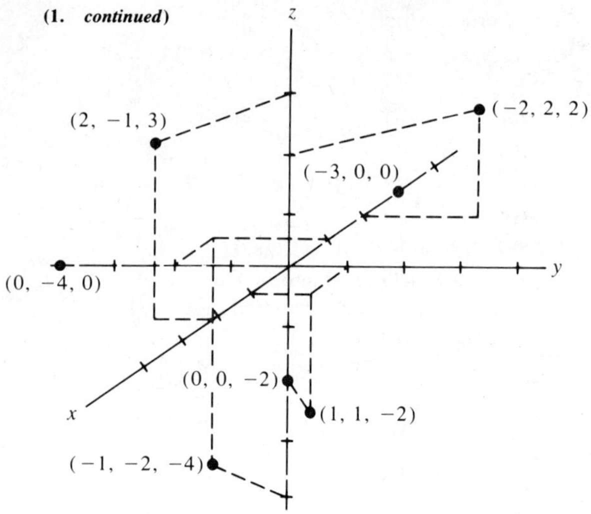

3.

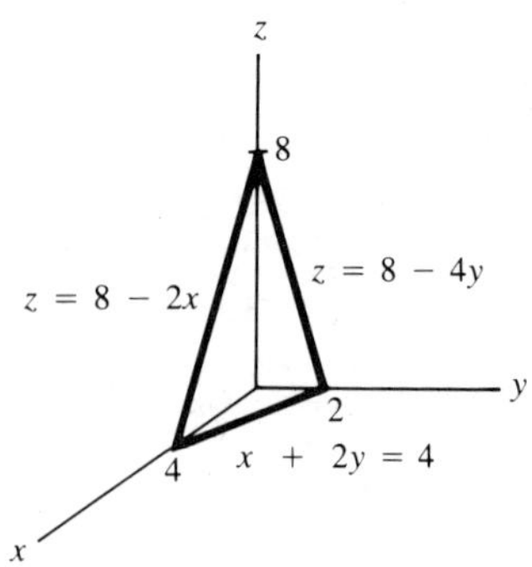

5.

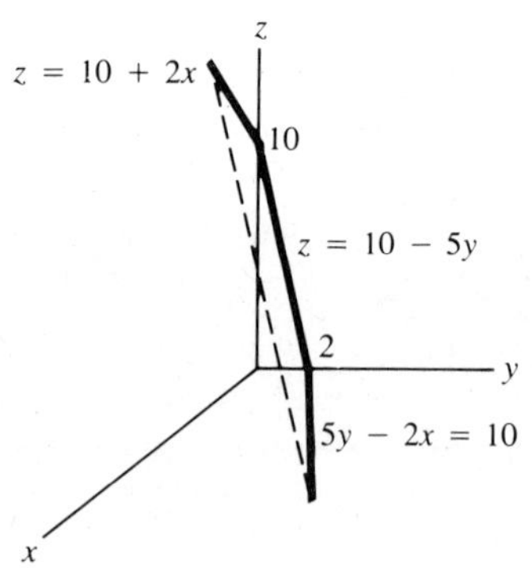

7.

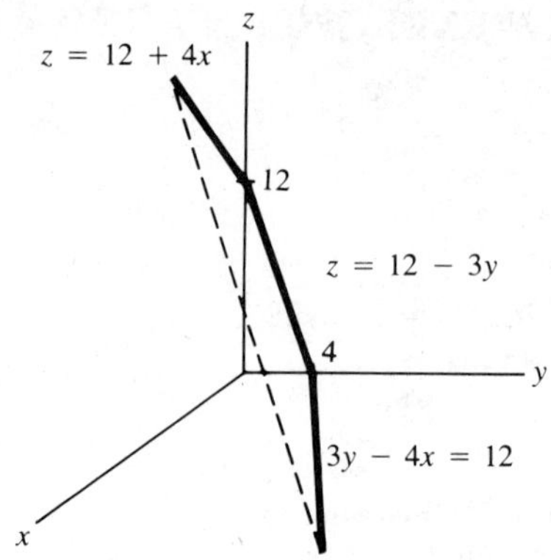

9.

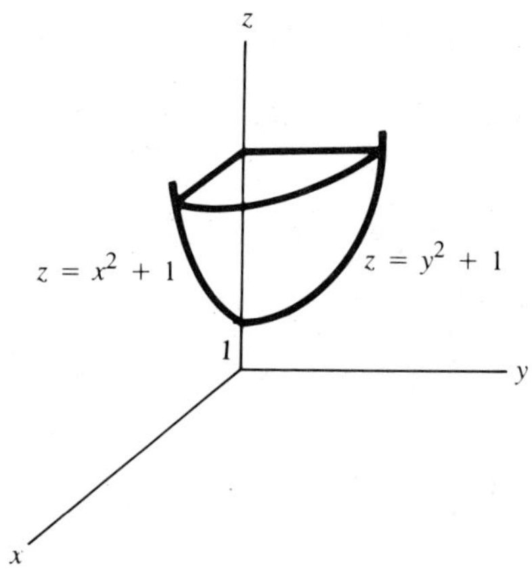

11.

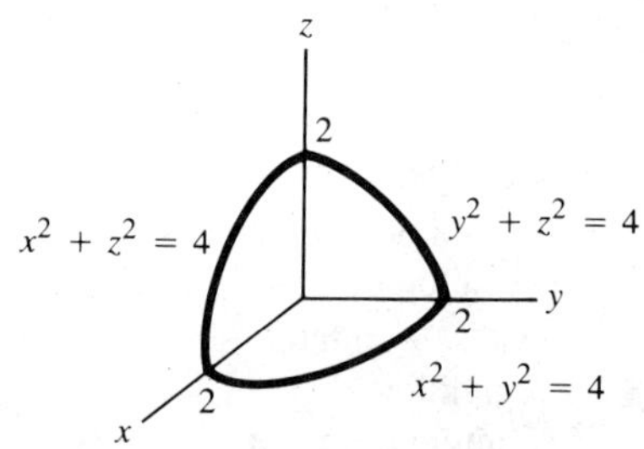

13.

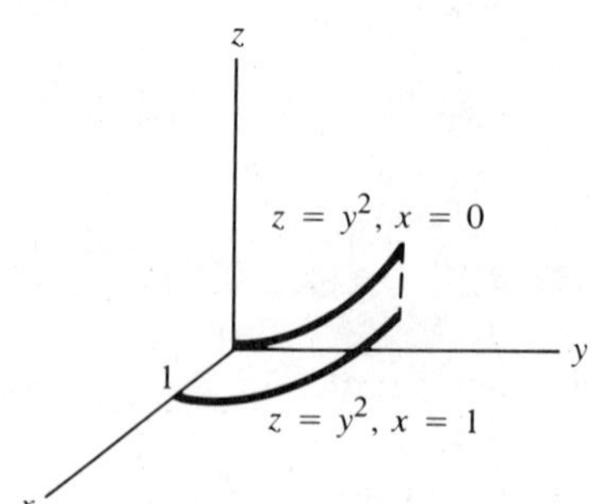

15.

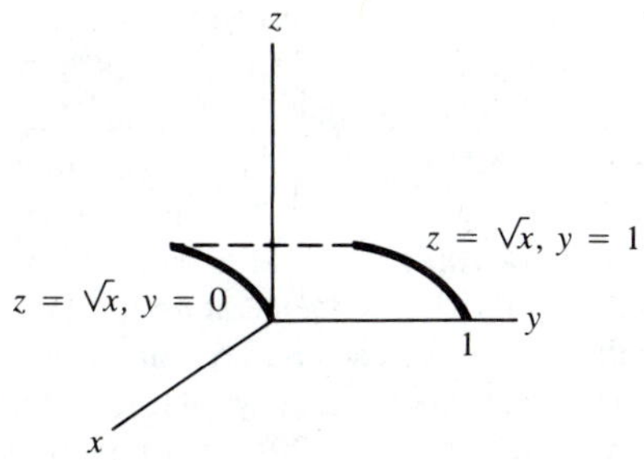

17.

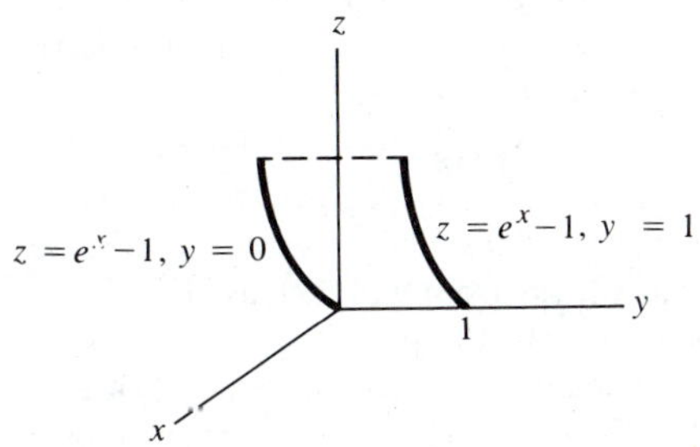

7.3 EXERCISES, PAGE 368

1. $f_x = 5, f_y = -7$

3. $\dfrac{\partial f}{\partial x} = 3 + \dfrac{4x}{y^2}, \dfrac{\partial f}{\partial y} = -\dfrac{4x^2}{y^3}$

5. $g_x = \sqrt{\dfrac{y}{2x}}, g_y = \sqrt{\dfrac{x}{2y}}$

7. $\dfrac{\partial h}{\partial u} = 4(2u + v)(u^2 + uv + v^3)^3,$
$\dfrac{\partial h}{\partial v} = 4(u + 3v^2)(u^2 + uv + v^3)^3$

9. $\dfrac{\partial z}{\partial x} = 2 \ln y, \dfrac{\partial z}{\partial y} = \dfrac{2x}{y}$

11. $\dfrac{\partial w}{\partial x} = \dfrac{3y}{(x + y)^2}, \dfrac{\partial w}{\partial y} = -\dfrac{3x}{(x + y)^2}$

13. $h_x = 2ze^{2xz}, h_y = -\dfrac{1}{y}, h_z = 2xe^{2xz} - \dfrac{1}{z}$

15. $f_x(0, 0) = 2, f_y(1, 1) = 7$

17. $\left.\dfrac{\partial f}{\partial x}\right|_{(8,3)} = \dfrac{1}{5}, \left.\dfrac{\partial f}{\partial y}\right|_{(8,3)} = \dfrac{3}{10}$

19. $\left.\dfrac{\partial w}{\partial x}\right|_{(1,-1,1)} = \dfrac{1}{\sqrt{2}} + 4, \left.\dfrac{\partial w}{\partial y}\right|_{(1,-1,1)} = -\dfrac{1}{\sqrt{2}},$

21. $f_{xx} = 0, f_{xy} = f_{yx} = 0, f_{yy} = 6$

23. $f_{xx} = 0, f_{xy} = f_{yx} = -\dfrac{4}{3y^3}, f_{yy} = \dfrac{4x}{y^4}$

25. $\dfrac{\partial^2 z}{\partial x^2} = \dfrac{48x^2}{y} + \dfrac{12y^2}{x^4}, \dfrac{\partial^2 z}{\partial y\,\partial x} = \dfrac{\partial^2 z}{\partial x\,\partial y} =$
$-\dfrac{16x^3}{y^2} - \dfrac{8y}{x^3}, \dfrac{\partial^2 z}{\partial y^2} = \dfrac{8x^4}{y^3} + \dfrac{4}{x^2}$

27. $f_{xx} = ye^x - \dfrac{2}{x^2}, f_{xy} = f_{yx} = e^x, f_{xz} = f_{zx} = \dfrac{1}{z},$
$f_{yy} = 0, f_{yz} = f_{zy} = 0, f_{zz} = -\dfrac{x}{z^2}$

29. $f_{xyxy} = e^{xy}(2 + 4xy + x^2y^2), f_{xyyy} = x^2e^{xy}(3 + xy)$

31. $f_{xx}(1, 1) = 10, f_{xy}(2, 2) = -3, f_{yx}(3, -5) =$
$-3, f_{yy}(-2, 7) = 0$

33. $f_{xyx}(1, 1) = 6, f_{yyx}(-1, -1) = 12$

7.4 EXERCISES, PAGE 374

1. **(a)** 8 **(b)** 24 **3.** **(a)** 6 **(b)** 3

5. **(a)** $C_x = 20, C_y = 14$;
if y is held fixed, each additional unit of A made will cost \$20; if x is held fixed, each additional unit of B made will cost \$14
(b) $R_x = 25, R_y = 20$;
if y is held fixed, each additional unit of A can be sold for \$25; if x is held fixed, each additional unit of B can be sold for \$20
(c) $P_x = 5, P_y = 6$;
if y is held fixed, each additional unit of A made and sold will contribute \$5 to profit; if x is held fixed, each additional unit of B made and sold will contribute \$6 to profit

7. $C_x(100{,}200) = 1200, C_y(100{,}200) = 1300,$
$R_x(100{,}200) = 1200, R_y(100{,}200) = 2000,$
$R_x(100{,}200) - C_x(100{,}200) = 0, R_y(100{,}200) - C_y(100{,}200) = 700$;
if y is held fixed at 200 units, the 101st unit of x will cost approximately \$1200 to make, bring in revenue of approximately \$1200, and contribute approximately \$0 to profit; if x is held fixed at 100 units, the 201st unit of y will cost approximately \$1300 to make, bring in revenue of approximately \$2000, and contribute approximately \$700 to profit

9. $V_x = -200, V_y = -100, W_x = -400,$
$W_y = -300$;
if price y of W is held fixed, an increase of \$1 in price x of V will cause quantity of V demanded

to decrease by 200 units and quantity of W demanded to decrease by 400 units; if price x of V is held fixed, an increase of \$1 in price y of W will cause quantity of V demanded to decrease by 100 units and quantity of W demanded to decrease by 300 units; products are complementary

11. **(a)** 800,000 units
(b) $f_x(1600, 10{,}000) = 250$, $f_y(1600, 10{,}000) = 40$; if y is held fixed at \$10,000, an additional worker-hour (the 1601st) could be used to increase production by approximately 250 units; if x is held fixed at 1600 worker-hours, an additional \$1 of capital (the 10,001st) could be used to increase production by approximately 40 units

13. **(a)** $f_x(50, 50) = e^{0.5} \approx 1.6487$, $f_y(50, 50) = 0.5e^{0.5} \approx 0.8244$; if y is held fixed at \$50,000, an additional 1000 worker-hours (beyond 50,000) could be used to increase production by approximately 1649 units per week; if x is held fixed at 50,000 worker-hours, an additional \$1000 of capital (beyond \$50,000) could be used to increase production by approximately 824 units per week
(b) $f_x(100, 100) = f_y(100, 100) = e \approx 2.7183$; if y is held fixed at \$100,000, an additional 1000 worker-hours (beyond 100,000) could be used to increase production by approximately 2718 units per week; if x is held fixed at 100,000 worker-hours, an additional \$1000 of capital (beyond \$100,000) could be used to increase production by approximately 2718 units per week
(c) $f_x(200, 200) = e^2 \approx 7.3891$, $f_y(200, 200) = 2e^2 \approx 14.7781$; if y is held fixed at \$200,000, an additional 1000 worker-hours (beyond 200,000) could be used to increase production by approximately 7389 units per week; if x is held fixed at 200,000 worker-hours, an additional \$1000 of capital (beyond \$200,000) could be used to increase production by approximately 14,778 units per week

15. $P_x(2, 20, 3) = 20e^{0.4} + 40 \approx 69.8$, $P_T(2, 20, 3) = 404$, $P_t(2, 20, 3) = 10{,}000$; if T is fixed at 20°, an additional gram of nutrient (the 3rd one) will cause the number of bacteria present at time $t = 3$ to increase by approximately 70; if x is held fixed at 2 grams, an increase of 1° in the temperature (to 21°) will cause the number of bacteria present at time $t = 3$ to increase by approximately 404; if x is held fixed at 2 grams and T at 20°, then running the experiment an additional hour (from $t = 3$ to $t = 4$) will cause the number of bacteria to increase by approximately 10,000

17. $f_x(100, 1) = 0.1$, $f_y(100, 1) = -10$; if the price of a ticket is held fixed at \$1 and the population of the city increases by 1000 (to 101,000), ridership will increase by approximately 100 persons per day; if the population of the city remains constant at 100,000, an increase of \$1 in the price of a ticket (to \$2) will cause ridership to decrease by approximately 10,000 persons per day

7.5 EXERCISES, PAGE 384

1. saddle pt: (0, 0, 0) **3.** rel max: (0, 0, 9)
5. rel min: (3, −2, −37)
7. rel min: (−3, 10, −486)
9. saddle pt: (0, 0, 0), rel max: (−1, 1, 1)
11. rel min: (5, 10, −48), saddle pt: $\left(\frac{1}{3}, \frac{2}{3}, \frac{76}{27}\right)$
13. saddle pt: (0, 1, −6), rel min: (±2, 1, −19)
15. saddle pt: (0, 1, 0), (−1, 1, 0), (0, 0, 0), rel max: $\left(-\frac{1}{3}, \frac{2}{3}, \frac{1}{27}\right)$
17. base 2 m by 2 m, height 1 m
19. $A(4, 6) = \$162$ **21.** $P_A = P_B = \$250$; no
23. 1 ton of fertilizer, $\frac{5}{6}$ ton of pesticide

7.6 EXERCISES, PAGE 391

1. min: $f\left(\frac{3}{2}, \frac{3}{2}\right) = \frac{13}{2}$ **3.** max: $f(3, 3) = -9$
5. max: $f\left(3, \frac{3}{2}\right) = \frac{9}{2}$ **7.** max: $f(0, -3, 3) = 9$

9. $f\left(\frac{6}{5}, \frac{8}{5}\right) = \frac{13{,}824}{3125}$ **11.** $f(\pm 4, \pm 4) = 29$

13. $f\left(\frac{1}{2}, \frac{1}{2}\right) = \frac{1}{4}$ **15.** 10, 10

17. 4, 16 **19.** $2\sqrt{2}$

21. cube of side $2r/\sqrt{3}$

23. max production is $f(300, 450) = 300^{1/4}450^{3/4} \approx$ 406.6 units; marginal productivity of money $= \frac{3}{160}\left(\frac{300}{450}\right)^{1/4} \approx 0.0169$ units; 1 additional dollar could be used to produce approximately 0.0169 additional units

25. $\approx 328{,}634$

7.7 EXERCISES, PAGE 403

1. $\frac{1}{2}$ **3.** 16 **5.** $\frac{1}{2}(1 \quad c^{-2})$

7. 3 **9.** -4 **11.** $\frac{\sqrt{2}}{30}$

13. 16 **15.** 4 **17.** $\frac{4}{3}$

19. $\frac{4}{3}$ **21.** $\frac{1}{12}$

7 REVIEW EXERCISES, PAGE 404

1. $\{(x,y) \mid x \neq y\}$; undefined, 1, 0, 27

2. $\{(x,y,z) \mid xy > 0\}$
(a) undefined **(b)** $e^{-8} + \ln 8$
(c) undefined **(d)** $e^{6} + \ln 4$

3. $R(p_1,p_2) = 200p_1 - 5p_1^2 + 750p_2 - 20p_2^2$; \$2920

4. \$11,000; contributes \$10 to profit

5.

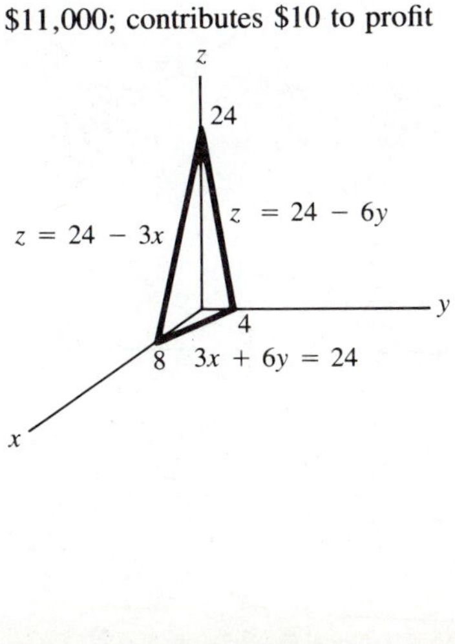

6.

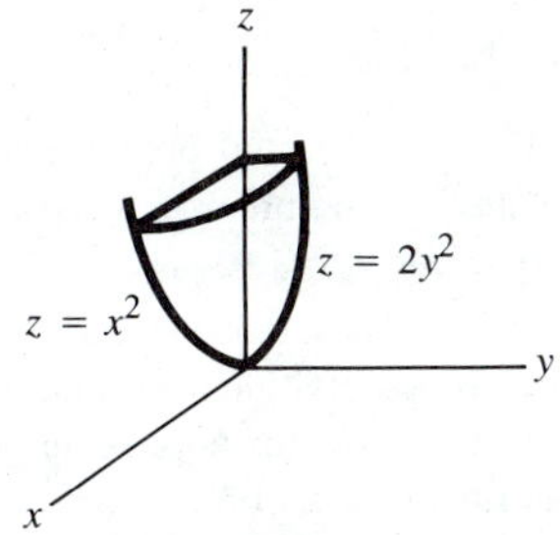

7.

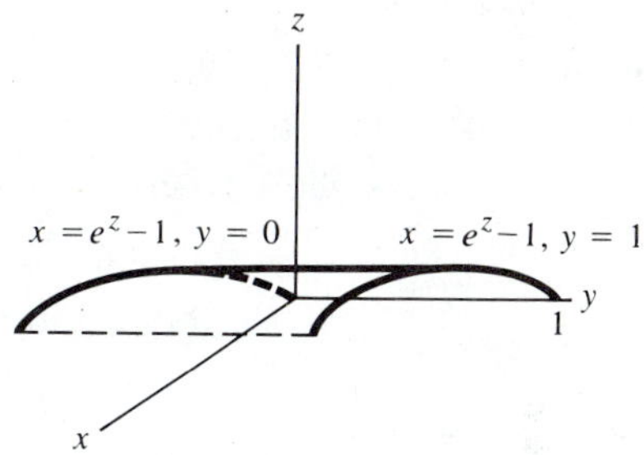

8. $f_x = 16xy - 2y^3 + 12, f_y = 8x^2 - 6xy^2 + 2,$ $f_{xx} = 16y, f_{xy} = f_{yx} = 16x - 6y^2, f_{yy} = -12xy$

9. $f_x(2, 1) = 42, f_y(-1, 0) = 10, f_{xx}(2, 2) = 32,$ $f_{yy}(-3, 4) = 144$

10. $g_{pq} = e^{pq}(pq + 1), g_{qqpp} = e^{pq}(p^2q^2 + 4pq + 2), g_{qppp} = q^2e^{pq}(pq + 3)$

11. $\frac{\partial z}{\partial x} = 12(x + y)(x^2 + 2xy - y^3)^5,$

$\frac{\partial z}{\partial y} = 6(2x - 3y^2)(x^2 + 2xy - y^3)^5,$

$\frac{\partial^2 z}{\partial x\,\partial y} = 12(x^2 + 2xy - y^3)^4(11x^2 + 12xy - 15xy^2 - 16y^3),$

$\frac{\partial^2 z}{\partial y\,\partial x} = \frac{\partial^2 z}{\partial x\,\partial y},$

$\frac{\partial^2 z}{\partial x^2} = 12(x^2 + 2xy - y^3)^4(11x^2 + 22xy + 10y^2 - y^3),$

$\frac{\partial^2 z}{\partial y^2} = 6(x^2 + 2xy - y^3)^4(-6x^2y - 72xy^2 + 51y^4 + 20x^2)$

12. **(a)** 82 **(b)** 83

13. $f_x(16, 1) = \frac{1}{32}, f_y(16, 1) = -4;$

if predators remain fixed at 1, an additional prey animal (the 17th one) would cause the number of wolves to increase by approximately $\frac{1}{32}$; if prey remains fixed at 16, an additional predator (the 2nd one) would cause the number of wolves to decrease by approximately 4

14. $R_x(25, 100) = \$20.10$, $R_y(25, 100) = \$25.025$; if y is held fixed at 100 units, an additional unit of x (the 26th) could be sold for approximately \$20.10; if x is held fixed at 25 units, an additional unit of y (the 101st) could be sold for approximately \$25.025

15. rel min: $(0, 0, -1)$

16. rel max: $(0, 2, 4)$, saddle pt: $(\pm 2, 2, 36)$

17. base is 2 m by 2 m, height is 2.6 m

18. $p_X = \$400$, $p_Y = \$250$

19. max: $f(1, 3) = 27$

20. min: $f(2, 8, 4) = 80$

21. 8, 8

22. max productivity is $f(10, 8) = 1000(10^{1/5}8^{4/5}) \approx 8365$; marginal productivity of money $= 2(10^{4/5}8^{-4/5}) \approx 8.3625$; an additional \$1 could be used to increase production by approximately 8.365 units

23. $\frac{92}{3}$ **24.** $\frac{10}{3}$

7 SUPPLEMENTARY EXERCISES, PAGE 411

3.

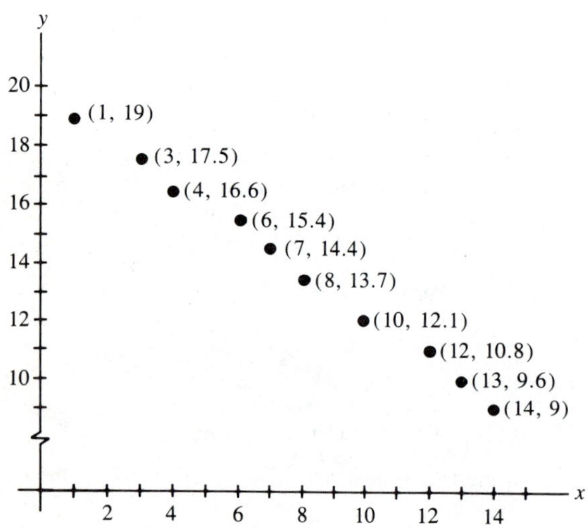

$-0.77X + 19.81$; time until tumor disappears is decreased by approximately 0.77 days

A.1 EXERCISES, PAGE 420

1. 90°, 180°, 270°, 360°

3. −45°, −90°, −180°, −135°

5. 450°, 540°, 30°, 150°

7.

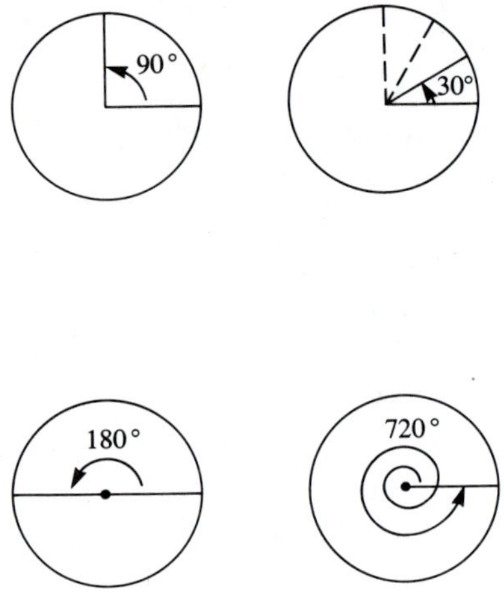

9.

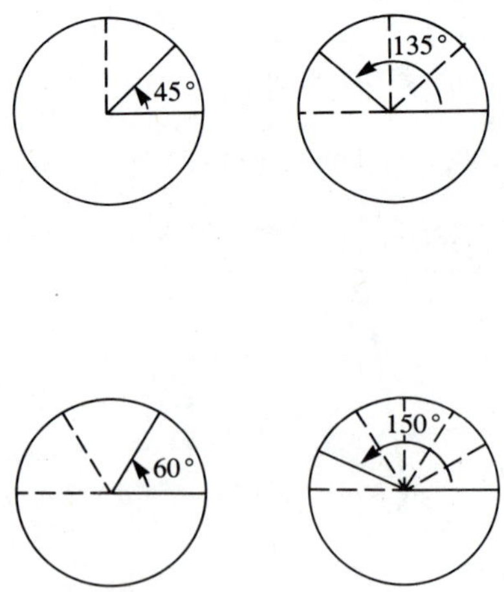

11.

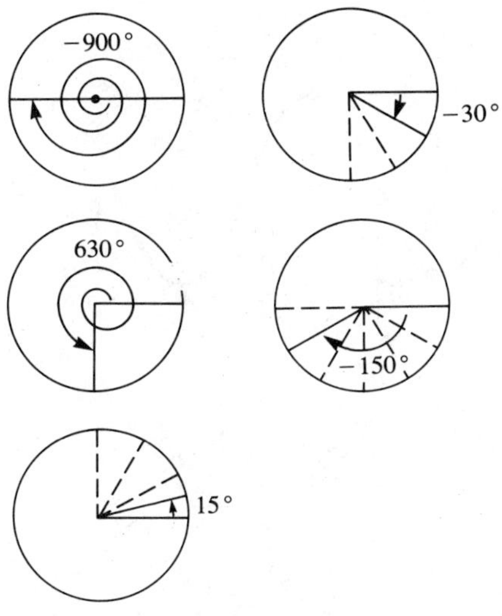

13. $\pi/2,\ \pi,\ \pi/4,\ 2\pi$

15. $5\pi/6,\ -7\pi/6,\ -\pi/2,\ 3\pi/4$

17. $-8\pi/3,\ 3\pi/2,\ -3\pi/4,\ 5\pi$

19.

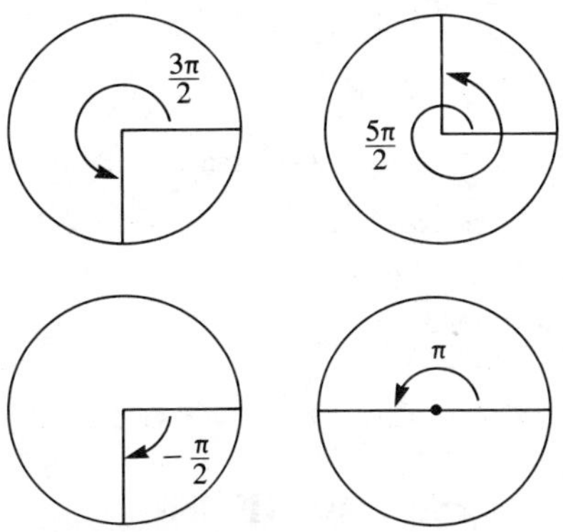

21.

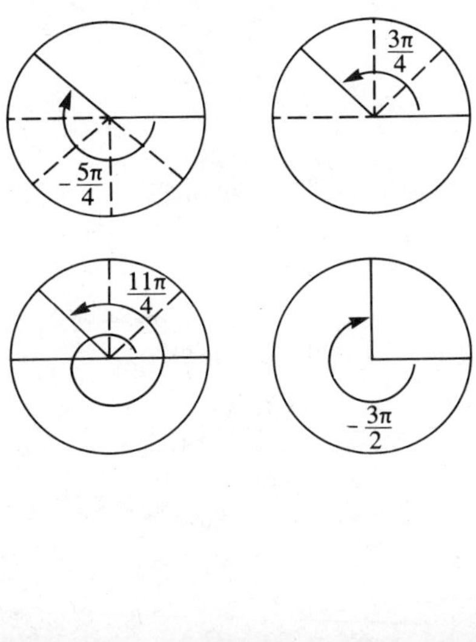

23.

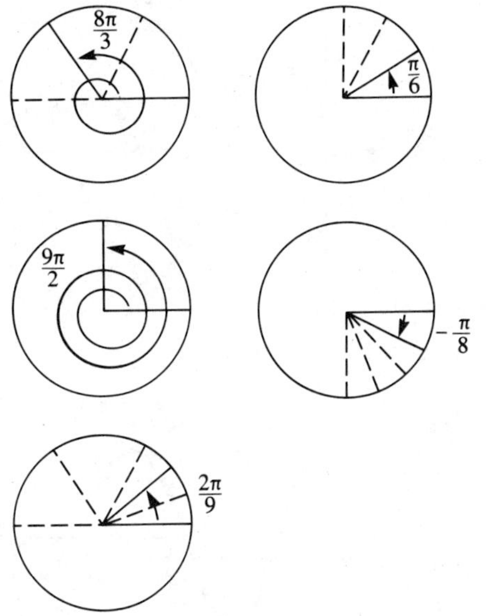

25. $\pi/2,\ \pi/6,\ 25\pi/36,\ -\pi/3$

27. $3\pi/4,\ -5\pi/6,\ 35\pi/18,\ 52\pi/15$

29. $-1080°,\ 80°,\ 900°,\ 67.5°$

A.2 EXERCISES, PAGE 437

1. $-1, 0, 0, 1$ **3.** $-1, 0, 0, 1$

5. **(a)** $n\pi,\ n = 0, \pm 1, \pm 2, \ldots$
(b) $n\pi/2,\ n = \pm 1, \pm 3, \pm 5, \ldots$

7. **(a)** $2n\pi,\ n = 0, \pm 1, \pm 2, \ldots$
(b) $n\pi,\ n = \pm 1, \pm 3, \pm 5 \ldots$

13. $\sin \dfrac{\pi}{12} = \dfrac{1}{2\sqrt{2+\sqrt{3}}}$, $\cos \dfrac{\pi}{12} = \dfrac{\sqrt{2+\sqrt{3}}}{2}$

15. **(a)** $-\sqrt{3},\ -2,\ 2/\sqrt{3},\ -1/\sqrt{3}$
(b) undefined, undefined, 1, 0
(c) 0, −1, undefined, undefined

17.

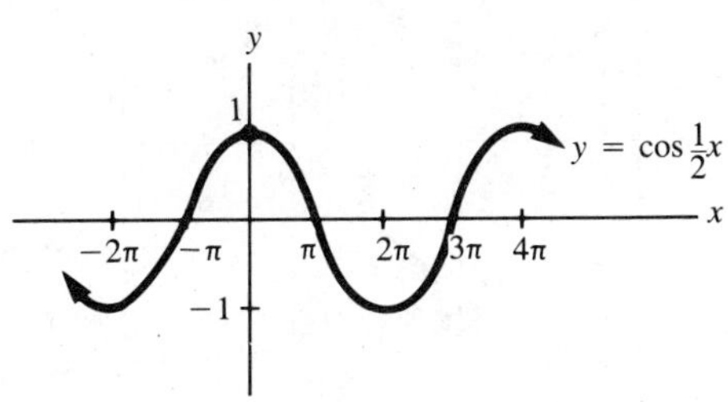

4π

19.

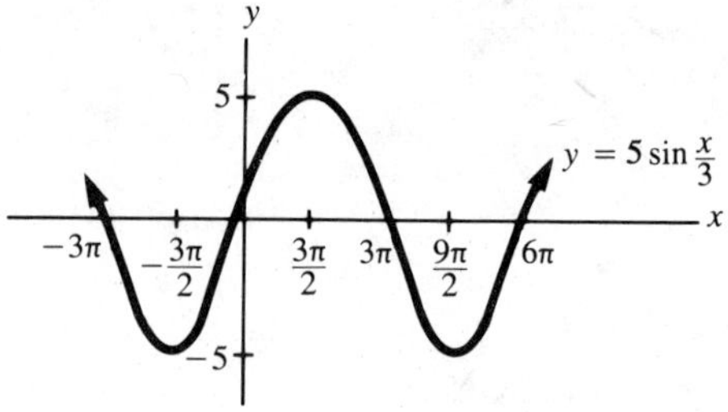

6π

21.

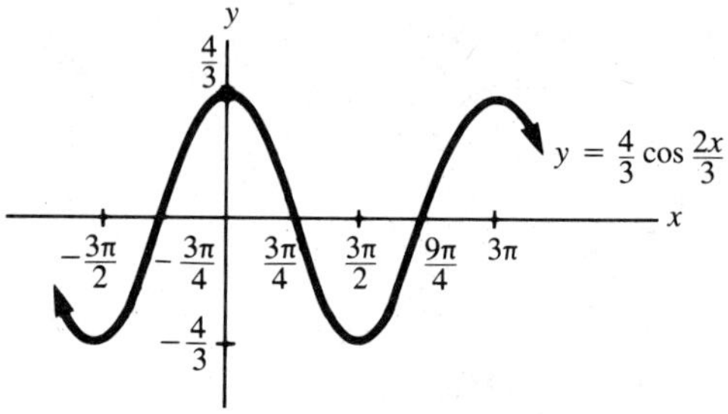

3π

23.

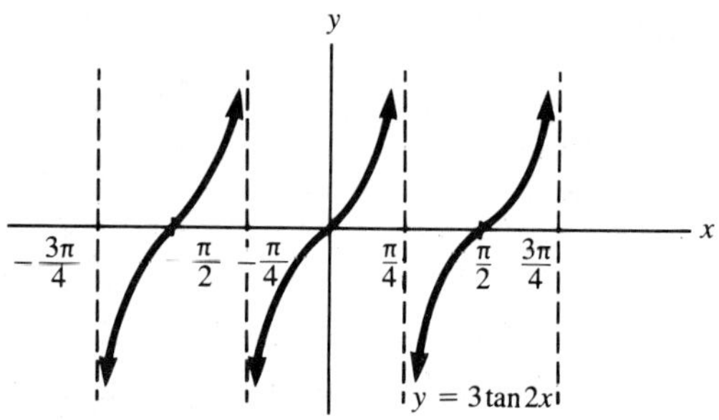

$\frac{\pi}{2}$

25.

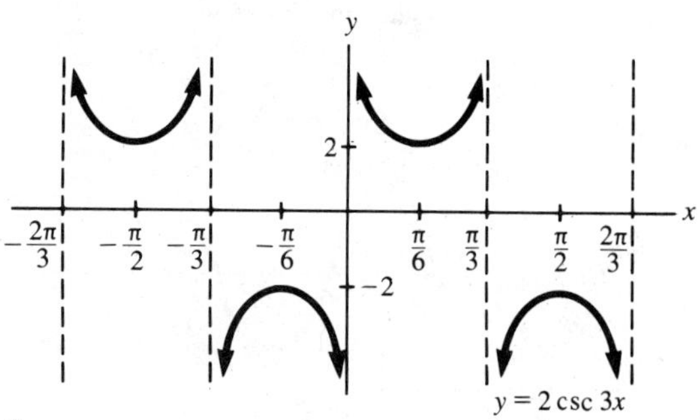

$\frac{2\pi}{3}$

27.

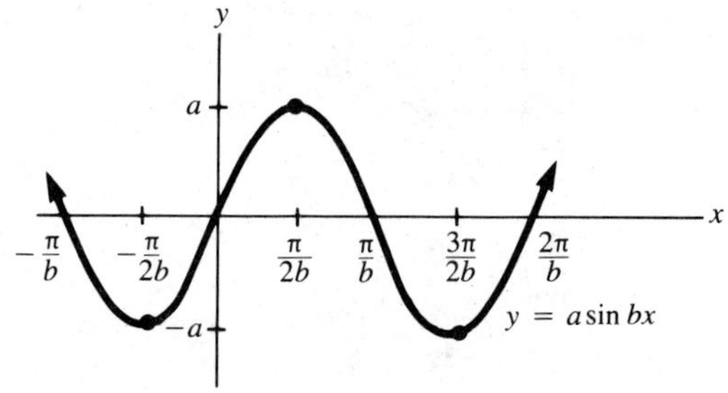

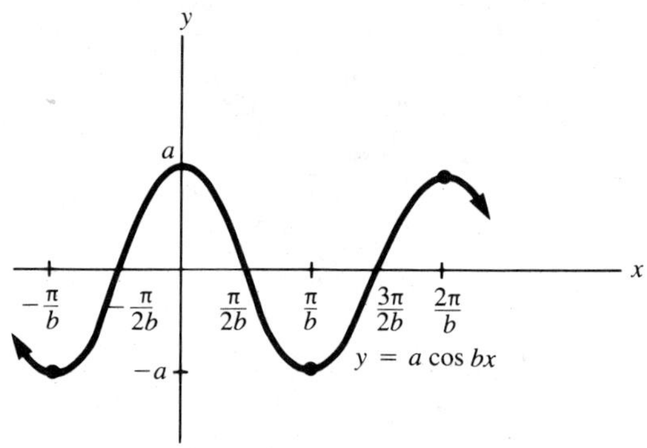

$\frac{2\pi}{b}$

29. $\sin x = 2\sqrt{2}/3$, $\cos x = \frac{1}{3}$, $\tan x = 2\sqrt{2}$,
$\sec x = 3$, $\csc x = 3/2\sqrt{2}$, $\cot x = 1/2\sqrt{2}$

31. $r = 4/\sqrt{3}$, $s = 2/\sqrt{3}$

33. $r = 2$, $s = 2\sqrt{3}$

35. $x = \pi/4$ radians

A.3 EXERCISES, PAGE 450

1. $\frac{dy}{dx} = -3 \sin x$ **3.** $\frac{dy}{dx} = 15 \cos 3x$

5. $\frac{dy}{dx} = 2(x + 4) \cos(x^2 + 8x)$

7. $\frac{dy}{dx} = 2(2 \cos x \cos 2x - \sin x \sin 2x)$

9. $\frac{dy}{dx} = -14 \cos 7x \sin 7x$

11. $\frac{dy}{dx} = 2 \sin x - 7 \sin^3 x + 5 \sin^5 x$

13. $\frac{dy}{dx} = -\frac{4 \cos^3 4x\,(4 \sin 3x \sin 4x + 3 \cos 4x \cos 3x)}{\sin^5 3x}$

15. $\dfrac{dy}{dx} = e^{-x}(2 \cos 2x - \sin 2x)$

17.

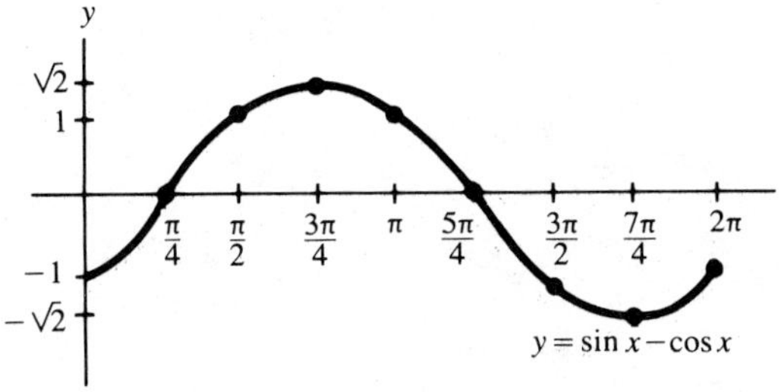

19.

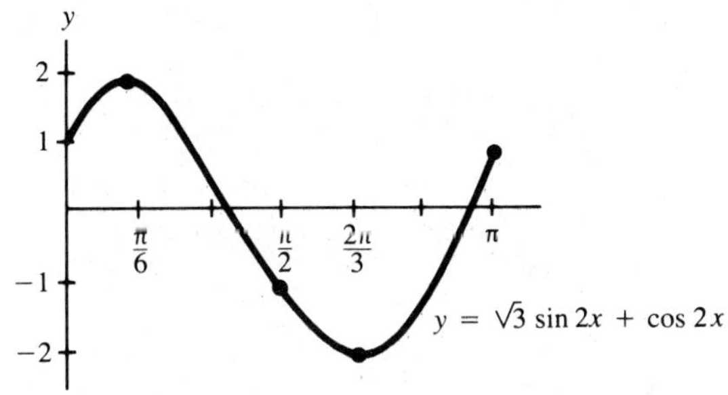

21.

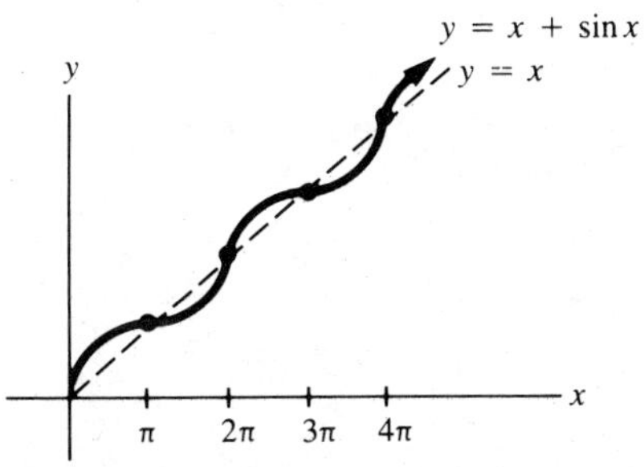

27. $\dfrac{dy}{dx} = 2x \sec x^2 \tan x^2$

29. $\dfrac{dy}{dx} = -e^x \csc^2 e^x$

31. $\dfrac{dy}{dx} = (2 \sec^2 x - 1) \sec x$

33. $\dfrac{dy}{dx} = 2 \sec^2 x \tan x$

35. $\dfrac{dy}{dx} = \sec x$

37. -0.05 rad/sec

39. transient: $T = 2e^{-5t}$;

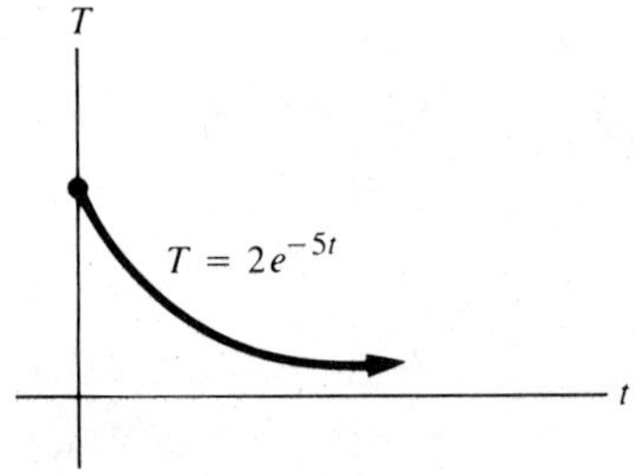

steady state: $S = 2\sqrt{3} \sin 4t - 2 \cos 4t$;

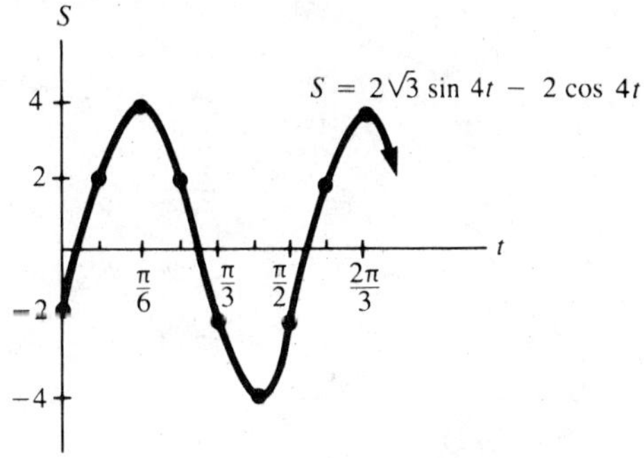

41. **(a)** $t = 0$: 1 in below rest position;
$t = \pi/6$: $2e^{-\pi/6} \approx 1.18$ in below rest position;
$t = \pi/4$: $e^{-\pi/4}/\sqrt{2} \approx 0.32$ in below rest position

(b) $t = 0$: velocity $= 5$ in/sec, acceleration $= -20$ in/sec^2;
$t = \pi/6$: velocity $= -5e^{-\pi/6} \approx -2.96$ in/sec, acceleration $= -10e^{-\pi/6} \approx -5.92$ in/sec^2;
$t = \pi/4$: velocity $= -5\sqrt{2}\, e^{-\pi/4} \approx -3.22$ in/sec, acceleration $= 5\sqrt{2}\, e^{-\pi/4} \approx 3.22$ in/sec^2

(c)

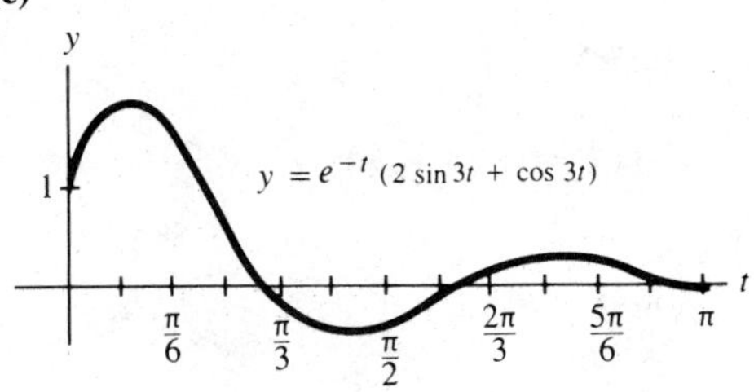

$2\pi/3$

A.4 EXERCISES, PAGE 459

1. $-\dfrac{1}{3}\cos 3x + c$

3. $\dfrac{1}{\pi}\cot(-\pi x) + c$

5. $\dfrac{1}{2}\sin \pi^2$

7. $-2\cos\sqrt{x} + c$

9. 0

11. $\dfrac{1}{8}\sin^4 2x + c$

13. $1 - e^{-1}$

15. $-\cos x + c$

17. **(a)**

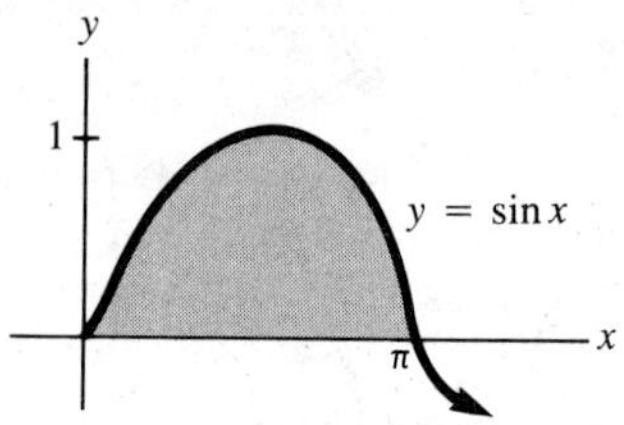

2

(b)

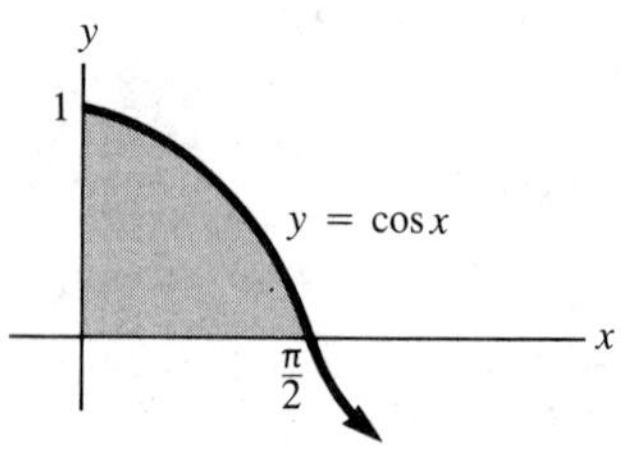

1

19. **(a)** $2/\pi$ **(b)** 0

21. $-\dfrac{1}{5}\ln|\cos 5x| + c$

23. $-\dfrac{1}{2}\ln|\csc x^2 + \cot x^2| + c$

25. 1

27. $-\dfrac{2\pi}{3} + \ln\left(\dfrac{2+\sqrt{3}}{2-\sqrt{3}}\right)$

29. $\dfrac{2}{\pi}\ln\left(\dfrac{\sqrt{2}+1}{\sqrt{2}-1}\right)$

31. $x\sin x + \cos x + c$

33. $\pi - 2$

35. $\dfrac{1}{2}(1 + e^{-\pi})$

37. $\dfrac{1}{2}(x - \sin x\cos x) + c$

APPENDIX REVIEW EXERCISES, PAGE 461

1. 22.5°, $\pi/8$; −150°, $-5\pi/6$

2. −840°, $-14\pi/3$; 720°, 4π

3. 210°, $7\pi/6$; 405°, $9\pi/4$

4. −75°, $-5\pi/12$; 100°, $5\pi/9$

5.

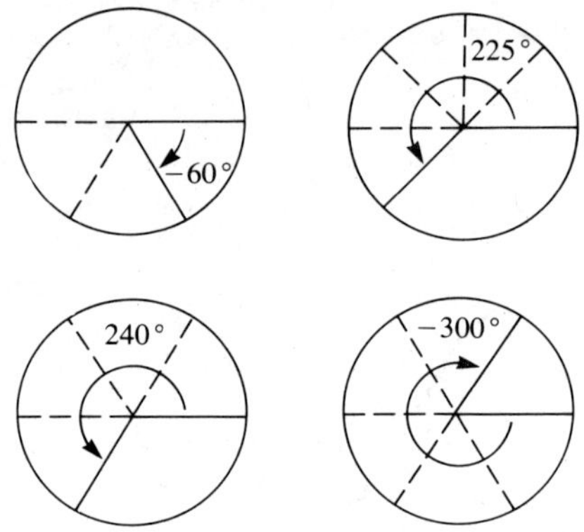

6.

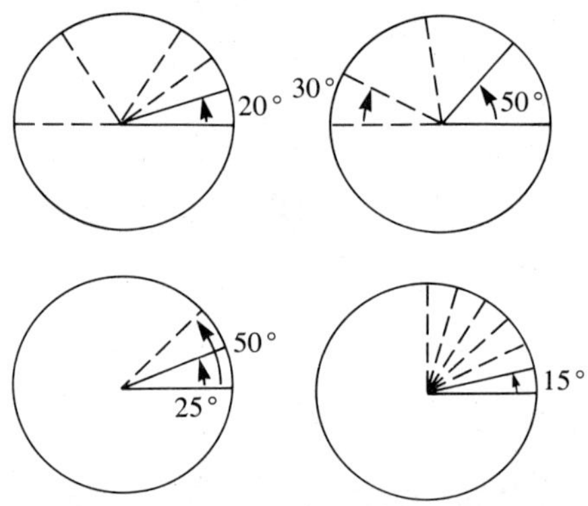

7.

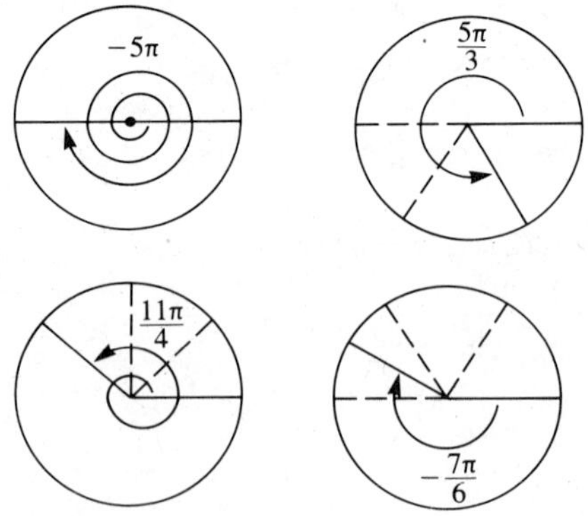

8.

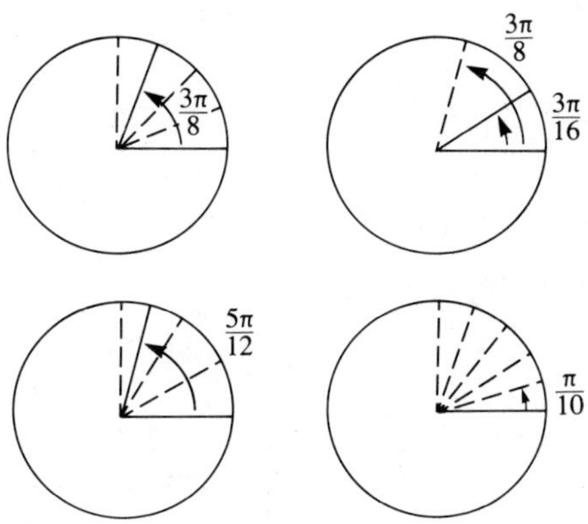

9. $3\pi/4,\ -4\pi/3,\ \pi/8,\ -\pi/45$

10. $240°,\ -225°,\ 12°,\ 585°/\pi \approx 186.2°$

11. $\sin x = -1$, $\cos x = 0$, $\tan x$ is undefined, $\csc x = -1$, $\sec x$ is undefined, $\cot x = 0$

12. $\sin x = -1/\sqrt{2}$, $\cos x = -1/\sqrt{2}$, $\tan x = 1$, $\csc x = -\sqrt{2}$, $\sec x = -\sqrt{2}$, $\cot x = 1$

13. $\sin x = 0$, $\cos x = 1$, $\tan x = 0$, $\csc x$ is undefined, $\sec x = 1$, $\cot x$ is undefined

14. $\sin x = \sqrt{3}/2$, $\cos x = 1/2$, $\tan x = \sqrt{3}$, $\csc x = 2/\sqrt{3}$, $\sec x = 2$, $\cot x = 1/\sqrt{3}$

15. $\sin x = 1/2$, $\cos x = -\sqrt{3}/2$, $\tan x = -1/\sqrt{3}$, $\csc x = 2$, $\sec x = -2/\sqrt{3}$, $\cot x = -\sqrt{3}$

16. $\sin x = 1/2$, $\cos x = \sqrt{3}/2$, $\tan x = 1/\sqrt{3}$, $\csc x = 2$, $\sec x = 2/\sqrt{3}$, $\cot x = \sqrt{3}$

17.

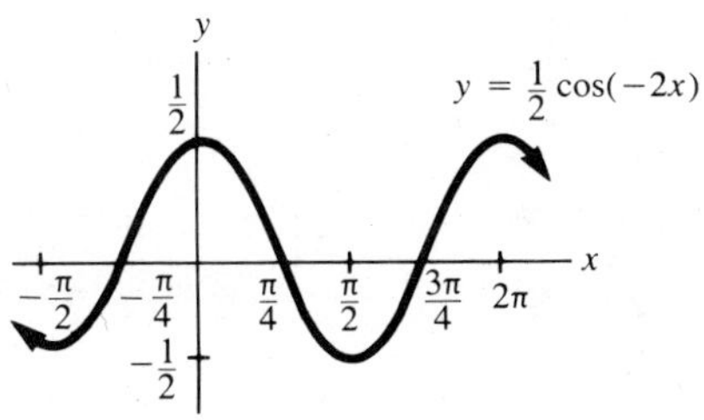

π

18.

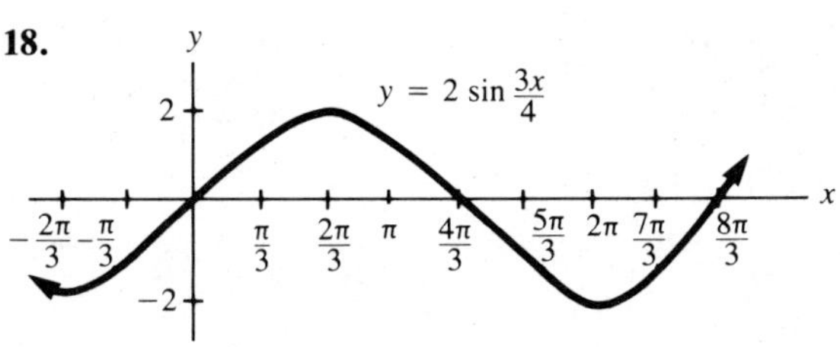

$8\pi/3$

19. **(a)**

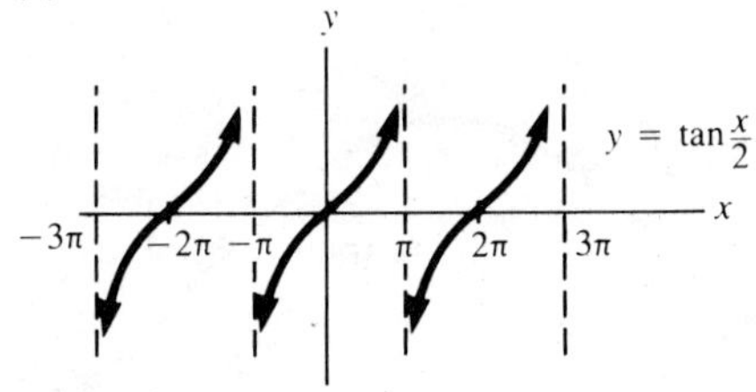

(b)

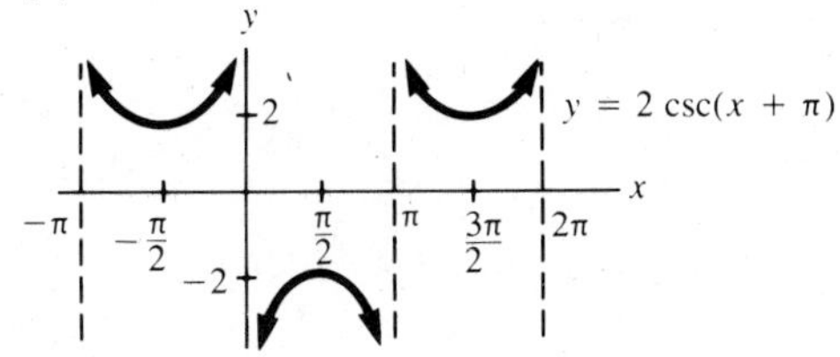

20. **(a)** $\sin x = 5/9$, $\cos x = 2\sqrt{14}/9$, $\tan x = 5/2\sqrt{14}$, $\csc x = 9/5$, $\sec x = 9/2\sqrt{14}$, $\cot x = 2\sqrt{14}/5$

(b) $a = 6$, $b = 3\sqrt{3}$

(c) $x = 5\pi/6$ radians

21. $\dfrac{dy}{dx} = 3\cos 3x$ **22.** $\dfrac{dy}{dx} = -\dfrac{\sin 2\sqrt{x}}{\sqrt{x}}$

23. $\dfrac{dy}{dx} = \dfrac{x\cos x - \sin x}{x^2}$

24. $\dfrac{dy}{dx} = \dfrac{1}{6}(3\cos 2x \cos 3x - 2\sin 2x \sin 3x)$

25. $\dfrac{dy}{dx} = e^{\tan x}\sec^2 x$ **26.** $\dfrac{dy}{dx} = -12x(\csc^2 3x^2)$

27. $\dfrac{dy}{dx} = \dfrac{2(2\sec^2 4x + \cot 2x \tan 4x)}{\csc 2x}$

28. $\dfrac{dy}{dx} = -12\cos 4x \sin^2 4x$

29. $\dfrac{dy}{dx} = 2\sec^2(\tan x)\tan(\tan x)\sec^2 x$

30. $\dfrac{dy}{dx} = \dfrac{\cos x - \csc x \cot x}{\sin x + \csc x}$

31.

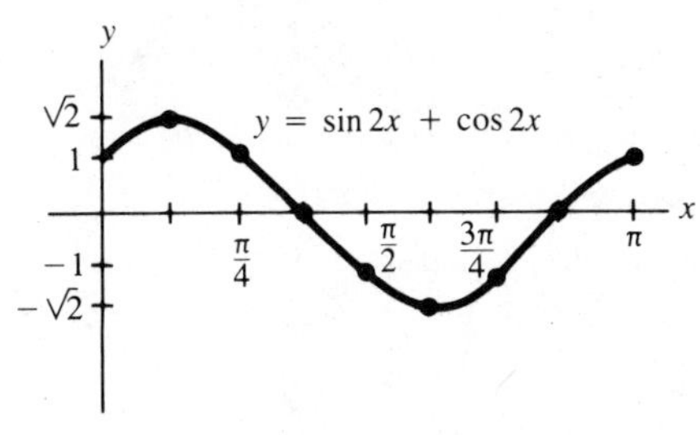

32.

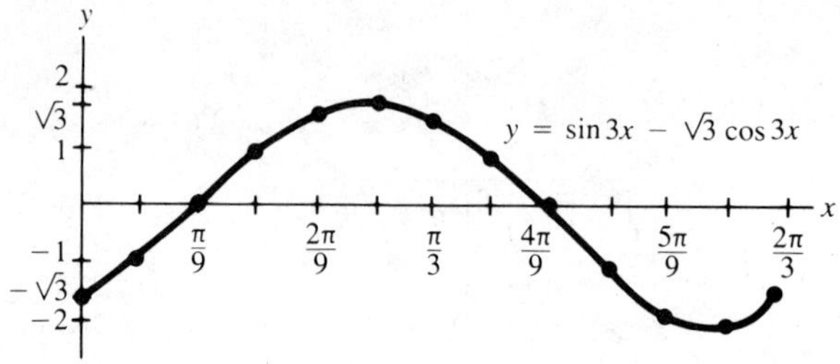

33.

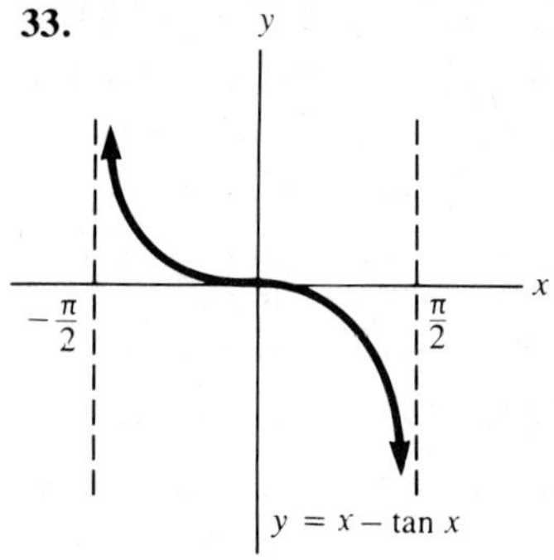

34. $\frac{3}{10}$

35. $-\cos x^2 + c$

36. $-\frac{2}{5}\csc 5x + c$

37. $-\frac{\cos^5 x}{5} + c$

38. $\frac{3}{2}\ln\sqrt{2}$

39. $\frac{1}{2}\sin 2x + c$

40. $\sin x\,[\ln(\sin x) - 1] + c$

41. $\frac{1}{2}e^{\sec 2x} + c$

42. $2\sin\sqrt{x} + c$

43. $\frac{3}{2}\left[\ln\left|\sec\left(\frac{\pi^2}{16} + 2\right) + \tan\left(\frac{\pi^2}{16} + 2\right)\right| - \ln|\sec 2 + \tan 2|\right]$

44. $-\frac{x\cos 2x}{2} + \frac{\sin 2x}{4} + c$

45. $\ln\left(\frac{6(\pi - 1)}{4\pi - 3}\right)$

46. $\frac{e^{3x}}{13}(2\sin 2x + 3\cos 2x) + c$

47. **(a)** 0 **(b)** $3\sqrt{3}/\pi$

48. $\frac{1}{2} - \frac{\pi^2}{72}$

Applications Index

Subject Index

1. $$\int u^n\,du = \frac{u^{n+1}}{n+1} + c, \qquad n \neq -1$$

2. $$\int \frac{1}{u}\,du = \ln|u| + c$$

3. $$\int e^{au}\,du = \frac{1}{a}e^{au} + c$$

4. $$\int b^{au}\,du = \frac{1}{a\ln b}b^{au} + c$$

5. $$\int \ln au\,du = u(\ln au - 1) + c$$

6. $$\int \log_b au\,du = u\left(\log_b au - \frac{1}{\ln b}\right) + c$$

7. $$\int \sqrt{u^2 + a^2}\,du = \frac{1}{2}[u\sqrt{u^2 + a^2} + a^2 \ln|u + \sqrt{u^2 + a^2}|] + c$$

8. $$\int \sqrt{u^2 - a^2}\,du = \frac{1}{2}[u\sqrt{u^2 - a^2} - a^2 \ln|u + \sqrt{u^2 - a^2}|] + c$$

9. $$\int \frac{1}{\sqrt{u^2 + a^2}}\,du = \ln|u + \sqrt{u^2 + a^2}| + c$$

10. $$\int \frac{1}{\sqrt{u^2 - a^2}}\,du = \ln|u + \sqrt{u^2 - a^2}| + c$$

11. $$\int \frac{1}{u^2 - a^2}\,du = \frac{1}{2a}\ln\left|\frac{u - a}{u + a}\right| + c$$

12. $$\int \frac{1}{u\sqrt{a^2 + u^2}}\,du = -\frac{1}{a}\ln\left|\frac{a + \sqrt{a^2 + u^2}}{u}\right| + c$$